The Art of Floral Design

The Art of Floral Design

Norah T. Hunter

Herb Mitchell
Contributing Author (Chapters 18-22)

I(T)P™

Notice To The Reader

Publisher does not warrant or guarantee any of the products described herein or perform any independent analysis in connection with any of the product information contained herein. Publisher does not assume, and expressly disclaims, any obligation to obtain and include information other than that provided to it by the manufacturer.

The reader is expressly warned to consider and adapt all safety precautions that might be indicated by the activities described herein and to avoid all potential hazards. By following the instructions contained herein, the reader willingly assumes all risks in connection with such instructions.

The publisher makes no representations or warranties of any kind, including but not limited to, the warranties of fitness for particular purpose or merchantability, nor are any such representations implied with respect to the material set forth herein, and the publisher takes no responsibility with respect to such material. The publisher shall not be liable for any special, consequential, or exemplary damages resulting, in whole or in part, from the readers' use of , or reliance upon, this material.

Cover design: Brian Yacur
Cover and section opener artist: Jenni Christensen

Delmar Staff
Publisher: Tim O'Leary
Senior Editor: Mark W. Huth
Developmental Editor: Nancy P. Belser
Production Editor: Brian Yacur

For information, address Delmar Publishers Inc.
3 Columbia Circle, Box 15-015
Albany, New York 12212

Printed in the United States of America
Published simultaneously in Canada
by Nelson Canada,
a division of the Thomson Corporation

1 2 3 4 5 6 7 8 9 10 XXX 00 99 98 97 96 95 94

Library of Congress Cataloging-in-Publication Data

Hunter, Norah T.
The art of floral design/Norah T. Hunter, with Herb Mitchell.
p. cm.
Includes bibliographical references and index.
ISBN 0-8273-5089-9
1. Flower arrangement. I. Mitchell, Herb. II. Title.
SB449.H86 1994 93-31901
745.92—dc20 CIP

Contents

Preface

The Art of Floral Design is a practical, student-oriented textbook that is easy to use and understand. This book is intended to make an understanding of, and appreciation for, the basic concepts of floral design more accessible to students. Topics are presented in clear, easily understood language. The purpose is to thoroughly teach the art of floral design, and is intended not only for the beginning floral student, but is a helpful resource for the amateur and professional as well.

ORGANIZATION AND APPROACH

The text is divided into five sections and is written in such a way that information is organized into self-contained chapters to allow individual instructors to use it in any sequence that suits their needs. Although no single book can cover all there is to know about a field of study, this text attempts to be a basic introduction to a great variety of floral design subjects.

Section I: Theory and Design supplies the reader with a background of floral design. Chapter 1 covers the history of floral design, helping the student to realize the importance of flowers throughout time—giving inspiration to contemporary floral artists. Through many museum reproductions of floral art from around the world, the student is better able to visualize the floral art of the ages. Specifically geared to the arrangement of flowers, chapters 2 through 7 cover design principles and elements, including in-depth discussions of harmony, unity, color, balance, proportion, scale, line, form, space, depth, texture, and fragrance. Chapter 8 thoroughly discusses tools, containers, and the mechanics of floral design. Although much of this chapter is basic in nature, the beginning floral student, as well as the experienced designer will benefit from sound principles of construction and mechanics.

Section II: Flowers and Foliage correlates flower anatomy and postharvest physiology to floral design. Understanding the structure of flowers and the postharvest physiology of cut flowers, presented in chapter 9, will aid the designer in the selection and proper care and handling, thoroughly discussed in chapter 10. Chapter 11 covers the forms of flowers and foliage, as they relate to floral design.

Section III: Basic Techniques and Styles presents a variety of central topics. Chapter 12 discusses the basic shapes of arrangements and provides numerous easy-to-follow instructions and illustrations. Historical information about floral and civic holidays in chapter 13 provides the student with a better understanding of the floral holidays—to glean from the many examples, illustrations, and photographs what is personally applicable. Basic to floral design, presented in chapter 14, is a knowledge of personal flowers—corsages and boutonnieres. Many clear illustrations are provided, helping the student learn construction techniques. Because fresh flowers are not always an option, chapter 15 discusses the alternative—everlasting flowers—covering both man-made and dried plant materials. A variety of drying and preservation techniques are presented to expand individuality in design.

Section IV: Beyond the Basics includes topics that are advanced in nature. Chapter 16 overviews oriental design styles and techniques, with appropriate illustrations, helping the beginner and advanced designer understand more clearly basic rules and applications. Advanced, contemporary designs and techniques are covered in chapter 17. Because wedding flowers and sympathy flowers are so important in the floral industry, both chapters 18 and 19 overview these styles and floral traditions with emphasis placed on the retail aspect of design.

Section V: The Floral Industry overviews central topics important for the student of floral design. Chapter 20 discusses the harvest and distribution of cut flowers—where they are grown and how flowers are moved from the grower to the final consumer. Chapter 21 overviews types of retail floral shops and discusses important selling techniques, employee roles, and shop layout. To become a talented floral designer, it is important to make a lifelong commitment to learning. Chapter 22 emphasizes continued education and discusses career opportunities. Names and addresses of national floral industry organizations and available trade publications are listed.

MAJOR FEATURES

It is hoped that this book will prove to be a catalyst to student learning and create imagination. Chapters contain organized and in-depth examples, illustrations, step-by-step construction techniques, and design ideas. The examples demonstrate the how-to's of floral design in a logical approach to beginning and finishing each arrangement. The text is well organized and written directly to the reader in a helpful, friendly fashion.

Illustrations

The Art of Floral Design is heavily illustrated with clear and helpful drawings that are a necessary component for teaching and understanding. Step-by-step instructions are adjacent to the detailed line drawings, explaining by example the techniques being presented in all aspects of floral design.

Photographs

The text includes many color photographs of floral designs, historical arrangements, various mechanics, and important floral industry subjects. Included are reproductions from national and international museums and libraries emphasizing the important historical-period styles.

Selected Key Terms and Glossary

Key terms are highlighted within the chapters as well as listed at the end of each chapter. An extensive glossary is included which alphabetically lists selected key terms. While not all key terms have been listed in the glossary, an attempt has been made to include those terms which are particularly unique and important, for a better understanding of floral design and the floral industry.

Questions and Activities

Along with selected key terms, at the end of each chapter are questions and suggested activities. These terms, questions, and activities will increase the level of understanding and internalization of chapter concepts.

Appendices

The Appendices contain extensive listings and illustrations of cut flowers and cut foliage, common to the floral industry. The 37 color plates and 53 detailed line drawings, make it possible to easily identify common cut flowers and foliage. Plants are organized alphabetically by botanical name and include family name, name origin, species, common names, seasonal availability, description of the plant, its uses in floral design, and specific care and handling techniques, and approximate vase-life. The common names of plants are listed and cross-referenced to the botanical names because many flowers and foliage are better known by their common names.

Syndicated Cartoons

Many syndicated Farside and Herman cartoons are included throughout the text. Keeping a sense of humor always helps the day go better.

Instructor's Guide

An Instructor's Guide is available, containing helpful teaching aids. Included are chapter outlines, learning objectives, student questions and solutions, suggested activities, and transparency masters.

About the Author

Doug Martin Photography

Norah T. Hunter is a faculty member at Brigham Young University in Provo, Utah, where she has taught students the art of floral design for more than a decade. She holds both Masters and Bachelors Degrees in Horticulture. Ms. Hunter is a member of the American Academy of Floriculture, and recently served as a committee member on the Redbook Florist Education Advisory Board writing curriculum for floral workshops and seminars.

After working in the retail business for over 15 years, Ms. Hunter, a freelance designer, established her own floral consulting company called Norah's Wildflowers. She lives with her husband and five children in Pleasant Grove, Utah.

We are delighted to have had the added expertise of Herb Mitchell as a contributing author for chapters 18–22 of *THE ART OF FLORAL DESIGN.* In 1981, Mitchell completed his credentials as a Certified Management Consultant, and for the past 17 years has had his own consulting business where he has extensive experience working with retailers, wholesalers, and manufacturers. He is a member of the Institute of Management Consultants, the National Speakers Association, the Society of Professional Journalists, and the Association of Psychological Type. In 1978 and again in 1984, Mitchell was selected as one of the three float judges for The Pasadena Tournament of Roses. In 1983, Mitchell was inducted in the Floriculture Hall of Fame in recognition of his outstanding contributions to the establishment, advancement, and improvements of Floriculture in America.

Introduction

Flower arranging is an art form that enhances the quality of life. From birth to death flowers and floral arrangements heal sadness and sorrow and give comfort to the living. Flowers give inspiration and hope in times of illness and despair. They express emotions from deep within the soul, that often cannot be expressed with words. Symbolic of life and vitality, fresh flowers celebrate the happy and joyous occasions of life. Floral arrangements are therapeutic, not only to the designer, but to the recipient as well. Flowers are sentimental and lovely gifts that help create memories.

It is my hope that you will discover within you the motivation to become the best you can be by stretching and expanding your skills, imagination, and desire to build upon your talents. Learning the basics and techniques of floral arrangement, understanding floral arrangement history, knowing how to incorporate the fundamental design principles and elements, and realizing there is always opportunity for growth and learning, are all essential to becoming a talented floral artist who is able to master nature within the confines of a container.

> To become a talented floral designer, you must make a commitment to lifelong learning. No designer ever knows everything there is to know about floral design. When you make a commitment to excellence—you pledge to be the very best you can possibly be. This requires becoming a student again and again—every time you have an opportunity to explore the new. Through this learning, you develop a constant appetite for change and growth.
>
> Herbert E. Mitchell

Last Flowers *by Jules Breton, 1890. Cincinnati Art Museum, Gift of Emilie L. Heine in memory of Mr. and Mrs. John Hauck*

The purpose of this book is to teach and inspire, and to help unleash your potential. Internalization of knowledge elevates you to new discoveries and opportunities to experience your creative talent. Realize the power you possess—a kind of stewardship—to share the beauty and vitality of flowers with others so they may enjoy the fullness of your talent through the art of floral design.

N.T.H.

Acknowledgements

Many people assist, directly and indirectly, in the process of writing a book. I am particularly grateful to Marcia M. Bales of Renton, WA, for her precise and engaging illustrations throughout this text. It has been a pleasure to work closely with her from the beginning to the completion. I wish to especially thank Valerie Holladay and Dorine Jesperson for their editorial assistance and never-rending encouragement.

Thanks also are extended to the following reviewers who thoughtfully considered endless pages of material and made helpful comments and suggestions:

Teresa P. Lanker, The Ohio State University
Holly Money-Collins, City College of San Francisco
James L. Johnson, Texas A&M University
Diane Noland, University of Illinois
C. Burton Knandel, Brainerd Technical College
Frank McElwain, Limestone High School
Bryan Starr, Jefferson High School
Lonnie D. Koepke, Broken Bow Public School
David Soucy, SUNY-Morrisville
William J. McKinley, Kishwaukee College

I am particularly grateful to Herbert E. Mitchell for his willingness to give professional expertise and help with the final chapters of the text in the event of the early arrival of my twin daughters. Special appreciation is extended to the many museums, floral companies, manufacturers, floral designers, and artists. The quality of this text is enhanced by the support, contributions, and willingness of so many companies and individuals and acknowlegement is given at each illustration; however, the following individuals and companies gave an extraordinary amount of support:

Marie Ackerman, American Floral Services, Inc., Oklahoma City, OK
Kate Wootton and *Frances Porterfield,* Florists' Review, Topeka, KS
James Moretz, American Floral Art School, Chicago, IL
Michele Malone and *Pamela Murphy,* The John Henry Company, Lansing, MI
Hermine Warringa, Flower Council of Holland, Netherlands
Marvella Crab, Society of American Florists, Alexandria, VA
Jennifer Sparks, American Floral Marketing Council, Alexandria, VA
Jenni and *Day Christensen,* Pleasant Grove, UT
Lynne Oliver, Flower Farm, Provo, UT
Nelson B. Wadsworth, Archival Photography, Logan, UT
Doug Martin, Doug Martin Photography, Provo, UT
Raymond Gendreau, Seattle, WA
Jerry Alhadeff, Evergreen Wholesale Florist, Inc., Seattle, WA
Mary C. Suggett, Universal Press Syndicate, Kansas City, MO
John and *Rhonda Kauwe* and *Gary Merritt,* Aloha Wholesale, Orem, UT
Stanley L. Welsh, Brigham Young University, Provo, UT
Kimball T. Harper, Brigham Young University, Provo, UT

Photos by Doug Martin: 2-7, Chapter 3, 4-6, 5-15, 7-16, 8-5, 8-6, 8-7, 8-8, 8-9, 11-20, 12-8, 12-15, 12-55, 19-17. All other photos unless otherwise indicated are by Norah Hunter.

I also wish to thank my editor, Nancy Belser, and the staff at Delmar, particularly, Mark Huth, and Brian Yacur for their confidence in the project, attention to detail, and valuable professional advice.

To my colleagues and students at Brigham Young University, I wish to express my gratitude and appreciation for their support and encouragement.

Finally, to my husband and family, for their sacrifice, prayers, and wholehearted support, I give my sincere thanks.

N.T.H.

Section 1

THEORY AND DESIGN

Chapter 1

A History of Floral Design

The art of floral design has a rich worldwide history. Floral artists of today inherit the floral art of the ages. Since flowers have long been used for decoration and adornment, the art of arranging flowers goes back to ancient cultures. We can learn a great deal from the past. Excavated remains, wood carvings, ***bas-reliefs,*** tapestries, rugs, wall paintings, and many other tangible artifacts tell us that flowers were brought indoors for enjoyment and have always played an important part in the lives of people (see figure 1-1).

Figure 1-1. Flowers and floral arrangements are often shown in Egyptian art, as in this painting from a tomb wall at Thebes of the 14th century B.C. Formal bouquets of lotus flowers and buds in orderly rows are typical in the Egyptian style of design.
Ipuy and his wife receiving offerings from their children. Copy of a wall painting from the Tomb of Ipuy. Egyptian Expedition of the Metropolitan Museum of Art, Rogers Fund, 1930. All rights reserved, The Metropolitan Museum of Art, New York.

It is exciting to learn the ways flowers were used or arranged by different groups of people or during different time periods. Floral artists are encouraged to study and learn about the history of floral design. As you study the various styles of ancient and modern times, you will gain inspiration, increasing your creativity and professionalism. You can increase your awareness and appreciation for floral arrangements simply by looking at historical paintings, engravings, tapestries, or other works of art, and studying what kinds of flowers and foliage were used, how the flowers have been arranged, and in what type of vase. If an interior setting is visible or people are present in the artwork, notice the style and accessories within the room and the clothing and hairstyles of the individuals. Often these will give you an overall feeling for a particular ***period style*** and help you remember different bouquet types more effectively than just reading historical dates. Various kinds of arrangements are often called period styles because they are representative of a certain time period.

As you gain knowledge of floral styles and traditions of past eras, you will be able to create compositions that express a unique feeling of another time and place. This skill is especially important when a floral composition must harmonize with a period-style room, or perhaps accent a historic museum or hotel lobby. Authenticity in the shape, style, container, and flower types will enhance these interior settings. A basic knowledge of period styles will also strengthen your design skills and confidence for flower shows and competitions.

This chapter is in no way comprehensive but it will give you a solid foundation of basic historical styles. This history highlights the major styles that have influenced and contributed to the floral design styles of today. There are ideas in every floral tradition that can guide you. Learning the history of floral design will spark your imagination and heighten your understanding and awareness of floral art throughout time.

FLOWER ART IN ANCIENT CIVILIZATIONS

The remains of many ancient cultures give proof that people have always placed importance on flowers and their use indoors. Certain ancient civilizations provide us with a clear record of how and when flowers were used. Others offer only a glimpse of their floral history. Important floral traditions and styles are highlighted from the Egyptian, Greek, Roman, and Byzantine periods, each offering significant recordings of the use of flowers.

Egyptian Period

History records through precious artifacts, wall paintings, and bas-reliefs, that in the ***Egyptian period*** cut flowers and foliage were used as decorations. Much evidence shows how the ancient Egyptians used cut flowers and placed blossoms, foliage, and fruits in a variety of vases. As in our day, flowers were a part of important life events and were used during religious ceremonies, for decorative purposes in the home, and for personal adornment.

The favored containers were of the wide-mouthed variety, although many vases were designed to hold flowers (see figure 1-2). Vases, jars, and bowls were made from a variety of materials including pottery, gold, slate, polished alabaster, and dark green diorite. The early Egyptians also produced a ware of finely ground silicate called ***faience.***

The floral art of Egypt was simplistic, repetitious, and highly stylized. Flowers and fruits were placed in carefully alternating patterns, resulting in regimented rows. The primary colors of red, yellow, and blue were

Figure 1-2. Although the usual Egyptian container was a basin or a wide-mouthed bowl, ancient Egyptians placed flowers and foliages in a variety of containers of extraordinary interest. Many containers were made of molded glass, pottery, metals and stone.

used predominately in floral designs, as were other vibrant color combinations. The lotus flower or water lily, was the flower of the Goddess of Iris and, therefore, was considered sacred. It is commonly depicted in Egyptian art and can be traced as far back as 2500 B.C. As illustrated in figure 1-3, lotus flowers and buds were often placed in low bowls in orderly rows. With this particular style, a pattern of similar floral groups—flower, bud, and foliage—were repeated in an arrangement around the rim of a container. It is known that bowls of flowers were set on the banquet tables, used in processions, and given to honor others.

Other floral decorations included chaplets for the hair, wreaths, ***garlands,*** and flower collars. For today's application of the Egyptian floral design style, materials commonly used by the Egyptians can be found in table 1-1.

Figure 1-3. Lotus flowers and buds were often placed in low bowls in repetitious orderly patterns.

TABLE 1-1: FLORAL DESIGNS CHARACTERISTIC OF THE ANCIENT EGYPTIANS

Floral Designs
- flowers placed in containers in orderly, alternating patterns
- flowers placed in spouted vases with no stem visible
- floral chaplets, collars, wreaths, and garlands

Popular Container Shapes
- wide-mouthed bowls; often sectioned for fruit, with flaring sides
- rounded basins
- spouted vases
- deep or shallow baskets

Popular Flowers
- water lily and lotus
- acacia
- rose
- violet
- lily
- narcissus
- poppy
- gladiolus
- lupine
- jasmine
- bachelor button
- morning glory
- strawflower

Examples of Foliage
- ivy
- myrtle
- oleander
- palms
- papyrus
- reeds
- olive
- laurel

Examples of Fruit
- figs
- gourds
- grapes
- peach
- plum
- pomegranates

Figure 1-4. Ancient containers and vases of the Greek and Roman times were not made originally for holding flowers but have influenced later vase shapes.

Greek Period

Even though many vases and containers remain from the ***Greek period*** as shown in figure 1-4, none were made solely for the purpose of holding flowers. Rather, flowers and petals were commonly scattered on the ground during festivals and used to make wreaths and garlands. We know from ancient writings that flowers were important in the daily lives of people, and the wearing of wreaths or ***chaplets*** on the head and garlands around the neck was quite common. Professional wreath makers were engaged to make them as gifts and as decorations for special occasions (see figure 1-5). Because the wreath was the symbol of allegiance and dedication, it was awarded in honor to athletes, poets, civic leaders, victorious soldiers, and other heroes. Garlands and wreaths were worn at weddings and used to decorate interiors and exteriors of homes, statues, and graves. Wreaths were so important to the Greek way of life that books were

Figure 1-5. Professional wreath makers made garlands and wreaths for decoration and for personal adornment.
Preparations for the Festivities *(Opus XXXIII) Sir Lawrence Alma-Tadema, British (1835–1912). Bequest of Madeleine Dahlgren Townsend. Sterling and Francine Clark Art Institute, Williamstown, Massachusetts.*

written about the proper etiquette for wearing them as well as the appropriate flowers, foliage, symbolism, and various styles.

The ***cornucopia*** we use today was introduced during the Greek period (see figure 1-6). Originally, it was placed in an upright position with flowers, fruits, and vegetables. Often called a ***horn of plenty,*** today it is commonly laid on its side with its contents overflowing and spilling out. As the symbol for abundance, the cornucopia is associated with a bounteous autumn harvest and with Thanksgiving.

Roman Period

During the ***Roman period*** wreaths and garlands were heavy and elaborate. Fragrant flowers with bright colors were favored by the Romans. Rose blossoms and petals were scattered lavishly on banquet tables, streets, and lakes during festivals and ceremonies.

A basket of mixed flowers appeared for the first time in a Roman mosaic now in the Vatican Museum. It shows a charming mix of roses, anemones, tulips, dianthus, and other flowers in a delightful combination of colors and forms. This is the first introduction of a truly naturalistic flower bouquet. See table 1-2 for a listing of floral styles and common flowers and foliages used during the Greek and Roman periods.

Figure 1-6. Wreaths and garlands were commonly worn for personal adornment or used in decoration. A popular container of the Greek period, the cornucopia, is still used today to symbolize abundance.

TABLE 1-2: FLORAL DESIGNS CHARACTERISTIC OF THE GREEK AND ROMAN PERIODS

Floral Designs
- wreaths and garlands
- loose flowers, petals

Examples of Flowers
- rose
- hyacinth
- lily
- iris
- honeysuckle
- narcissus
- violet
- poppy
- daisy
- fragrant spring blossoms (apple, quince, etc.)
- crocus
- fragrant herbs

Examples of Foliage
- myrtle
- laurel
- ivy
- olive
- oak

Examples of Fruit
- grapes
- cones
- acorns
- berries
- pomegranates

Figure 1-7. A rare Grecian example depicts leafy olive branches arranged in a decorated terra cotta vase.

Figure 1-8. Early Byzantine mosaics reveal that the floral styles of the time were symmetrical and conical, distinguished by height and symmetry, as shown in this conical, foliage design.

Byzantine Period

The floral styles during the ***Byzantine period,*** which takes its name from the city Byzantium, an Eastern Roman Empire, are evident through different mosaics of this period. Symmetrical, stylized tree compositions were introduced during this time (see figure 1-8). Containers were filled with foliage to resemble conical trees. These designs were distinguished by height and symmetry.

The floral traditions of the Greek and Roman periods continued; however, the garland was constructed differently. It was often made of narrow bands of flowers or fruit alternating with foliage and given a twisted, spiral effect.

EUROPEAN PERIODS

There are several important periods of floral history that have influenced the art of design from ***European periods.*** Your design capacity will be augmented as you study and recognize various period styles. All great artists can learn and gain insight from the past to enhance their own techniques. A professional floral artist should be knowledgeable of the various design styles and should be able to recreate the past in the art of floral design. Important floral traditions and styles are highlighted from the Middle Ages, Renaissance, baroque and Dutch-Flemish period, French period, English-Georgian period, and the Victorian era.

Middle Ages

Very little is known about floral art during the ***Middle Ages,*** the period of time between the ancient and modern times in European history, about 476 to 1450 A.D. We do know that flowers, both wild and cultivated, were an important part of everyday life and used in food, drink, and medicine. Fragrant flowers were highly favored for strewing on the ground, freshening the air, and for making wreaths and garlands for personal adornment and decoration (see figure 1-9).

Figure 1-9. In the Middle Ages, fragrant flowers were sewn tightly together to form wreaths and garlands. Medieval gardens as this served many different functions; including, protection, socializing, and the growing of herbs, food, and scented flowers.
Emilia in her Garden *by the Master of the Hours of the Duke of Burgundy, c. 1465. Bibliothek National, Vienna/The Bridgeman Art Library, London.*

Renaissance

The ***Renaissance*** refers to the period in Europe after the Middle Ages. Beginning in Italy in the 14th century, the Renaissance is associated with a resurgence and flourishing of the arts. During this time period, art represented religious history. Flowers in vases are often shown in paintings from this time period as great emphasis was placed on flower symbolism. For example, the rose represented sacred or profane love. The white lily (*Lilium candidum*) was the symbol of chastity and fertility. Because it appeared in many Annunciation paintings, it soon became known as the Madonna lily (see figure 1-10).

Figure 1-10. Pink and white roses are placed in a classic Renaissance vase. The white lily, common in many Renaissance paintings depicting the Annunciation, symbolized chastity and fertility.

The Annunciation; *MASTER of the BARBERINI panels; Tempera on wood; c. 1450; National Gallery of Art, Washington; Samuel H. Kress Collection.*

Several design styles are typical of the Renaissance period. A single stem of white lily in a humble jug (see figure 1-11) is seen throughout art of this time symbolizing chastity and fertility. Flowers were also arranged in vases so that only the blossoms were visible. Stems were covered creating a massed, symmetrically stiff conical arrangement (see figure 1-12). Even though the flowers were usually compactly arranged, each flower stood out with distinction because of the variety of bright colors and forms that were used. The art of making wreaths and garlands continued during this period.

During the Renaissance, the art of arranging was patterned after the classical Greek and Roman styles, and many types of containers were used to hold flowers. Elaborate metal containers with well-formed bases, stems, necks, and handles were commonly used (see figure 1-13). Illustrated in a gardening book entitled *Flora overo Cultura di Fiori,* published in Rome in 1638, are pictures of containers fashioned especially for the arrangement of flowers (see figure 1-14). One such container made it convenient to insert flowers in the holes of a removable lid. Notice how the arrangement is made from a variety of different flowers with all the flowers' heads facing out, forming a tight, symmetrical bouquet. For a detailed listing of design styles, containers, and plant materials common to the Renaissance period, see table 1-3.

Figure 1-12. Mixed bouquets of the Renaissance period completely covered stems, showing only the blossoms, making a massed, symmetrically stiff display.
Courtesy The Professional Floral Designer.

Figure 1-11. A single stem of white lily placed in a crude jug or ewer is typical 15th century Renaissance floral display.

TABLE 1-3: FLORAL DESIGNS CHARACTERISTIC OF THE RENAISSANCE PERIOD

Styles of Design
- mass, symmetrical arrangements (oval, conical, triangular, and circular)
- tight designs with short-stemmed flowers
- fruits and vegetables often combined with flowers
- garlands and chaplets
- loose flowers and petals

Containers
- urns made of pottery, marble, bronze, and Venetian glass
- blown glass vases, beakers, and goblets
- simple bowls, jugs, and jars of stone, pottery, and terra cotta
- metal pedestal vases, vessels, and trays

Examples of Flowers
- many garden flowers
- carnations and pinks
- lily
- violets
- rose
- primrose
- iris
- anemone
- campanula
- stock
- lily of the valley
- marigold
- monkshood
- lupine
- pansy
- poppy
- narcissus

Examples of Foliage
- ivy
- olive
- laurel
- boxwood
- myrtle

Examples of Other Plant Materials
- vegetables and fruits
- cones

Figure 1-13. Made of metal, pottery, and glass, Renaissance-period vases were often ornate, formal, and designed for the arrangement of flowers.

Figure 1-14. Containers often had removable lids with holes in them, as shown in this pottery vase. These lids made it easier to insert flowers and thus create floral bouquets.
Flora Ouero Cultura di Fiori, Giovanni Battista Ferrari; Rome 1638. The Pierpont Morgan Library, New York. PML 33421.

Figure 1-15. Floral arrangements of the baroque period were asymmetric, massed, and overflowing with bold sweeping lines, displaying dramatic contracts of large and small, and dark and light, as shown in this engraving by Jean Baptiste Monnoyer. Museum of Fine Arts, Boston.

Baroque and Dutch-Flemish Styles

The era following the Renaissance during the 17th and 18th centuries in Europe is named the ***baroque*** period. During this period, economic conditions changed dramatically. Art was no longer just for the Church or the nobility. It now became accessible to the previously ignored middle class. As a result, paintings from this period show flower arrangements in everyday situations rather than just in religious settings.

As a reaction against the severe classic styles, this period is characterized by elaborate ornamentation and curved rather than straight lines, as shown from an engraving of flowers in a basket in figure 1-15. Interiors were fantastically overdecorated and many accessories were gaudily ornate.

Flower arrangements depicted in paintings during the baroque period are typically massed and overflowing. Some are symmetrically oval while many display a rhythmic asymmetrical balance, as with the S, or Hogarthian curve, named after the English artist William Hogarth (see chapter 12).

We gain greatest insight into the baroque style of arrangement through flower paintings by the ***Dutch-Flemish styles*** of the artists of the 17th and 18th centuries. Many artists fled the strife and religious persecution of the South to settle in the northern provinces. However, because of the religious reform that took place in the Netherlands and Flanders (now known as Belgium) creative expression became a reality for these artists. The painters of this period have left a legacy of art for grouping mixed flowers in large, flamboyant arrangements.

One such early artist, Ambrosius (the Elder) Bosschaert (1573–1621) is regarded as perhaps one of the most influential painters in the history of floral genre painting. One of his famous paintings (see figure 1-16) shows a symmetrical composition in which each flower is clearly presented with little overlapping. As with Bosschaert, the majority of Flemish artists such as Jan (Velvet) Brueghel (1568–1625) and even Jan van Huysum (1682–1749) a century later, combined fruits, bird nests, shells, insects, and other decorative objects at the base of these lavish bouquets. Another major distinguishing characteristic of this period is the great variety of flowers within one bouquet. Tulips were a part or many bouquets as were roses, peonies, iris, lilies, and poppies.

Figure 1-16. This famous, lovely painting shows a symmetrical composition with little or no overlapping flowers, characteristic of the early Dutch-Flemish style. Each flower is clearly evident. Typical are the striped tulips combined with a great variety of flowers—anemones, ranunculus, dianthus, columbines, iris, roses, narcissus and marigolds. A scattering of shells, insects, and other accessories are characteristic of the Flemish setting.

Flowers in a Niche, *Ambrosius Bosschaert (1614–1654). Photograph C. Mauritshuis, The Hague, Holland, inv. nr. 679.*

It is important to understand that, in reality, many of these early artists never actually arranged the flower bouquets or even painted their pictures using an actual arrangement as a model. Instead, the artists painted a composite bouquet from individual drawings and from visualizations in their minds, which explains why short-stemmed flowers are often positioned high and flowers from all seasons are placed side by side.

Before this period and during the early part of this era, common containers for flowers included the bottle-shaped, handled jugs and vessels, ***ewers*** (water pitchers), and other objects of ordinary daily use. As the 17th century progressed, however, containers for holding flowers became more diverse in shape and included glassware, ***terra cotta (earthenware)***, and porcelain. When imported Chinese porcelain vases in blue and white became the rage for flower arrangers, Dutch craftsmen developed a cheaper container known as Delft (named after the city in the Netherlands where ***delftware*** originated) to fill the need for this popular container.

The artists of the later part of the 18th century emphasized many traditional design elements, including depth, texture, form, line, and color. In contrast to early paintings, later artists treated these elements, especially depth, more seriously by showing the backs and profiles of flower heads. Flowers were also overlapped for an even greater feeling of depth and space. The symmetrically oval floral arrangement shape during this period gradually changed to a soft, asymmetrical triangle as shown in figure 1-17.

See table 1-4 for a partial listing of design shapes, container types, flowers, foliages, and accessories common during the baroque period, particularly the Dutch-Flemish style of design.

Figure 1-17. In contrast to the early Dutch-Flemish style of arrangement, the artists of the later part of the 18th century emphasized texture, form, depth, and space. Floral displays became more asymmetrical, informal, and naturalistic. Flowers were overlapped, with their backs and profiles visible to create a greater feeling of depth and space.

TABLE 1-4: FLORAL DESIGNS CHARACTERISTIC OF THE BAROQUE AND DUTCH-FLEMISH STYLES

Styles of Arrangement
- massed, full, and drooping designs; predominantly asymmetrical
- Hogarth (S-curve) style of design
- casual mixed bouquets

Containers
- massive and sturdy
- classic metal and stone urns
- Chinese and Japanese porcelain vases, bowls, and flasks
- glass and pottery jugs, vases, and plates

Partial List of Flowers
- great varieties of tulips
- carnations
- cyclamen
- foxglove
- stock
- hellebore
- hollyhock
- hyacinth
- iris
- larkspur
- lilac
- lilies
- lupine
- marigolds
- nerine
- peony
- poppy
- roses
- sunflowers
- snowball

Partial List of Foliage
- leaves of flowers emphasized
- coleus
- olive
- bold leaves of hosta, castor bean, and canna plants

Examples of Accessories
- sprays of flowers loose on the table
- fruit and vegetables placed in or near the bouquet
- birds nests with eggs
- shells
- jewels and coins
- butterflies, flies, beetles, spiders, ants, snails, grasshoppers, etc.
- snakes and lizards
- nuts, berries, and cones
- figurines, cups, plates

Figure 1-18. Characteristic of the floral arrangement style popular during the French period, this lavish mix of flowers and foliage in a two-piece Sevres ornamental porcelain bowl supported by an engraved base, displays elegance. Gladiolus, snapdragons, ranunculus, iris, tulips, lilies, heather, fillers, and ivy all radiate outward in a characteristic triangular fan-shape. *Mauritshuis—The Hague.*

Figure 1-19. Floral arrangements, often fan-shaped or circular, and massed, were frequently placed on mantles, console tables, and pedestals to decorate interior chambers. N'ayez pas peur, ma bonne amie, Jean Michael Moreau, le jeune, French (1741–1814). Engraved by Helman. Proof state of place used in Le Monument du Costume.
The Metropolitan Museum of Art, New York, Harris Brisbane Dick Fund, 1933. (33.6.18).

French Period

The ***French period,*** also known as the ***Grand era,*** flourished during the 17th and 18th centuries in France. It is associated with court life, beginning with Louis XIV in France and continuing with his many successors. Even though the influence of the Dutch-Flemish period was evident in most floral arrangements of the French style, the emphasis was on classic form, refinement, and elegance, as compared to the often overdone flamboyance of the Dutch-Flemish period.

During this period, the arts expressed the baroque magnificence and luxury of the aristocracy. Floral arrangements, commonly fan-shaped and massed, were often used as decorative accents in interiors (see figure 1-19). Designs were large in scale and massed with flowers, but arrangements were never overdone, flamboyant, or overly massive. The French style can generally be described as having a distinctive overall shape of a fan or triangle with radiating stems.

Several types of containers were popular during the Grand era (see figure 1-20). Generally vases were highly ornate, giving a feeling of elegance and tasteful extravagance. For a partial listing of design styles, container types, flowers, and accessories typical of the French style, see table 1-5.

Figure 1-20. French floral vases during the Grand era were elegant and highly ornate. Many were made of fine porcelain.

TABLE 1-5: FLORAL DESIGNS CHARACTERISTIC OF THE FRENCH PERIOD

Styles of Design
- tall, triangular, fan-shaped designs
- round and crescent shapes
- smaller designs, often shorter than the container

Typical Containers
- elegant and highly ornate
- epergnes made of ceramic, crystal, and silver
- flasks, goblets, vases, wall pockets made of glass, ceramics, and porcelain
- classic urns made of faience, metal, porcelain
- spout or five-finger vases and bricks and square vases fitted with holes for flowers made of ceramics, metal, and faience
- basket, cornucopia, and shell-shaped containers made of ceramics, porcelain, and metal

Popular Flowers and Foliage
- roses
- lilacs
- tulips
- narcissus
- lilies
- pansy
- larkspur
- lily of the valley
- marigold
- hyacinth
- carnation
- ranunculus
- poppy
- acacia
- snapdragon
- aster
- flowering branches
- ferns

Accessories
- porcelain figurines
- fans
- sheet music
- candelabra and candlesticks
- baskets with tatting bobbins and thread

English-Georgian Period

The baroque period in England during the 18th century is often referred to as the ***English-Georgian,*** or simply the Georgian period, named after the English rulers King Georges I, II, and III.

The history of the English-Georgian period includes many comments about the importance of the fragrance of flowers. Fragrance was the first prerequisite to a bouquet, for it was believed that their perfume would rid the air of contagious and infectious diseases. In order to have the perfumed flowers always close by to safeguard them from pestilence, the English created the ***nosegay,*** a small hand-held bouquet to carry the sweet scents (see figure 1-21). Nosegays also gave relief from the smells of unsanitary surroundings in a society with less-than-modern standards for bathing and cleanliness. These small hand-held bouquets

Figure 1-21. The nosegay was originally designed by the English, who believed that carrying a small perfumed bouquet of flowers and herbs would rid the air of infectious diseases.

Figure 1-22. This British etching shows a charming English woman adding tulips to the mixed bouquet of lilies, irises, roses, and snowball flowers (Viburnum). The flower bouquet, large in proportion to the container, massed, and oval was commonly made in a two-handled metal container.

Spring, *Wenceslaus, Hollar, British, 17th century, 1641. Etching. The Metropolitan Museum of Art, New York, Harris Brisbane Dick Fund, 1947. (47.100.514).*

are often called ***tussie-mussies,*** sometimes spelled tuzzy-muzzy. The word tuzzy refers to the old English word for a knot of flowers.

These small fragrant bouquets were first used solely for fragrance but soon became a fashion trend. Women of the Georgian period wore flowers in their hair, around their necks, or on their gowns, either at the waist, shoulder, or in the ***décolletage*** (the neckline or top of a dress cut low to bare the neck and shoulders).

Most English-Georgian arrangements were formal and symmetrical, often tightly arranged with great varieties of flowers (see figure 1-22). During this time period, the custom was introduced of filling the drawing room fireplace with flowers during the months when heat was unnecessary. The flower-filled container, when set within the fireplace, was called a ***bough pot.***

Figure 1-23. English-Georgian floral containers including Wedgwood, stoneware, and porcelain, of which many were made with holes and special openings for the arrangement of flowers. Also popular were metal (especially silver) urns, handled cups, and epergnes. Wall vases, often called wall pockets, made of ceramic were also popular.

Flowers were arranged in an extensive variety of containers. **Wedgwood** (named after the English potter Josiah Wedgwood about 1759), a fine ceramic ware popular during this time period, reflected the English taste for ancient Greek and Roman designs. Wedgwood containers were manufactured to hold flowers; many were made with special holes and openings, which served to hold the stems of flowers in place to keep arrangements stiff and formal (see figure 1-23).

The English are often credited with introducing miniature arrangements of a few blossoms in bud vases. Even though small, these dainty arrangements included a wide variety of flowers and colors typical of the large bouquets (see figure 1-24). Many historians suggest that the English were the first to use the formal centerpiece as we know it today. See table 1-6 for a list of design styles, container types, and popular flowers characteristic of the English-Georgian period.

Figure 1-24. The English created tiny, dainty arrangements that, nevertheless, used a wide variety of flowers and colors as did large bouquets.

TABLE 1-6: FLORAL DESIGNS CHARACTERISTIC OF THE ENGLISH-GEORGIAN PERIOD

Styles of Design
- small tussie-mussie bouquets
- large mixed flower arrangements
- small mixed vase arrangements
- dried bouquets
- fan-shaped bough pot arrangements (chimney flowers)

Typical Containers
- Wedgwood containers
- posy-holder vases with five or more fingers made of ceramics
- urns and stem-cups with handles made of pewter, silver, and ceramics
- jars, jugs, bottles, and vases made of ceramics and glass
- wall pockets made of ceramics
- enclosed bricks with holes for flowers made of delftware

Partial List of Popular Flowers
(These flowers are additions to those listed previously under baroque and Dutch-Flemish styles and the French period.)
- flowering branches and catkins
- veronica
- snowdrop
- scabiosa
- phlox
- geraniums
- clover
- hibiscus
- passion flower

Victorian Era

The ***Victorian Era*** was named after Queen Victoria, who reigned in England from 1837 to 1901. During this time, a variety of architectural styles and furnishings were designed bearing her name. Victorian interiors were lavishly decorated in heavy colors and patterns. Some historians suggest that this era made the most significant contribution to establishing the use of plants and flowers in daily life.

A time of great enthusiasm for flowers, plants, and gardening, the Victorian era is important in the history of floral design because flower arrangement was taught and recognized as an art. In fact, the Victorian era records the first attempt to establish rules for flower arranging. A proliferation of magazines and

Figure 1-25. A three-tier epergne design.
The Professional Floral Designer by Herb Mitchell

Figure 1-26. The great source of inspiration for 19th-century arrangements lies in French paintings. Full and overflowing, this mixed bouquet of peonies, roses, lilacs, iris, and other garden flowers expresses beauty and overabundance.

Spring Bouquet, *1866, Pierre-Auguste Renoir, French (1841–1919). Courtesy of the Fogg Art Museum, Harvard University, Cambridge, Massachusetts. Bequest of Grenville L. Withrop.*

The Victorian style of floral arrangement can generally be described as grouping large masses of flowers, foliage, and grasses together to create a compact design with few voids and no obvious center of interest. Usually these arrangements were round or oval in shape. Flowers on short stems were generally packed tightly into containers. Inspiration for the Victorian style of floral arrangement comes from paintings of the time, with abundant varieties of flowers spilling from vases (see figure 1-26).

Containers typical of the Victorian period were highly ornate and showy (see figure 1-27). Containers of all shapes and materials were used to hold abundant masses of flowers.

The Victorian era was primarily one of sentiment, when the language of flowers was thoughtfully and carefully studied. The nosegay, introduced during the English-Georgian period, reached great popularity.

Figure 1-27. Elaborate, hand-painted porcelain, metal, and ceramic containers typify the entire Victorian period.

books about the art of flower arranging appeared. Many seriously studied the techniques of flower arrangement. Society required cultured young girls be taught not only to arrange flowers and make tussie-mussies, but also to grow, preserve, press, draw, and paint flowers. Another floral pastime included making artificial flowers from an array of materials such as shells, wax, feathers, hair, textiles, and beads.

Figure 1-28. Before the invention of the flower holder, tied nosegays were short-lived and cumbersome during such activities as eating and dancing. The bouquet of flowers together with the holder have several names including tussie-mussie and nosegay. The holders are called posy bouquet holders, posy holders, bouquet holders, porte-bouquets, and bouquetiers (BUK-tje).

A young man admiring a lady would bring her a hand bouquet of flowers with a romantic note alluding to the symbolic message of the particular flowers in the nosegay.

Proper Victorian ladies appeared at social gatherings carrying nosegays of fresh flowers. These delicate bouquets were not only used as air fresheners, but the flowers and foliage conveyed special sentiments as well.

While these nosegays were fashionable and demure, they were short-lived and cumbersome during such activities as eating or dancing. To overcome these problems, the flower holders were invented (see figure 1-28).

Hand bouquet-holders, or ***posy holders,*** were made from various precious metals, steel, and alloys. Other materials used to make holders included ivory, glass, painted porcelain, amber, tortoiseshell, and mother-of-pearl. Holders were often decorated with jewels, pearls, and small mirrors, and engraved with elaborate designs (see figure 1-29). Many holders had two small chains: one chain had a ring attached, to be worn on the finger, and the other chain had a pin attached for securing the flowers.

Flowers were arranged, then tied, wrapped in damp moss, and inserted in the bouquet holder. Most bouquet holders had a long pin attached that fit through perforations in the holder, crossed through the bound stems, and came out a hole in the opposite side and often secured with a screw.

Two distinct types of nosegay designs—formal and informal—were placed in the bouquet holders. The formal style had concentric rows of flowers with a rose or other fragrant, symbolic flower in the center. The rows of flowers and foliage formed rings around the top flower. The other style was informal, with a more casual mixing of fragrant blossoms.

Figure 1-29. Bouquet holders reveal exquisite workmanship, evident in this highly ornate holder studded with semi-precious stones. This elaborate holder features a small chain with a pin attached for securing flowers in place while dancing, and another chain with a ring attached intended to slip over the finger.

Photo and design by James Moretz, American Floral Art School, Chicago, IL.

Figure 1-30. Jewelers rivaled one another in making bouquet holders. This resulted in an enormous variety of workmanship in shape, decoration, and style. Accumulating the largest and most valuable collection of its kind, William Kistler, former owner and director of the American Floral Art School, collected Victorian bouquet holders over a span of three decades.

Photo and design by James Moretz, American Floral Art School, Chicago, IL.

Figure 1-31 and 1-32. During the 18th century, bosom-bottles made of metal and glass and able to hold a small amount of water, were invented to keep tiny bouquets of flowers from withering, when tucked centrally into the décolletage of a dress. Also popular during the 19th century (the Victorian era), tiny holders were sometimes worn at the waist, securely fastened with a brooch, or securely pinned among hair curls.

Photo and design by James Moretz, American Floral Art School, Chicago, IL.

Holders were also invented to hold and keep flowers fresh while worn by both women and men. Flowers worn in the décolletage were placed in ***bosom bottles*** to keep the flowers fresh and lovely (see figures 1-31 and 1-32). Small bottles were also fashioned to hold tiny clusters of flowers for men to wear. For a partial listing of design styles, container types, popular flowers, and accessories, see table 1-7.

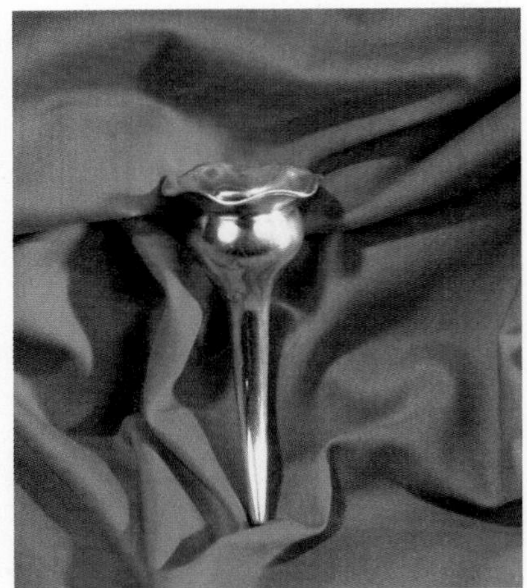

Figure 1-32 A.

Figure 1-32 B.

TABLE 1-7: FLORAL DESIGNS CHARACTERISTIC OF THE VICTORIAN ERA

Styles of Design
- tight masses of flowers; often softened by foliage at the design edges
- two- or three-tier flower arrangements made in epergne containers combined with fruits and vegetables
- hand-held bouquets
- dried flowers; pressed flowers under glass
- artificial flowers made from shells, feathers, paper, hair, etc.

Typical Containers
- popular container materials include metals, porcelain, ceramics, glass, and alabaster
- epergnes
- trumpet-mouthed vases
- wall pockets
- tuzzy muzzy holders
- quaint and curious vases made in innumerable shapes
- urns and jugs

Examples of Flowers
- flowers rich in texture, solid blooms, and unusual and interesting patterns were favored.
- streaked, fringed, and spotted flowers; especially carnations, pinks, tulips, camellias, lilies, chrysanthemums, peonies, poppies, roses, and dahlias
- bleeding heart and fushia
- freesia
- gardenia
- geraniums
- hydrangea
- lilac
- morning glory
- narcissus
- salvia
- stephanotis
- stock
- sweet pea
- tuberose
- violets

Examples of Foliage
- ferns
- grasses
- plumed grasses

Accessories
- figurines, fans, and shells
- Victorian greeting cards
- paperweights and oil lamps
- knick-knacks
- stuffed birds, butterflies, etc.
- dried flowers

ORIENTAL INFLUENCE

Contrasting considerably to the European style of arrangement, the ***oriental style*** of flower arrangement has greatly influenced the development of art of floral design.

Floral styles of both China and Japan share similar philosophy of nature but can be easily distinguished from each other. Both styles place emphasis on the individual form, texture, and color of plant material. Great importance is placed on line and negative space. The following paragraphs give a brief overview of Chinese and Japanese styles. For an in-depth look at the art of oriental arrangement, see chapter 16.

Figure 1-33. A Chinese arrangement, although appearing unstructured and naturalistic, requires careful thought and planning.

Chinese Style

China has long been known as "The Flowery Kingdom." Glimpses of the ancient ***Chinese style*** are preserved in paintings, prints, and scrolls. Many of these art pieces tell us how cut flowers were used. Although Chinese floral designs are less stylized and the technique of arrangement is less studied the Chinese art of flower arrangement has always been a respected art form and worthy of attention.

Japanese Style

The ***Japanese style,*** in contrast to the Chinese style, is highly formalized and follows strict rules of construction. The Japanese style was adapted from the ancient Chinese art and is steeped in tradition and symbolism. Generally, cut flowers and accessory materials are placed in arrangements in a manner reflective of how they are found in nature. Natural growth patterns are considered and plant material is groomed and pruned to perfection (see figure 1–34).

Figure 1-34. Japanese arrangements emphasize simplicity and line.

AMERICAN STYLES

America has few pictures or artifacts recording its history of flower arrangement from 1620 to the middle of the eighteenth century. Although earlier pilgrims and settlers brought with them a European heritage of floral design, their lifestyle was austere and plain; plant life was considered in terms of food and health rather than decoration. Life was hard for these newcomers to America, and their focus was no doubt survival in a harsh land, with little time for what they may have felt was unnecessary.

Early-American Style

The first settlers who came to New England were not part of aristocracy. They were Puritans seeking religious freedom. The ***Early-American style*** demonstrated that these colonists were avid gardeners, and their attention focused on providing plants and herbs for food and medicine. Once settled, early colonists placed wildflowers, grains, and grasses in everyday jugs, simple pottery pitchers, and pewter and copper kettles and pans, since they were not able to bring fancy and decorative floral vases with them to American (see figure 1-35).

Figure 1-35. Early Americans used a variety of simple and charming containers; pitchers, jugs, cups, and kettles made from pottery, copper, and pewter.

Colonial Style

When Williamsburg was established as the capital of the Virginia colonies, active trading began to take place. Many people and artistic influences were joined together creating diverse styles. The period between 1714 and 1780 is known as the Colonial Williamsburg period, and the floral style is named the

colonial style. The typical floral arrangement of this period is a massed, rounded, or fan-shaped bouquet that is casual, open, and homespun in style (see figure 1-36). Dried flowers and grasses were often included with fresh, cultivated flowers. Fruit and flowers were often placed on the dining table around the floral centerpiece. For a detailed listing of design styles, containers, plant material, and accessories common to the colonial style, see table 1-8.

Figure 1-36. Due to active trading and economic development, colonial vases show more diversity than the Early-American containers. Many were manufactured specially for the arrangement of flowers, as with the 18th-century, five-fingered posy holder and the rectangular brick vase.

TABLE 1-8: FLORAL DESIGNS CHARACTERISTIC OF THE COLONIAL STYLE

Styles of Design
- rounded, symmetrical bouquets
- fresh and dried flowers used in combination
- bouquets of one type of flower with a filler flower

Popular Containers
- bowls and baskets made of pottery, reeds, silver
- bricks, square and rectangular, enclosed with holes for flowers
- bough pots made of pottery
- canisters and oval tins
- handled cups
- epergnes made of ceramics and silver
- jars, jugs, and pots made of delftware, earthenware, stoneware
- urns made of earthenware, metal, and ceramics
- miscellaneous vases made of ceramics
- wall pockets made of ceramics

Flowers
- many plants under cultivation
- favored were roses, lilies, tulips, and carnations
- geraniums
- hyacinth
- narcissus
- lilac
- snapdragon
- stock
- babies breath, Queen Anne's Lace, and other delicate fillers
- wildflowers
- daisies
- dried flowers
- marigold
- fragrant herbs
- sunflowers
- rhododendron

Other Plant Materials
- grasses
- gourds
- nuts
- fruits and vegetables
- seed pods
- ferns

Accessories
- figurines
- candles and candlesticks
- Bible, books
- period portrait

Neoclassicism—Federal and Greek Revival

The ***Federal period,*** equivalent to the English-Georgian period in England, lasted from about 1790 through 1825. It is known as the political, social, and decorative formation era in America following the Revolutionary War. The style of floral arrangement during this period in America was distinct, reflecting the ***neoclassic style.*** The tendency was to break with anything reflecting English dominance. The shape of arrangements was often pyramidal or fan-shaped, influenced by the French design style. Roses were a favorite flower of this period, as were geraniums and trailing ivy.

Figure 1-37. The compact nosegay and the tight bouquets, softened with foliage, are characteristic of the Victorian style in America. Children of Israel Griffith. *(1799–1875) by Oliver T. Eddy, about 1844. Gift of John Buck in memory of Alverda Griffith Buck. Collection of the Maryland Historical Society, Baltimore, Maryland.*

The ***Greek revival*** (1825–1845) similar in style to the Federal period, was the final phase of neoclassicism. Large and elegant plantation mansions were lavishly deco-rated for extravagant parties and dinners. Arrangements were generally large and symmetrical in neoclassic containers.

American Victorian

The Victorian period in America is often called the ***Romantic era*** and was popular from about 1845 to 1900. Floral styles during this time in America copied the European Victorian styles. Interest in flower arrangement

Figure 1-38. Epergnes, characteristic of the Victorian era in Europe, as well as America, were popular vases for holding fruit and flowers. Bouquets of this type were often used as centerpieces.

Figure 1-39. Floral vases of Art Nouveau (1890–1910) were inspired by curving lines of nature; organic motifs of flowers, foliage, vines, and animals. The vase in the center left reflects new fern fronds ready to uncurl.

is evident with the beginning of garden clubs in 1862 and the founding of the Society of American Florists in 1864.

The ***epergne*** container remained popular, often decorated with fruit and flowers. The favorite epergne had a trumpet form, for tall flowers, rising out of a stemmed bowl base for fruit or low-spreading flowers (see figure 1-38).

FLORAL ARRANGEMENT STYLES

#2

Various floral arrangement styles have emerged in more recent times as well. As architectural, interior, fashion, and other aspects of design change, so do floral arrangement styles. Often floral arrangements reflect societal or emotional transformations as well. The following are just a few of the numerous floral arrangement styles that have emerged since the early 1900s, many of which are still popular today.

Art Nouveau

The ***Art Nouveau period*** lasted approximately twenty years from 1890 to 1910 and was popular in both Europe and America. This period was one of the most original manifestations of the human creative spirit. Art Nouveau is the term used for defining the style of art that had its main theme of curvilinear lines, often patterned after nature in the shape of plants of flowers as well as the human form.

Containers, as shown in figure 1-39, although made in different materials and shapes, were all inspired and patterned after curving lines. To repeat the style of these decorative containers, floral bouquets were curving and often asymmetrical to lend a more natural, rhythmic, and graceful appearance. Today, floral bouquets inspired by the Art Nouveau period are often arranged asymmetrically with curving flower stems and branches spilling abundantly (see figure 1-40).

Figure 1-40. The asymmetrical, cascading, waterfall style is deeply rooted to the Art Nouveau period.

Figure 1-41. The Art Deco style is characterized by strong, streamlined, geometric lines, forms, and patterns. The strong vertical lines of the vase and calla lilies are contrasted with the curving lines of the curly willow branch, bear grass, and strelitzia leaves, forming a stylized, geometric design.

Art Deco

The term ***Art Deco*** is derived from the 1925 Paris World's Fair exhibition called *Exposition Internationale des Arts Decoratifs et Industriels Modernes.* This style at the time was referred to as moderne of modernistic and was popular during the 1920s and 1930s in Europe and the United States.

A blending of several influences including the ancient Egyptian and Aztec cultures, the jazz age, the new industrial age, and other aspects of society are reflected in this style of design. This modern style was characterized by strong, streamlined, geometric lines, forms, and patterns including zigzags, pyramids, and sunburst motifs. The containers and floral arrangement styles displayed in figure 1-42 show geometrically bold and modern patterns representative of the Art Deco period.

The descriptive term Art Deco or ***le style 25*** as others call it, was not actually used to describe this modernistic style until the 1960s when the style reemerged.

Figure 1-43. Freedom of expression was clearly evident in the 1950s with bold foliages, flowers, and patterns reflecting form and idea. The oriental influence of line and the use of driftwood were vogue.

Free-form Expression

The 1950s brought a new approach to the arrangement of flowers—a modern, natural appearance. Floral arrangements became more expressive with both a feeling of movement and a feeling of freedom. It allowed for ***free-form expression.*** Linear designs, patterned after oriental styles, had a new look. The incorporation of driftwood and other figurines into floral designs were popular. As shown in figure 1-43, the use of flowers with bold forms, such as anthuriums, prevailed. Textural differences became apparent with the use of coarse, knotted driftwood juxtaposed with smooth, textured flowers and foliage. The dominant curvilinear lines of the driftwood and other elements rendered a feeling of movement.

Figure 1-42. Left: Surplus glass bricks with holes cut in them became trendy during the 1940s. Right: An abstract, angular arrangement of bird of paradise. These wonderful, bizarre designs during the 1930s reflected the influence of Art Deco—abstract, angular, and unusual. It was revived in the mid-1960s.

Geometric Mass Design

Popular during the early 1960s and early 1970s were tight geometric bouquets (see figure 1-44). These arrangements combined mass and line, and included both symmetrical as well as asymmetrical forms in stiff vertical or horizontal patterns. Mixed bouquets were popular as were tight gladiolus

Figure 1-44. Tight geometric forms, both symmetrical and asymmetrical, prevailed in the 1960s and early 1970s.

bouquets. Compote containers were commonly used to hold these arrangements.

Changing Contemporary Styles

Contemporary styles of floral arrangement change with the times. ***Contemporary*** refers to the time of the present or of recent times. Contemporary arrangements are also called modern because they generally are different than whatever style had been previously popular. (For a further look at contemporary styles see chapter 17.)

REVIEW

Flowers have, since the beginning of time, reflected the lives of people. Floral designs have been used for special times and they will continue to be a part of life's important occasions. The art of modern floral design is the blending of all the previous eras of historical flower arranging as handed down from generation to generation. It is the summation of all the finer and lasting qualities of each period. Almost every period in floral arrangement history has left its influence on the floral art of today.

In order to attain professionalism, you must be aware of color trends, interior design, and fashion not only of past floral periods but also of the present. Each generation sets a new pace of floral styles reflecting the way people live and think during a particular period.

TERMS TO INCREASE YOUR UNDERSTANDING

American styles
Art Nouveau period
Art Deco
baroque
bas-reliefs
bosom bottles
bough pot
Byzantine period
chaplets
Chinese style
colonial style
contemporary
cornucopia
décolletage
delftware
Dutch-Flemish style
Early-American style
earthenware
Egyptian period
English-Georgian period
epergne
European periods
ewers
faience
Federal period
free-form expression
French period
Japanese style
garlands
Grand era
Greek period
Greek revival
horn of plenty
le style 25
Middle Ages
neoclassic style
nosegay
oriental style
period style
posy holders
Renaissance
Roman period
Romantic era
terra cotta
tussie-mussies
Victorian era
Wedgwood

TEST YOUR KNOWLEDGE

1. What are practical reasons for knowing the history of floral art?
2. What are some factors that dictate change in floral arrangement styles throughout periods of time?
3. What are some floral traditions of the ancient civilizations?
4. Describe the influences of the European and the oriental style on the American style of design.

RELATED ACTIVITIES

1. Select pictures of flower arrangements from interior and home furnishing magazines and catalogs. Identify period-style arrangements and containers.
2. Choose a favorite period style. Plan the supplies needed and construct an authentic replication.
3. Make an English-Georgian style bud vase.
4. While visiting a museum, notice floral bouquets in paintings and prints from earlier periods. Speculate by the shape, style, and container the appropriate time period.
5. Gather together a few accessories that might typically be used in a Dutch-Flemish floral design. Make a Flemish-style bouquet and incorporate your accessories.

Chapter 2

Design, Harmony, and Unity

Figure 2-1. This centerpiece for a formal tea, reflects a theme and style. The parts, including the silver compote, the delicate roses, bouvardia, and heather, and the shape and style of the composition all support a similar style.

The dynamics of life are expressed through the art form of floral design. As in any art, the floral designer embellishes the form with personal interpretation. This interpretation is referred to as ***style.*** The word style in floral art may be used to describe a characteristic manner of construction by a floral designer.

Many factors contribute to the style of a floral composition. As an artist, the floral designer follows guidelines in order to communicate the idea that is being conveyed through the design. As with other forms of art, ***floral art*** is a form of expression that relies on the principles of ***harmony*** and ***unity*** to create pleasing designs. In order to increase your awareness of design and style, this chapter will discuss thoroughly the design process and the concepts, or principles, of harmony and unity. These principles, or guidelines, are the foundation upon which the floral designer bases the expressions that are conveyed through the designs. Once this foundation is established, the floral designer is free to explore many different avenues in order to establish a particular style.

DESIGN PROCESS

The word ***design,*** when used as a verb, means to plan and carry out a project in a skillful way. An arrangement of parts, details, forms, colors, and other elements should produce a complete and artistic unit—often called ***composition*** and design. The basis of design is that it is a ***planned process;*** it is not mere chance or haphazard luck that results in a pleasing composition.

A floral designer plans and organizes what others will look at and enjoy. Although there are many formulated rules of design—the standard do's and don'ts of making arrangements—these rules should be regarded as guidelines to assist you in the construction process of making successful floral compositions. It will be helpful for you to have a knowledge of the process of design as well as a working knowledge of the rules, or guidelines, of construction.

Three activities can stimulate the design process, or more specifically, increase your design ability; thinking, observing, and practicing. All three components are interrelated, not sequential, steps. When consciously repeated, this design process will inspire artistic design and discover solutions to various design situations.

Thinking

Great floral compositions cannot be created mindlessly simply by impulsively throwing flowers, objects, and various foliages together. It is necessary to think about the requirements of a particular floral arrangement. What are the visual effects, style, and theme? What will be the placement? What are the physical limitations—size, color, shape? When is the design needed? And where will it be used? Generally, design problems arise simply because the specific requirements needed for a floral composition are not fully understood.

Thinking of design solutions is particularly important when communicating a theme or message. For example, a holiday design for Halloween will portray a distinct theme of colors, flowers, and accessories. A bouquet for a special occasion, such as the birth of a new baby, will also display a unique theme. An arrangement made for a design competition generally follows a given theme and conveys a message to the viewer. It is helpful to consider various possible solutions that could be appropriate for a desired theme. As ideas come to mind, write them down and sketch these possibilities. Sometimes discussing ideas with others will help us choose the best idea or combination of ideas that will result in a design that conveys a theme, and, better still, communicates positively with the viewer.

Observing

We can educate ourselves by observing carefully the work of other floral artists of today as well as those of the past. Studying the floral art from all periods and of different countries is also instructive and enlightening. As designers, we benefit from what other floral artists have done and are doing. We learn by observing others.There is a great deal of visual information available in floral magazines and books. Visiting floral shops will also give you insight to floral design and composition as with attending floral demonstrations and symposiums. Looking at the work and style of other floral designers will give you ideas for finding solutions to your design situations.

As you observe the designs of others, analyze what you see. When you are drawn to a design, consider what captivated your attention. Consider what techniques or visual design elements the designer used to make an impression. On the other hand, if you see an arrangement that is disturbing or unfavorable, don't disregard it as unworthy of attention, but stop and analyze to figure out what it is that bothers you. By examining every composition you see, your design awareness will become greater, your confidence will increase, and your own style of design will develop. Careful observation of the work of other floral artists provides a basis to create your own original floral designs.

Practicing

Thinking and observing are only parts of the design process. The well-known cliché "practice makes perfect" is a reminder to the importance of hands-on work. Gaining experience in the actual construction process and finding solutions to design problems are vital to the entire process of design.

Experimenting and exploring on your own with flowers and foliages is important and will give you experience in the design process. However, attending hands-on workshops and floral classes, and working in floral shops will give you essential practice in finding solutions for design situations. Practicing the art of floral arrangement will give you experience, as you gain more and more experience as a floral artist, you will also increase in confidence in expressing your own style of design.

Figure 2-2. The parts of a floral composition include a container and mechanics, the flowers, the foliages, and accessories. Other important considerations for creating a theme or style include the placement or location and purpose or function for the bouquet.

STYLE AND COMPOSITION

Many famous floral artists often develop a unique or distinct style. Style can also refer to the type or expression of a design itself. The container, flowers, foliages, the completed shape, and other factors together help to determine the style of a design. Style can relate to a time period, such as the Renaissance or the Victorian era, or it can relate more specifically to a region, such as Japan or Europe. Style can also relate to the purpose, function, or message of a floral arrangement. Sometimes a certain mood or theme may be needed. Asking yourself "what style is needed?" before beginning construction of a floral bouquet, will expedite the design process and the gathering of supplies to make a floral composition.

Composition is defined as putting together a whole by combining parts or elements into a unified, harmonious whole (see figure 2-2). Often in floral design, we think of a composition as a completed arrangement. A composition is made from parts, individual objects, and even ideas that are organized into a unified whole. Each element contributes effectively to the whole to complete a single theme or idea.

In forming a floral composition and creating a style of design, it is necessary to select the parts according to the purpose for which the arrangement will be used. Every arrangement will have a different set of design problems. Theme, color, size, shape, message, placement, and other factors must all be considered before making a floral arrangement.

Before choosing the parts for a floral arrangement—the container, flowers, foliages, and accessories—it is important to consider each of them carefully. The following questions relative to the size and shape of the completed design and its final placement are important considerations in planning and constructing a floral design: For what purpose is the arrangement going to be used? Does the arrangement need to be made for a special occasion, or does it need to match a predetermined theme? What is the size of the room and the location in the room in which the design will be placed? From what distance will the arrangement be viewed? At what eye level will the composition be viewed? (See figure 2-3.) These are just a few considerations important to the creation and organization of parts into a floral composition.

Figure 2-3. Many factors help determine the design style. Eye level is an important consideration.

Figure 2-4. These arrangements lack a sense of harmony shown by flowers, foliages, and containers. The sizes, textures, colors, forms, patterns, and style need to harmonize with one another to create a desired style.

Container and Mechanics

The choice of a container may depend upon physical factors such as the size of the bouquet, the size of the flowers, the mood and color for a design, the shape of the design, and the placement for a bouquet. The container must harmonize with the other parts of the design as well as to the background setting and the mood to be achieved. The mechanics of the design may also be a factor relating to harmony. If the container chosen for a particular arrangement is clear glass, the mechanics or foundation must not be seen; otherwise, unsightly mechanics distract from the aspects of harmony in design. (For more information on containers and mechanics, see chapter 8.)

HARMONY

Harmony is the wise selection of individual parts to form a floral composition. Harmony is often associated with music, when individual sounds from instruments or voices combine together to make a beautiful and pleasing composition. However, the parts must be in tune with one another and should combine into a pleasing and orderly whole. Harmony refers to the effect of the proper relationship between all the various elements in a composition.

The principle of harmony is an essential ingredient in floral design; a lack of harmony is disturbing (see figure 2-4). A determined theme and purpose for a floral arrangement will enable you as a designer to select materials for a harmonious design more easily.

Harmony of parts is evident when the same color is used throughout a design. Other elements, such as form and line, can also be repeated. To study the structure of a floral composition as it relates to harmony, first it is necessary to look at the requirements for the bouquet. You might ask yourself the following questions to get a feeling for harmony: What mood and theme is needed—Formal, sophisticated, informal, natural, cute, funny, period style, contemporary, special occasion, season, or holiday? What size, shape, texture, and color are needed for the various parts to match this theme? After selecting the parts, do they all go well together? Are they "in tune" with one another? Do added accessories enhance the theme? (See figure 2-5.)

Figure 2-5. A feeling of harmony is exhibited in this buffet design. The design style, shape, flowers, and added accessories enhance and support a football or homecoming party theme.

Figure 2-6. All the component parts of a floral composition should somehow harmonize with one another. It is vital that the flowers in a design are compatible to each other in order to support an overall design style.

Flowers

The flowers you select for a floral composition must be in harmony with each other. Their sizes, colors, textures, and forms, must all be considered carefully (see figure 2-6). They do not all need to be the same size or color, but they do need to suggest a similar mood or theme. For example, in order to create a country-garden theme, the flowers generally need to look like they come from a country garden. Flowers such as zinnias, marigolds, delphiniums, asters, sunflowers, etc., support this theme and blend harmoniously with one another. Any exotic or tropical theme is supported with the use of flowers with exotic, tropical forms. Heliconia, anthurium, bird of paradise, and other similar flowers are harmonious to a bold tropical theme.

If a container is predetermined, the flowers must be chosen to harmonize with its shape, texture, color, and size. Often however, flowers are chosen first, before a container, which gives more freedom to the selection of flowers for a theme or style.

Foliage

The foliages selected for a floral composition must harmonize with the flowers as well as with each other. Their sizes, forms, colors, and textures must be considered, and each should be selected to support and harmonize with the desired theme.

Accessories

When used for ***theme*** bouquets or special occasions, accessories are an important addition to creating a desired style or conveying an important message (see figure 2-7). Whether the accessories are placed within the bouquet or set near the floral bouquet on the table to form a larger area of composition, their purpose is

Figure 2-7. Accessories are often important additions in floral arrangements. The candles, grapes, plums, and kiwi in these 85th birthday party designs repeat colors and enhance an overall harmonious theme.

Figure 2-8. Harmony is ultimately achieved when the arrangement blends, and looks correct in style and placement with its setting or location.

Figure 2-9. Unity in floral design refers to the completed arrangement or the "whole" being more important, or predominant, over the individual "parts." This elaborate, tropical centerpiece demonstrates unity as a result of harmonious parts.

to enhance the total theme. As you choose your accessories, be careful that they do not detract from the flowers and the style of a composition. Do not let accessories override the importance of the flowers and foliages.

Placement

A knowledge of where the floral composition will be placed is helpful in the selection of parts (see figure 2-8). The size of the room and the location in that room where the floral design will be placed is important for knowing the relative sizes of the container and flowers. For example, a large room generally requires a larger bouquet with larger flowers and a larger container in order to maintain balance and proportion.

Also important is a knowledge of the interior colors, patterns, and style of the room in which the arrangement will be placed. With the room interior in mind, you can choose the parts of a floral design to complement and harmonize with an existing style.

You can practice creating harmony by envisioning floral arrangements to go with a certain set of circumstances. For example, what materials can you envision for a springtime arrangement to be delivered to a young teenage girl in the hospital who has just had knee surgery? What size, shape, and mood will harmonize with the purpose, style, and setting? Asking questions to match the criteria of a floral design will enable you to select the parts for a design with harmony in mind.

Figure 2-10. A lack of unity exists when an arrangement can be divided into chunks or separations.

Figure 2-11. Three important ways of achieving unity are proximity (combining flowers into one design), repetition (repeating similar elements such as color throughout a design), and transition (providing a gradual change from one part of a design to another).

UNITY

Related to harmony, but slightly different, is the idea of unity. The word unity is defined as the state of being one, united or complete in itself. Unity means that the whole composition must be predominant over its individual parts. In order to achieve unity, it is vital that you view the floral design initially as a unit, not as a combination of parts. As illustrated in figure 2-10, if the arrangement has chunks or separations, it is not completely unified.

Unity is oneness of purpose, thought, style, and spirit. Complete unity of a floral arrangement is a conscious effort, involving the skill of a floral designer in the selection of parts, as well as the purpose and placement for the bouquet. Unity in floral design is best expressed through proximity, repetition, and transition (see figure 2-11).

Proximity

A fairly easy way to achieve unity in a floral design is to organize flowers and foliage into one container. Generally, by placing the flowers close together, the varying textures, sizes, shapes, and colors of the flowers result with a type of unity. It may not always be harmonious, but a certain unity does result. For example, the ***Mille de Fleur*** designs (meaning "thousands of flowers"), which incorporate many diverse flower varieties and colors in one composition, somehow blend together to make a unified whole.

Repetition

A valuable and widely used method for achieving unity is through repetition, in which some element of the floral arrangement is repeated throughout the design to relate the parts to each other and to the whole. Repetition of the same color throughout a bouquet is the easiest way to achieve immediate unity. Other elements such as texture, flower type and shape, size, and line angles may also be repeated to enhance the feeling of unity in a floral design.

Transition

Another way to achieve unity is through transition, also called ***continuation***. This method takes a little more planning. Transition refers to a gradual change from one element to another; elements are arranged and organized in a pattern that provides continuous eye movement.

Transition is best achieved by having intermediate colors, sizes, shapes, textures, and spacing within a design. For example, color is an easy way to achieve transition. The orange flowers shown previously in figure 2-11 visually connect the red flowers with the yellow. Eye flow is smooth, and a sense of unity results.

Figure 2-12. An overwhelming desire for harmony and unity can result in an arrangement that is uninteresting, dull, and boring.

Unity with Variety

Too much blending of parts can be monotonous and boring. The cliché "Variety is the spice of life" can be a useful guide to floral design. In our desire to seek harmony and unity, it is possible to create a composition that is visually dull, as shown in the uninteresting arrangement in figure 2-12.

Variety can be achieved by using certain techniques that alleviate boredom. For example, you may repeat colors, but use different values and intensities. Or you may repeat shapes but at the same time use different sizes, or use similar flower types, in different colors. Adding a strong focal point or a striking accent area within a bouquet can also relieve monotony. Another way to add interest is to vary the outside shape of a floral arrangement, so that it is not geometrically stiff (see figure 2-13).

Figure 2-13. A geometrically stiff design appears more relaxed and natural with some variation in the outside perimeters. A unified appearance with added variety is achieved.

REVIEW

The procedure of design is a planned process that involves thinking, observing, and practicing. These activities will stimulate the design process and your sense of style by increasing your awareness and experience. The creation of a certain style is directly related to the parts or elements selected to form a floral composition, which is the organization of parts into a harmonious whole. You can achieve this harmony and unity in your designs through the conscious, wise selection of parts. These parts—which include the container, flowers, foliage, and accessories—will together support a desired theme, mood, or style, and result in a harmonious and unified design.

TERMS TO INCREASE YOUR UNDERSTANDING

composition
continuation
design
floral art
harmony
Mille de Fleur
planned process
proximity
repetition
style
theme
transition
unity

TEST YOUR KNOWLEDGE

1. What are some ways to enhance and increase your design skills?
2. What are some popular styles in floral arrangement today?
3. Name the parts of a floral composition.
4. What are the similarities and differences between harmony and unity?
5. How can unity in floral design be easily achieved?

RELATED ACTIVITIES

1. Plan a theme and setting. Make an arrangement suitable in shape, size, color, and style.
2. Visit a retail shop and notice the display arrangements. What themes do you notice?
3. As a class activity, divide into groups. Each group will select a different theme or style. Plan container possibilities, as well as flowers, foliage, and accessories that will be needed to support the selected theme.

Chapter 3
Color

Color is perhaps the most important element of design. As shown in figure 3-2, it might very well be the only visual element of a flower arrangement that anyone ever notices. Often color is the ultimate decision for the high school student ordering prom flowers. It is a major decision for the bride-to-be in planning flowers for the wedding. And it is of utmost concern for the fastidious designer using flowers to accent the interior of a home or business. Therefore, a knowledge and understanding of color as well as its properties, effects, and schemes is beneficial. The floral designer who is well versed in the use and function of color will be able to use it freely and effectively.

Figure 3-1. Vivid Color. Flowers and foliage with various tints and shades of the primary, secondary, and tertiary colors make up this striking "color wheel" design.

The fundamental basis of color theory is that color is a property of ***light.*** In reality, light is the source of all color. Without light, color does not exist. The property of light with regard to color was illustrated by Sir Isaac Newton in the 17th century. He passed white light through a prism, which caused the light to separate into a rainbow of colors because of varying wavelengths.

Figure 3-2. As it is generally noticed first, color is often the most important visual element in a floral design.

Color theory is an important and complex science that goes beyond the scope of this book. However, to understand color and its use in floral design, you must be familiar with its properties and psychological effects, its use in creating successful design principles, and the various color schemes.

COLOR PHENOMENON

Because color is a property of light, colors are not constant. As light changes, so do colors. Flowers and objects under artificial lighting alter when they are moved outside to natural sunlight.

Colors can also change according to their surroundings and juxtaposition with other colors. A significant fact to remember in floral design is that a color is not important in itself. Color cannot be isolated from its surroundings. The adjacent colors in a bouquet and the background colors in a bouquet all affect each other. The eye perceives color in relation to its environment.

PROPERTIES OF COLOR

Although color is difficult to define, every color has three basic properties: ***hue, value,*** and ***intensity.***

Hue

The varying wavelengths created as white light is passed through a prism produces vast visual differences. Hue is the property that gives the color its name, such as red, blue, yellow, and green. This property sets each color apart from all the others.

Each color on the wheel shown in figure 3-3 is often called a hue. The basic ***color wheel*** of twelve hues can be divided into three categories, (see figure 3-4). The ***primary colors*** are red, yellow, and blue. These three colors are often referred to as ***foundation colors*** because from these all other hues on the wheel can be created. (The simplest method of seeing how colors are "made" is to mix paints.) The second group of colors are referred to as ***secondary colors.*** These colors are made up of a combination of two primary colors, resulting in orange, green, and purple. The third group, called ***tertiary colors*** or ***intermediate colors,*** is created by mixing a primary with an adjacent secondary color on the wheel. The six tertiary colors get their names from their "parent" colors, resulting in hyphenated words such as red-orange, yellow-green, and blue-purple.

Value

Value is another property of color (see figure 3-5). Value is a measurement of the amount of light reflected from a colored object. Value refers to the lightness or darkness of an individual hue. In paints, value can be changed by adding black or white to a hue. Adding black darkens a color and produces what is called a ***shade.*** White lightens a color and produces a ***tint.*** Value, much like color itself, can vary with surrounding hues. For example, red roses in a black vase against dark foliages will appear lighter in color than the same red roses adjacent to pink and white flowers in a white vase, in which the roses will appear darker in color.

Intensity

Another important property of color is intensity, sometimes called ***chroma*** or ***saturation.*** Intensity refers to the brightness or concentration of a color. A low intensity or neutralized version of a color is a ***tone.*** Dulling a pure hue to make a tone is accomplished by adding gray to a hue or mixing a hue with the opposite or complementary color on the color wheel. ***Complementary colors*** placed next to each other intensify one another, making each other brighter.

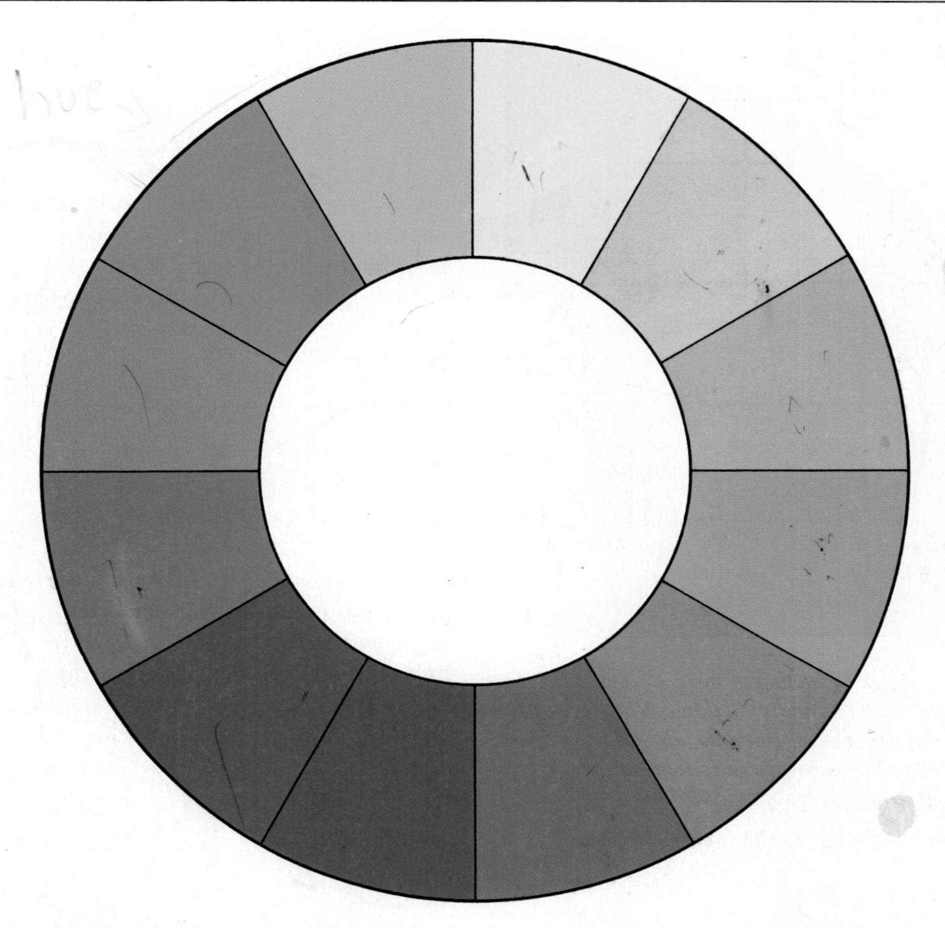

Figure 3-3. The most common organization of twelve hues, patterned in a wheel, shows color relationships.

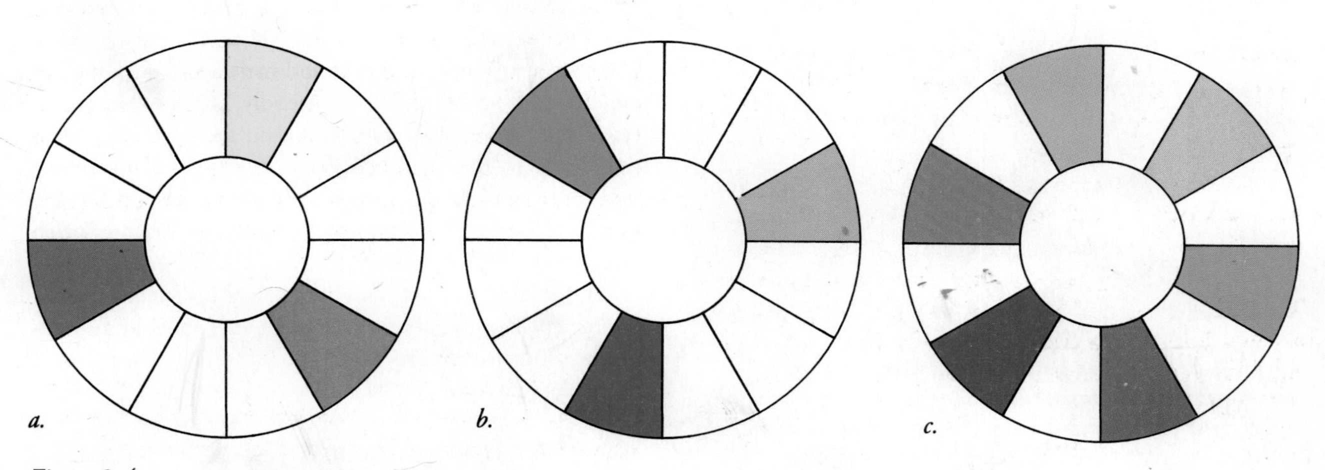

Figure 3-4.
a. Three primary colors—red, yellow, and blue—are foundation colors.
b. Three secondary colors—orange, green, and purple—are made from a combination (not always equal) of two primary colors.
c. Six tertiary colors—red-orange, blue-green, etc.—are intermediate colors created by mixing a primary color with an adjacent secondary color.

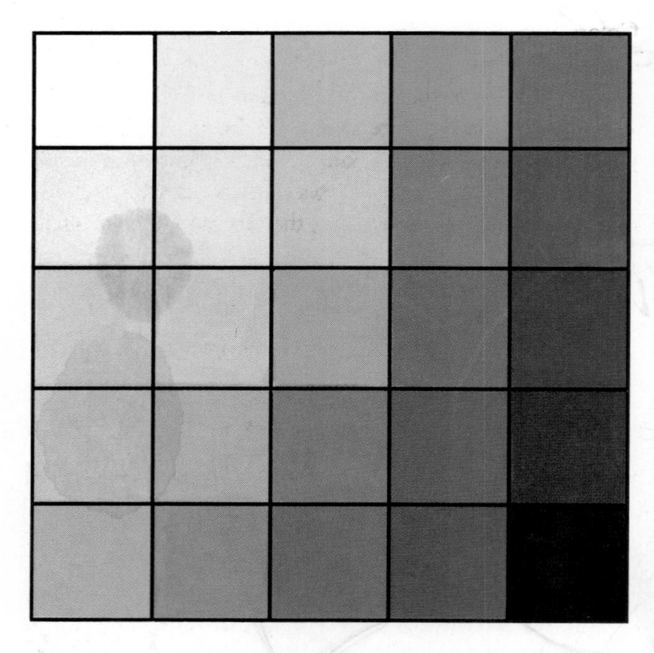

Figure 3-5. Value, which is another property of color, refers to the lightness or darkness of an individual hue. Various value levels are called tints and shades. Intensity, also called chroma or saturation, refers to the brightness or concentration of a color. Various intensity levels are often called tones.

Figure 3-6. Colors can be easily divided into two groups—warm and cool. Because of learned associations, we often identify color as either warm or cool. Hence, yellow (sunshine) and red and orange (fire and heat) are warm colors. Blue (water and ice), green (grass and plants), and some purples (shaded snow and mountain shadows) are cool colors.

PSYCHOLOGICAL EFFECTS

Colors can set the mood of an arrangement and create emotional impressions. The simplest way to classify colors according to their psychological effect is to divide them into two groups—warm and cool (see figure 3-6).

Warm Colors

The ***warm colors*** are red, orange, yellow, and various colors containing these three hues. They are termed warm because of their association with warm and hot things, like the sun, heat, and fire. These colors are active, cheery colors, generally evoking warm and happy feelings. Flowers and objects in warm colors dominate an arrangement, look larger, and appear to advance and move toward the viewer. Because warm colors look larger, they can be easily seen from a distance. Warm colors are generally informal and blend easily with each other. However, large amounts of warm colors can be psychologically irritating to some viewers.

Cool Colors

The ***cool colors*** are blue, green, purple, and colors containing these three hues. These colors are called cool because of their association with cool things like grass, water, and ice. These colors are generally restful, peaceful, and soothing. They are associated with quiet and melancholy feelings and are, therefore, considered "less friendly." Objects and flowers in the cool colors recede and tend to move away from the viewer so they look smaller than they are. Since they fade into the background, cool colors often cannot be seen from a distance. These colors may appear formal and often can display a lack of unity.

A blending of warm and cool colors can create a visual excitement because of the balance of advancing and receding qualities (see figure 3-7). Thus, the use of warm and cool colors increases depth in a floral design.

Even though purple is generally categorized with the cool colors of blue and green, it is easily swayed into displaying either cool or warm properties simply by the colors adjacent to it. As shown in figure 3-8, the iris in both arrangements are fairly similar in color; yet in the vase on the right where the iris are juxtaposed with reddish purples, they display cool properties. In the vase on the left, the purple iris are adjacent to varying blue flowers, so the iris look redder and warmer.

In addition to being warm or cool, each color has other characteristics that invite varied emotional responses. Colors have various emotional associations and they vary from one society to another. For example, if you are feeling sad and blue or maybe green with envy, or feeling cowardly yellow, or perhaps you're seeing red for some reason and you're really in a black mood, it won't help to use purple language or tell white lies! Instead, get yourself some flowers. No matter what their color, they are sure to cheer you!

Figure 3-7. Antithesis of Warm and Cool. A blending of warm and cool colors in a floral arrangement can create visual excitement while enhancing a feeling of greater depth. The yellow daffodils and acacia, orange ranunculus, and red tulips all advance toward the viewer, while the blue delphinium and cornflower, purple statice, and green stems and leaves tend to recede.

Figure 3-8. Color Phenomenon. The visual effects of purple vary greatly and depend upon surrounding colors. The purple iris in both designs are similar in color, but when juxtaposed with the cool blue iris, delphinium, statice, and cornflower, as in the vase on the left, the iris appear redder (warmer). When surrounded by reddish liatris, godetia, heather, statice, and freesia, the iris shifts and becomes more blue (cooler).

Individual Colors

Since everyone has favorite colors, it is generally easier to make an arrangement using preferred colors. Likewise, it is more difficult and frustrating to make a floral design using colors that are less favorable. In addition to individual preferences, colors also are known to evoke moods and feelings and appeal to our emotions. For this reason, a knowledge

Figure 3-9. Pure Elegance. Varying textures and forms are more apparent in this all-white bouquet of lisianthus, larkspur, waxflower, and Queen Anne's lace.

Figure 3-10. Fiery Flowers. This composition of red-hot gerbera, anthurium, protea, telopea, and red-veined croton leaves embodies strength and dominance.

of the varied emotional responses of individual colors can help you as a designer to use color effectively.

White

Although white is not found on the color wheel, it is a color common in floral work. Nature offers many white or near-white flowers that blend easily with other colors and make adjacent colors look cleaner and livelier. White is a perfect neutral, which, because of its reflective qualities, adds brightness and contrast. Varying textures are heightened in an all-white bouquet, which is sometimes called ***achromatic*** (without color). White is simple, portraying both elegance and sophistication (see figure 3-9).

Red

Red is a lively, stimulating color that embodies strength and dominance, as shown in figure 3-10. Because of its bold and dramatic characteristics, it should be used with care. The advancing warm quality of red can result in an overcrowded look. A good design rule to remember when combining red with other colors is to leave enough space between red flowers and objects so they have room to "visually grow." Red in only small amounts is showy and says, "Look at me." It demands attention and often can become overpowering when used with other colors, which often get visually pushed aside. It takes some practice and experience to mix red tastefully with other colors. Because the complement to red is green, using green foliage will intensify the red flowers.

Pink

A tint of red, pink, is not a color on the standard color wheel, although it is common in many flower types. Pink successfully combines with many colors and is often enhanced by the use of stronger contrasting colors. A mint or light green is the natural complement to pink. A delicate pink clearly portrays romance and femininity. Intensity and value levels are greatly varied in pink flowers as shown in figure 3-11. Bright and deep-colored pinks draw more attention than do the subdued or light, pastel pinks.

Figure 3-11. Romance Unparalleled. Pink flowers are available in a wide range of intensity and value levels as shown in this composition of gerbera, lily, nerine, lisianthus, freesia, statice, sweet William, and mini carnation.

Orange

The warmth of red and the liveliness of yellow combine together to make the radiance of orange. Like red, orange is a stimulating color but not as visually demanding. Orange compels attention and adds brightness (see figure 3-12). An intense orange hue is often reserved for autumn and Halloween bouquets. More common are the tints (coral, salmon, peach, etc.) and shades (rust, brown, etc.) of orange that blend effectively with many colors. Blue is the complement to orange on the color wheel. Tints and shades of these two hues can skillfully be combined to form lively combinations.

Figure 3-12. Energetic Orange. Orange compels attention and adds brightness as shown by all the flowers in this composition. Bird of paradise, crocosmia, lily, rose, celosia, Euphorbia fulgens, Chinese lantern, croton leaves, and berries form a lively, warm composition.

Yellow

Yellow reflects a great deal of light. Because of this quality, it is vibrant and highly visible. Yellow suggests cheerfulness and sunshine, as pictured by the sunflowers in figure 3-13. Yellow is viewed as a "friendly" color, often dispelling gloom. Adding small amounts of yellow to an arrangement

Figure 3-13. Sunny Sunflowers. Yellow suggests cheerfulness and sunshine, evidenced by the brilliant sunflowers. The value and intensity levels of the lilies, yarrow, freesia, solidaster, and button mums, as well as the flower forms, contrast with each other, alleviating boredom and maintaining interest.

adds spirit and can perk up an otherwise dull design. However, when used alone at full intensity, some viewers may find yellow arrangements monotonous and even annoying. The true complement to yellow is purple, and flowers in these two colors are often combined to create springtime designs. An extremely versatile color, yellow will blend well with many other colors and schemes.

Green

The cheerfulness of yellow and the coolness of blue combine to make green. In fresh arrangements, green is the natural background, as readily seen in foliages and flower stems (see figure 3-14). Green is a cool color that is generally soothing and restful. Because green containers do not attract unwanted attention, they are more usable than any other color. Green flowers are available but are rare.

Figure 3-14. Uncommonly Incredible. Green flowers and variegated foliages add interest to floral designs. The various greens of the gladiolus, Euphorbia marginata, and foliages combine together to form a cool and relaxing design.

Blue

Blue is generally peaceful, quiet, and cool as shown in figure 3-15. Blue varies greatly under different lighting; many blue flowers are actually purple. Blue flowers recede and fade into the background when viewed from a distance. Be careful when using large quantities of dark, deep shades; they can have a depressing psychological effect.

Figure 3-15. Heavenly Blues. The blue delphinium, iris, cornflower, and statice combine together, forming a peaceful, visually soothing design.

Purple

The coolness of blue and the warmth of red blend to form purple or violet, a rich and dramatic hue shown in figure 3-16. Because purple is a combination of the two extremes in emotional temperature, it can be cool or warm

Figure 3-16. Purple Pizazz. Purple is a dramatic, rich color with a wide range of value, intensity, and red or blue levels. In this composition, dark purple gladiolus, lisianthus, and iris contrast with lighter-colored liatris, freesia, and aster.

depending upon many factors; including lighting, background, juxtaposition with other colors, and the percentage of red and blue in the make-up of the hue itself. This phenomenon is clearly illustrated in figure 3-8. A purple can appear warm or cool depending upon neighboring colors.

COLOR IN DESIGN

Color can be used effectively to create many of the principles of design, such as balance, depth, focal point, rhythm, harmony, and unity. A knowledge of how these design principles can be incorporated through the use of color is a useful tool for making successful and pleasing floral compositions.

Balance

Color can be used to achieve or maintain visual ***balance,*** especially in asymmetrical designs. Using colors with bright intensity or dark value levels near the rim of the container, near the base of an arrangement, will lend a sense of visual balance. The brighter or darker the color, the less quantity is needed because of the attention these colors demand. Greater quantities of tints and subdued colors should generally be used when contrasted with an area of extreme brightness or darkness providing areas for the eye to rest.

Depth

Depth can be maximized by using a combination of warm and cool colors within an arrangement. Mixing warm colors with cool causes warm colors to advance and come forward and cool colors to recede, which enhances visual depth.

Focal Point

Color is often the most effective way to provide a ***focal point*** in a floral arrangement. Other methods of creating a focal point may include the use of a larger flower or an interesting flower shape, but color is the element that demands attention. Bright, intense colors immediately attract the eye, as do dark, deep colors. Because of their advancing properties, warm colors will attract attention over cool colors. In a floral design, you can create a focal point simply by using a contrast in color.

Rhythm

The use of similar colors throughout a bouquet is a simple way of attaining ***rhythm.*** When you use the same or corresponding colors at the focal point, and then use them again throughout a design, you invite eye movement. Another way of achieving eye movement or rhythm is to use a gradual transition from one color to another.

Harmony and Unity

Harmony and unity are both achieved more readily by the repetition of color throughout a design. You can unify the parts of the arrangement and create a harmonious overall look with color.

COLOR SCHEMES

Often in floral design work, it is easy to get locked into choosing favorite colors or well-known combinations. Remember to expand your color selection, become familiar with various color schemes, and consciously plan the visual effects you intend for finished bouquets. Learning about color and the color wheel will enable you to approach unusual color combinations and create an unlimited number of them freely.

The color schemes of fresh flower designs are determined generally by the flower and accessory colors. The green already present in the foliages and in the flower stems generally does not count as one of the hues forming a scheme unless it is consciously planned by the designer. ***Neutral colors*** such as white, black and gray may be added without changing the name of a scheme. Color schemes derived from the color wheel are either ***related, contrasting,*** or ***discordant.***

Related Schemes

Related schemes, such as monochromatic and analogous, use a common color to act as a unifying element. These schemes are generally pleasing and harmonious, but they can sometimes be boring.

Monochromatic (one color)

A ***monochromatic scheme*** (mono meaning one and chroma meaning color) involves the use of a single hue. A danger associated with monochromatic schemes may be a monotonous design. To help alleviate boredom, vary the individual hue to incorporate tints, tones, and shades to include a wide range of values and intensities (see figure 3-17). The visual effect should be harmonious with unity generally prevailing. These schemes can be enhanced by the use of varying textures and patterns. The floral centerpiece pictured in figure 3-18 uses red as its base hue. Tints and shades soften the boldness of red and provide visual rhythm and unity. Other examples of monochromatic floral arrangements in this chapter can be seen in figures 3-9 through 3-16.

Figure 3-18. Monochromatic Harmony. While red is the base hue in this monochromatic design, the pink tints and maroon shades help to soften its boldness.

Figure 3-17. Monochromatic color schemes are developed from a single hue and may include a wide range of values and various degrees of intensity.

Analogous

The word ***analogous*** means similar in function, origin, and structure. When applied to the color wheel, analogous refers to adjacent or neighboring hues. These schemes are created using several colors that appear next to each other on the color wheel. And because in nature real flowers do not always appear in the exact hues from the wheel, the analogous scheme may be modified to include tints, tones, and shades of adjacent hues as shown in figure 3-19. The analogous floral arrangement in figure 3-20 uses blue-purple, purple, and red-purple in varying shades and tints.

Contrasting Schemes

Contrasting schemes are more visually exciting because the colors used to form these schemes are from distant parts of the wheel. Of these, the most easily recognized are complementary and triadic. Since these schemes use unrelated colors, you must determine correct color proportions and proper color placement so that your designs will not look "thrown together."

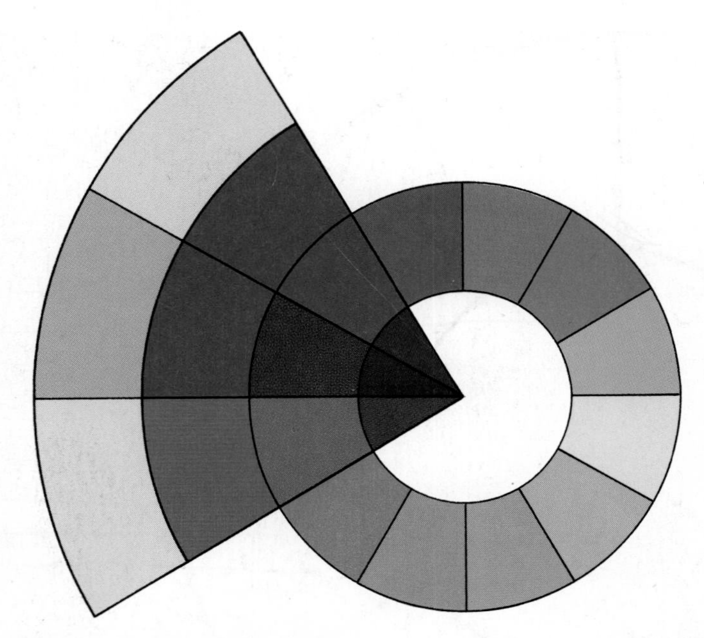

Figure 3-19. Analogous color schemes are developed from several (commonly three) colors adjacent to each other on the color wheel. Harmony is easily maintained with these schemes because they usually have one color in common.

Figure 3-20. Analogously Gorgeous. Various tints and shades of red-purple dendrobium and liatris, purple china aster, trachelium, statice, blue lace flower, and larkspur, and blue-purple delphinium blend harmoniously together.

Complementary

A ***complementary scheme*** uses two hues that lie directly opposite from one another on the color wheel. When juxtaposed, these hues intensify and complement one another. Some easily recognizable complementary schemes are red and green, yellow and purple, and blue and orange. Flowers in varying tints, tones, and shades can be used to provide further interest as shown in figure 3-21. Floral arrangements using complementary schemes are lively. They involve the use of a warm and cool color, which causes emotional excitement and enhances visual depth. The floral arrangement pictured in figure 3-22 uses cool blues with warm oranges to create an interesting color pattern.

Split Complementary

Another form of a complementary scheme is the ***split complementary scheme***. This is a three-color scheme composed of any hue, plus the two hues adjacent to its complement. An easily identified split complementary floral arrangement uses yellow, but instead of using the opposite color of purple, it uses the two hues that are adjacent—blue-purple and red-purple. These colors contrast less than the direct complement, which adds variety and interest (see figure 3-25a).

Triadic

Derived from the word triad, a ***triadic scheme*** means "involving three." When applied to the color wheel, the three colors are equidistant or equally spaced, from one another. Because triads use colors that are unrelated, they are difficult to work with. Great care is needed in choosing flowers and accessories to make these designs palatable. Tints and shades of a triad can soften the sometimes harsh effects

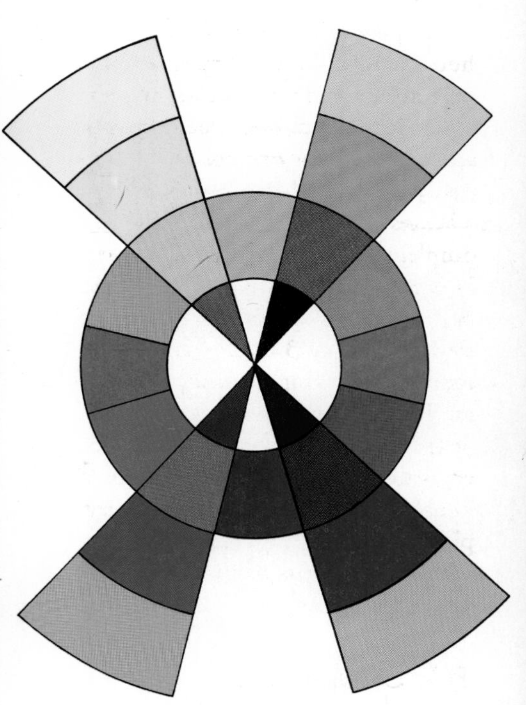

Figure 3-21. Complementary or contrasting color schemes are developed from colors that are opposite to each other on the color wheel. Values and intensities may vary for increased interest.

Figure 3-22. Striking Combination. The warm, orange gladiolus, lilies, safflowers, and croton leaves are intense when juxtaposed with cool, blue delphiniums.

of the unrelated colors (see figure 3-23). The most common and easily identified triadic scheme uses the primary colors of red, yellow, and blue, as shown in figure 3-24.

Discordant Schemes

Discordant schemes, such as double complement and tetrad, use four different colors that are widely separated on the color wheel. Discordant schemes are rarely considered to be a valid option in creating successful compositions; however, these schemes offer new territory for the experienced designer. These schemes use unrelated hues that have no affinity for one another and usually clash when placed side by side. In order for them to appear dynamic rather than ostentatious, the values, intensities, and proportions should be varied. If this is done, few fresh flower combinations will be truly unpleasant. The ***double complement scheme, alternate complement scheme,*** and ***tetrad scheme*** are clearly energetic color combinations since they use four extremely dissimilar colors (see figure 3-25b, c, d).

Double Complement

The double complement scheme uses four colors (two pairs of complements), such as yellow and purple combined with yellow-orange and blue-violet.

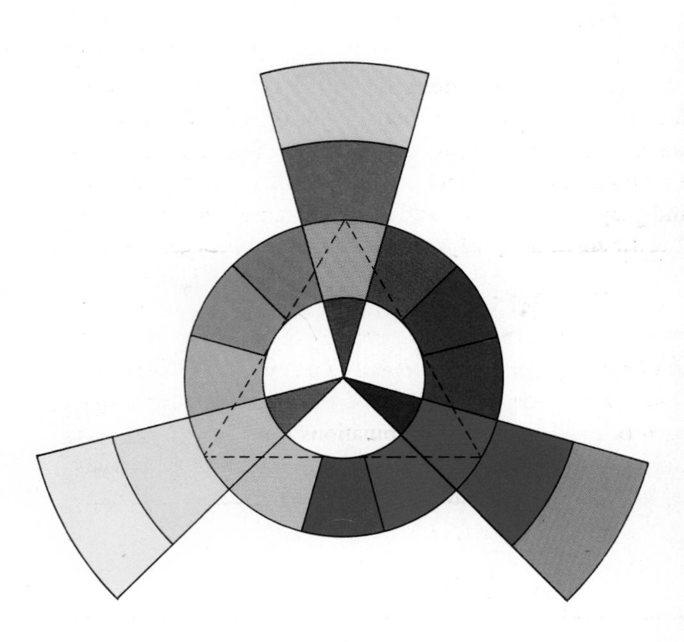

Figure 3-23. Triadic color schemes are developed from three colors that are equally spaced on the color wheel. Tints and shades help to soften these color schemes.

Figure 3-24. Primary Colors. Red anthuriums and croton leaves, blue delphinium, and yellow yarrow, solidaster, and croton leaves form this lively, unusual triadic design.

Because of the variety of hues incorporated into one bouquet, this scheme offers a great deal of visual stimulation.

Alternate Complement

This scheme incorporates a triad, plus one direct complement to one of the three hues. For example, if the triad of red, yellow, and blue is selected, combine any one of the complements of a color in that triad, such as purple (which is the complement to yellow). Color proportions and values are best varied in order to provide rest for the eye.

Tetrad

The word tetrad means a group or set of four. When applied to the color wheel, the four hues used are equidistant from each other. Selecting colors to make a tetrad combination is almost impossible without the helpful aid of a color wheel. These combinations involve the use of a primary, a secondary, and two tertiary colors. For instance, if the color red (a primary color) is selected, its complement green (a secondary color) is also incorporated. The other two colors forming the tetrad are blue-purple and yellow-orange (tertiary colors) because they are equally spaced from the first two colors selected. These schemes use such different and unrelated colors that the end result is eye-catching and contemporary.

COLOR INSPIRATION

In addition to using various schemes derived from the color wheel, other methods can inspire color selection for your floral designs.

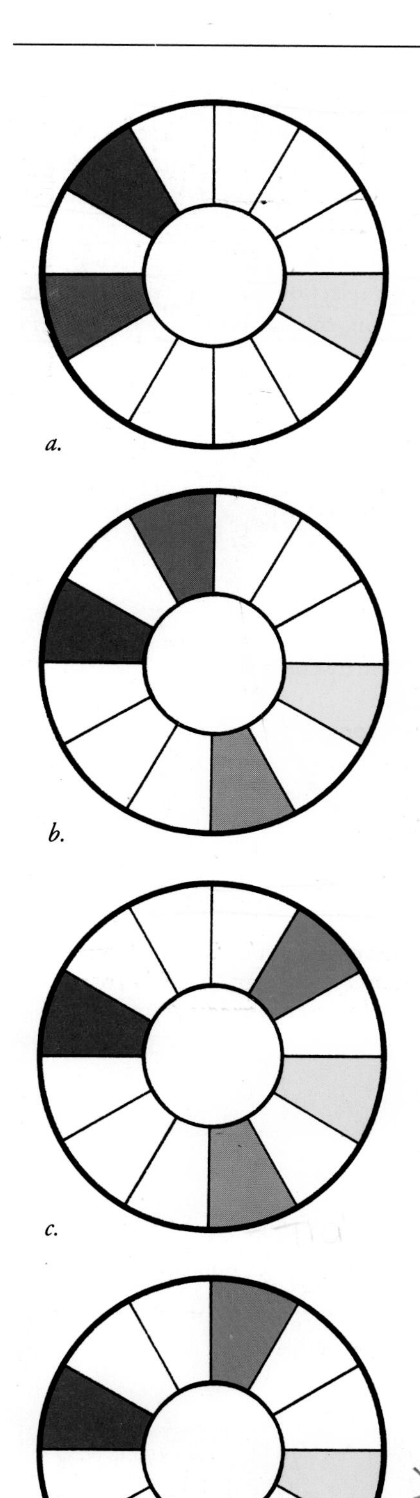

Seasons or Holidays

Seasons or holidays can and do have a definite impact on color selection. Certain holidays inspire various color combinations. For instance, red and green are the traditional choices for Christmas designs. Red, pink, and white are popular for St. Valentine's Day bouquets. Black and orange are associated with Halloween. Pastels and purple and yellow are common springtime combinations, and autumn is known for its array of yellows, oranges, browns, and reds.

Special Occasions

Special occasions can certainly dictate color choices. Baby arrangements are often made with flowers in soft pinks and blues, whereas homecoming corsages might be made to match bold school color combinations like black, silver, and purple. Gold may be chosen to decorate tables for a 50th wedding anniversary party. And, of course, it is crucial to match wedding theme colors and party dress colors.

Symbolism

Flower colors have symbolic meanings and at times, the giver of the flowers may wish to convey a certain message that is associated with the color itself. Red roses symbolize "true love" while yellow flowers suggest "friendship" or "cheer." As you choose flowers for your floral designs, often color and its symbolic message may dominate your choice of flowers.

Favorite Colors

One guide in color selection is simply to choose the favorite colors of the giver or recipient. Several color studies reveal that men prefer flowers in bold, strong colors like red, yellow, and blue. Women generally prefer the feminine colors of pink and purple. As mentioned earlier, everyone has certain color preferences and dislikes that will influence color choices.

Existing Colors

Color selection will be influenced by the ***existing colors*** of the interior where the design will be placed. For instance, colors can be chosen for a centerpiece with the tablecloth, napkins, china, even the food in mind. The size of a room and the distance at which an arrangement will be viewed can also help in color selection. And lastly, the color of a predetermined container or accessory is an important consideration in color selection for flowers.

Figure 3-25.
a. *Top: A split-complementary color scheme uses three colors—any hue plus the two hues adjacent to its complement.*
b, c, and d. *All discordant schemes use four colors:*
b. Double-complement scheme is developed using two pairs of complements.
c. Alternate-complement scheme uses a triad and one direct complement.
d. Tetrad color scheme is developed from four colors that are equally spaced on the color wheel.

REVIEW

Because color is perhaps the most important element in design, it is essential for you to gain a knowledge of its properties and psychological effects. Color can be used to achieve balance, depth, emphasis, rhythm, harmony and unity. The visual success of a floral design is generally dependent upon color and color relationships. Whether colors that form schemes are classified as related, contrasting, or discordant, it is beneficial to know the visual effects, advantages, and disadvantages. Choosing various colors for the arrangements you design may be based on a scheme derived from the color wheel or other factors may lead you in your color selection, such as; the season or holiday, a special occasion, color symbolism, and adjacent background colors.

TERMS TO INCREASE YOUR UNDERSTANDING

achromatic
alternate complement scheme
analogous
balance
chroma
color wheel
complementary colors
complementary scheme
contrasting scheme
cool colors
depth
discordant scheme
double complement scheme
existing colors
focal point
foundation colors
hue
intensity
intermediate colors
light
monochromatic scheme
neutral colors
primary colors
related scheme
rhythm
saturation
secondary colors
shade
split complementary scheme
tertiary colors
tetrad scheme
tint
tone
triadic scheme
value
warm colors

TEST YOUR KNOWLEDGE

1. In floral designs, what are some of the emotional responses to warm colors ? to cool? to individual colors like yellow, red, and blue?
2. Which visual elements do achromatic and monochromatic schemes rely upon to make them more interesting?
3. While color schemes may be derived from the color wheel, what else can inspire color selection for floral designs?
4. What are good rules to remember for discordant color schemes to make them palatable?
5. How can color create focal points, depth, balance, rhythm, and unity?

RELATED ACTIVITIES

1. Visit a retail florist. What color schemes are used in floral arrangements?
2. Visit a wholesale florist. Notice the great varieties of flower color. What blue and green flowers are available?
3. Choose a color scheme from a color wheel (one in which you have not had much experience). Plan and construct an arrangement incorporating the scheme you have chosen.

Chapter 4

Balance, Proportion, and Scale

Figure 4-1. In addition to being visually balanced and lovely, this floral composition of delphinium, iris, China aster with aspidistra, ivy, and eucalyptus pods is also mechanically balanced and self-supporting.

B***alance*** and ***proportion*** are closely related in floral design. If an arrangement lacks balance, it often lacks proper proportion. If proportions are not correct, imbalance generally results. Proportion is the comparative relationship between the compositon parts with respect to the finished size of an arrangment. Visual balance is the result of a desirable relationship among parts. The various elements—container, flowers, foliages, and accessories—need to "fit" one another visually as well as physically.

We often use the word proportion to describe the size relationships between the various parts of a composition. The words proportion and scale are related in that they both refer to size. Scale is another word for size. In order to determine size, we need some standard of reference. Size or scale is most often determined by the surrounding environment where the floral arrangement will be placed.

BALANCE

Balance is the quality of a floral arrangement that gives a sense of equilibrium and repose. Generally, we have a need for balance and can sense when there is imbalance. We tend to avoid dangerously precarious ladders, shelves, branches, and anything else that poses a physical threat. Although imbalance in floral design generally does not pose a physical danger, it is disturbing to us as balance is needed to achieve a sense of order. Balance exists as both an actual ***physical balance*** and a perceived ***visual balance;*** in fact, they often overlap each other. Both are essential to the success of a floral composition.

Physical Balance

Physical balance is often called ***mechanical balance.*** This means the floral arrangement needs to be mechanically sound to independently stand on its own, to be stable and self-supporting. If a design easily falls over, as shown in figure 4-2, it does not possess sound mechanical balance.

In order to provide actual balance, containers must be the proper size, weight, and shape for the flowers they will hold. Physical or mechanical balance is often achieved by placing an equal amount and weight of flowers and foliages on each side of a container The materials are distributed evenly so that the container will support them.

Visual Balance

Closely related to physical balance is visual balance. Even though a floral design might possess sound mechanical balance, if it doesn't look balanced, it appears unstable and is visually disturbing.

Figure 4-2. A design that easily falls over, or looks like it will easily tip over, does not possess sound mechanical or visual balance.

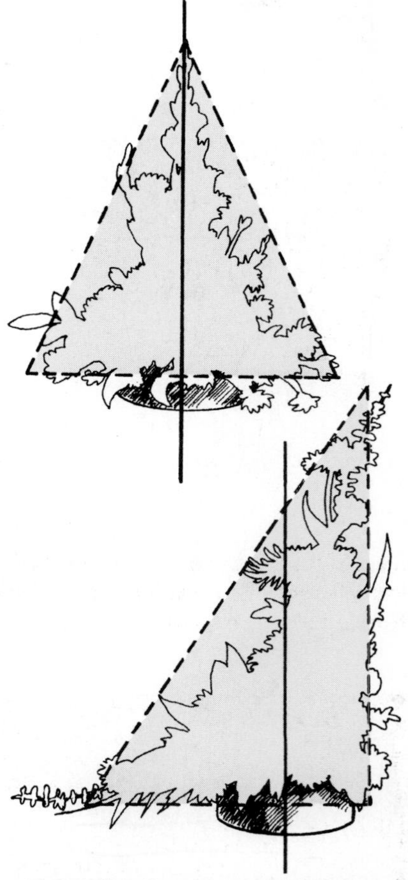

Figure 4-3. In symmetrical balance, one side of a floral design mirrors the other. In asymmetrical balance, when divided in half, the two sides are dissimilar. The different sides, however, have equal visual weight or equal eye attraction, which creates visual balance.

Every floral arrangement relies on visual balance to provide a sense of equilibrium. Visual balance is achieved when the arrangement gives a feeling of stability. There are four types of visual balance: ***symmetrical balance, asymmetrical balance, radial balance,*** and ***open balance.***

Symmetrical Balance

Symmetrical balance is the simplest type to recognize and create. Symmetrical balance is achieved when

Figure 4-4. Left: Often called formal balance, perfect symmetry, as in this rose arrangement, is static and unmoving. Right: Asymmetrical balance, as in this tropical design, appears more natural and casual. It is often described as informal balance.

Figure 4-5. This arrangement, though symmetrical in balance, is much too perfect, rendering a contrived and fake appearance.

Figure 4-6. Near Symmetry. This arrangment of gerberas, lilies, nerine, heather, and genista, although not symmetrical in placement, is arranged in a near-symmetry pattern to form a symmetrical design shape.

identical flowers and foliages are arranged and repeated in the same position on either side of an imaginary central vertical axis; one side is a mirror image to the other (see figure 4-4). Symmetry displays immediate balance and conveys a restful, peaceful sensation. Appearing formal and dignified, this type of balance also displays strength and stability. However, these arrangements often contrast with nature, and sometimes appear stiff and contrived (see figure 4-5). A solution to a contrived appearance is to place flowers in a ***near symmetry pattern,*** which is more common: rarely do floral arrangements repeat exactly the same elements on both sides. Flowers and foliages on opposing sides may vary, but because they are so nearly alike, they do not change the impression of symmetry as seen in figure 4-6.

Asymmetrical Balance

A floral design using asymmetrical balance is a more complex effort. It is achieved by placing unequal visual weight on each side of a central vertical axis. More subtle than symmetrical balance, it is often referred to as ***optical*** (connected with the sense of sight) or ***occult*** (hidden and concealed) balance. Even though these floral designs look more natural and less contrived, more thought and imagination is actually required in creating them. Something on each side of the floral composition must have equal eye attraction to provide balance. Asymmetrical arrangements remain interesting for a longer time because flowers and foliages of different shapes, sizes, and colors may be arranged in an unlimited number of ways.

Asymmetrical balance is based on equal eye attraction of dissimilar elements. For instance, one element that attracts immediate attention is color. A contrast in value and intensity is often used to provide equilibrium. Several large light-colored flowers on one side of an arrangement can be balanced by just one small, yet darker or brighter flower on the other side. Generally, dark shades appear heavy, so they can be used to help strengthen balance by being placed low in a floral composition.

Another method of achieving a sense of balance is through the use of contrasting ***shapes*** and ***textures.*** One small, interesting, exotic flower or a coarse-textured flower can balance a larger area.

The seesaws or teeter-totters in figure 4-7 illustrate an important idea of balance by ***position.*** An adult can balance with a small child simply by sitting closer to the center or fulcrum. Likewise, flowers can balance each other by their position. A large flower positioned closer to the focal point or center of interest can balance a smaller flower or object on the perimeter of a design (see figure 4-8). Accessories and figurines can also be added or placed next to a floral arrangement to help give a sense of equilibrium, as shown in figure 4-9.

Eye direction can also help achieve equilibrium. Facing flowers in a certain direction invites the viewer's eye to follow in that direction. Asymmetrical balance can also be achieved through the use of strong lines and the angling of stems (see figure 4-10). Eye direction and the angling of stems must be carefully planned, but once asymmetrical balance is achieved viewer interest is maintained.

No rule of measurement tells us how and where different flowers and objects must be placed in order to achieve asymmetrical balance. The point at which balance is achieved must be felt by each individual floral designer. With practice, you will sense proper distribution of dissimilar elements, and you will recognize when equilibrium exists.

Figure 4-7. An adult can balance with a small child by moving closer to the center of a seesaw. This simple example demonstrates a well-known principle of physics: two items of unequal weight can be brought to balance or equilibrium by moving the heavier object inward toward the center or fulcrum.

Figure 4-8. Balance by position is shown in this arrangement, as the large lily at the container rim balances with small gladiolus buds toward the design's outer edge.

Radial Balance

Another type of balance is radial. All the elements of a floral design radiate or circle out from a common central point like the spokes of a wheel or the rays of the sun (see figure 4-11). In floral design, radial balance can be combined with symmetrical or asymmetrical balance, depending upon whether the focal point is positioned in the middle or off center.

Figure 4-9. Accessories and figurines may be used to achieve a sense of balance.

Figure 4-11. Another type of balance common in floral arrangement is called radial balance. Like the rays of the sun, the spokes of a bicycle wheel, the petals of a daisy, the lines of a fan, or the pattern of a shell, all the elements radiate or circle out from a central point. This balance is not entirely separate from symmetrical or asymmetrical balance, but merely a fine distinction of one or the other.

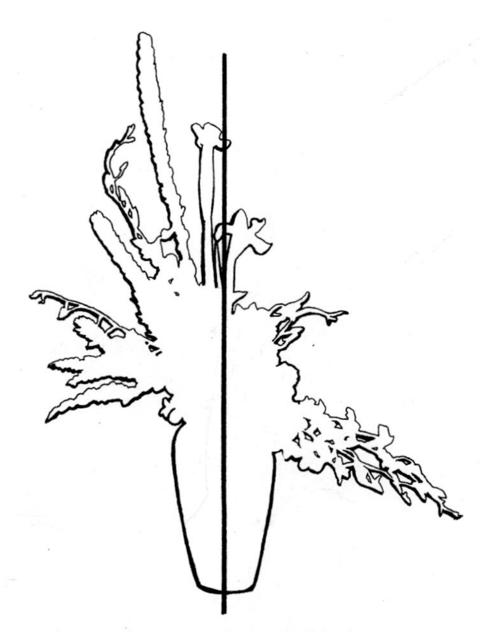

Figure 4-10. Asymmetrical balance may be achieved simply by line directions and the angling of stems as shown in the silhouette of this design. More material is used on the left side, but the downward angling of dendrobium orchids on the right side gives equal eye attention, lending a sense of balance.

Incorporating a strong radial balance in a floral design provides an immediate and obvious focal point. Radial balance is often displayed in bridal bouquets, where all elements visually and physically extend outward in all directions from a central location (see figure 4-12).

Open Balance

Open balance is another type of balance, common in contemporary designs (see chapter 17). Designs classified as having open balance are neither prominently symmetrical or asymmetrical. Open balance uses material throughout a design in a more relaxed and unstructured manner (see figure 4-13). Open balance does not follow any format or rules like the other types of visual balance, but it does rely on a perceived balance—one that you and the viewer sense.

PROPORTION

Proportion, often inseparable to the idea of balance and equilibrium, generally describes the relationship of one part of a floral design to the other parts as well as to the whole. The early Greeks discovered a few secrets of good proportion, which can be directly applied as you make floral designs.

The Greeks found that a rectangle or oblong with its sides in a ratio of 2 to 3 was a standard for good proportion. This particular shape is named the golden rectangle (see figure 4-14).

Another proportion discovery is called the ***golden section.*** This involves the division of a form or line. The ratio of the smaller section to the larger section is the same as that of the larger to the whole. The progression of numbers based

Figure 4-12. Often bridal bouquets display radial balance, all elements extending out from a common point.

Figure 4-14. According to a Greek discovery, a rectangle with the sides in a ratio of 2:3, called a golden rectangle, is a standard for good proportion.

on the ratio of the golden rectangle is 2, 3, 5, 8, 13, 21, and so on. Each number in the series is the sum of the two preceding numbers. For instance, 2 to 3, 3 to 5, and 5 to 8 are each about the same ratio and, therefore, equally pleasing. The golden section can be used in floral design, especially in determining how high the flowers should be to the container (see figure 4-15). A 3:5 ratio where the container equals 3, the flowers equal 5, and the whole design equals 8, is a good rule to go by.

Often we hear that a good rule to follow to determine how high flowers should extend is to simply double the vase height as shown in figure 4-16. This

Figure 4-13. Many contemporary, abstract, and oriental designs display an open balance. This type of balance is not prominantly symmetrical, asymmetrical, or radial, but instead relies on a perceived balance.

Figure 4-15. The Greek golden section may be used in floral arrangements where the ratio of vase to flowers is 3:5, or roughly 5:8.

Figure 4-16. The old rule that says simply to double the vase height to determine how high flowers should extend may work at times but generally results in flower arrangements that are stubby and display poor proportion.

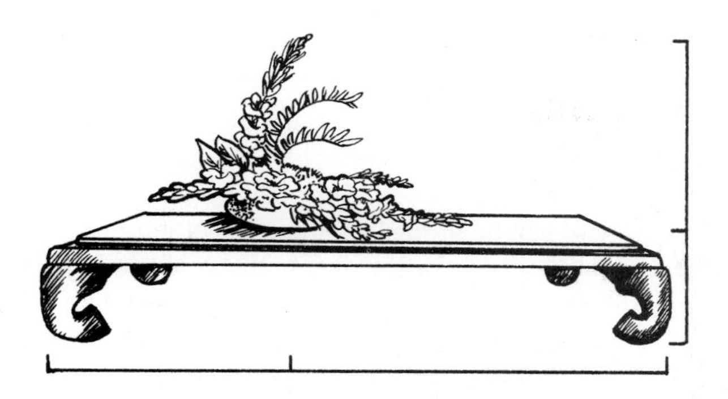

Figure 4-17. Another Greek discovery, the golden mean refers to the division of a line somewhere between one-half and one-third its length. This division is more pleasing and natural as shown by a floral design that is placed on a coffee table slightly off-center.

rule might work sometimes, but for the most part, a 2:2 or 3:3 ratio displays poor proportion and is somewhat disturbing.

Another proportion discovery, called the ***golden mean,*** is also a dependable guideline for use in floral design. This rule has application anytime something can be divided in half. Instead of dividing an area exactly in half, divide the area instead somewhere between 1/2 and 1/3. For instance, the golden mean can apply to the placement of a floral arrangement on a table as shown in figure 4-17. Rather than placing the bouquet right in the center of the table, it appears more pleasing and natural when placed slightly off-center.

SCALE

Scale is another word for size and refers to the overall size of an object compared with other objects; a flower, container, or accessory is only large or small when in comparison with something else. Parts of a floral composition need to be in scale with one another, and the size of the entire bouquet needs to be in scale with its surroundings.

Flowers to Container

The container is an important part of scale and proportion because its size and shape help determine the size and shape of the entire design. A small delicate container generally dictates that petite and dainty flowers should be used to fill it. Likewise, a large heavy vase looks best with massive or showy flowers.

Flowers to Flowers

Just as flowers need to be in scale with the container, the flowers within the same composition must be similar in scale with each other. Huge flowers can overshadow tiny blossoms. Extreme size differences in flower sizes may be distracting.

Flowers to Foliage

The size of foliage you select to go in a floral bouquet should be in scale with the flowers; this will complement them. The various foliages in a composition should also be in scale to one another. Just as extreme flower sizes may be distracting, so can extreme foliage sizes. For instance, lily of the valley foliage becomes lost and unimportant next to bird of paradise foliage.

Figure 4-18. Scale refers to the overall size of a floral composition or its parts compared with other objects or their parts. The elements of a floral composition must be scaled to each other. As shown here, the large ti leaves and container are overwhelming and out of scale to the single small rose and cluster of delicate baby's breath. Likewise, the large incurve (football) mums distract from the tiny lily of the valley flowers.

Figure 4-19. Floral arrangements should fit on a table or other flat surface both physically and visually, and at the same time be in proportion to the surrounding area.

Arrangement to Surroundings

There are several considerations of scale and proportion regarding where floral designs will be placed. One important consideration is the size of the table or area where the arrangement will be sitting. An arrangement must fit on a table physically and visually and at the same time be in proportion to the surrounding area (see figure 4-19).

Another important consideration is the size of the entire room in which the floral design will be placed. A floral arrangement is perceived in relation to the area around it. If a bouquet is large, it will become even larger when placed in a small room. The room will appear smaller; the bouquet larger.

A large and grandiose arrangement of heliconia and protea made for a spacious hotel lobby would certainly be cramped in a hospital room. Likewise, a tiny, delicate bud vase of roses and baby's breath in scale for a small office would look ridiculous in a large auditorium.

Floral bouquets should be in scale to their surroundings. Chapels and auditoriums are generally large and, therefore, call for larger floral designs that can be viewed by many people at great distances (see figure 4-20). In con-

Figure 4-20. In order to fit scale and proportion, floral bouquets placed in chapels, auditoriums, and other large rooms or areas must also be large. The large size will permit people from a great distance to view the floral bouquets.

trast, a bouquet for a patient in a small hospital room does not need to be big enough to be seen across a great distance by large masses of people (see figure 4-21).

Other important elements to remember for scale and proportion are texture and color. Shiney or coarse-textured containers, coarse-textured flowers, and foliages will appear larger because they attract the eye. Likewise, bright, dark, and even warm colors generally demand more attention and look larger than dull, light, and cool colors.

"It was a choice between one of these or a dozen skimpy little roses."

Figure 4-22. Herman © 1983 Jim Unger. Reprinted with permission of Universal Press Syndicate. All rights reserved.

REVIEW

The success of a floral composition is dependent upon many considerations, which include balance, proportion, and scale. You will realize just how closely related these principles are if you will deliberately consider them as you construct your floral compositions. Before you begin construction on a floral bouquet, consider where the arrangement will be placed. Such factors as the size of the room and the size of the table or area where the arrangement will be sitting are both important scale considerations for deciding how large your arrangement should be. The container should be chosen so it will be in scale and proportion to the whole of the design. Flowers, foliages, and accessories must be in scale to the container, to each other, and to the room size.

Once your supplies are selected, beware of what type of visual balance is needed for your bouquet. Often the location where the arrangement will be placed, the purpose of the bouquet, and customer preference all dictate the type of balance most appropriate. Even if there is no criteria to be met or you are making the design for yourself, it helps to be aware of what type of balance you want before beginning construction. Your arrangement must not only look balanced, it must be mechanically sound and stable. If it is unbalanced in any way, it will not fulfill its purpose of being pleasing and restful to its audience.

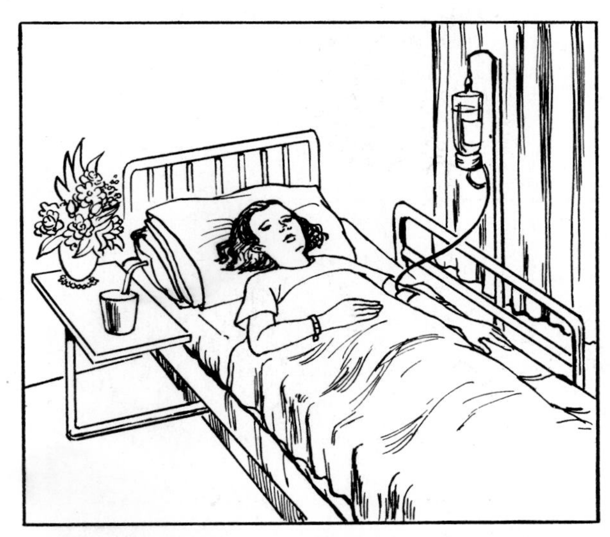

Figure 4-21. A floral bouquet placed on a small table in a small room, close up to the viewer, does not need to be large. Rather, it should be small and unobtrusive.

TERMS TO INCREASE YOUR UNDERSTANDING

asymmetrical balance
balance
golden mean
golden rectangle
golden section
mechanical balance
near symmetry balance
occult
open balance
optical
physical balance
position
proportion
radial balance
scale
shapes
symmetrical balance
textures
visual balance

TEST YOUR KNOWLEDGE

1. What are advantages of each type of visual balance? Are there any disadvantages?
2. How are proportion and scale related? How do they differ?
3. When can the golden mean be applied in floral arranging?

RELATED ACTIVITIES

1. Visit a retail shop. Notice floral arrangements on display and identify the types of visual balance.
2. While at the shop, notice proportions and the scale of individual parts of floral arrangements. Critique the proportions.
3. Plan a floral arrangement. Determine first the type of balance needed, and choose the parts according to proportion and scale guidelines.

Chapter 5

Focal Point and Rhythm

Your goal as a designer is to create, with flowers, a composition that will attract immediate attention. It should "speak" to the viewer. Something in the design should say, "Hey look at me! I'm important." Other elements in the design then keep the attention of the viewer by saying, "Now, follow me. I'll take you on a picturesque journey."

The best way to capture interest is to establish a point of ***emphasis,*** or ***focal point.*** This attracts attention. The rhythm of the design encourages the viewer to look further by providing a visual path to follow. Creating a focal point and incorporating rhythm in your floral designs provides the viewer with greater visual satisfaction.

Figure 5-1. Traditional focal points are located at the rim of a floral composition. In symmetrical designs, the focal point is placed in the center, whereas asymmetrical designs, the focal point is generally placed off to one side.

FOCAL POINT

In traditional floral designs, the center of interest and visual activity is called the focal point or ***focal area.*** In many floral compositions it is necessary and helpful to plan out a focal point before you begin. When the flowers or other material used to create the emphasis differs from the rest, a focal point will usually result because a ***contrast*** is created. A focal point is important because of its uniqueness to the rest of the composition. The focal point dominates and calls attention to the entire arrangement. Because of its strong emphasis, the focal point compels the viewer to look at this location first within the design.

Location

Traditionally in most arrangements, the focal point is located low in the design, near the container rim. In symmetrical designs, the focal point is centralized; in asymmetrical designs, it is placed off center (see figure 5-1). For visual balance, the main lines of the design generally converge at the focal area.

The focal point can also be placed in an intermediate area within the design to connect two vastly different parts of an arrangement. The different sections in the composition rely on a dominate feature to make the entire design complete and unified (see figure 5-2).

In contemporary designs, the focal point can be placed virtually anywhere. The area of interest is most effectively created by ***isolation.*** When one element is isolated or sits apart from all others, it becomes the focal point (see figure 5-3).

There can be more than one focal point in a floral composition, although secondary points of interest or ***accents*** do not require as much attention as the main emphasis area. A focal point highlighted with several accents can be very exciting; however, remember that too many focal points and accents cause confusion. The viewer does not know where to focus. If everything in the floral arrangement is emphasized, nothing is visually important.

Figure 5-2. A focal area (as with these iris) may also be placed in an intermediate area of an arrangement helping to connect two vastly different areas of the design.

Ways to Achieve Emphasis

Establishing a focal point in a floral arrangement can be achieved through a variety of methods. Generally, a focal point is created when one element differs from all the others in a design. This is ***emphasis by contrast.*** It is the difference that creates the focus.

Color

Color contrast is the simplest way to achieve a focal point (see figure 5-4). A change in value, intensity, or the hue itself immediately attracts attention. Since dark shades carry more visual weight than do light tints, flowers in deep shades should generally be placed low in the design while flowers in light tints should be placed at the perimeter. This helps to clearly establish a traditional focal point, at the same time providing visual balance.

Flowers that have bright and intense colors are instantly spotlighted. They demand attention and can be used to create strong areas of attraction. The duller-colored flowers when used in larger quantities, provide contrast, making intense colors even brighter.

You can create an area of emphasis simply by using a different color from the rest of the arrangement. Any color by itself, when juxtaposed with a different color throughout the rest of the composition, will catch the viewer's attention.

Warm colors when contrasted with cool colors can also be used effectively to create an area of emphasis. Flowers in these colors visually advance toward the viewer and demand attention.

Figure 5-3. The focal area may be set apart from the rest of the arrangement, as with these bird of paradise flowers placed in the upper edges of the design.

Figure 5-4. A contrast in color is the simplest way of creating a strong focal point. As shown here, a contrast in value, intensity, or the hue itself immediately attracts attention. Warm colors attract attention quicker than do cool colors.

Size

Another way to achieve a focal point is by ***contrast in size.*** Larger and more open blooms can be used to draw the eye just by their weightiness. When contrasted with smaller and less-opened blossoms, the size of a larger flower will provide emphasis (see figure 5-5).

Shape and Pattern

Interesting flower shapes, called ***form flowers,*** command respect and attention. Flowers such as heliconia, bird of paradise, and anthurium are meant to be focal flowers. Many look best when isolated at the perimeter of a design so their silhouettes can be fully appreciated. Others, such as large single orchids and protea work well at the traditional focal area (see figure 5-6).

Unique foliage can also entice the viewer. Foliage such as croton, diffenbachia, monstera, and cyperus leaves with interesting color patterns, shapes, sizes, or textures can create emphasis or help to accent an existing focal area (see figure 5-7).

Figure 5-5. Large, opened flowers compel attention and create emphasis when contrasted with smaller, less opened blossoms.

Figure 5-6. Worthy of Attention. *The interesting form, larger size, and vivid colors of the cattleya orchid creates a center of interest.*

Figure 5-7. Unique foliage may be used alone to create a center of interest.

Spacing

Spacing can also be used to draw the eye. As illustrated in figure 5-8, an area with closer massing of flowers appears heavy and thus is compelling. As flowers move out toward the composition edges, larger areas of space create a contrast.

Texture

The texture of some flowers, for example, a banksia flower with its leaves appeals to the viewer (see figure 5-9). Other materials with coarse surface textures, such as those found on wood, mosses, and baskets, have more visual weight as well. These unique and interesting textures attract attention, especially when contrasted with fine or smooth surfaces, such as glass and metal.

Accessories

Many types of ***accessories*** can be added to create a center of interest (see figure 5-10). Bows, candles, figurines, and novelty items are

Figure 5-8. Massing flowers tightly together can create a focal point.

Figure 5-9. Although somewhat subtle, the element of texture can be used to create an area of emphasis. Coarse textures (as with this banksia) attract attention when contrasted with fine or smooth surfaces.

frequently incorporated into floral designs. The occasion, season, and purpose of an arrangement help to determine the use and placement of various accessories.

Line Direction

Line direction can help strengthen an area of emphasis. Lines radiating out from and converging into the focal point will draw the viewer into the center of interest. However, remember to avoid crossing stems, which causes confusion (see figure 5-11).

Directional Facing

Another important technique that can be incorporated is called ***directional facing*** (see figure 5-12). The direction the flowers are pointed toward can be a tool

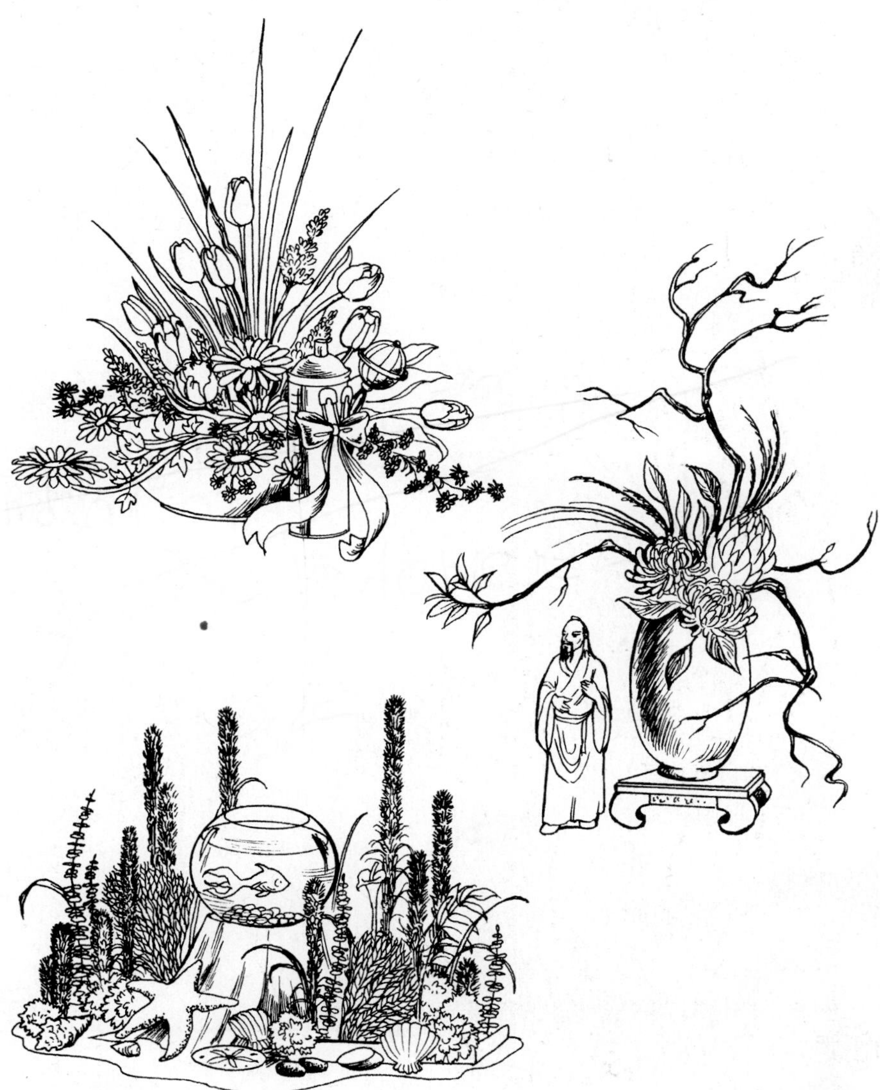

Figure 5-10. Accessories may be added to floral designs to create areas of emphasis and enhance a theme or style.

Figure 5-11. Radiating lines help to strengthen a focal point, while crossing stems are disturbing and cause confusion.

Figure 5-12. The direction the flowers are facing can be a tool to accent the focal area. As shown here, stems and flower heads may radiate out from the area of interest at the rim of the container or they may face inward toward the focal point.

to accent the focal area. Flowers throughout the entire design can radiate out away from the center of interest or they may face inward toward the focal point. The viewer's eye is guided by the ways the flowers are pointed.

Framing

One techinque to further enhance a focal area is often referred to as ***framing.*** Encircling the entire composition with a branch, stem of foliage, or other linear material, draws the viewer into the center to look more closely at the flowers placed within (see figure 5-13).

Isolation

When one flower or grouping of floral material is isolated or separated from the rest of the compositon, the viewer's eye is automatically drawn to that area (see figure 5-14). This technique is referred to as ***emphasis by isolation.*** Placing certain elements in a detached and independent location gives visual importance.

Figure 5-13. Framing is one technique used to enhance a focal area.

Figure 5-14. When one flower or a group of flowers or other floral material is isolated or separated from the rest of the design, an area of emphasis is created by isolation.

Figure 5-15. Absence of Emphasis. *Since the use of a focal point is so common, sometimes attention can be achieved by simply not using one, as in this exotic arrangement of bird of paradise, leucospermum, protea, ginger, anthurium, kangaroo paw, orchid, and lycopodium. There is no dominant flower in the arrangement that allows the viewer to be intrigued by a pattern of widely different forms, colors, and textures.*

Many of these methods for achieving focal points overlap, and often several of them will occur as you organize flowers, foliage, and accessories into a floral composition. For instance, many tropical flowers command attention not only by their striking form but by their intense color, interesting pattern, large size, and unusual texture as well.

Remember that the focal point must not be too overwhelming or unfitting to the design. It must remain part of the composition. However, an area of emphasis is not necessary in every composition. Some floral designs do not require a focal point. Many designs are intriguing merely by a pattern of widely differing elements in which all have equal emphasis (see figure 5-15).

RHYTHM

The term ***rhythm*** is defined as a flow or movement characterized by the regular recurrence of elements or features. In a work of art, it is also the effect of ordered movement attained through repeated patterns and graceful spacing. The word rhythm is easily understood when associated with music. The flow and movement can be ***adagio*** (slow and leisurely) and ***legato*** (smooth and flowing style with no noticeable interruptions), while some musical rhythm is ***allegro*** (fast and cheerful) and ***staccato*** (having distinct and abrupt breaks). As with music, the concept of rhythm expressed through floral compositions creates a sense of movement that stirs emotion (see figure 5-16).

Figure 5-16. Rhythm in floral design is similar to rhythm in music. Rhythm, or eye flow, may be slow and smooth, as in adagio and legato. Other times our eyes may move at a quicker pace, jumping from one element to another, as in allegro and staccato.

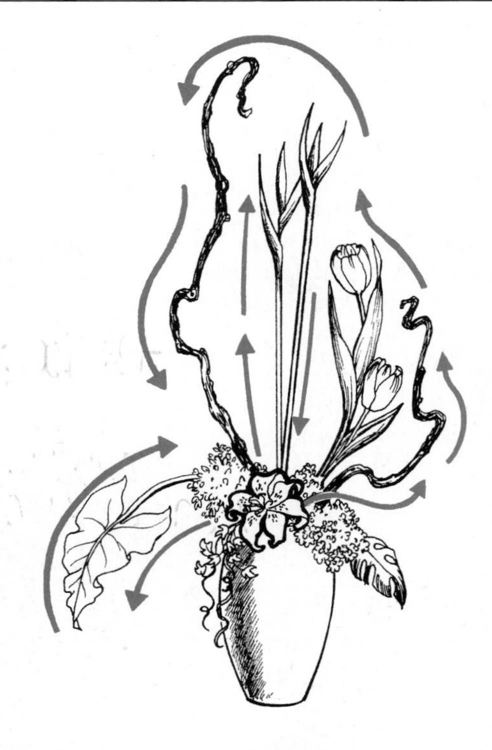

Figure 5-17. The goal of rhythm in a floral composition is to direct attention away from the focal area out to all parts of the design.

In order to achieve rhythm, it is vital that you gain an awareness of the placement of flowers, foliage, and accessories in relation to the focal area. To create visual success in a floral composition, you want to take the eye away from the focal point out to all parts of the design. Your goal is to create a visual pathway that leads the viewer's eye around and through the design. The pathway should then direct the viewer's eye back to the area of emphasis to rest at the focal point (see figure 5-17). This visual movement is created by the rhythm of the design.

Repetition is the basis for achieving rhythm, a concept that is all important in a floral composition. Repeating similar or slightly modified elements throughout a design will cause feelings of motion and energy. The rhythmic pattern then causes an emotional response in the viewer. Such responses can be slow, relaxing, and calm, or fast, happy, and abrupt.

In addition to repetition, rhythm is also achieved through the use of ***radiating lines*** and transition. Often, all three methods are used simultaneously to create an even greater impact of visual flow and eye movement that originates and converges at the focal area.

Figure 5-18. The viewer is compelled first to the bright yellow sunflower at the base of the design, afterwhich the other sunflower and yellow solidago provide an instant pathway. Other elements of color, line, form, and texture carry the eye throughout the design.
EPI, Excelsior Plastics Industries and Robert Andersen, AIFD—Andersen's Florist, Redlands, CA.

Repetition

Rhythm is easily expressed through the repetition of similar colors, shapes, textures, and lines. Color repetition (especially a bright, intense color) has the most impact in getting the eye to move from the focal point to other areas of the composition. The viewer automatically looks for similar colors in the composition to direct visual movement (see figure 5-18). Repetition of shapes can also cause eye movement by repeating flower types and forms throughout a composition. A more subtle method of achieving rhythm is through the repetition of similar textures. Interesting surface characteristics, for example pubescent flowers or shiny leaves, will attract attention, and when repeated throughout a design, will increase eye movement. Another important tool for attaining rhythm is the repetition of similar ***lines*** (see figure 5-19). Eye motion is invited most easily by the use of curved lines, while repetition of horizontal lines will result in a slower, more relaxed motion. Vertical and diagonal lines, when repeated throughout a composition, cause rapid eye movement from the focal point out to the perimeters and back.

Radiating Lines

Radiating lines are those that ***branch*** out from a central location to form a focal point (see figure 5-20). Placement of flowers and foliage in a radiating fashion will quickly take the viewer's eye out to the blossoms at the perimeters and then back to the focal point several times in succession. However, if stems cross one another, eye flow is interrupted, rhythm ceases, and the composition lacks movement.

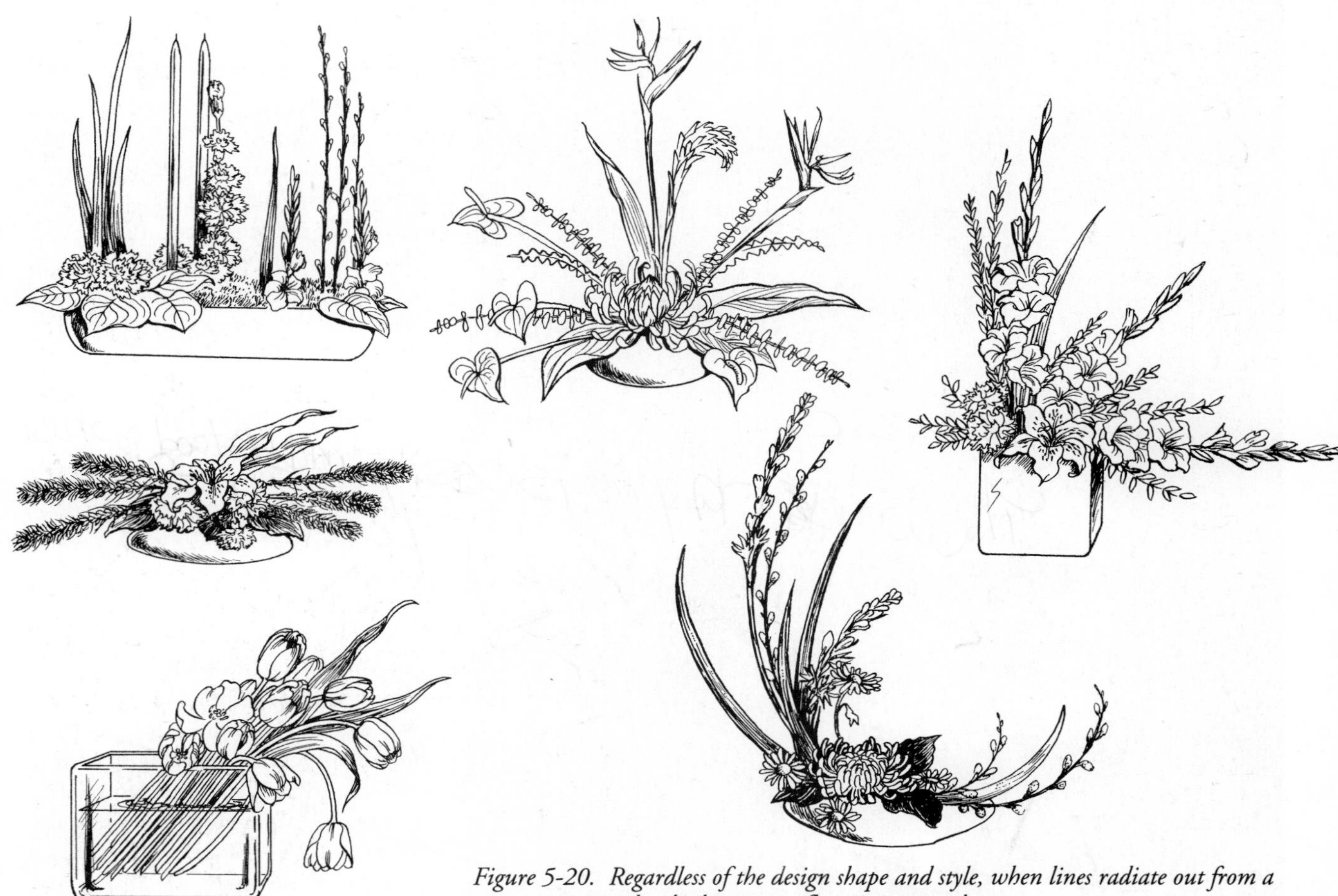

Figure 5-20. Regardless of the design shape and style, when lines radiate out from a common point, the rhythm, or eye flow, is increased.

Figure 5-19. Whether vertical, horizontal, diagonal, or curved, the repetition of similar lines throughout a design creates rhythm.

Transition

The word ***transition*** is defined as a passing from one condition or form to another. In music, transition occurs through a shift from one key to another. In writing, transition means to connect one part to another, often a very different part. In floral design, transition is sometimes called ***sequencing.*** The materials in a design move in a progressing pattern of change. A transition in color, value, intensity, size, form, spacing, or texture creates a rhythmic, visual pathway.

Where flower colors are dramatically different in sections of a bouquet, the eye will travel from focal point to perimeter, or from one side to the other, more smoothly if there is a transitional color or intermediate shade, tint, or tone. By using intermediate colors and values, the eye will move steadily throughout the design (see figure 5-21).

Gradual but irregular spacing from the focal point out to the margins often helps to create rhythm. Figure 5-22 illustrates how smaller spaces at the base in between flowers creates the focal point. The spaces then become progressively larger toward the top and edges of the arrangement. A rhythmic movement results.

Figure 5-21. This impressive, unique design illustrates how transition creates rhythm. The colors gradually change from the dark red of the amaryllis and kalanchoe to the orange gerbera, the lighter orange carnations and peach kalanchoe, then to the yellow ranunculus and roses. The size of the flowers, as well as their heights also gradually change from one extreme to another.

In addition to careful spacing, texture can also be used to help create rhythm. If just one color is used in a floral composition, varying textures become emphasized and are more easily noticed. A gradual transition from a rough texture to one that is smooth will create a sense of motion.

Figure 5-22. Rhythm is created by gradual but irregular spacing from the focal point toward the outer margins.

REVIEW

Knowledge of the design principles of focal point and rhythm will help you more fully understand how to captivate and keep attention. An area of emphasis, or a focal point, gives dominance to a floral design. A focal point within a floral composition may be achieved by a number of creative methods. Characteristics of the flowers themselves, such as color, size, and form, or the method of arrangement, such as spacing, framing, or isolation, can create a center of interest. Many arrangements rely on the use of a focal point for visual success. Others do not necessarily need a single area of focus. Rhythm in a floral composition is the quality that gives a sense of motion. It is a feeling created by actual movement of the eye as it follows line or pattern from the focal point out to the design edges and then back again to rest at the focal point. Creating an area of emphasis and incorporating rhythm in your floral designs will provide the viewer with increased visual pleasure.

TERMS TO INCREASE YOUR UNDERSTANDING

accessories
accents
adagio
allegro
branch
color contrast
contrast in size
directional facing
emphasis
emphasis by contrast
emphasis by isolation
focal area
focal point
form flowers
framing
isolation
legato
line
line direction
radiating lines
repetition
rhythm
sequencing
spacing
stacatto
transition

TEST YOUR KNOWLEDGE

1. What are some ways of creating focal areas in floral designs?
2. Why is the concept of rhythm important to a floral design?
3. What are some unique accessories that may be incorporated into arrangements to draw the eye?

RELATED ACTIVITIES

1. While visiting a retail floral shop or reading a floral magazine, study the floral arrangements you see. Determine if an area of emphasis exists and analyze what type of rhythm is used. Share your findings.
2. Plan and make a floral arrangement with a strong focal point and obvious rhythm.

Chapter 6

Line, Form, Space, and Depth

The elements of line, form, space, and depth have been grouped together in one chapter because each one relies on the other three for importance; all are essential components of floral design. The proper and skillful blending of these elements will result in a dynamic and distinctive floral arrangement (see figure 6-1). As you learn more about these elements, you will understand what part each of these play in the creation of a pleasing floral composition.

Figure 6-1. This fresh arrangement offers a distinct, dynamic pattern through the use of line, form, space, and depth. The lines of the container are repeated by the various foliage and curving branch. Similar elements are placed in groups, emphasizing forms; a contrast in space is evident; and depth is enhanced by angling and overlapping.

LINE

An impressive tool for the floral artist, ***line*** can give distinction and importance to your designs. Lines can express various moods and feelings simply by their direction. Inherent in all floral compositions, line provides shape and structure. It provides a pathway for the eyes to follow throughout a composition, which increases rhythm and visual enjoyment. Lines expressing motion can be effectively incorporated into floral designs in three ways—***actual lines, implied lines, and psychic lines.***

Figure 6-2. Flower stems, curved branches and foliage, and the container edges provide actual lines that are clearly delineated.

Actual Line

Actual lines are plain to see. The eye can easily go from one place to another because it is following a real, existing line. For instance, as shown in figure 6-2, the flower stems and curving branches are real lines that radiate out from the focal point. Our eyes easily follow the lines. The outer edges of a container and accessories also provide actual line. Actual lines set up the skeleton of a design and provide structure for the arrangement.

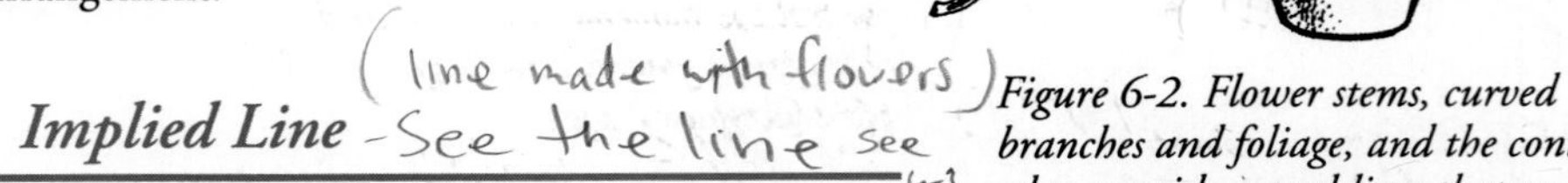

Implied Line

An implied line also provides a pathway for the eye to follow, but no line actually exists. These lines are created by a series of repetitious elements, as in a child's dot-to-dot puzzle. As you look at a floral design, you visually create lines between similar flowers and elements. As illustrated in figure 6-3, our eyes connect the same flowers together—forming more visual pathways.

Figure 6-3. With implied lines, our eyes connect similar flowers, as in a dot-to-dot, creating more lines, or pathways, to follow.

Psychic Line

A ***psychic line*** is one that does not exist, yet we ***feel*** a line between two elements. Although no actual lines or even repetitious elements form an implied line, we mentally connect two objects. Psychic lines are created in floral arrangements simply by placing flowers or materials in a way that directs the eye. As shown in the floral composition in figure 6-4, a psychic line is formed by the flowers that seem to look at each other. This facing or pointing flowers and objects in a certain way to form a line is referred to as ***directional facing.***

Often all three types of lines can be used to create visual movement, as seen in the floral design in figure 6-5.

Line Direction

Not only is the type of line important to visual satisfaction, but its direction can express varying feelings. Line directions are classified into four patterns: ***vertical lines, horizontal lines, diagonal lines, and curved lines.***

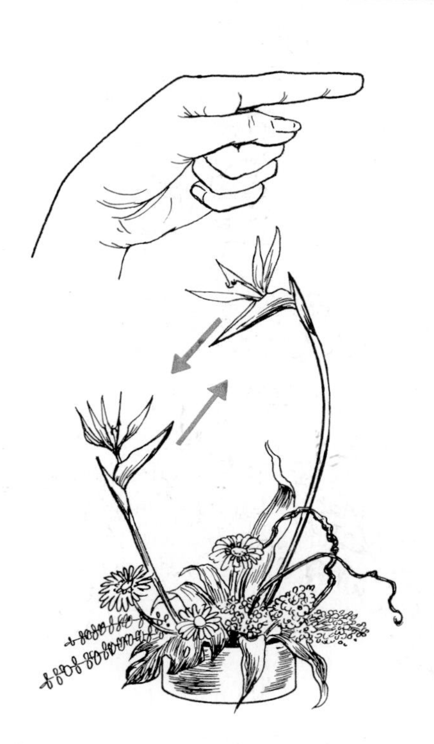

Figure 6-4. One element points to another (as in the bird of paradise flowers), and the eye connects the two forming a psychic line.

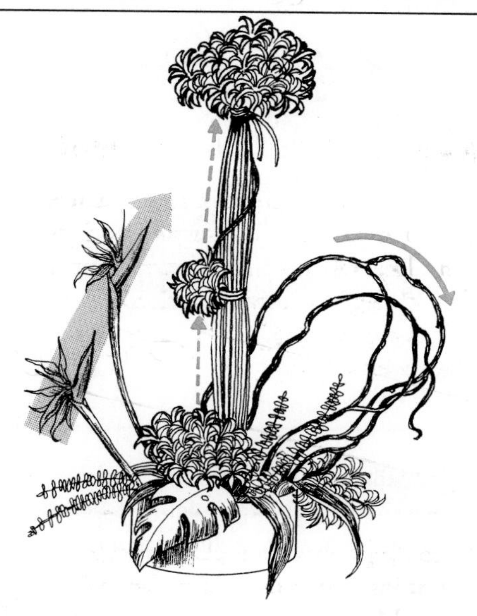

Figure 6-5. Actual, implied, and psychic lines can be used together to create visual movement. To a large extent, you can control eye movement around and throughout your floral designs by using various types of line.

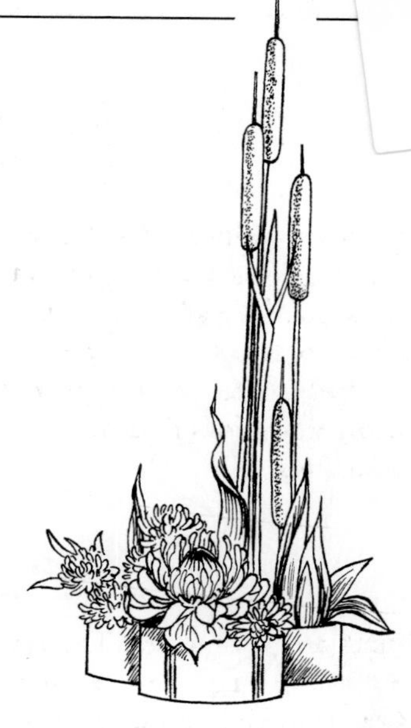

Figure 6-6. Vertical lines (shown here by the cattails) command visual attention and suggest intense energy.

Vertical Line

Vertical lines stress height and suggest power and strength. They often give a feeling of formality and dignity. Vertical lines cause continuous eye movement from the focal area up to the top edges of a design and then back down to the focal area. The exaggerated vertical lines in figure 6-6 demand visual attention and suggest stamina and intense energy.

Horizontal Line

Horizontal lines stress width rather than height. Since they seem peaceful and calm compared to other line directions, they imply restfulness. Because they are generally parallel with the container rim and the surface of the table, they help provide a sense of stability. As shown in figure 6-7, horizontal lines provide eye movement but at a much slower and leisurely pace than other line directions.

Diagonal Line

As shown in figure 6-8, diagonal lines are dynamically energetic. These slanting lines cause more eye movement, especially when juxtaposed between vertical and horizontal lines. Because diagonal lines suggest motion and excitement, you will want to use them sparingly in a composition. Too many can prove to be overwhelmingly busy and confusing.

Figure 6-7. Suggesting quiet and repose, horizontal lines provide eye movement at a much slower and more leisurely pace than other line directions.

Figure 6-8. Diagonal lines suggest great dynamic energy and motion.

Curved Line

Curved lines, like diagonal lines, suggest motion but in a softer, more comforting way. As you look at the illustration in figure 6-9, it doesn't take long to view the entire design because the curves cause your eyes to move throughout the composition quickly. Curved lines add interest and gentleness, especially when used with other line directions.

Within a Frame

Visually frame a floral composition in a vertical rectangle as shown in figure 6-10. This technique will help you determine which lines of your floral design are ***static lines*** and which are ***dynamic lines***. Vertical and horizontal lines (those parallel to the imaginary frame sides) are static. Curving or diagonal lines are ***counterpoint*** or contrasting to the imaginary frame sides; they are dynamic. Conversely, visually framing a floral composition in a circle or oval frame causes the curving and diagonal lines to be static (repeating the lines of the frame) and the vertical and horizontal lines to be dynamic (counterpoint to the frame sides). A blending of both static line and dynamic line in a floral composition provides visual energy by creating visual ***tension*** among contrasting vertical, horizontal, diagonal, and curving lines.

Static Line

Lines in a composition that are parallel to imaginary frame sides are called static (see the top illustration in figure 6-11). These lines, within the vertical and horizontal sides of the frame are somewhat inert, displaying a lack of visual energy. In some designs, these lines can almost appear boring and motionless unless they are exaggerated.

Figure 6-9. Curving lines in a floral design lead the eyes to move throughout the arrangement quickly.

Dynamic Line

Lines in a composition that are not parallel to the sides of an imaginary frame are called dynamic, as shown in the bottom illustration of figure 6-11. These lines provide opposition to the majority of the lines in the design and create visual excitement by expressing motion. Dynamic lines are active and add vibrancy to common floral designs.

Since line is a component of every floral design, it is important for you to become familiar with the various line types and directions and to understand how to use line to create eye movement. A well thought-out and controlled line will give your floral designs distinction. Remember, however, that too much line movement, whether deliberate or accidental, can cause confusion and instability.

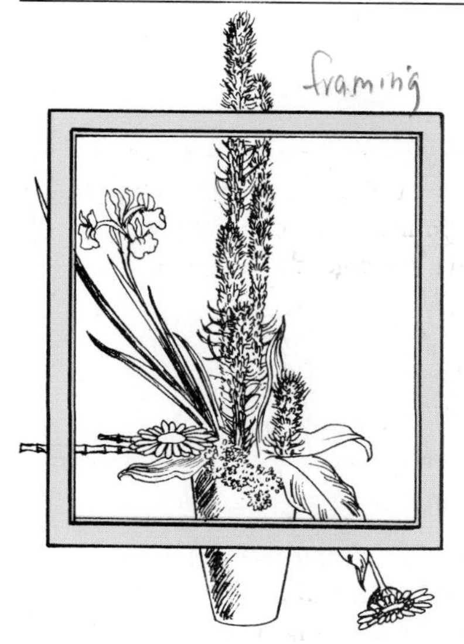

Figure 6-10. Lines that are static or dynamic may be easily identified by imagining or actually placing a frame around a completed design.

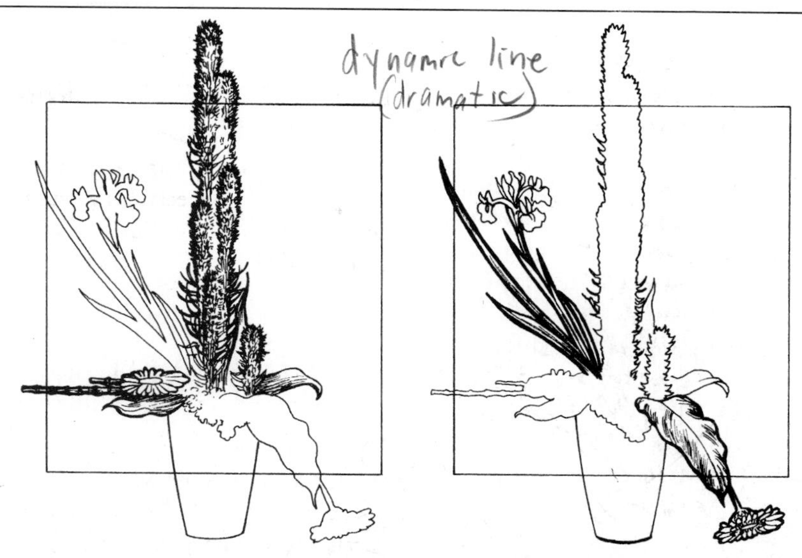

Figure 6-11. Top: Lines parallel to imaginary frame sides are static and unmoving. Lower: Lines that are not parallel to the sides of an imaginary frame are dynamic and serve to create visual excitement by expressing motion.

FORM

In design, ***form*** refers to the three-dimensional aspect of an object. A more common, and probably more precise, word for form, is shape. Because a floral composition is basically the arrangement of different forms, it is essential for you to see shapes. If you look at a floral design, the container, flowers, foliage, and accessories all have certain shapes (see figure 6-12). To have a variety of shapes within one design is more visually satisfying than to have only the same shape repeated throughout (see figure 6-13), which can be monotonous and uninteresting. Skillfully combining various forms can enhance the visual success of a floral design. This is because

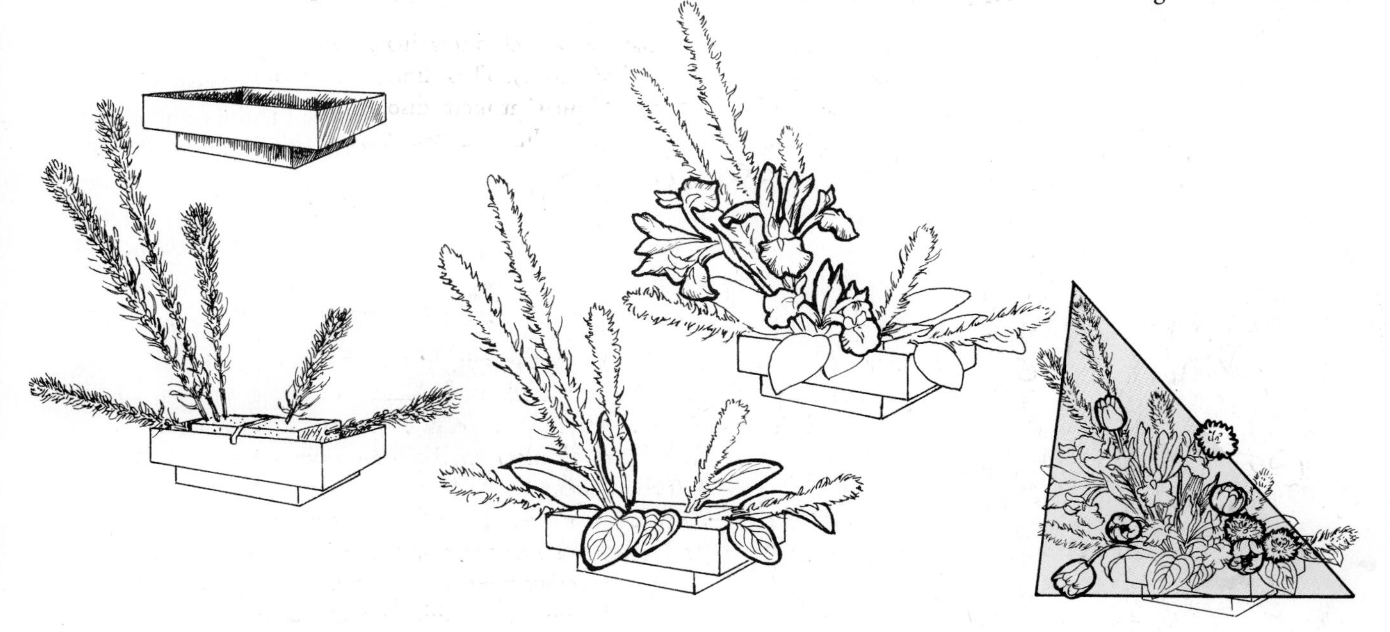

Figure 6-12. A variety of forms in one floral composition provides more interest: a rectangular container, linear liatris, oval maranta leaves, form iris, and rounded tulips, and carnations combine in a triangular design.

many principles of design such as focal point, rhythm, harmony, and unity often rely on form for increased visual quality.

Distinctive flower shapes, called form flowers, can easily create a focal point. These flowers can attract attention just by their odd or unique forms. Some foliage that have interesting shapes are also named form. In order to allow their individual shapes to be fully appreciated, it is important to leave enough space around form flowers. (See chapter 11 for an in-depth look at form flowers and foliage.)

The rhythm of a design often relies on the forms of the individual components as well as their arrangement. Depending on the composition shape and style, different flower and foliage forms can be skillfully mixed with each other throughout a design. Similar shapes can also be grouped together for emphasis and contrast (see figure 6-14).

Figure 6-14. When similar shapes are grouped together, emphasis is placed on form. Differences in shape are more readily noticed.

Figure 6-13. A design that repeats only one shape is less visually exciting than one that contains a variety of shapes. The liatris arranged in a tall container form a linear, vertical design, that is rather uninteresting.

Remember that the container shape and the flower and foliage shapes need to be in harmony with each other. Often, when combining extremely different shapes, try emphasizing and using more of one form and lesser amounts of other forms. This will help unify and give a theme to your entire composition. For instance, a tall cylinder container is an excellent choice for tall line flowers, foliage, and sticks where the emphasis is on line and shape. However, other shapes in lesser amounts will add interest and increase visual enjoyment. Remember that the success of any design depends on the proper blending of shapes. (See also chapters 8, 11, and 12.)

SPACE

Space is an important design consideration that is often overlooked, but it is vital to give line and form significance. Flowers and containers are three-dimensional forms that occupy space. The area where forms are placed is sometimes called ***positive shape*** or ***positive space.*** If you look at an entire composition, imagine a frame around it, the area inside the frame and between the forms is often called ***negative shape*** or ***negative space*** as shown in the shaded portions of figure 6-16. Negative shapes are necessary in order to make the positive shapes look more important and interesting.

Figure 6-15. As shown in this design, space gives line and form significance. Similar elements are grouped with each other (emphasizing shapes), yet it is the contrast in negative space between flowers that helps create interest. Although the amaranthus is massed at the base of the design, negative spaces remain between individual flower spikes. The iris, liatris, and various foliage rely on elongated negative space and voids, resulting in a contemporary and dynamic pattern.

Figure 6-16. Negative shape or space (the shaded portions) are vital for making areas of positive shape or space stand out.

A ***void*** is an open space or break in a design that connects an area of positive shape with another. Generally, it is created by a single stem, line, or a grouping of clean stems or lines (see figure 6-17). Voids will give further impact to the positive as well as negative spaces of your designs.

A gradual increase of space from the focal point out to the design edges enhances rhythm and eye movement. You can create this focal point simply by spacing flowers closer together in one area, usually near the rim of the container. The eye is drawn to the mass of flowers (see figure 6-18).

Remember that the flowers in your designs need "breathing room." Some ***space*** around each flower gives individual forms importance, no matter what their shape. If flowers are too tightly arranged throughout your entire composition, they will look cramped and uncomfortable. Do not be tempted to completely fill in your designs.

Figure 6-17. A void in a floral arrangement is an open space or break that connects an area of positive shape with another. (The anthurium flower is connected to the rest of the arrangement by its own clean stem.)

Figure 6-18. Tight massing of flowers with little space between them may be used to create an area of emphasis. A gradual increase of space from the focal point out to the design edges enhances rhythm and eye movement.

DEPTH

Because most artists use two-dimensional mediums such as paint on canvas, they are concerned with achieving ***depth.*** However, the tools of the floral designer, such as flowers and container, are three-dimensional. For this reason, the floral designer is generally not as concerned about depth. However, in order to prevent a two-dimensional, flat design, it is vital to create an overall feeling of depth, particularly in one-sided arrangements.

Artists who work on a two-dimensional canvas employ several methods of creating the illusion of depth. Two important artistic techniques for creating illusions of depth or perspective are the use of angled lines and the overlapping of objects. You can use these same techniques, ***angling of stems*** and ***overlapping flowers,*** and apply them directly as you arrange flowers. You can also use ***size,*** color, and value of materials to increase the illusion of depth.

Angling of Stems

A simple technique to give one-sided designs a fuller, more balanced look is to exaggerate the angles of the flower and foliage stems. As illustrated in figure 6-19, a one-sided arrangement will have increased depth and better balance if the tallest background stems are angled backwards and the lowest foreground stems are angled downwards in front of the container. The stems between should be gradually angled for smooth transition and rhythm. Proper angling of stems will create a more balanced and beautiful design.

Overlapping

A slight overlapping of flowers and foliage will also enhance depth. When flowers hide parts of other flowers, being placed "on top" or in front of each other, a greater sense of depth results (see figure 6-20). Varying the height, width, and depth of flowers and foliage will also create a more natural appearance. If you arrange the flowers so all the blossoms are the exact same distance from the traditional focal point area, your arrangement will appear flat and contrived.

Figure 6-20. A feeling of depth becomes greater as flowers and foliage slightly overlap one another.

Size

Another method artists employ to give the illusion of depth is through the size of objects. When artists paint a landscape scene, in order to give a flat canvas miles of distance and great depth, objects in the foreground are drawn exceptionally large in proportion to the objects furthest away, which are made smaller. You can use this same technique in one-sided floral designs to maximize depth. Arrange large flowers low and in the foreground of your composition, and place the smallest flowers at the back of your design. Remember to have a gradual transition in size from large to small. This technique can create a dynamic visual pattern (see figure 6-21).

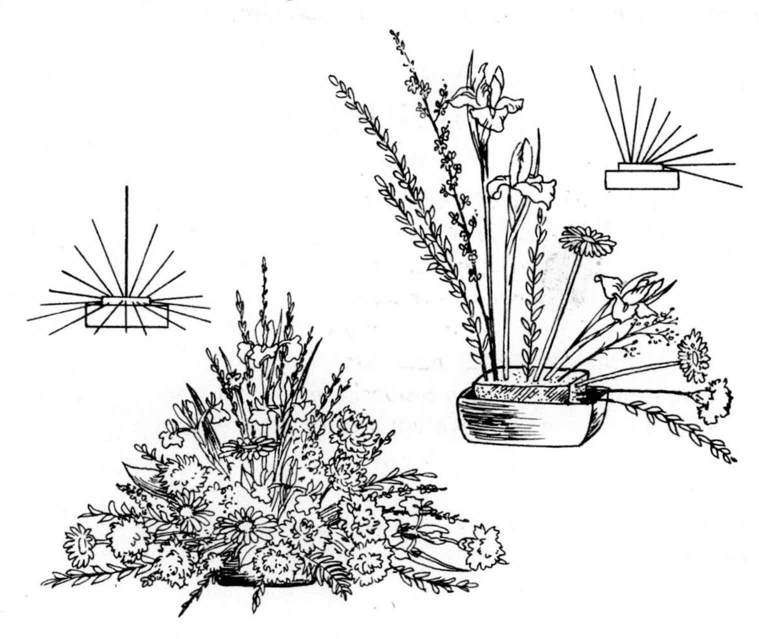

Figure 6-19. Top: In one-sided designs, depth can be easily increased by angling the tallest stems slightly backwards and the lowest, foreground stems downward in front of the container. Lower: A gradual angling of stems throughout a bouquet, enhances depth and increases rhythm.

Color and Value

Another way you can increase depth in one-sided arrangements is to place bright and dark colors low and in the front of your designs. Use flower colors that are duller and lighter further back in the composition (see figure 6-22.) This technique is based on ***atmospheric perspective,*** which means using color and value to show depth. This simple phenomenon can readily be seen as you view a distant land-

Figure 6-21. Another tool used to increase the feeling of depth in one-sided designs is to place larger flowers low and in front with a gradual decrease in size towards the back.

scape. The objects closest to you, for instance the meadow and flowers, will be lowest and brightest in color. As distance increases, the colors of the far-off meadow and flowers become increasingly subdued and muted. Value contrast is also apparent as the atmosphere lightens and neutralizes colors that are far away. The more distant an object, the lighter it appears.

Figure 6-22. Placing lighter and more subdued colors at the back of an arrangement with the brighter and darker colors low in front will enhance the feeling of depth.

REVIEW

Each of these design elements—line, form, space, and depth—is necessary in constructing a successful floral composition.

Inherent in every floral arrangement, line provides shape and structure, height, width, and depth. Lines provide pathways for the eye to follow. The use of line results in increased eye movement and rhythm to form striking and visually dynamic designs. The skillful use of line will give your compositions strength and distinction.

Form refers to shape. A completed flower arrangement is a certain shape and is actually a composite of various smaller shapes. The container, flowers, foliage, and accessories all have their individual shapes, which are essential as a variety of shapes in a single design increases interest and visual satisfaction.

Space gives line and form importance. Negative spaces and voids in floral compositions are necessary to give distinction to the parts of a design. Enough space also needs to be left around flower heads so they do not look cramped and uncomfortable.

Creating a sense of depth is essential in floral arrangements, especially in one-sided designs, so they do not look flat. Many techniques can be used to enhance the feeling of depth. Proper depth will also aid in the overall balance of a design.

TERMS TO INCREASE YOUR UNDERSTANDING

actual line
angling of stems
atmospheric perspective
counterpoint
curved lines
depth
diagonal lines
directional facing
dynamic lines
feel
form
horizontal lines
implied line
line
negative shape
negative space
overlapping
positive shape
positive space
psychic line
space
static lines
tension
vertical lines
void

TEST YOUR KNOWLEDGE

1. What are some emotional responses to the different line directions—vertical, horizontal, diagonal, and curved?
2. What are factors that make a line dynamic or static?
3. What are the techniques most artists use to achieve a feeling of depth? How can these same techniques be used by floral artists?

RELATED ACTIVITIES

1. Make a floral design that uses not only actual lines but implied and psychic as well.
2. Sketch a line design. Plan the negative shapes.
3. Make a one-sided design appear less flat by exaggerating the stem angles and overlapping flower heads.

Chapter 7

Texture and Fragrance

Figure 7-1. An interesting variety of coarse textures, both visual and tactile, are used in this distinct composition. The prickly pin cushions (leucospermum), velvety kangaroo paw, rough lotus pods, tightly leaved rosette echeveria plant, and guava foliage with woody stems, all combine pleasantly together in an unpolished pottery container.

While the design elements that involve the sense of sight are important considerations for achieving great floral designs, ***texture*** and ***fragrance*** are elements that are often overlooked. When thought of at all, they may be considered as optional extras, but they can be important design aspects in your floral compositions. A sense of texture appeals not only to one's sense of sight, but also to one's tactile senses. Your knowledgeable use of texture will bring character and distinction to floral compositions.

Fragrance is another powerful element in your floral designing. By adding an extra dimension, fragrance increases one's awareness and enjoyment of a floral design. A floral composition that not only appeals to the sense of sight, but also to touch and smell, enhances visual pleasure through a more complete sensory satisfaction.

TEXTURE

Texture refers to the surface characteristics of objects or elements. Texture is the ***visual*** and ***tactile*** exterior quality of everything in a floral design, including the container, flowers, foliage, and accessories (see figure 7-1). Although we might not actually feel a certain texture within a floral composition, our memory provides a sensory reaction. Visual texture arouses tactile sensation. An example of this is the texture of clouds. Some clouds may appear as soft and fluffy as cotton. Although in reality they are a mass of tiny, condensed water droplets or ice crystals suspended in the atmosphere, the visual sensation of billowy white clouds evokes the softness of cotton.

Similarly, various light and dark patterns created from different surfaces give visual signals so that we enjoy the differing textures vicariously. We often describe textures as rough or smooth, shiny or dull, coarse or fine. Sometimes when referring to the concept of texture, only coarse and rough surfaces are thought of as important in flower arrangements (see figure 7-2). However, it is the successful combination and blending of various textures that increases depth and visual interest.

The concept of texture is especially important when it comes to selecting the materials for each floral design you create. In order for the element of texture to make a powerful statement, you will need to carefully and consciously select the parts of your compositions. A proper blending of similar-type surfaces will result in harmony and unity. However, texture is heightened by the juxtaposition of varied and opposite surfaces. A variety of surfaces from the container, flowers, foliage, and accessories can be used to enhance and accent the overall design effect (see figure 7-3).

Figure 7-3. Many floral arrangements achieve visual success through the use of extreme textural differences. Smooth callas, shiny ornaments, a smooth pillar candle, and a reflective container contrast with coarse conifer foliage, prickly holly, and ruffly carnations.

Figure 7-2. Containers with nonreflective surfaces and bumpy surfaces, such as baskets and unglazed pottery, are homespun and informal. Many flowers and foliage seem coarse because they are prickly, fuzzy, bumpy, or busy in appearance. Accessories such as branches, pods, and cones provide rough textures that add interest.

Containers and Accessories

Throughout history, smooth and shiny polished surfaces have been symbolic of wealth and high status while rough, coarse, or dull surfaces have characterized peasantry and informality. All surfaces display a particular character, and each possesses a unique beauty. Containers that are smooth and highly reflective, such as glass, brass, silver, and polished ceramic, often suggest elegance and formality. In contrast, containers having coarse and rough surfaces, such as baskets, pottery, and unfinished wood, are more natural and informal.

Accessories in varying textures can be used to emphasize and help give distinction to your floral compositions. Adding accessories that have smooth or rough textures can be used to make your designs more formal or casual. The reflective and smooth surfaces of shiny ornaments, figurines, glass stones, mirrors, and some candles often add a touch of opulence (see figure 7-4). On the other hand,

Figure 7-4. Smooth, reflective containers, such as glass and polished ceramic, often suggest formality. Many flowers, foliage, and accessories also have the ability to reflect light, because of their smooth surfaces, which often convey a sense of richness.

Figure 7-5. Texture is an important element in the selection of flowers for arrangements. Different textures will have various visual appeals. The textures of banksia, celosia, and carnation render different sensory results than do the textures of anthurium and calla.

Figure 7-6. Spiky margins suggest coarse texture.

the textures of sticks, bird nests, moss, and dried plant materials create a very different feeling.

Remember that the textures of containers and accessories should be in harmony with the textures of the flowers and foliage. If the surfaces of the container and accessories do not harmonize and blend with the flowers, they often compete for attention. As a result, the flowers become less important.

Flowers and Foliage

All flowers and foliage possess texture. However, the textures of some may be more apparent and have differing effects. A coarse and hairy protea will have a different visual impact than that of a smooth and shiny anthurium. The texture of a velvety calla lily will have different sensory appeal than that of carnations or a coarse and ruffly celosia (see figure 7-5).

Coarse textures that compel attention can be used to create a focal point, and leaves of varying textures can certainly be used to enhance and accent a focal point. The total composition needs to be analyzed as you choose individual flowers and foliage.

Texture will have a greater impact in floral arrangements that are all white or monochromatic because texture does not have to compete with contrasting and showy color schemes. Likewise, color schemes that are sometimes boring can be made more interesting through the use of varied textures.

Abstract and freestyle oriental designs often place emphasis on texture. Many are successful because of their extreme textural differences (see chapters 16 and 17).

Foliage with their varying textures can stir the senses and compel attention. Since ***pubescent*** leaves or spiky ***succulents*** draw attention, they can be used to support a focal area (see figure 7-6). The textures of ti leaves or shiny galax leaves provide a striking contrast to Christmas foliage or to the fuzzy texture of ming fern (see figure 7-7). Remember also that although some foliage have a smooth and shiny surface, they appear heavy in texture if they have striped or varied color patterns (see figure 7-8).

As you prepare your floral designs, remember that the element of texture becomes less important the further away it is to the viewer. The closer we get to a floral composition, the rougher and more varied the surfaces of flowers and objects become. Much like viewing something through a microscope, texture becomes more obvious the closer we scrutinize.

It is important to harmonize the textures of flowers, foliage, containers, and accessories within one composition. Although textural variety is desired, too many intermediate textures can weaken a design. A repetition of ***extreme contrasts*** in textures, however, adds interest and beauty.

Figure 7-7. A combination of foliage textures give variety and distinction to designs—smooth ti and galax leaves, coarse and fuzzy ming fern, and conifer foliage.

Figure 7-8. Although the surface of some foliage is actually smooth, various striped or color patterns create a sense of texture (as found in diffenbachia, croton, and calathea).

FRAGRANCE 

Fragrance can be an element of floral design just as much as form, texture, and color, but too often we consider it nice to have, but not necessary. However, fragrance can be an integral part of many compositions since it adds yet another pleasurable dimension to our enjoyment of flowers (see figure 7-10). An appeal to our sense of smell heightens our awareness and increases our sensory enjoyment. Even more important, when we expect to breathe in a wonderfully sweet fragrance to match the beauty of a floral design, and no fragrance is present, we are often disappointed no matter how lovely the flowers are to our eyes.

For thousands of years, man has regarded natural fragrances as being clean and pure. In earlier days, fragrance was regarded as the most important aspect of flowers because it was believed that fragrance had the power to ward off and guard against disease.

Figure 7-9. Fragrance is an element of floral design just as are form, texture, line, and color.

Figure 7-10. Fragrance in floral arrangement has long been important. This 16th-century, allegorical engraving represents one of the five senses. To depict the sense of smell, the artist used a floral bouquet containing sweet-scented flowers, some of which are lilies, roses, tulips, and dianthus.

The Sense of Smell *by Frans Floris, Netherlandish 16th-century engraving. The Metropolitan Museum of Art, Harris Brisbane Dick Fund, 1928. [28.4(55)].*

Fragrance can take us back through time and space; fragrance triggers our memory (see figure 7-12). The nose contains the organ of smell in which scents are detected by the very fine and sensitive, hair-like ends of the ***olfactory*** nerve. The nerve endings pick up various scents and transmit this information to the brain, which matches it against stored information from past experience. This is how smells are perceived and identified; we associate certain scents with past experiences.

Figure 7-11.

Herman © 1989 Jim Unger. Reprinted with permission of Universal Press Syndicate. All rights reserved.

We all have favorite scents we enjoy and certain other scents we dislike. This is often due to an associated event, place, or person from our past. For example, the combined

Figure 7-12. Fragrance often serves to trigger our memory.

Figure 7-13. Some fragrances (in this case, stock) may be pleasing to one, annoying to another.

scents of stock and chrysanthemum may evoke sadness or depression and general dislike if a death and funeral are associated with them. Obviously, fragrances with a negative memory attached are often harder to enjoy.

Likewise, fragrances that bring about positive or happy memories are usually pleasant and enjoyable. For example, the delicate scents of lily of the valley and daffodils may elicit pleasant thoughts of new life in a spring garden.

Sometimes, however, our reaction to a certain scent may be inexplicable. Because of our individual fragrance sensitivities, a certain floral scent, such as that of a gardenia or hyacinth, may be sweet to one person, and to another overwhelmingly nauseating. As with color, we all have certain favorites and preferences and certain dislikes (see figure 7-13).

A basic understanding of the nature of scent is important and helpful. The scent from all fragrant plants is produced by an essential oil found either in the flowers or in the leaves. In flowers, this essential oil is called ***attar*** and is mostly stored in epidermal cells in petals or petal substitutes. Made up of a number of chemical compounds, the formula for each is unique. Flowers, of course, use their scents to attract and lure insects, birds, and other pollinators.

By contrast, the purpose of scented foliage is not to attract insects but to repel them. The scent from foliage is used as a protection against disease and harmful predators. Many of the compounds found in foliage are strongly antiseptic and smell like medicine. The smell of spiral eucalyptus, for example, may be wonderfully exhilarating for some or terribly odorous for others.

Strong fragrances should generally be avoided in designs for small hospital rooms or small office spaces. Any fragrance can turn odorous and become annoying if it is close and in a limited space (see figure 7-14). Allergies to certain fragrances will also limit the choice of flowers for a bouquet.

The scent of flowers and foliage can be bewitching, yet often the most neglected and least understood part of floral arranging. The fragrance of flowers varies from sweet and delicate to spicy and bold. Take the opportunity to incorporate fragrance in your selection of flowers.

Figure 7-14. Any fragrance may turn odorous and become annoying, especially in a small, closed area.

"How's your hay fever, Mother?"

Figure 7-15.
Herman © 1987 Jim Unger. Reprinted with permission of Universal Press Syndicate. All rights reserved.

Figure 7-16. Many flowers offer fragrance, as shown here by this small sampling of fragrant flowers.

Figure 7-17. line drawing identification of Figure 7-16. Some of the fragrant flowers available:
a. acacia, b. eucalyptus, c. genista (Cytisus canarienis), *d. rose, e. paperwhites* (Narcissus) *and waxflower* (Chamelaucium uncinatum), *f. daffodils* (Narcissus) *and freesia and stock* (Matthiola incana) *and forsythia, g. tulips, nerine, and heather* (Erica), *h. sweetpeas* (Lathyrus), *i. gardenia, j. lilac* (Syringa), *k. sweet William* (Dianthus), *l. carnations and spray carnations* (Dianthus), *m. diosma* (Coleonema pulchrum), *n. stephanotis, o. bouvardia , p. stock and prunus branches*

REVIEW

Floral arrangements can gain more sensory importance through the use of texture and fragrance. Both texture and fragrance are often overlooked as elements of floral design. They should be carefully considered when you are selecting materials for your designs. Without textures and surface variations, floral arrangements can appear flat and uninteresting. Texture adds character and gives distinction to a floral composition. Likewise, fragrance adds to the enjoyment of flowers by adding a new dimension. A floral bouquet that appeals to more senses than just sight will result in increased visual and sensory pleasure.

TERMS TO INCREASE YOUR UNDERSTANDING

attar
extreme contrasts
fragrance
olfactory
pubescent
succulents
tactile
texture
visual

TEST YOUR KNOWLEDGE

1. Name some coarse-textured flowers and foliage.
2. Which color schemes are enhanced through the use of texture?
3. Why is a variety of textures important?
4. What are some of the pros and cons of fragrance in floral designs?

RELATED ACTIVITIES

1. While visiting a retail floral shop or a wholesale operation, notice texture extremes in flowers and foliage. What are some of your favorite flower and foliage textures?
2. What are some of your favorite flower fragrances? Are there any explanations as to why you like certain fragrances and dislike others?
3. Plan and make an arrangement that incorporates both fragrance and extreme contrasts in texture.

Chapter 8

Tools, Containers, and Mechanics

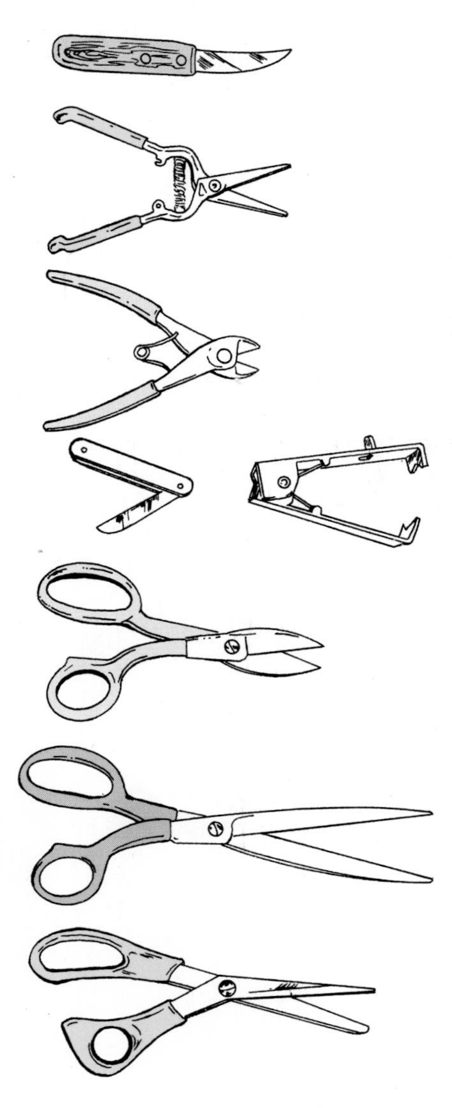

Figure 8-1. Basic hand tools for floral work include knives, clippers, wire snips, strippers, wire scissors, ribbon scissors, and utility scissors.

The success of any floral arrangement is dependent on how it is constructed. Using the proper tools, choosing the correct container, and knowing how to put all the parts of a composition together are important in the arrangement of flowers. This chapter will help you have a better understanding of the basic tools necessary to construct floral arrangements. It will be beneficial for you to learn about the various shapes, styles, and types of containers so you can achieve harmony and unity in your designs. A thorough knowledge of the materials available and of construction techniques will build your confidence and increase your skill in the art of floral arrangement.

HAND TOOLS

Certain hand tools are essential for design work, as shown in figure 8-1. If readily accessible, a few tools will make the process of arranging flowers much easier and much more time efficient. It is important to purchase high quality tools and use each as intended. Maintain your tools by keeping them sharp and clean in order for them to work efficiently and last a long time.

Knives

A knife is perhaps the most important tool to a designer because all flowers must be cut before being inserted into a design. A wide variety of knives are available for floral work. Pocket knives are commonly used by florists because they are sharp with quality blades and convenient because they fold up when not in use. Most other floral knives look like kitchen paring knives. The type of knife you will use largely is a matter of personal preference. Be sure to choose one that fits comfortably in your hand.

First attempts at using a knife might be awkward, but with practice it will become second nature. Soon cutting stems any other way will seem cumbersome. (See chapter 10 for more information on the proper use of a knife.)

Clippers

Many different styles of clippers or floral shears are available. These are often used by designers for cutting or trimming thicker flower and foliage stems as they allow the cutting action to take place in both directions without pinching the stems.

Wire Cutters

There are several types and styles of wire cutters. Wire snips that have a spring action handle and a short blade can easily cut any thickness of floral wire, corsage stem, and other heavy material. Utility wire scissors are generally serrated and can cut heavier materials, florist wire, and plastic. They have longer blades and can cut across chicken wire more easily than the short-bladed wire snips.

Strippers

Different types of hand strippers are available. Hand strippers, often called ***rose strippers*** or ***de-thorners,*** are used for removing thorns and leaves from stems. It is important not to strip tender bark from stems. This practice can harm flowers.

Scissors

Sharp, quality scissors are essential. Ribbon scissors have a long, slender blade and should be used only for cutting ribbon, fine netting, and fabric in order to maintain their sharpness. It is also helpful to have utility scissors available for trimming foliage and cutting other materials.

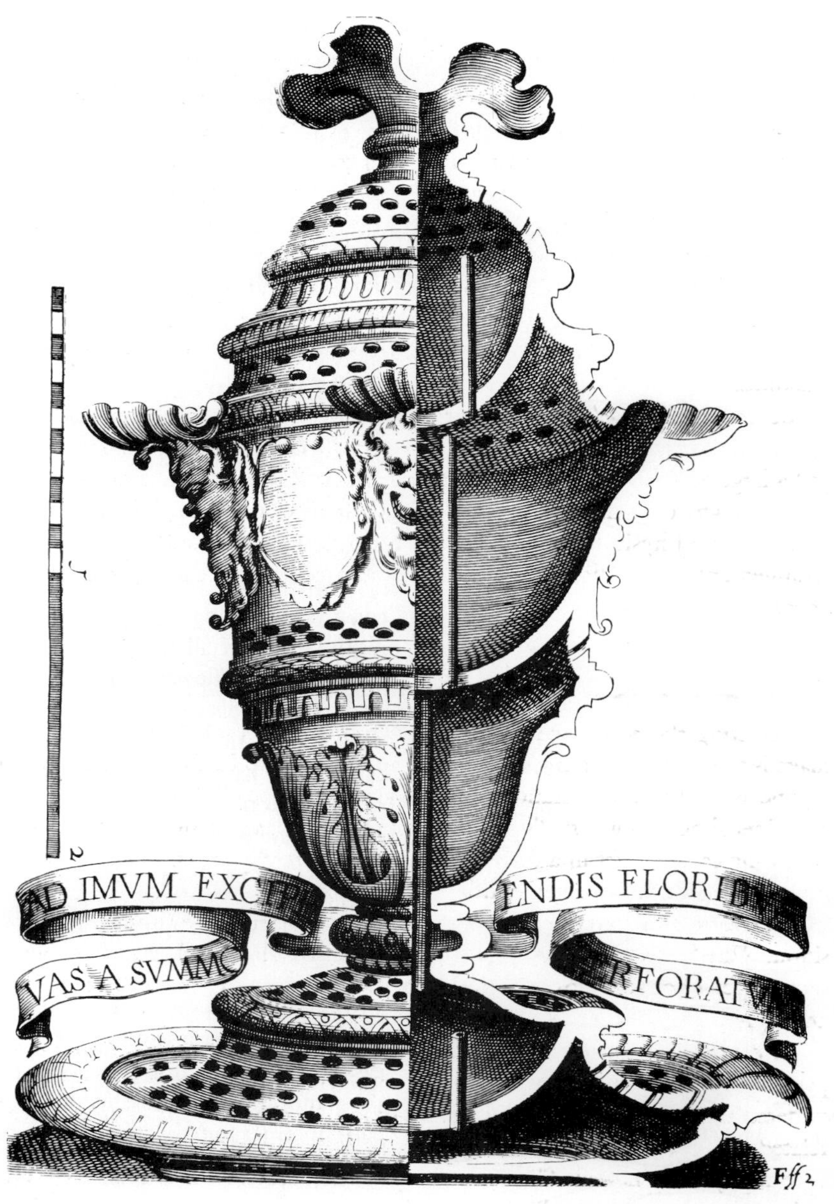

Figure 8-2. This cutaway drawing of an elaborate vase is from Ferrari's "Flora overo Cultura di Fiori," Rome, 1638. Short-stemmed flowers could easily be inserted through the holes and kept fresh with a unique watering system.
The Pierpont Morgan Library, New York. PML 33421.

CONTAINERS

A multitude of containers are available, which come in a wide variety of shapes, sizes, styles, materials, and prices. Choosing a container not only depends on your choice of flowers, but also where the design will be placed. The container should be compatible with the entire design and setting, not only visually but physically as well.

Function

Most containers from ancient times were not made uniquely to hold flowers; rather they were designed primarily for other purposes, for instance, the storage of food and water. It wasn't until later that containers were made for the particular function of floral design. As seen in figure 8-2, the ease of floral arrangement has long been a concern. Some of these containers even had holes and unique water-filling systems for flowers.

The container is an all-important part of any floral composition. It is the foundation and starting point for construction of the entire design. If the foundation is inferior in any way, the floral arrangement usually reflects that weakness.

Several factors make a container suitable for the arrangement of flowers. Containers for fresh flowers must hold water, with a deep reservoir for water, and a large area for the stems. The opening where flowers extend from must be wide enough to fit individual stems or large masses of stems. The container must also be physically heavy enough to support the actual weight of the arrangement. It should also function in helping to conceal any mechanics of construction.

Figure 8-3. The container chosen should fit the design shape and style.

Shape and Size

Selecting the correct container shape and size for each arrangement you make is essential for creating designs that appear unified. In choosing a container, one of the first considerations must be given to shape. The shape of a chosen container should flow visually with the completed arrangement (see figure 8-3). Harmony cannot exist in a design if the container is the wrong shape or out of proportion with the arrangement.

The form of a container will often dictate the form of the finished composition. Whether you need a vertical cylinder, a rounded bowl, or a horizontal rectangle, choose a container that will complement the shape of the arrangement.

A container must be in scale with the arrangement and be the proper proportion to the flowers and setting. The container size should not overwhelm or underestimate the flowers and foliage it holds. The size of a container is closely related with function, because a container must not only visually hold the flowers but physically as well. The size of a container must allow it to easily hold the weight of its flowers and foliage.

Larger, heavier flowers—for example, incurve (football) mums and gladiolus—need a container that is big enough not to tip over. In contrast, tiny sweetheart roses and gypsophila do not demand a large container and are more suited to a smaller vase.

Style

The container greatly influences the style and mood of a floral arrangement. The decorative appeal of a container should be in harmony with the style of an arrangement. The container should be suited to the flowers as well as the decor where the arrangement will be placed. Not only is the container shape and size important, but, so are color, pattern, and texture, as they help determine mood and theme.

Certain container shapes can set a period-style theme for an arrangement: for instance, a large classical metal urn will display a certain style, perhaps of lavish opulence. In con-

Figure 8-4. The proportion of the container to the entire arrangement can vary considerably.

Figure 8-5. Glassware offers containers for flowers that are simple and affordable and yet elegant.

trast, the style of arrangement made in a low, wooden tray or a small, porcelain teacup is very different.

The color of a container should be compatible with the entire arrangement. Neutral container colors are tints and shades of green, tan, gray, and brown. Black and white containers also can serve as neutral container colors; however, you must use caution with certain flower colors. For instance, a black container might not harmonize with springtime pastels, and prove instead to be too stark a contrast. A plain white container may compete and conflict with autumn yellows, oranges, and reds. Coordinated container colors that repeat flower colors can also be used, as the repetition of similar colors helps to create a sense of harmony and unity. Whatever container color is chosen, it should help accent and give importance to the flowers.

Some containers offer a choice of decorative patterns. These are useful for novelty or holiday designs, where the pattern helps to execute a theme. Highly ornate containers also have a place in floral design and are appropriate for more lavish settings.

The textures of the flowers, foliage, and accessories will help determine the container texture that is most appropriate. Reflective containers, such as shiny metals or fine porcelain, usually appear more formal and often suggest a luxurious style. Nonreflective surfaces like coarse textured baskets and wooden crates display a more casual mood and often suggest a rugged, earthy style.

Figure 8-6. Plastic containers offer versatility in shape, size, style, and color.

Containers are an integral part of floral arrangements. However, the amount of the container that is seen will vary with the design style. As shown in figure 8-4, the proportion of the container to the entire arrangement can vary considerably, depending upon the decorative qualities. Sometimes, the container can be the visual emphasis of a composition, while at other times, the container may be totally hidden and is solely functional.

Common Materials

Many choices of container materials are available. The most widely used containers are glass and plastic. Baskets are also a popular container choice. In addition, various types of ceramic containers are also commonly used. For a more sophisticated, formal look, brass and other metal containers are sometimes used. A look at the arrangement requirements, and often budget as well, will dictate your container choice.

Glassware

Glassware is common, affordable, and versatile. Glass containers are available in a vast selection of shapes, sizes, and styles. Glass offers a way to present flowers that is simple and affordable, yet displays high quality and elegance. Since light can pass through it, another dimension of beauty is added.

Floral arrangements made in clear glass containers do present a few minor challenges to designers. All of the inner workings of the arrangement are readily seen, so it is important to use little or no mechanical aids. Visually, it looks best not to use floral foam or shredded styrofoam in clear vases. However, glass stones or marbles sometimes are used to hide mechanics and help support and keep stems at various angles. To keep all the flowers positioned, use a vase with a small opening or use a plastic or tape grid at the top of the vase. Or better still, make a natural grid with foliage. (See chapter 13.)

Since it is vitally important to keep the water in the vase clean and pure, use a floral preservative that will noy cloud the water. Remember to remove any foliage from stems that will be in the water. Soggy foliage is unsightly and accelerates the growth of bacteria. Utility or everyday glass containers are available in clear, green, frosted, and white. Some manufacturers offer glassware in subtle hints of peach, pink, green, and blue.

Plastics

Plastic containers come in a wide variety of shapes, sizes, styles, and colors. No longer are plastic containers considered necessarily cheap and cheap-looking, since plastic can take on the appearance of different materials including glass, brass, and ceramic.

Plastic containers have several advantages, one of which is cost, since plastic vases are the most cost-effective container material. Most plastic containers are unbreakable and are suited for the arrangement of flowers.

There is a plastic container that will meet the needs of most arrangement moods and styles.

Because plastic containers are lightweight, extra weight must be added to them to ensure stability when arrangements contain tall and heavy flowers . A variety of materials may add weight for stability such as rocks, sand, gravel, or plaster of Paris. Use of these is appropriate for fresh or artificial arrangements.

Baskets

Baskets are available in a wide range of styles, sizes, shapes, and fiber combinations. They are generally inexpensive. They come in various colors and textures and can be easily altered with spray paints.

Baskets are inherently charming and their coarse textures offer a homespun appearance. Appropriate for many different flower combinations throughout the year, baskets can help create different looks.

While working with baskets is a pleasure, they do present occasional challenges. For instance, some baskets present stability problems if they are not woven with a sound horizontal base. Baskets must also have a liner to hold floral foam and water. Simply putting foil or clear wrap in the bottom of the basket is not sufficient to keep stems from poking through and water leaking out. Use a plastic bowl or commercial liner that will hold the floral foam and water. Liners are available in a range of sizes. If the exact size liner is not available to fill a basket, use a slightly smaller one and anchor it with tape or glue.

Figure 8-8. Ceramic containers can be plain and simple or decorative and unusual.

Figure 8-7. Baskets come in a wide variety of color, textures, and styles.

Ceramics

Ceramic containers are available in many styles and types. Produced by heating clay, ceramics include earthenware (otherwise known as pottery), stoneware, and porcelain. Glazes applied to pieces before and after heating create a wide variety of surface finishes and textures. A coarse earthenware pot will present a different design option to that of a highly polished porcelain vase.

Many ceramic pieces look like fine art with stylized lines and form. Some vases offer beautiful decorative designs and patterns. These are generally more expensive and reserved for high style and elaborate designs.

There are also everyday ceramic containers, many of which are novelty or special occasion containers and mugs. These, used together with flowers, emphasize whatever theme you choose to follow.

Figure 8-9. Brass containers with their highly reflective qualities offer a touch of opulence.

Brass

Brass generally has a highly reflective surface that signifies formality and elegance. Although brass containers come in several shapes, sizes, and styles, they generally are expensive and reserved for special orders.

With brass and other metallic containers, there is a possibility that floral preservatives will not work because of their chemical interaction with the metal. Also, the metal may corrode as a result of contact with water and chemicals. It is, therefore, best to use a plastic liner inside the container. This will allow the preservative to remain active and at the same time help the brass remain clean, shiny, and untarnished.

A container is often chosen after flowers and foliage have been selected. When you choose a container that will complement a design, remember that it needs to relate in color, texture, or shape to the arrangement so that harmony and unity can exist. If the container may be of primary importance, however, select flowers and foliage according to the container's style, size, shape, or color.

MECHANICS OF ARRANGEMENT

The stability and quality of a floral composition depend on its ***mechanics of arrangement,*** which refer to the methods of construction. In order to create high-quality, successful designs efficiently, it is important for you to gain a practical knowledge of arrangement techniques.

Establishing the Foundation

In order for you to arrange flowers in a container (and have them stay where you want them within the design even after they are delivered to someone) your arrangement requires a secure and functional foundation. The most popular and efficient method of construction is with the use of ***floral foam.*** Other time-honored methods include ***pin holders, grids,*** and the ***lacing of foliage.***

Floral Foam

Floral foam is a porous material designed specifically for the arrangement of flowers. A membrane covers the floral foam, helping to hold in the moisture. To be of use, the floral foam for fresh arrangements must first be saturated with water. These foams are green and most common in block or brick shape.

Soaking Foam

Floral foam must be fully soaked before fresh flowers may be inserted. The best method of soaking foam is to let it float freely in a clean sink or bucketful of preservative solution. There should be enough water to fill

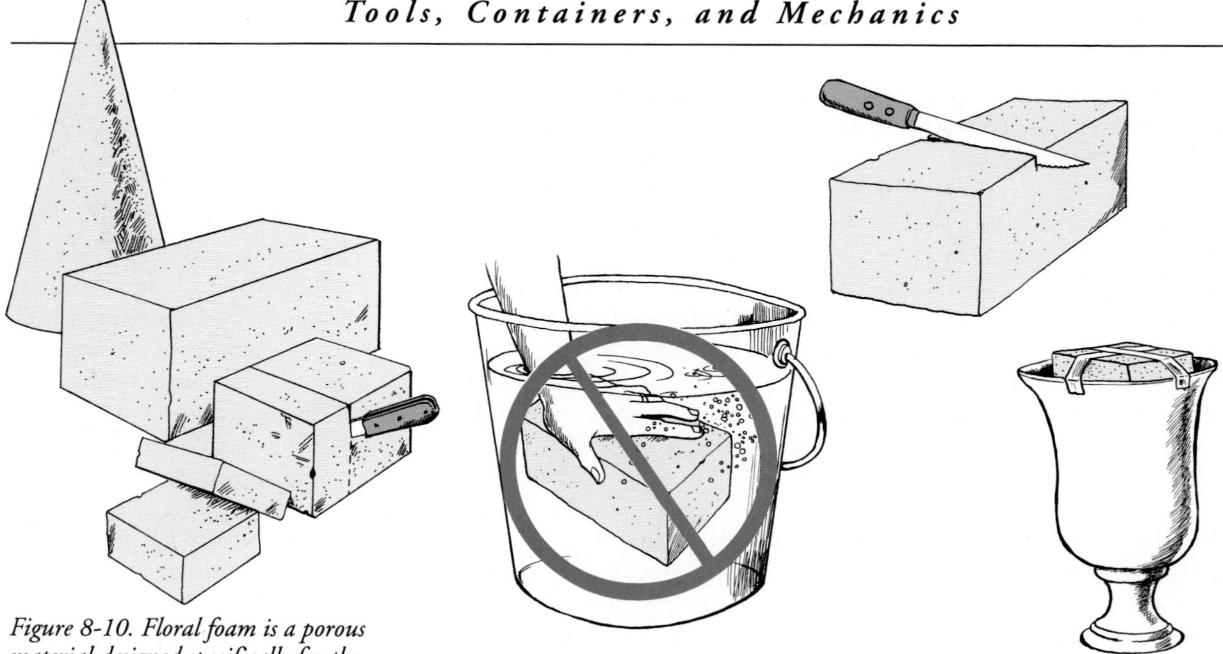

Figure 8-10. Floral foam is a porous material designed specifically for the arrangement of flowers to keep them fresh longer. It is available in several shapes and can be easily cut into desired shapes and sizes.

Figure 8-11. Floral foam must be fully soaked before flowers can be inserted; however, it should not be forced downward to speed up saturation. After being saturated, foam should be cut with a knife. Waterproof tape may be used for added security.

the block and allow it to float when it is fully saturated. Each block holds about two quarts of water.

Foams have various absorption rates, so it is best to let the foam soak on its own. This allows for total saturation. It usually takes one or two minutes for blocks to fully soak. Never force foam downward to speed up the saturation process as shown in figure 8-11. Leave the foam in the water until it has stopped bubbling. As water rapidly penetrates the foam, air pockets may form and become trapped inside. You can check for air pockets by inserting a knife blade into several places in the foam. This allows trapped air to escape and the foam to saturate more fully.

After a block of foam is saturated, it can be cut and placed into a container. Then flowers can be inserted. A second method is to cut smaller portions of foam before soaking, and then soak these smaller sections individually. Do not reuse foam once it has been used for a design previously. It will not have the same water-holding capacity, and flowers will wilt and die prematurely. This is also true for foam that has been soaked and allowed to dry without ever using it.

Cutting Foam

It is best to cut the floral foam with a knife. If you try to simply snap and break the foam or use scissors or wire to cut it, you will have a jagged mess. From the main block, cut off the amount of foam needed to fill a container. You should have enough room between the block of foam and the container sides to fill with water. If the block is solid against all the container sides, a water well needs to be made. This is done by trimming away a small corner section of the block or poking a stick or pen down through the foam edges so that water may be easily added to keep the container full of water (see figure 8-12).

Foam can extend beyond the container top about an inch and still draw water up into that portion that is not actually sitting in water. (Also, if the foam is 1/2 to 1 inch above the container top, this will allow room to extend stems horizontally or slightly downward.)

Securing Foam

If the foam is loose enough to shift at all in the container, flowers can easily become damaged or worse, the whole arrangement can fall out of the container during handling or delivery. There are several ways of securing foam into containers: ***waterproof tape, anchor pins,*** or ***pan-melt glue.***

Waterproof tape is available in 1/4- and 1/2-inch widths and comes in clear, green, and white. Also called ***anchor tape,*** it is used to hold the floral foam securely in the container, giving the entire arrangement stability. The tape is stretched across the top of the foam and then pressed onto the container sides. All tape must be hidden, so use only the amount of tape that is absolutely necessary to secure foam. The tape will not actually stick to the wet floral foam but adheres to the clean and dry surface of the container.

Anchor pins are easily attached to the bottom of a container with glue, tape, or floral clay. Saturated floral foam can then be placed on top of the anchor pin as shown in figure 8-13. The prongs will securely hold the foam in place, although several anchor pins may be required to hold a full block of foam.

Pan-melt glue can also be used to attach foam to plastic containers and liners. In order for the glue to adhere, however, the foam must be dry. The precut piece of foam must be dipped into the pan-melt glue, making sure all four corners are dipped. Only a thin coat of glue is needed and will adhere better if excess glue is scraped off. The glued side of the foam is then placed in the bottom of the container. The glue needs a minute to cool to fully adhere, afterwhich the foam can be saturated. To saturate foam, it is best to dip the container with the attached dry foam sideways in preservative solution. The water will soak into the foam from one side to the other as shown in figure 8-14.

The primary advantage to using pan-melt glue to secure floral foam is that many containers can be prepared ahead of time for a busy holiday or large party. The prefilled containers with dry foam can be soaked as they are needed.

Alternatives to floral foam are pin holders, chicken wire, tape grids, and foliage grids (see figure 8-15).

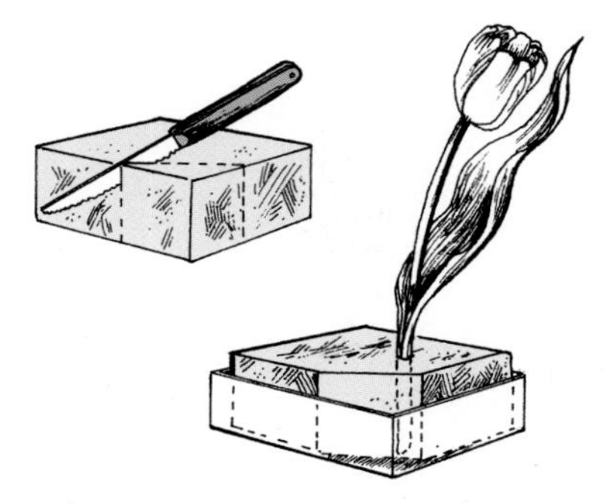

Figure 8-12. Floral foam is commonly used to create a foundation for flowers. When using foam, it is important to create a water well by trimming away a small corner of the block.

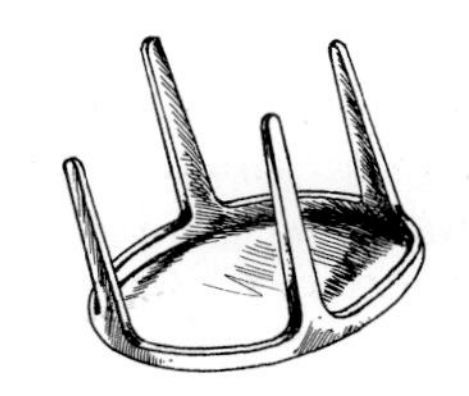

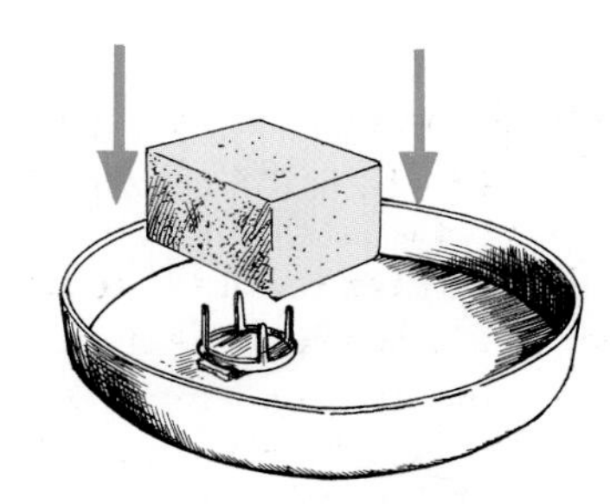

Figure 8-13. Anchor pins can be attached to the bottom of a container with glue, tape, or floral clay, after which saturated floral foam can be secured to the anchor pin.

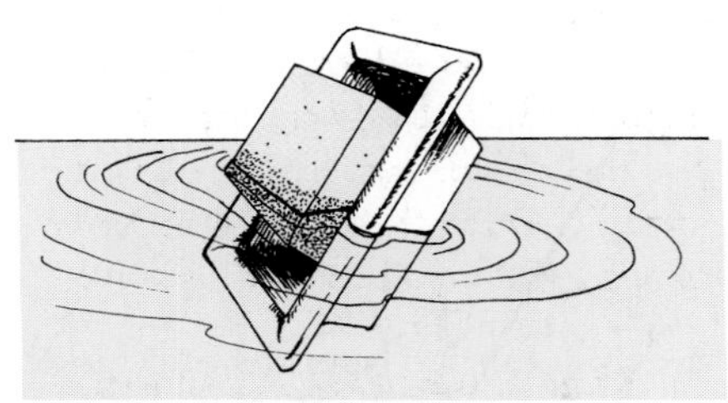

Figure 8-14. To fully saturate foam that has been glued into a plastic container or liner, dip the container into preservative solution sideways. The solution will gradually soak into the foam from one side to the other.

Pin Holders

Pin holders are also known as ***needlepoint holders*** and ***frogs.*** A pin holder has a heavy metal base with many closely spaced, sharp-pointed pins sticking upright. The pin holder is usually secured inside a container with floral clay. The flower stems may then be arranged directly onto the pins or wedged between them. Available in different sizes, pin holders are ideal for oriental, modern, and high-style designs with few flowers.

To make arrangements with pin holders, more time is needed; often, it is a challenge to keep the arranged flowers securely placed. Due to their expense, floral shops generally do not use pin holders for everyday customer designs.

Grids

A grid on the top of a clear glass vase will make it much easier to arrange flowers. It allows the natural beauty of stems to be seen without unsightly mechanics (see figure 8-17). Grids can be made by running 1/4-inch waterproof tape across the top of the container as shown in figure 8-18. In order for the tape to stay in place, the container must be clean and dry.

Many glassware manufacturers sell plastic grids that fasten to the container opening. These are convenient time-savers but are not available in all sizes for every glass vase.

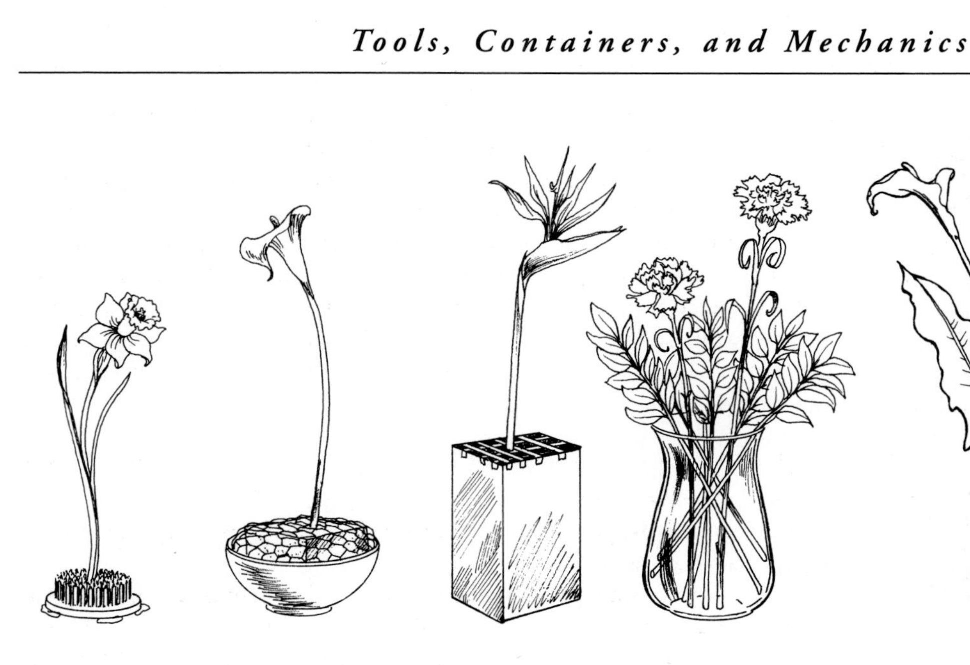

Figure 8-15. Alternatives to floral foam for securing flowers include the use of a pin holder, chicken wire, tape grids, and foliage grids.

Figure 8-17. The use of grids or the lacing of foliage is better in glass vases than floral foam where the mechanics of foam are visible. Furthermore, as stems poke through foam, foam pieces may shed and float around in the water, causing an unsightly mess.

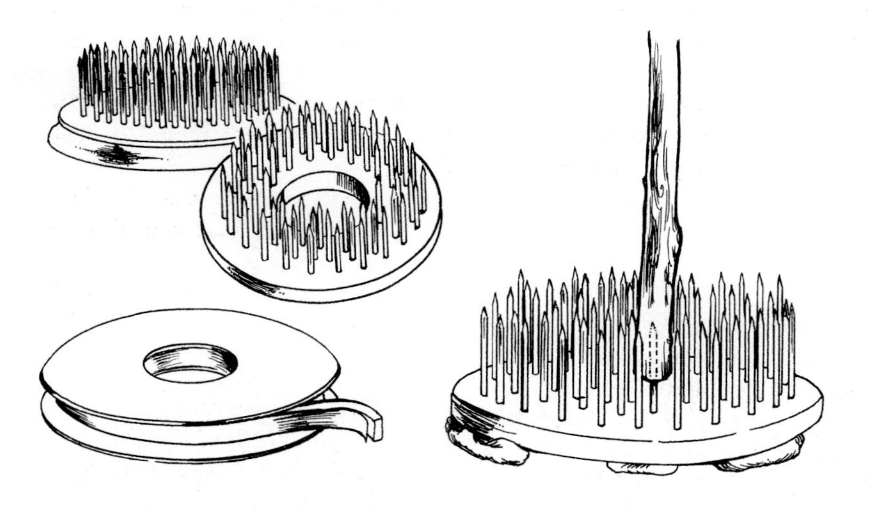

Figure 8-16. Pin holders, available in various shapes and sizes, are usually secured with floral clay. Flower and foliage stems can be arranged in between the sharp pins or directly on top of the pins, as shown.

Grids can also be made with chicken wire, which has long been used to help flowers stay in place. The edges of a piece of chicken wire are pressed over the edges of the lip of a container. The wire grid can then be taped in place with clear waterproof tape, or ***beauty clips*** may be used to fasten the chicken wire to the container rim. Care must be used so the wire ends do not leave scratches in the glass.

The flower and foliage stems must be cleaned of debris, and excess foliage must be removed from stems before they are inserted into the preservative solution. Remember that plant debris that floats in the vase is not only unpleasant to look at, it accelerates the growth of bacteria.

Foliage Grid

A grid can easily be made by interlocking foliage stems. This quick and simple, time-honored technique

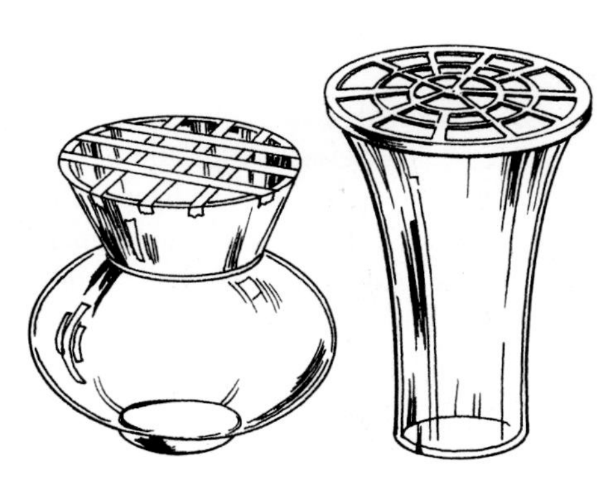

Figure 8-18. Convenient snap-on plastic grids are available for many glass vases. Grids can be made by running 1/4-inch waterproof tape across the top of the container.

is often referred to as lacing. Because only foliage is used to create the grid, there is no worry about concealing mechanics.

Leatherleaf is commonly used to create grids. However, many other foliage types or a combination of them can be used, including salal, fern, and huckleberry. The number of foliage stems required is determined by the size of the vase opening and the number of flowers that will be inserted.

The first step is to clean off all foliage stems and remove excess lower leaves. The grid can be made entirely in your hand or placed one stem at a time directly into the vase. As shown in figure 8-19, start with one piece of foliage and add another by inserting the second stem between the leaves or fronds of the first. Each foliage stem that is added to the grid is interlocked with another. Continue adding foliage until a tight interlocking grid is formed. Flowers and fillers can then be inserted and will stay securely positioned within the bouquet.

Wire

Wire has many different uses in floral design work. The types of wire available are ***straight wire, paddle*** or ***spool wire,*** and ***chicken wire.*** Once in a design, all types of wire must be disguised.

Straight wire is sold most commonly in 18-inch length pieces and is available in silver and green (enamel). The painted green wire is preferred because it blends with stems and is not as difficult to disguise. Also, it does not rust as easily as the silver wire when in contact with water.

Wire is given a gauge number according to its diameter or thickness. The lower the number, the thicker the wire. As the gauge number increases, the wire becomes thinner or finer. Common floral wire gauges range from very heavy #16 to delicate #30.

Straight wire is used to strengthen weak stems, support heavy flower heads, and bind various materials. It is also used to make corsages, boutonnieres, hair pieces, bridal bouquets, and many other floral items.

Paddle wire and spool wire are continuous wires wound on wooden paddles or spools. They are available in a variety of gauges. This wire is used when

Figure 8-19. Leatherleaf is often used to create foliage grids, although other foliage types such as salal, fern, and huckleberry, can be used. To create a foliage grid, insert stems within the vase to form a tight interlocking grid through which flower stems can be inserted.

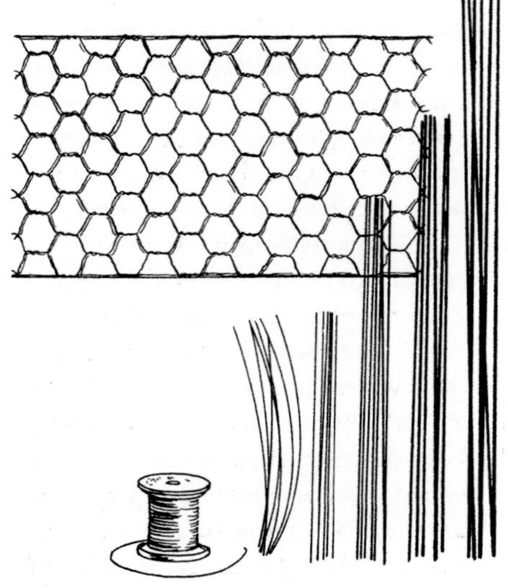

Figure 8-20. Available in straight wire lengths, on a spool, and in the form of chicken wire, wire has many different uses in floral design.

pieces longer than 18 inches are needed. It is commonly used to bind large materials, create garlands, and secure foliage wreaths.

Chicken wire is available in rolls of 12- to 18-inch widths. Cut pieces of chicken wire may be used to form grids, or pieces may be secured on top of floral foam to help support heavy, thick stems when inserted into the foam. It is also used to create topiary balls, cones, or other unusual or ornamental shapes, giving support to the floral foam. Any time floral foam is used without a container, a layer of chicken wire can be wrapped around the foam, giving further support to foam and flowers.

Wiring Fresh Flowers

Flower stems are often wired for several reasons. Wire can strengthen weak stems, keep flower heads upright, straighten crooked stems, or add curve to straight stems. There are several different ways to wire flowers. The flower head shape, stem thickness, and the reason for wiring a stem are all factors in your decision as to which wiring technique is most appropriate.

There are three basic wiring techniques; ***pierce, hook-wire,*** and ***insertion.*** As shown in figure 8-21, wire can be inserted vertically into the top of the stem (or at the base of the flower petals) and then wound spirally down the stem between leaves. This method is often used for roses and carnations.

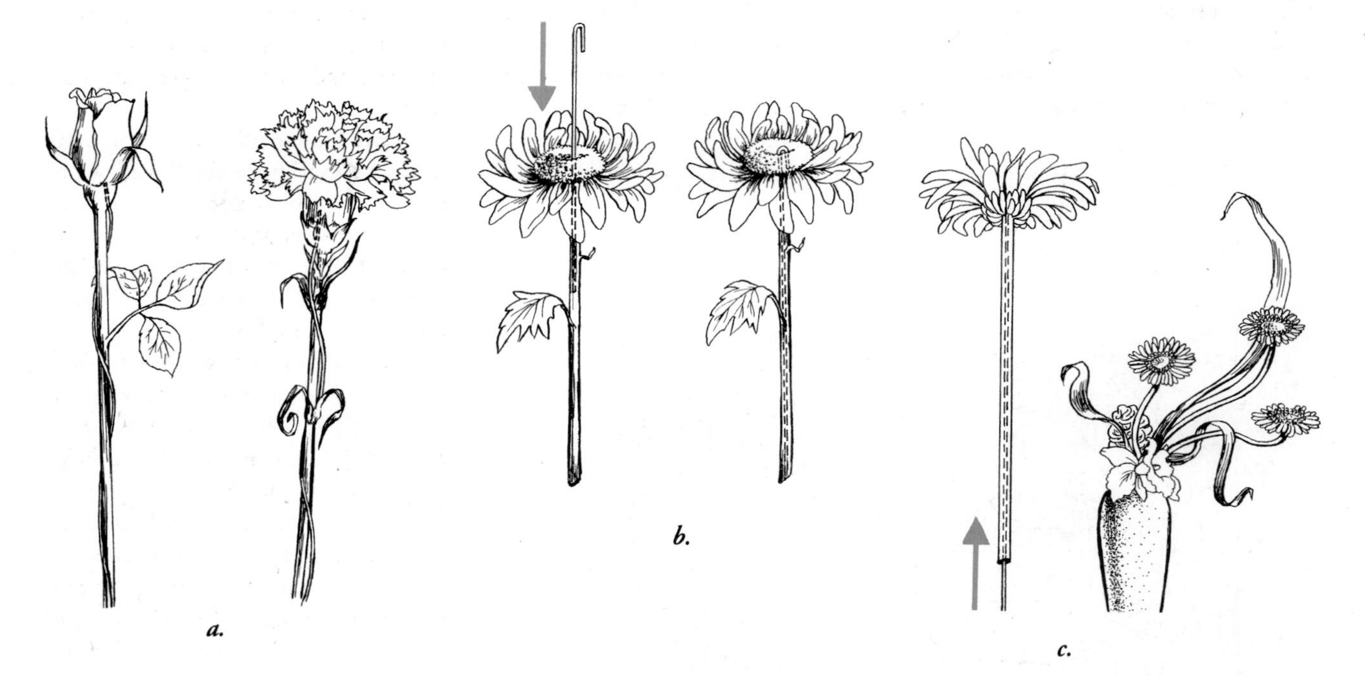

Figure 8-21. Three basic wiring techniques for fresh flowers include the following: [from left (a) to right (c)] (a) roses and carnations are often supported by inserting wire into the top of the stem or base of the flower and winding it vertically down the stem, (b) chrysanthemums, when in need of extra support, are wired by forming a hook at the top of the wire, then inserting the straight end of the wire into the center of the flower, called the hook-wire method, and (c) many flowers, such as gerbera, are given extra support by inserting a wire up the base of the stem of these and other hollow or fleshy-stemmed flowers to straighten stems or to create dynamic curves.

Another wiring technique, often referred to as the hook-wire method, is used commonly for weak flowers with flattened heads. First, a hook must be formed at the top of the wire. The straight end of the wire is then inserted in the top of the flower and pushed or pulled down far enough for the hook of the wire to be hidden. Often the end of the wire pokes through the stem after it passes through the flower. If this happens, leave the wire parallel with the stem or wind it vertically down the stem.

Hollow or larger fleshy-stemmed flowers will be better supported by insertion wiring. A wire is inserted up through the base of the stem all the way to the top. It is best to gently and slowly push the wire up so it will not poke through the stem or flower. Any excess wire at the stem base should be cut off. Gerberas are often wired this way for support. Dynamic curving lines can be created by gently molding the wired stem to the desired shape.

Glue Adhesives

Adhesives contribute to overall design security, keeping individual parts securely anchored and in place. Numerous glue and tape options are available that will help save time and labor. Many different adhesives or glues are used in a floral shop. The most common glues include pan-melt glue, ***glue guns*** with ***glue sticks, liquid floral glue,*** and ***spray glue.***

Pan-melt glue is available in solid pillows, blocks, and chips. It is usually melted in a small electric frying pan heated to about 275 degrees. Once the glue has melted to the thickness of molasses, materials can be dipped into the glue and attached to clean and dry surfaces. Pan-melt glue will hold securely even with changes in moisture and temperature, unlike other glues. When allowed to cool, the glue will harden in the pan, which can be stored until the next use. Only a few minutes are needed to reheat the glue to a usable consistency, and new glue pieces can be added when the glue level becomes low.

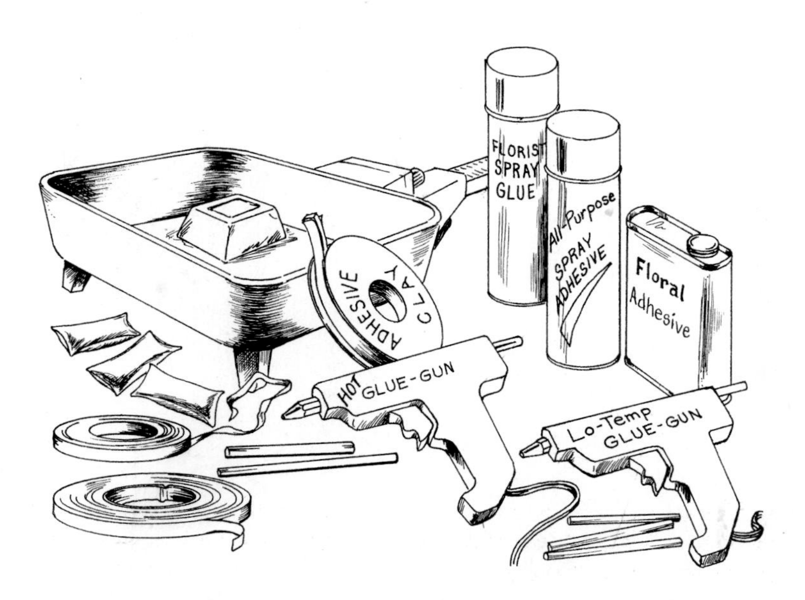

Figure 8-22. Many kinds of adhesives, including glues and tapes, help floral designers save time and labor.

Electric glue guns that heat glue sticks are commonly used for floral design work. The sticks of glue are easily inserted into the back of the gun. When the trigger is pulled, melted glue, which comes out of a very hot tip, can be directly applied to the surface of an object. Two gun types are available, ***hot melt*** and ***low temperature.*** The glue sticks for each type of gun differ in shape and are not interchangeable.

The advantage of glue guns is that the glue can be easily applied with control and versatility. However, the glue from glue guns is sensitive to temperature changes. At cold temperatures, less than 40° F, the glue separates and will no longer adhere to surfaces.

Liquid floral adhesive is manufactured specifically for the floral industry. It is a type of rubber cement that will not harm fresh flowers. This type of glue takes longer to dry than other glue types but is not sensitive to temperature or humidity.

Several types and densities of spray or aerosol glues are available. Floral sprays will not harm flowers. Some glues are manufactured to help keep petals intact, while others can be used in combination with glitter. Heavier nonfloral sprays are generally not used on fresh flowers but are commonly used for other design materials such as containers, ribbon, and permanent materials.

Clays and Tapes

In addition to these glues, other adhesives commonly used include

floral clay, floral tape, anchor tape, and ***double-face tape.*** To ensure adhesion, the surfaces to be bonded must be clean and dry.

Floral clay is a putty-like substance. It is available in strips or rolls of green or white with waxed paper between each layer, to prevent the layers from sticking to each other. This clay is moisture resistant and can be used to anchor pin holders and other items that will be in water. It is also commonly used to secure styrofoam into containers.

Floral tape, also called ***green tape,*** actually is not a regular adhesive tape; however, it will cling to itself as it is stretched. It is a paraffin-coated paper available in a rainbow of colors with green tints and shades most common. It is used to wrap wire, bind materials, construct corsages and boutonnieres, and many other floral items, including bridal bouquets.

Anchor tape, also called waterproof tape, ***bowl tape,*** or ***pot tape,*** is available in clear, green, and white in 1/4- and 1/2-inch widths. It is frequently used to secure foam blocks into containers. It is also used to make grids for the arrangement of flowers.

Doubleface tape is sticky on both sides and is temperature and moisture resistant. It is used in many design capacities, including securing anchor pins and connecting containers together, such as a liner in a basket.

Pins, Picks, and Water Tubes

Many other items are used in the mechanics of design. Several types of ***pins, picks,*** and ***water tubes*** are commonly used to help in construction of various floral designs.

Greening pins, often called ***philly pins*** and ***fern pins,*** are used to secure moss into foam in both fresh and everlasting designs (see figure 8-23). They can also be used to help anchor other materials into arrangements.

Corsage and boutonniere pins are available in various sizes with different head types. They are generally used to secure flowers to clothing but can also be used to pin flowers and ribbon to styrofoam.

Bank pins are straight heavy pins with flat heads. They are commonly used for pinning flowers, ribbon, and other materials to styrofoam.

A ***dixon pin*** is made up of two small wooden picks attached to each other with a thin, pliable wire. These are used to help anchor materials into designs.

Wooden picks are available in different sizes. One end tapers to a point while the other end has a pliable wire attached to it. These picks are used to help secure and anchor items into a floral design. They are also used to secure bows into potted plants and can be used to lengthen stems and accessories.

Long green wooden sticks, often called ***hyacinth stakes,*** are available in various lengths. They may be used to support heavy blooms or foliage in potted plants and fresh floral designs. They are also commonly used to extend stems and accessories or to help secure accessories into designs.

Steel picks available in several lengths are used in a steel pick machine. By pulling a lever on the machine, a pick is instantly attached to dry or silk flower stems (see chapter

Figure 8-23. Green pins are used to secure moss onto foam, as well as to firmly place other materials into arrangements.

15). These picks can also group delicate, everlasting stems together for easier insertion into floral foam. A pick machine should never be used on fresh flowers and foliage.

As shown in figure 8-24, a variety of tubes and funnels are available. Water tubes usually have a rubber or plastic lid that prevents the preservative solution from leaking. A flower can be poked through the center hole in the lid to receive a constant water supply. Depending upon what they are needed for, the shape of the tubes will vary. Pointed tubes are ideal for poking flowers into floral foam, plants, or dish gardens. The rounded tubes are commonly used for cut flowers that are wrapped or boxed that would otherwise be without water supply. Smaller tubes can be hidden inside floral tape for corsages and other floral pieces.

Funnels can be used for flowers with weak stems that would otherwise bend or break if poked into floral foam. Funnels can also be used to help lengthen short flower stems. Hyacinth stakes can easily be taped or glued to funnels for increased height of flowers within a bouquet.

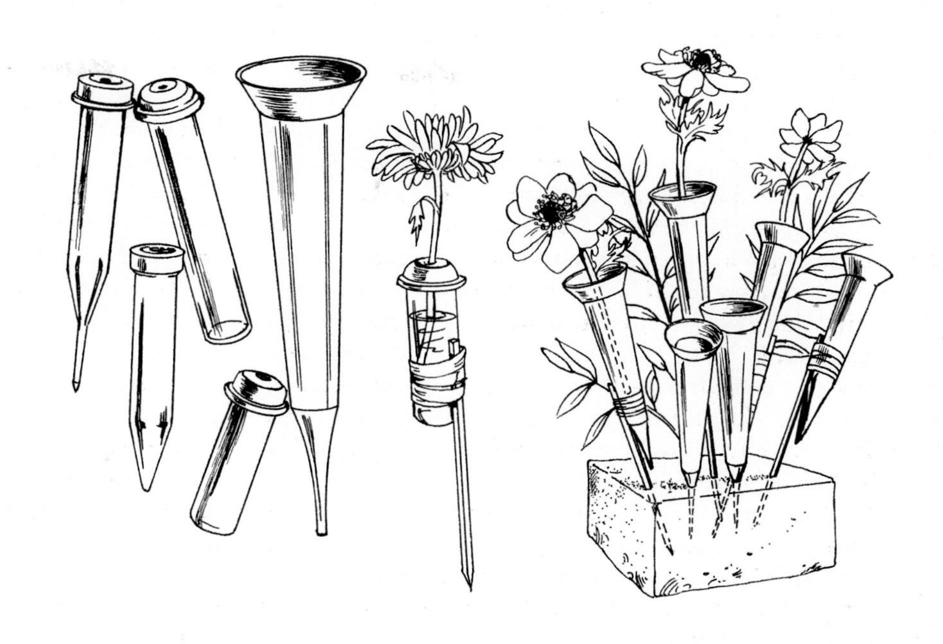

Figure 8-24. A variety of tubes and funnels are available to designers. Individual flower stems can be kept fresh with tubes holding preservative solution. Funnels can be used to give height to a single flower as well as a group of flowers in a bouquet.

Sprays

Floral aerosol and pump sprays can be grouped into three categories: ***paints, tints,*** and ***specialty*** or ***accessory sprays.*** While all are manufactured for use with fresh flowers and foliage, they can also be used on permanent flowers, containers, and accessories. Remember to follow the directions printed on the can, especially when you are spraying fresh flowers that can be easily damaged or ***frozen*** if you hold an aerosol can too closely while spraying.

While you are spray painting, or ***air brushing,*** move slightly both the aerosol can and the object being sprayed. This will allow for a light and even coating that won't drip or run. Also, it is best to test the spray first on a paper towel or newspaper. This will help prevent blotches.

Floral spray paints are available in a wide variety of colors. Since these paints are heavy pigments, they can dramatically change the surface color of an item (see figure 8-25).

Floral tint sprays are lighter pastel colors. These sprays offer a way of subtly altering the appearance of a flower, foliage, container, or accessory. Tints can easily be mixed with one another or ***double sprayed*** for exciting color results. For example, a pink tint may be sprayed on top of a yellow tint, resulting in a lovely peach hue.

A number of specialty or accessory sprays are helpful in design work. Some of these are used to help lengthen the life of the flower or add luster to flowers, foliage, and accessories; others are more commonly used on containers and accessories.

Sprays that help lengthen the life of cut flowers include ***anti-transpirants, sealers,*** and ***glues.*** The purpose of an anti-transpirant is just what it says: A spray that reduces water loss due to transpiration. Especially popular for corsage and wedding designs, these sprays help to minimize water loss in flowers and foliage. These are available in aerosol and pump sprays as well as dips.

Sealer aerosols can be sprayed on fresh flowers to seal pores, slowing water loss as well, but are more commonly used on dried materials to help prevent the shattering or dropping of flower parts.

Figure 8-25. Gold spray paint on the sabel palm, palmetto, cones, and dried berries in this design combine with yellow roses, solidaster, and juniper. The painted elements harmonize with the brass container and table top, so that the entire composition portrays unity and elegance.

Special spray glues are made for use on fresh flowers. A light coating on the back of the flower can help keep flower petals from shattering. These sprays are especially helpful on standard incurve (football) chrysanthemums, and can also be used on dried and silk materials for various design purposes.

Other specialty sprays include glitter glue, spray glitter, and shiny, iridescent, or lacey finishes. Glitter glue can be sprayed directly onto many surfaces, and immediately after application shimmering dusts or glitters can be sprinkled onto the sprayed surface. Spray glitter is an aerosol that already has glitter in the spray, allowing the glue and glitter to be applied in one step. Iridescent or pearl sprays add an opalescent finish. These do not change the existing color of a surface. These sprays can give new life to baskets and dried materials. Other aerosols can add a high gloss or porcelain finish. For sprays that offer lacey, antique, or white-washed finishes, it is important to follow the instructions on the can label to achieve the desired effect for a certain surface.

Ribbon

Ribbon adds a decorative touch to floral designs. Bows, loops, and streamers are popular in corsages, bridal bouquets, and floral arrangements. Bows are also used to dress up potted plants. Originally, ribbon was used in floral designs to tie or bind flower stems together. Now it can be used in many different ways, and is it used in virtually every aspect of floral design and decorating work. It will be helpful for you to become familiar with common widths, fabrics, styles, and uses of various ribbons. Knowing how to make decorative bows for floral work is essential for all designers.

Ribbon is available in many widths. As shown in figure 8-26, the different widths are referred to by number. Widths can vary slightly depending upon the ribbon material and manufacturer.

Colors, fabrics, and styles of ribbon appear limitless, changing frequently with current industry trends. The least expensive ribbon that is most commonly used is ***satin***

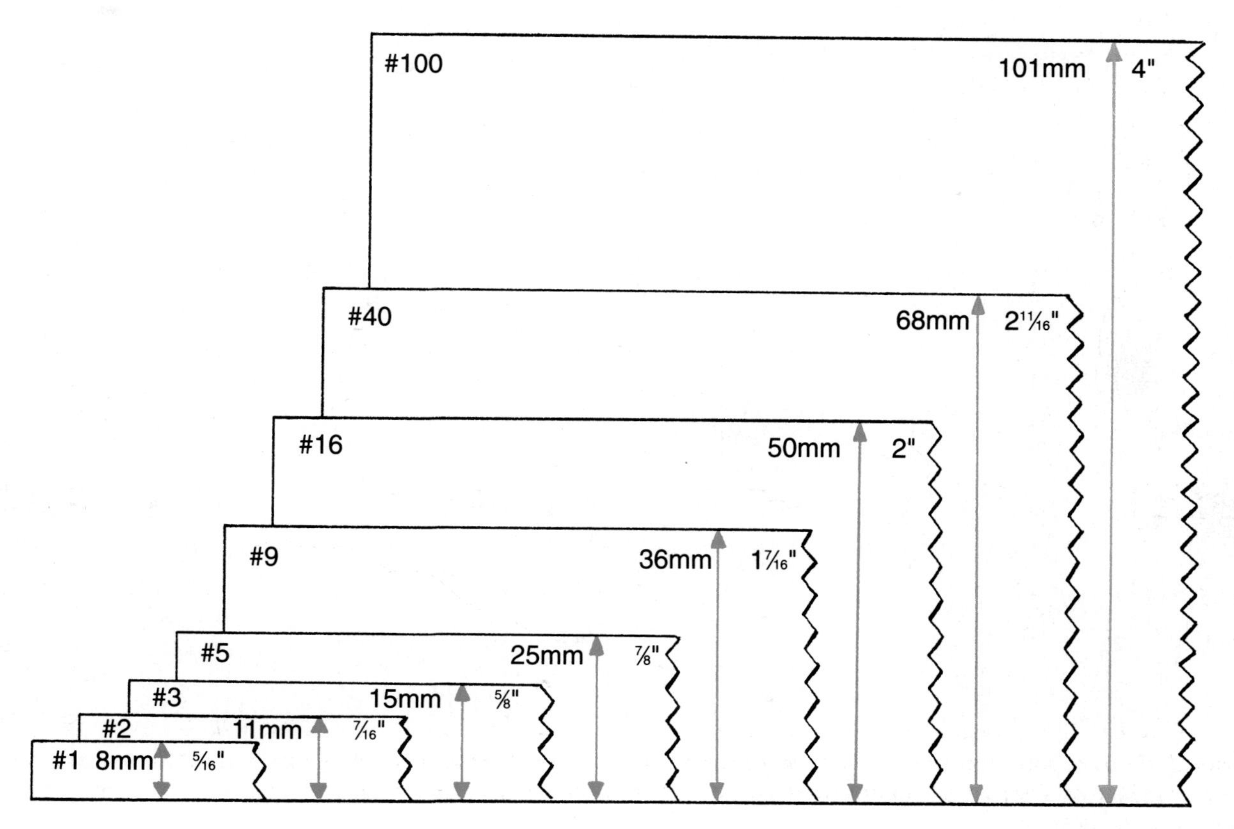

Figure 8-26. Ribbon widths with corresponding numbers.

acetate. It comes in an almost infinite spectrum of colors and can be used in a number of ways. It is the most cost-effective ribbon type.

Other ribbons that also have a place in design include wired ribbon, paper ribbon, shimmering or metallic ribbon, decorative lace and print ribbons, and novelty and seasonal ribbons. Whether you are choosing ribbon for use in a table centerpiece, wall hanging, bridal bouquet, or corsage, choose and place the ribbon properly to support the overall theme and style of a design.

The most common ribbon sizes are #3, #9, and #40 (see figure 8-27). Ribbon #3 is commonly used to accent corsages, bridal bouquets, small plants and planters, bud vases, and everyday floral arrangements. Ribbon #9 is commonly used to decorate plants in six-inch pots or larger. It is also used in larger designs, decorations, and presentation bouquets. Ribbon #40 works well in large designs (such as casket sprays and those placed on funeral easels) and other larger scale designs and decorations.

Figure 8-27. Ribbon #40 is commonly used in funeral work or other large designs. Ribbon #9 is commonly used to decorate potted plants and large bouquets. Ribbon #3 is often used to decorate everyday designs.

Often referred to as an accessory item within a floral composition, ribbon has the power to give a design a special flair. Alone it can create a focal point, or it may be used to help enhance a focal point. It can also be used to help achieve a sense of visual balance. It is important, though, never to overpower a design with ribbon. Of course, many designs are more successful without ribbon and bows. The size, color, style, and placement of bows, loops, or streamers must be planned. The right ribbon can help achieve harmony and unity in a design. The wrong ribbon can destroy a beautiful arrangement.

As you trim the end of ribbon streamers, cut the ribbon so it appears decorative. For instance, if you have ribbon streamers hanging from a bow in any design, a cut made perpendicular to the sides of the ribbon often will fray and be distracting. Rather than cut the streamer straight across, it looks more decorative to cut it at an angle, or you might even try something new as shown in figure 8-28. Before cutting, simply fold the ribbon in half from side to side to achieve various effects. Or you might knot the ends of the ribbon streamers for another design effect. This gives a little more emphasis and texture to the streamers.

A bow placed in any floral composition should be in scale and look proportionally correct to the entire design. Bows are generally placed at or near the container rim, or at the point where all lines and stems converge, helping to form a traditional focal point.

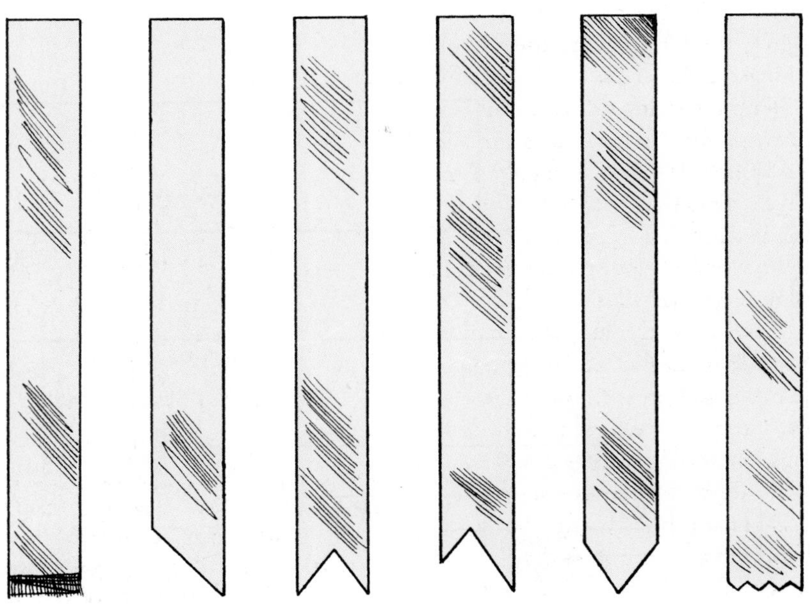

Figure 8-28. Rather than cutting a ribbon straight across, which can easily cause fraying, other effects, such as angle cutting, can give a more decorative flair.

Making a Bow

After learning the steps of bow-making, with practice you will become confident and proficient. The basic steps shown in figure 8-29 can be easily applied to all ribbon widths and fabrics. Most ribbon has a front, or finished, side and an underside. As you make the loops to the bow, it is necessary to twist the ribbon one half turn each time a new loop is formed, in order for the finished side to face outward. Once the bow is formed in your hand, bind it together with wire. The wire keeps the loops positioned and the wire ends can be used to anchor the bow within a design. Often bows can be made with ribbon that is still on the bolt. This way, there is no waste. For the fancier, more costly ribbon, it may be more effective to measure the ribbon first. Once cut, the bow can only be so big, which puts a limit on expense.

Whether you are left- or right-handed, place the ribbon end in your less-dexterous hand so that your more-coordinated hand will be controlling the long end of the ribbon and making the loops.

With the facing side of the ribbon outward, first form a loop around your thumb. It must be slightly gathered or pinched together behind the thumb. This loop will be the center of the bow and all loops and streamers will be connected to it.

Next, twist the long streamer of ribbon once, to put the facing side to the front, and form a loop off to one side. Continue twisting and forming loops, alternating their formation on each side of the center loop. The width of the ribbon and the bow's ultimate placement in a design, and the size of the design will all help dictate the sizes of loops. Several layers can be formed and the loops can

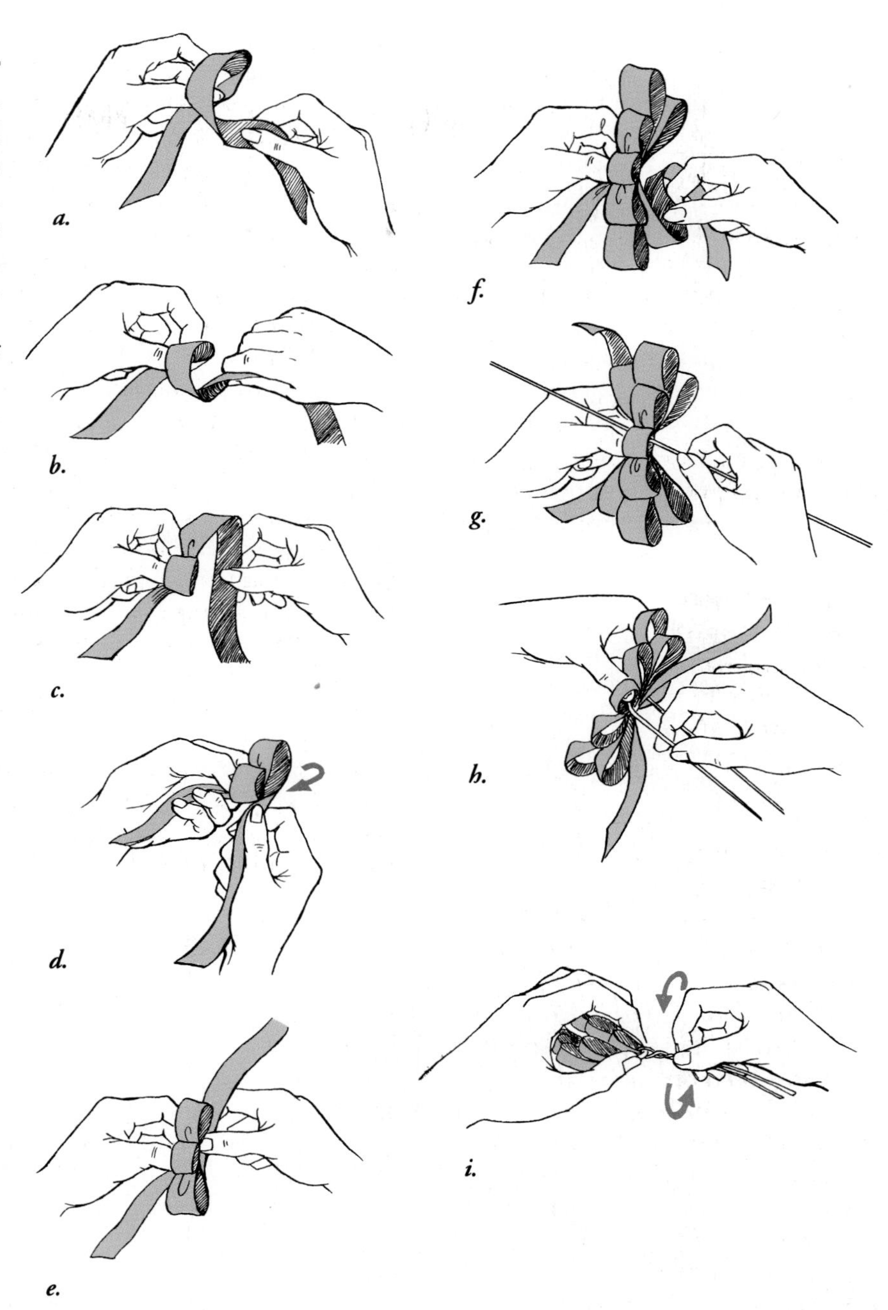

Figure 8-29. To make a bow: (a) Form a loop around your thumb. (b) Twist the long streamer of the ribbon once to put the facing side to the front. (c) Form a loop off to one side. (d) Twist the ribbon. (e) Form a loop to the opposite side, twist ribbon. (f) Continue forming loops, alternating their formation on each side of the center loop. (g) With an equal number of loops on each side of the center loop and streamers on different sides, insert a wire into the center loop.(h) Fold wire ends down together. (i) Twist the wire ends tightly.

increase in size with each layer. In deciding how big to make the bow, remember that the purpose of a bow is to add accent and enhance a focal point, not overwhelm and dominate a composition.

Finally, when an equal number of loops are on each side of the center loop and the streamers are on different sides as shown, insert a wire (the gauge will vary according to the ribbon width and size of the complete bow) into the back of the center loop. Fold both ends of the wire toward the back, and twist the wire tightly around all layers of the ribbon. The loops can then be spread apart, if desired, to form a more rounded or fuller shape. The wires attached to the bows may then be anchored in floral designs.

Accessories

Oftentimes novelty or seasonal accessories are added to floral arrangements. Common floral accessories are miniature plush toy animals or similar novelties, candles, and balloons. These decorative items help to strengthen or emphasize a certain mood or theme. Whenever accessories are added, they must be placed into a bouquet securely so they won't fall out or become damaged in any way. It is also important to make sure that none of the fresh flowers and foliage become damaged by the accessory.

Plush Toy Animals and Novelties

Plush toy animals are often added to floral arrangements to set a happy mood or send a message to the recipient. Usually they are placed in an arrangement to provide the focal point. They should be set low in a basket or other container, so that the entire arrangement will look visually and mechanically sound. These plush animals must be protected from water damage that might easily happen during handling and delivery. Arrange foliage over the foam where the toy will be placed, so that the toy will sit on foliage rather than wet floral foam. To protect the plush toy further, a piece of cellophane may be wrapped around the lower half of the toy. Use a piece of ribbon or raffia string to tie the cellophane in place. Then, use a heavy-gauge wire, a hyacinth stick, or a pick to secure the toy animal into the arrangement. Several techniques may be used. The shape, size, and flexibility of a plush toy will help you determine which technique will hold it most securely. Sometimes a heavy wire can be wrapped and hidden around its neck, stomach, or leg, and the wire can be inserted directly into the floral foam or attached to a stick for added security (see figure 8-30).

Figure 8-30. To protect a plush toy against water damage, wrap a piece of cellophane around the lower half and tie with ribbon. A hyacinth stake can be hidden in the back and the toy can be secured in the arrangement. Or a wire can be twisted around a portion of the toy and inserted directly into the foam or attached to a hyacinth stake for added security.

Many small accessories are available and often used to decorate arrangements, establish a theme, or celebrate a holiday. Many of these are hard plastic novelties such as bunnies, chicks, and eggs for Easter, or witches, pumpkins, and ghosts for Halloween. Remember that these accessories should not dominate a floral arrangement but provide accents or a focal point.

Most novelties have a wire or pick attached so they can be easily anchored into a floral design. Some, however, are not manufactured with stems or they may lose their stems during storage from season to season. It is easy to make a new stem or pick for the hard plastic novelties. As shown in figure 8-31, simply hold the end of a heavy wire over a candle flame. Once the end of the wire has heated up, immediately insert it into the base of the plastic novelty. If that alone is not secure enough or the novelty tends to spin

Figure 8-31. To make a new stem or pick for a plastic novelty toy, hold the end of a heavy wire over a candle flame. Immediately insert the hot wire end into the base of the novelty. Add a small amount of glue for added security.

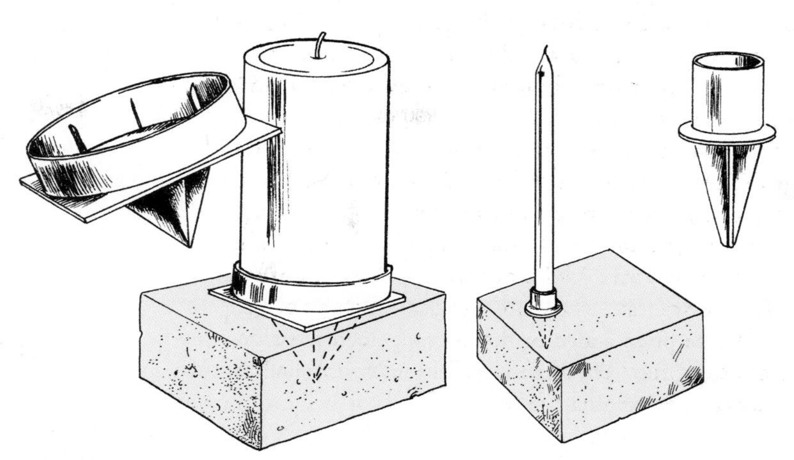

Figure 8-32. Several types of plastic holders are available to anchor candles into foam. Large pillar candle holders are convenient and save time.

on the wire due to its weight, add a tiny bit of glue where the wire pokes into the novelty. The wire can then be cut to the correct length for insertion into the floral design.

Candles

Candles generally signify formal or semiformal occasions and contribute greatly to the beauty of a table and a floral arrangement. Their soft glow spreads glamour that cannot be supplied by other types of lighting. Candles must harmonize with the floral composition and the entire setting, repeating the color of something that is on the table, whether it is flowers, cloth, or plates. In order to be an integral part of the setting or arrangement, the color, size, and style of candles must harmonize.

Candles should be slow burning and dripless. Most candles that are purchased with a clear wrap are high quality, meaning they burn slowly and do not drip excessively. To help them burn more slowly, cool the candles in a refrigerator for several hours before lighting them.

The height and type of candles depend upon the size of the table and the arrangement as well as the occasion. For instance, at a seated dinner it is best to avoid having the flame of the candle at eye level; either above or below is more comfortable for guests.

Whenever candles are used within a floral arrangement, it is vital that they are secure. Several types of plastic holders are made for various diameters of candles. Choose a holder that fits the diameter of the candle as shown in figure 8-32. The spike of the holder will anchor the candles securely into the floral foam, and the use of candle holders allows more foam space for flower stems, as illustrated in figure 8-33.

Candles can also be poked directly into floral foam. First, trim taper candles with two angled cuts at the base for easier insertion. Push the candle straight

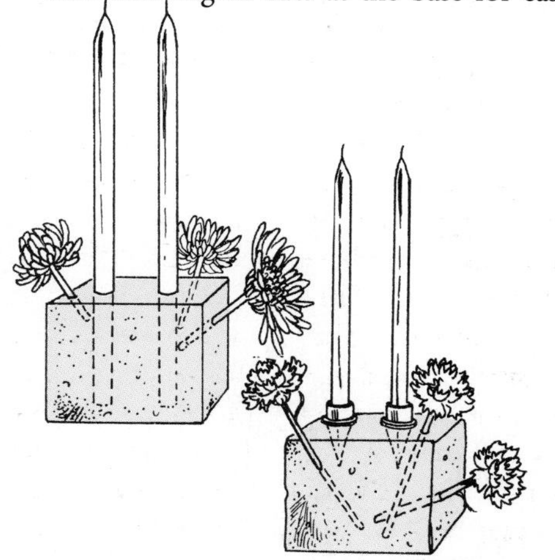

Figure 8-33. The use of candle holders allows more floral foam space for flower stems.

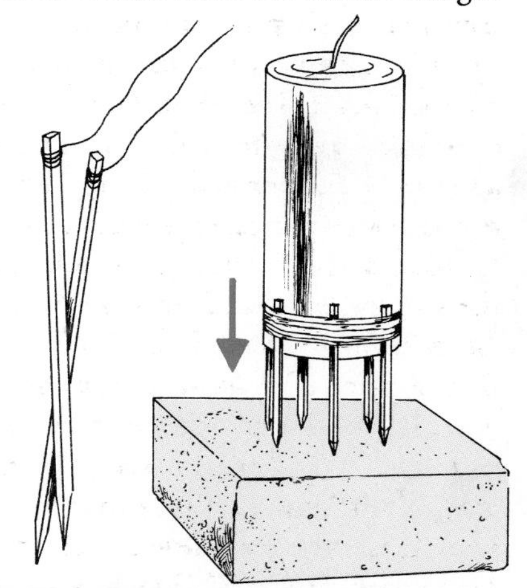

Figure 8-34. Large pillar candles may be secured into a design by taping several wooden picks around the base of the candle and inserting the picks into the foam while the heavy candle rests on top.

down slowly into the floral foam. Be careful not to twist candles during insertion, as this will cause a hole that is too big and the candle will be less secure.

Without a candle anchor, a larger and thicker pillar candle can be secured by taping several wood picks around the base of the candle, as shown in figure 8-34. The picks are then secured into the floral foam, two or three inches deep, and the heavy candle rests on top of the foam. Never try to push the large candle into the foam. The picks will allow the needed security.

Balloons

Balloons are often added to floral arrangements, making any floral bouquet more festive and fun. They can also help unify a theme through their color or message. Whether balloons are filled with air or helium, they must be securely attached. Properly anchoring balloons into a design is important not only for the sake of the balloons but also for the security of the flowers in the bouquet.

As shown in figure 8-35, helium balloon strings are best secured with a wired wooden pick. Once the wood pick is inserted into the floral foam, it will swell because of the moisture, keeping the balloons securely anchored.

Figure 8-35. Helium balloons can be secured into floral designs with wired wooden picks. Simply lay one to three strings across the top of the wood pick as shown. Next, wrap the wire around the ribbons and pick several times. Then pull the short ends of the ribbons up next to the wire. Continue wrapping the wire around the ribbons. The wooden pick will swell when inserted into the wet foam, keeping the balloons secure in the arrangement.

Packaging

Cut flowers are often wrapped in paper or cellophane or tucked neatly into a flower box. Plants are often packaged with decorative foil (poly foil) and cellophane. Distinctive and creative packaging of flowers and plants can make a spectacular presentation. However, in addition to being decorative, it is important that any type of packaging be functional and secure as well.

Wrapping Cut Flowers

Designers must have a knowledge of how to wrap loose cut flowers (see figure 8-36). Even a single rose can look like a masterpiece when wrapped by a skilled designer. Several paper products are available for wrapping cut flowers including wax tissue paper, clear or decorative cellophane, or paper. Roses and other wilt-sensitive flowers should first be placed in water tubes. Stems may also be poked into a small block of wet floral foam that has been placed in a small plastic bag. When wrapped cut flowers do not have a water supply, the recipient needs to be aware that the flowers must be unwrapped and cared for as quickly as possible. It is best to provide a care card and sachet of floral preservative with all cut flowers.

Boxing Cut Flowers

Flowers can also be arranged in a decorative floral box; these boxes are available in different sizes and styles. As shown in figure 8-37, a bed of foliage or shredded waxed corsage grass is laid down first in a clear plastic box. When using a cardboard floral box, wax tissue must be laid down first to protect the box from moisture. Next, follow the steps of

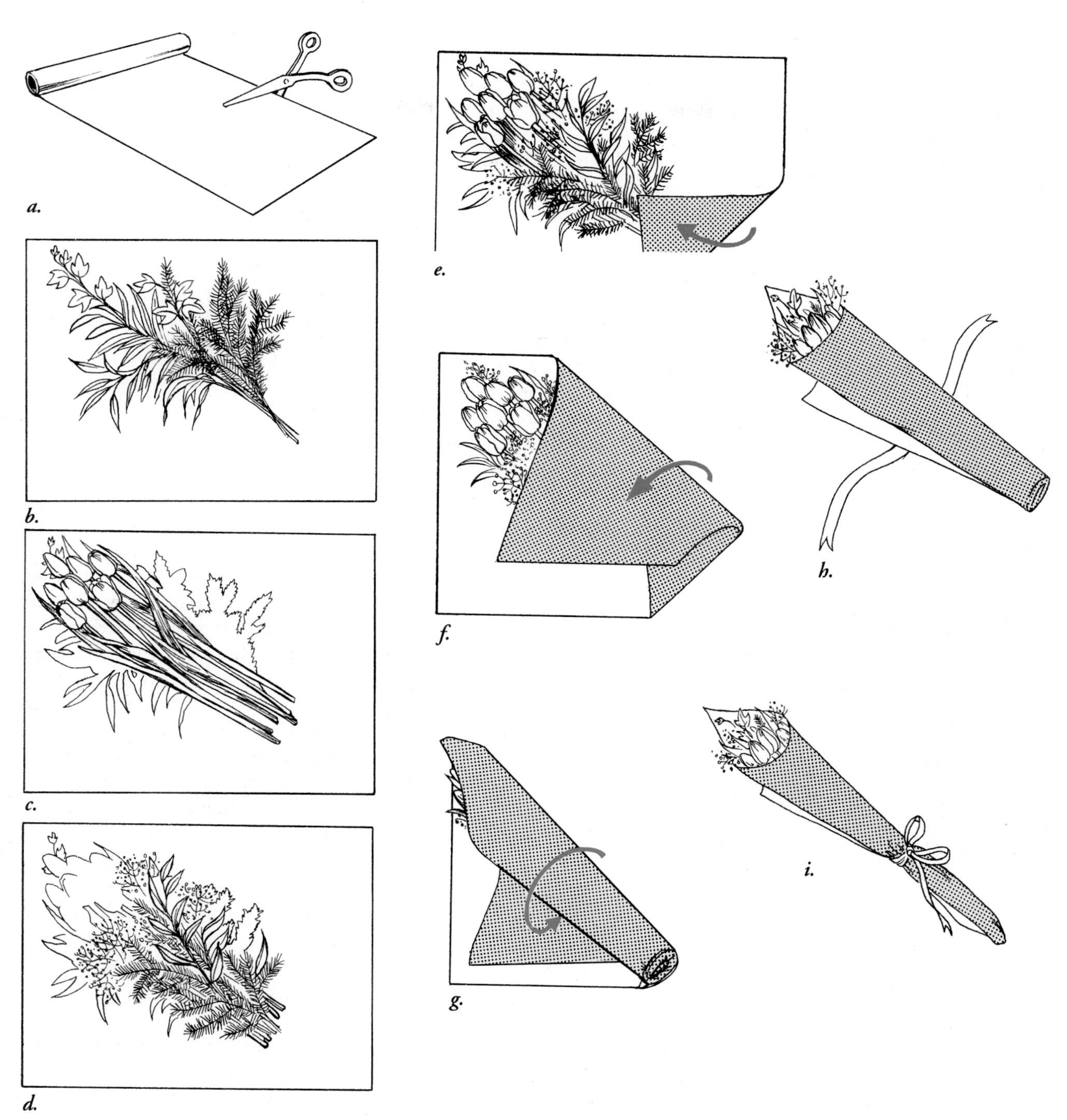

Figure 8-36. Steps of wrapping cut flowers: (a) Cut a square or rectangular piece of cellophane wrap or paper from the roll or use a precut piece of tissue. Layers of tissue and cellophane can also be combined. (b) Lay foliage diagonally on the piece of cellophane, tissue, or paper. (c) Position flowers on top of foliage. (d) Place filler flowers and more foliage on top to give a fuller look. (e) Once the desired flowers and foliage are layered on the cellophane, tissue, or paper, fold the lower corner over the base of the stems. (f) Next, fold one side of the piece of wrap over on top of the flower stems. (g) Roll the flowers and wrap together to form a tight package, as shown. (h) Slip a piece of ribbon under the packaged flowers. (i) Tie the ribbon around the wrap to secure the flowers inside and to make a lovely presentation. A care tag and sachet of floral preservative can be tucked down inside with the flowers.

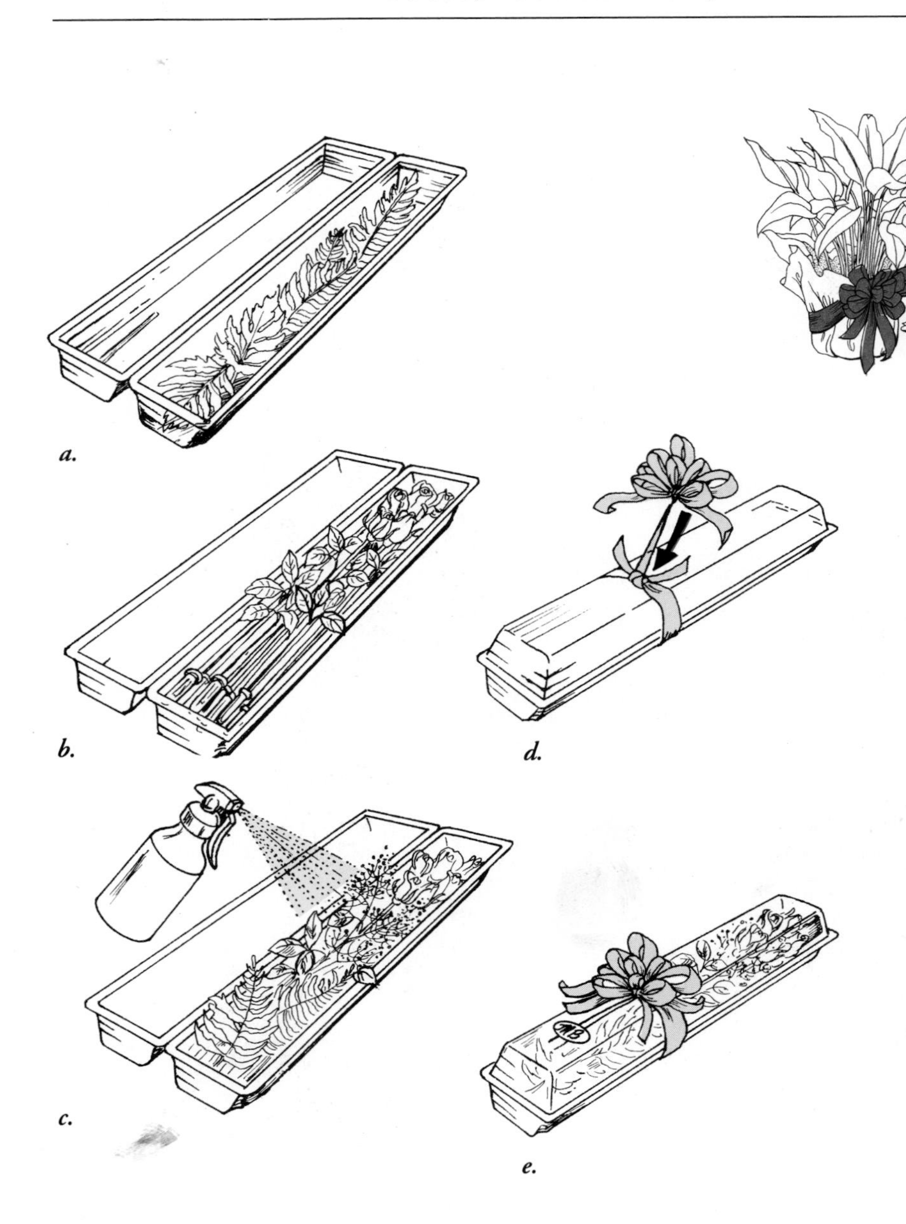

Figure 8-37. Steps of boxing cut flowers: (a) In a clear plastic box, first lay down a bed of foliage. If using a cardboard box, first insert tissue paper to protect the box from water damage. (b) Next, layer roses or other flowers so the flower heads are positioned in a clustered presentation. Rose stems must be placed in individual water tubes with preservative solution, or flower stems can be inserted into a small block of foam that has been wrapped in cellophane or is in a plastic bag and concealed under foliage. (c) Add a layer of filler flowers and foliage. Mist with water, to help keep cool and turgid during delivery. Care tag and preservative sachet may be tucked into the box after misting. (d) Close the box, making sure all foliage and flower parts are tucked neatly inside. Tie a piece of ribbon around the box. Attach a bow on top, which can be later added to the design. (e) Slide the wire stems under the ribbon and secure with a decorative sticker.

Figure 8-38. Many types and colors of poly foil, cellophane, and ribbons are available to decorate potted plants. Baskets and other containers can be used as well to hide the plastic pot.

placing flowers and foliage in the box, misting them, and tying the box with a ribbon for a decorative presentation.

Wrapping Potted Plants

Plants growing in plastic pots can be cleverly decorated in a variety of ways. A popular and decorative method of hiding the plastic pot and making a plant more showy is to wrap it in florists' foil. A variety of types and colors can be chosen to match a season or holiday and then coordinated with a ribbon and bow. Practice wrapping plants with foil by following the steps shown in figure 8-39. Many types of baskets, terra cotta containers, or up-side-down "hats" are also available for plants.

Wrapping Flower Arrangements and Plants

Floral arrangements and potted plants are sometimes covered with clear wrap or paper. This extra covering helps insulate and protect the flowers and plants while they are outdoors during delivery in inclement weather.

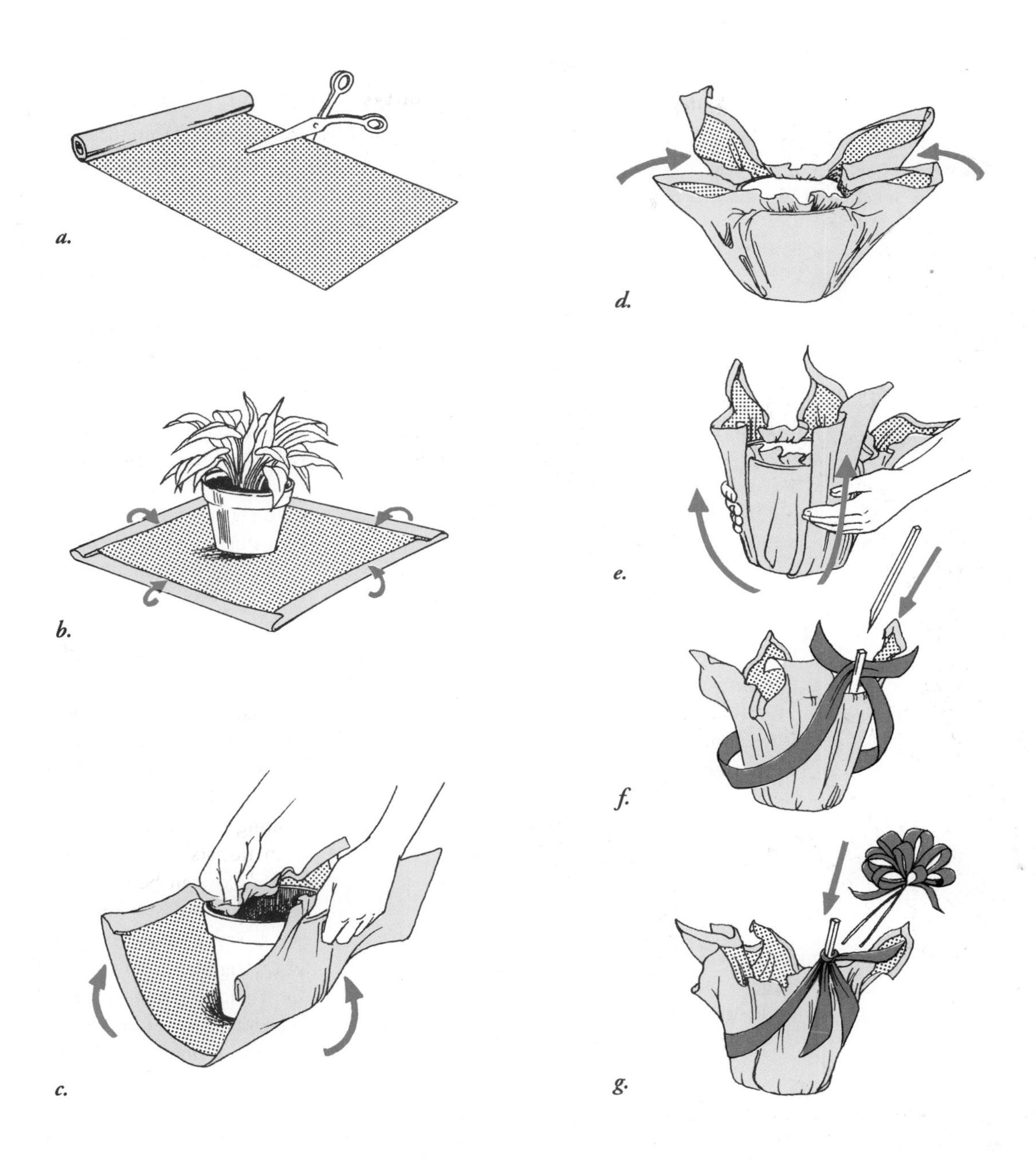

Figure 8-39. Steps of wrapping potted plants: (a) Cut a square piece of decorative poly foil from the roll. (b) Fold the cut edges inward for a cleaner, more decorative look. Place the potted plant in the center of the poly foil. (c) Bring opposite sides of the poly foil up and pinch into the inner rim of the plastic pot. (d) Next, bring the remaining two sides up and pinch them into the inner rim of the pot as well. (e) Smooth the poly foil against the pot for a smoother look. (f) Insert a wooden pick into the soil, against the container, on the front side of the plant. Wrap a piece of ribbon around the pot to keep the poly foil secure to the pot and tie it around the wooden pick. (g) Add a decorative bow next to the wooden pick. A plant care tag may be added with the bow. Slit a small hole in the bottom of the poly foil, underneath the pot, to allow for water drainage.

REVIEW

Beautiful, successful, and long-lasting floral designs do not just happen. The success of a floral arrangement is dependent upon its mechanics. Successful compositions are created by using the proper tools and selecting the appropriate container and accessories to harmonize with the style of the bouquet. It is helpful to know the tools needed as well as several different techniques for construction. As you learn and practice various techniques, your design ability will increase.

TERMS TO INCREASE YOUR UNDERSTANDING

accessory sprays
air brushing
anchor pins
anchor tape
anti-transpirants
bank pins
beauty clips
bowl tape
ceramics
chicken wire
de-thorners
dixon pin
double-face tape
double sprayed
fern pins
floral clay
floral foam
floral tape
frogs
frozen
glue guns
glues
glue sticks
green tape
greening pins
grids
hook-wire
hot melt
hyacinth stake
insertion
lacing of foliage
liquid floral glue
low temperature
mechanics of arrangement
needlepoint holders
packaging
paddle wire
paints
pan-melt glue
philly pins
picks
pierce
pins
pin holders
poly foil
pot tape
raffia string
rose strippers
satin acetate
sealers
specialty spray
spool wire
spray glue
spray paint
straight wire
tints
water tubes
waterproof tape
wire gauge
wooden picks

TEST YOUR KNOWLEDGE

1. What factors determine if a container is functional?
2. What are a florist's basic hand tools? Name the functions of each.
3. What are some advantages and disadvantages of different container materials?

RELATED ACTIVITIES

1. Construct a tape grid or lace foliage to form a grid in preparation for a floral arrangement.
2. Practice wrapping cut flowers and plants. Use your creativity and think of some alternative ways of dressing up potted plants.
3. Practice making bows. Determine how much ribbon is needed to make a bow.
4. Look at a variety of empty containers. What type of arrangement can you envision in each?

Section 2

FLOWERS AND FOLIAGE

Chapter 9

Nomenclature and Postharvest Physiology

Figure 9-1. Although flowers appear dramatically different, as shown here, their named parts are actually quite similar to one another. Inflorescence patterns (the arrangement of more than one flower on a stem), though seemingly diverse, are interrelated.
Courtesy of Evergreen Wholesale, Seattle, WA. Photo by Raymond Gendreau.

In order for you to fully understand care and handling procedures and techniques, it will be helpful for you to gain an awareness of basic flower and leaf structure, as well as the parts and types of flowers and foliage. Learning about ***postharvest physiology***—the functions and vital processes of flowers and foliage after they are cut—will enable you to better understand the methods of care and handling techniques for cut flowers and foliage.

The purpose of this chapter is to help you arrive at a broader knowledge of the plant material available. Rather than try to memorize all the botanical and physiological terms and concepts, use the information as a reference. The more you work with fresh flowers and foliage, the more familiar you will become with their structure, parts, and differences.

FLOWER NOMENCLATURE

All flowers can be easily classified according to their basic shape for design purposes. As you make arrangements, the varying shapes of flowers, or the inflorescence type, will directly influence the style, shape, and texture of your designs. (See chapter 11 for a further look at flower shapes, placement, and function within a floral composition.)

Although a wide variety of flowers exists and they often appear drastically unique, each flower has much in common. It will be helpful for you to learn and understand the basic parts or the ***anatomy*** of flowers. In your conversations with other floral industry professionals as well as in your reading of reference books and educational materials, you will find a knowledge of ***nomenclature*** (names of parts) useful.

Flower Parts

A ***botanically complete flower*** is a flower that has four main parts called ***sepals, petals, stamens,*** and ***pistils.*** The stem tip bearing all these flower parts is called the ***receptacle.*** Refer to the lilies shown in figure 9-2 to help you understand basic structure. Definitions of the Greek or Latin words will give you a better understanding of the shape or function of these individual parts.

Sepals

In some flowers, such as lilies and tulips, both the sepals and petals are brightly colored and difficult to tell apart. A sepal is one of the outermost flower structures that usually encloses the other flower parts in the bud. The word sepal (SEE-pul) comes from the Latin word *sepalum* (a covering). Sepals collectively are called the ***calyx*** (KAY-liks), from the Greek word *kalyx* (a husk or cup).

Sepals are commonly leaflike and green, but sometimes they are brightly colored like petals. When the sepals and petals appear identical, as in tulips, they both are often called ***tepals,*** or collectively the ***perianth.*** The word perianth comes from the

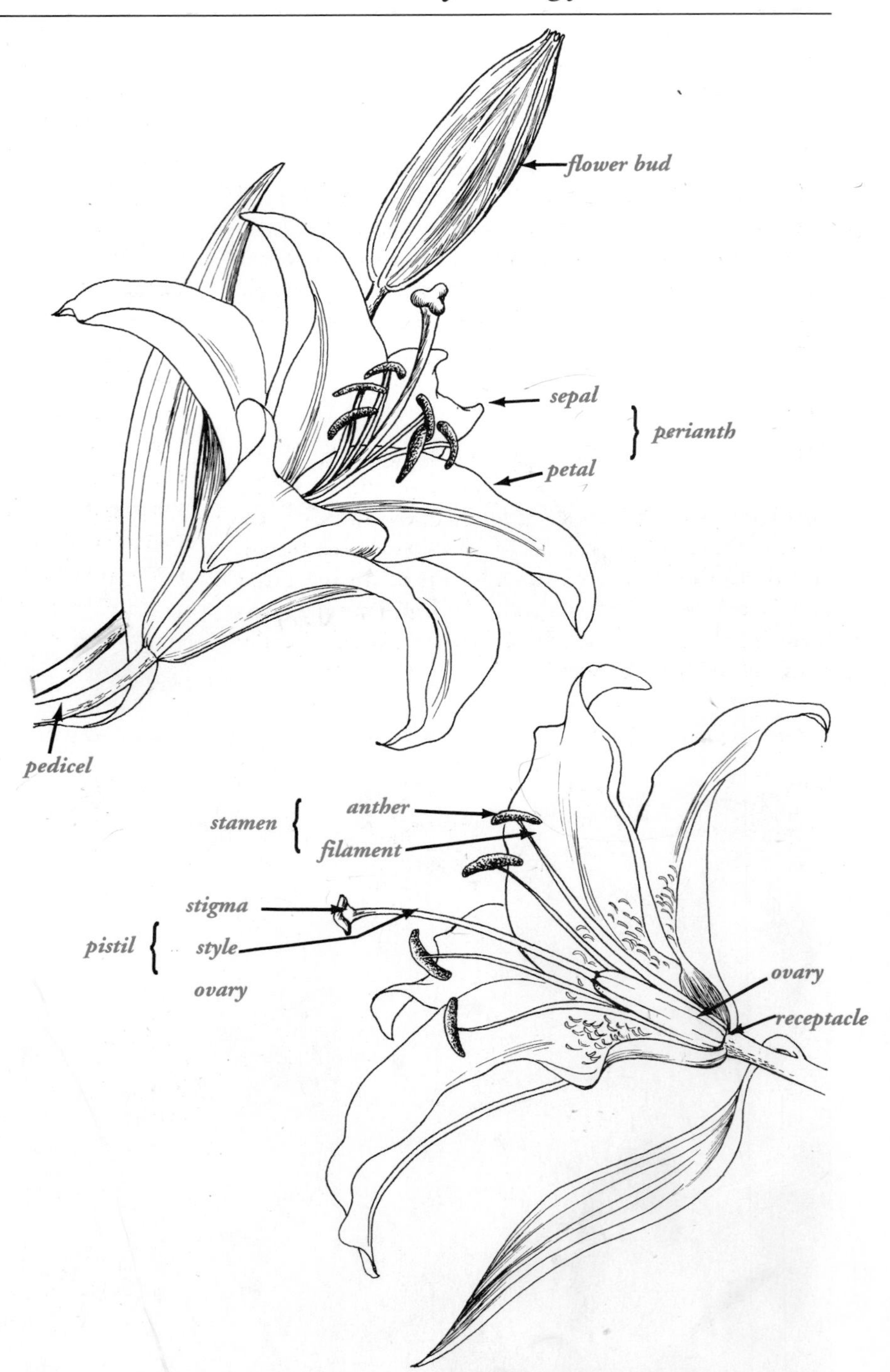

Figure 9-2. Parts of a lily flower: (Top lily) *An intact flower. In some flowers, such as lilies and tulips, both sepals and petals are brighly colored, and the perianth parts are often referred to as tepals. The sepals are attached to the receptacle below the petals. Collectively, the sepals form the calyx, and the petals form the corolla.* (Lower lily) *A partly dissected flower with two tepals and two stamens removed to reveal the ovary. The pistils, collectively called the gynoecium, consist of the ovary, style, and stigma. The stamens collectively are called the androecium.*

Greek words *peri* (around) and *anthos* (flower), literally meaning "around the flower."

Petals

Petals are usually conspicuously colored. The word petal comes from the Greek word *petalos* (outspread). Petals are collectively called the *corolla* from the Latin word corona (crown). The petals, generally the flower's showpiece, are normally positioned between the sepals and the inner flower parts.

Stamens

The Latin word stamen (thread) is similar to the Greek word stemon (standing). The stamens are the threadlike extensions that stand upright from the perianth. They are the male reproductive parts of the flower and are collectively called the ***androecium.*** The word androecium (an-DREE-see-um) comes from the Latin word *andros* (man) and the Greek word *oikos* (house), which literally translates to "house of man."

The stamen usually consists of two parts: the ***anther*** and the ***filament.*** The word anther comes from the Greek word *anthos* (flower). It is the ***pollen***-bearing portion of the stamen. The filament is the stalk of the stamen bearing the anther.

Pistils

Pistils, collectively called the ***gynoecium*** (ji-NEE-see-um), are the female reproductive parts of the flower and occupy a central position within the flower. The word gynoecium comes from the Greek words *gyne* (woman) and *oikos* (house) meaning "house of woman." The gynoecium may consist of a single pistil, as with the lily, or it may consist of several pistils. The basic unit of construction of a pistil is the ***carpel,*** which is a modified seed-bearing leaf. A pistil may consist of a single carpel or of two or more carpels partly or completely joined together, enclosing the ovules.

Each pistil usually consists of three parts: the ***stigma,*** the ***style,*** and the ***ovary.*** The stigma is the pollen-receptive part at the top of the pistil. The style (from the Greek

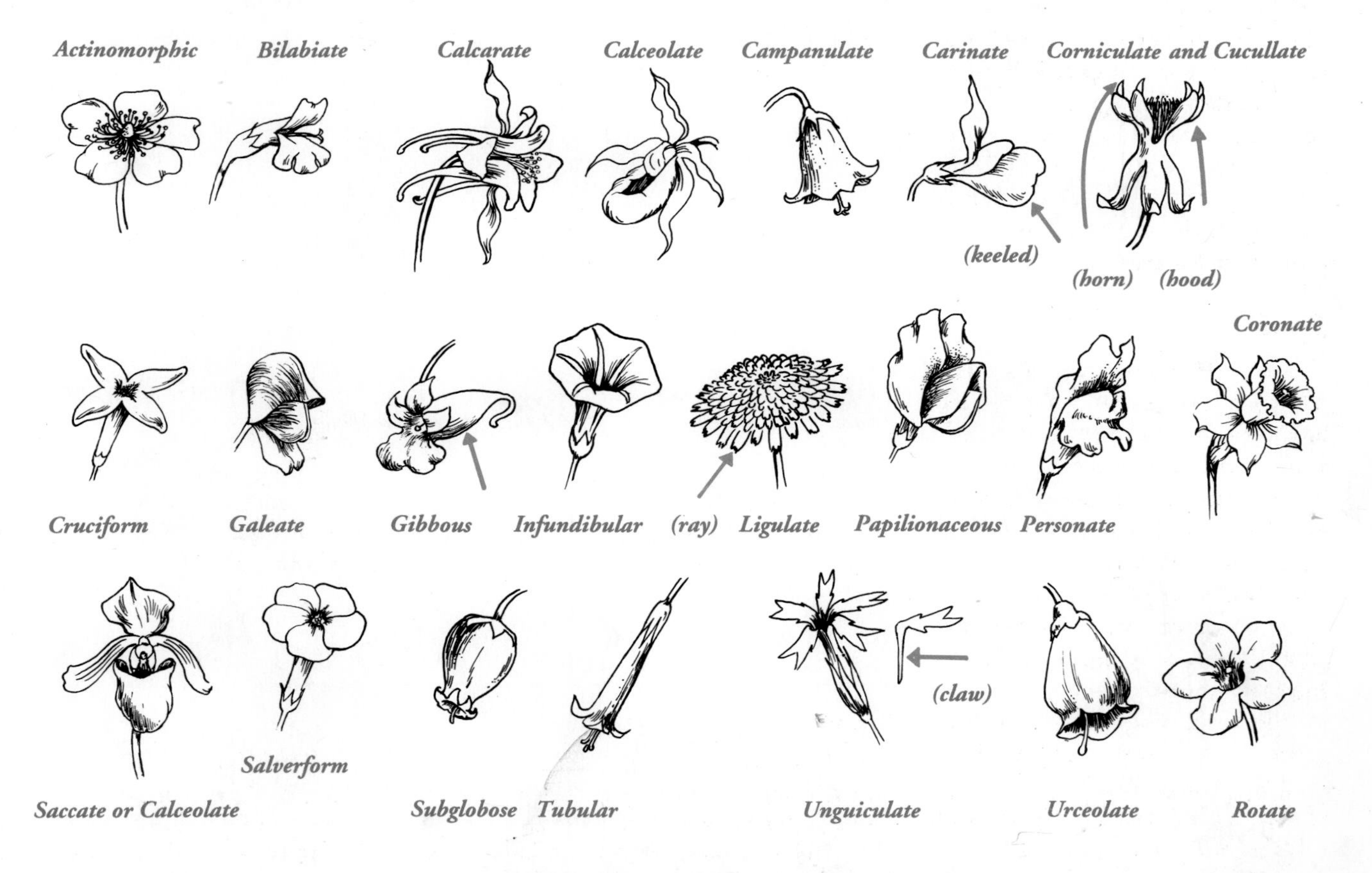

Figure 9-3. Perianth structural types are shown. The perianth is made of the calyx and corolla.

word *stylos,* meaning column) is the slender column of tissue that arises from the top of the ovary. The ovary (from the Latin word *ovum* meaning "an egg"), at the base of the pistil, is the enlarged portion containing one or more ***ovules*** or immature seeds.

Perianth Structural Types

The perianth is a collective term for the calyx and corolla. Various types of perianth shapes are shown in figure 9-3. As you work with fresh flowers, notice the various forms of flowers and ***florets*** that make up an entire flower. It is important that you realize and appreciate the vast differences among flowers and begin to appreciate those differences, as various shapes will suit different design needs in your floral compositions.

Solitary Flowers

Some flowers, such as roses and tulips, are called ***solitary, axillary,*** or

Figure 9-4. Tulips, roses, and daffodils are popular solitary flowers.

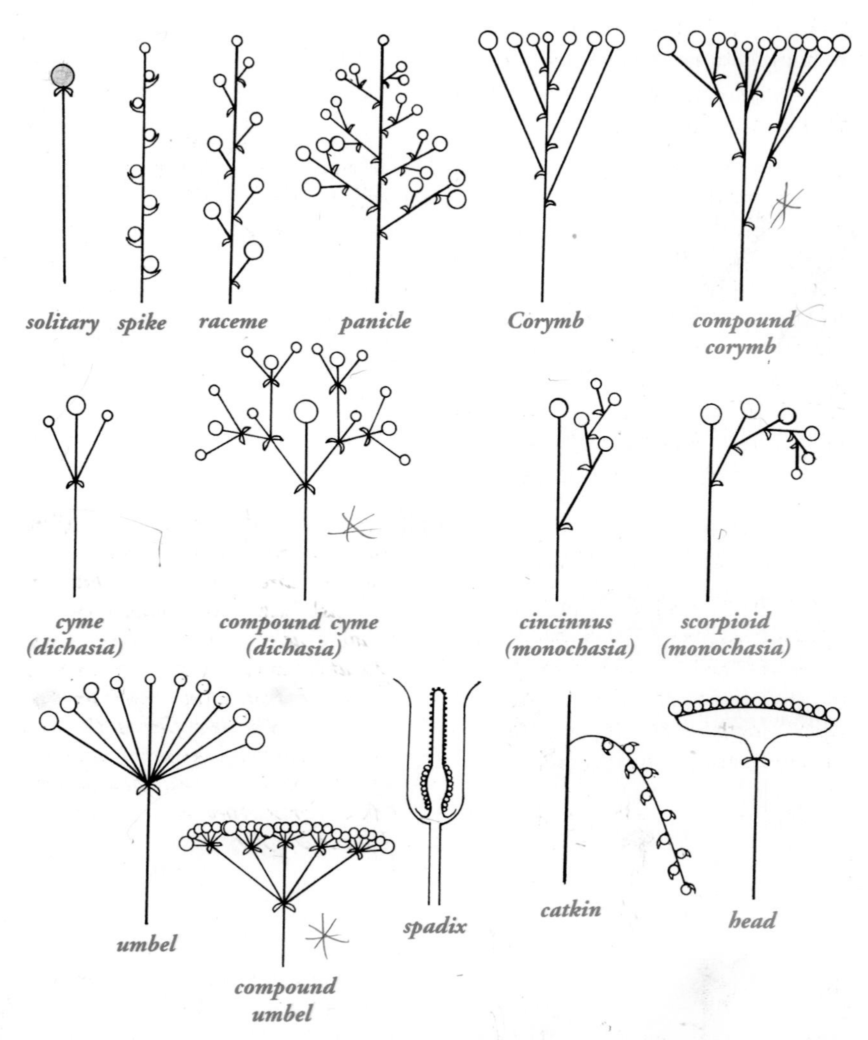

Figure 9-5. An inflorescence is any arrangement of more than one flower on a stem. Some of the more common inflorescence patterns are shown here.

terminal flowers because they form singly on upright ***stalks*** (see figure 9-4). Whether a solitary flower or a branching inflorescence, if a flower appears without ordinary leaves, as in *Narcissus* or *Nerine,* the flower is said to be ***scapose*** (a leafless flower stalk).

Inflorescence Patterns

An ***inflorescence*** (in-flow-RES-sens) is a flower cluster with a definite arrangement of flowers on a ***floral axis*** or stem. The inflorescence may be simple and easily recognized or it may be a highly complex structure, difficult to name or classify at a glance. An inflorescence may be ***determinate*** (central or top flower opening first) or ***indeterminate*** (outer or lower flowers opening first).

The main supporting stalk of the entire inflorescence is called a ***peduncle.*** The stalks supporting single flowers, or florets, are called ***pedicels*** (see figure

9-6). An inflorescence usually has modified leaves, called ***bracts,*** or reduced leaves, from the axils where the flowers originate.

Spike

A ***spike*** is an elongated inflorescence with a main stem. Its flowers are ***sessile,*** that is, unbranched and attached directly to the main stem without pedicels (see figure 9-7). Many so-called spikes are not actually this inflorescence type but only resemble it, and so they are termed ***spikelike.***

Raceme

A ***raceme*** (ray-SEEM) is an elongated inflorescence much like a spike except the florets each have their own stalks or pedicels, which are generally of equal lengths. Racemes are spikelike in appearance (see figure 9-8).

Figure 9-7. Variations of spike inflorescence patterns with sessile flowers (attached directly to the stem without pedicels). Liatris *is a unique spike inflorescence in which the florets open first at the top of the stem. In contrast, the florets of* gladiolus, *which are one-sided spikes, open first at the base of the stem. Although appearing uniquely different than the typical spike pattern,* freesia *florets appear on the upper side of a curved spike.*

Figure 9-8. Many raceme inflorescence patterns are spikelike, much like pedicelled flowers, such as dendrobium and delphinium. Stock (Matthiola) *and snapdragons* (Antirrhinum) *are terminal racemes.*

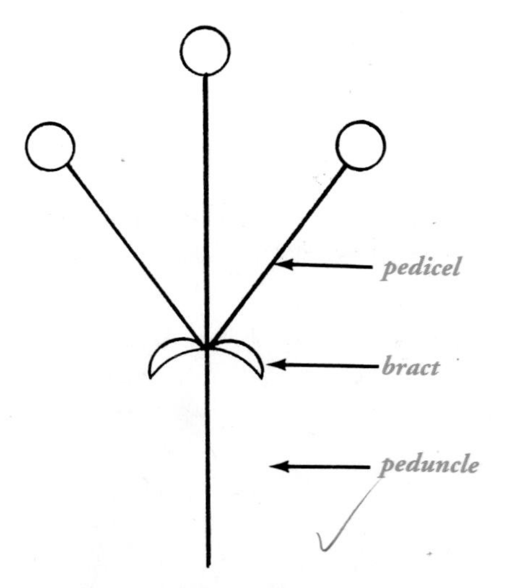

Figure 9-6. The main supporting stalk of an entire inflorescense is called a peduncle. The stalks supporting single flowers are called pedicels. An inflorescence may have bracts, which are modified leaves from the axils where flowers are produced.

Panicle

A ***panicle*** (PAN-i-cul) is a loose, irregularly branched flower cluster, a compound raceme. Most flowers classified as panicles have a central axis with branches that are themselves branched (see figure 9-9).

Corymb

A ***corymb*** (KOR-im) has a flat top or slightly convex shape. Corymbs have a main vertical axis and pedicels or branches of unequal length as shown in figure 9-10. The individual pedicels generally come from various alternate sides of the main stem, as with a raceme, but form a flat-top cluster. A ***compound corymb*** is a branching simple corymb.

Cyme

A ***cyme*** (rhymes with lime) is an inflorescence that is broad and often flat-topped. Cymes have many forms and are ***dichasial*** or ***monochasial.***

A dichasial cyme, or dichasium (die-KAY-zee-um), has two opposite divisions or branches that arise below a terminal flower. This simple inflorescence is a common unit that when repeated produces other more complex branching patterns called both ***compound cymes*** and ***compound dichasia*** (see figure 9-11).

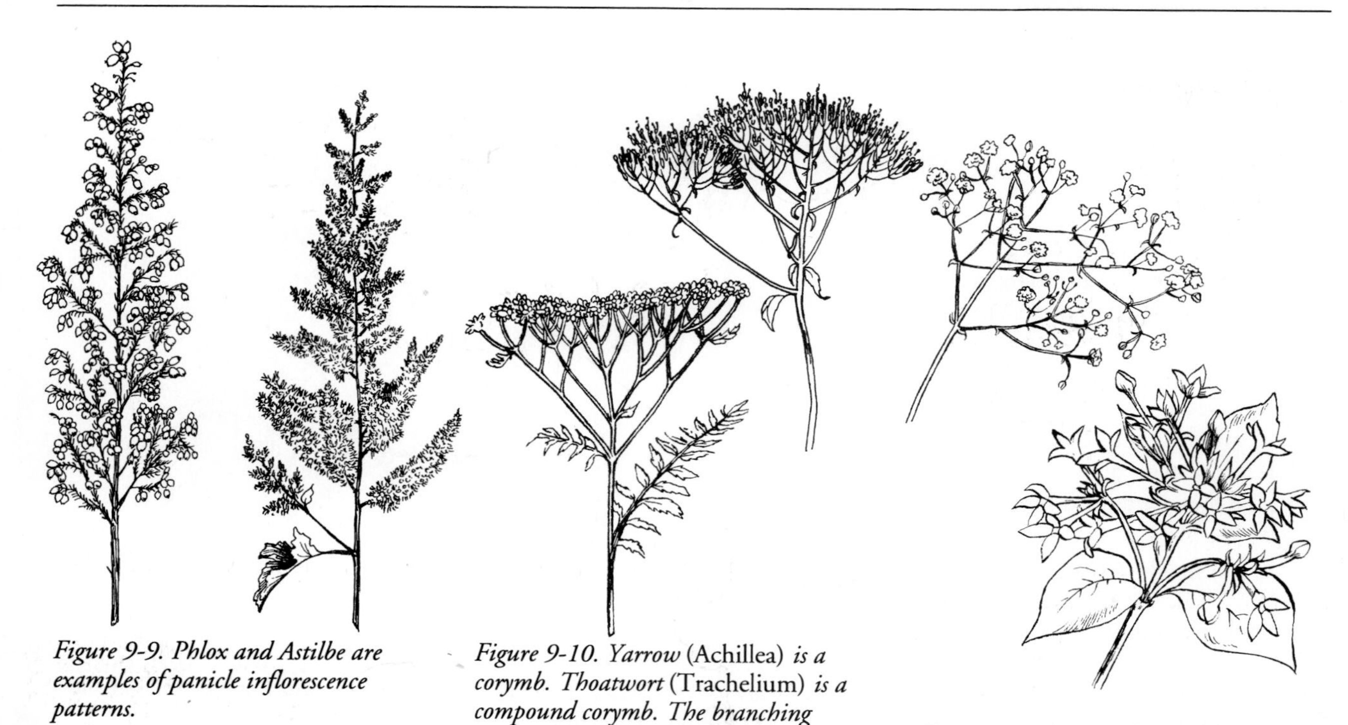

Figure 9-9. Phlox and Astilbe are examples of panicle inflorescence patterns.

Figure 9-10. Yarrow (Achillea) *is a corymb. Thoatwort* (Trachelium) *is a compound corymb. The branching pattern of corymbs are similar to racemes, except the pedicels are unequal in length to form a flat-topped clusted.*

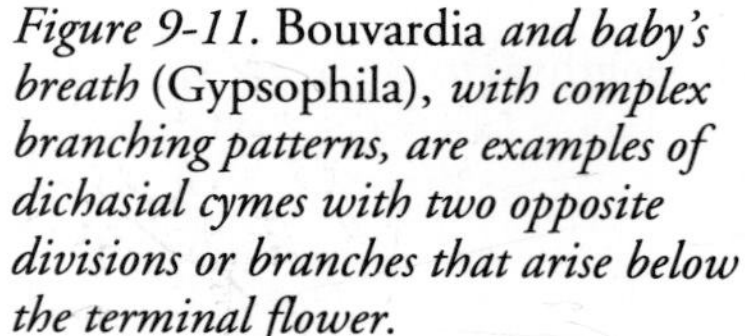

Figure 9-11. Bouvardia *and baby's breath* (Gypsophila), *with complex branching patterns, are examples of dichasial cymes with two opposite divisions or branches that arise below the terminal flower.*

A ***monochasial cyme,*** or ***monochasium*** (MON-oh-kay-zee-um), has one division. A repetition of branching flowers on further lateral branches creates various types of cyme inflorescence patterns including ***helicoid*** and ***scorpioid*** (see figure 9-12).

A helicoid cyme, or ***bostryx,*** is an inflorescence coiled in a bud and superficially resembles a raceme. The oldest, largest flower is considered the terminal flower; florets continue to branch out on one side, resulting in an interesting and unique form.

A scorpioid cyme, or ***cincinnus,*** is a coiled cyme similar to a helicoid cyme, but the flowers or branches develop alternately to the left and right, in a zigzag pattern, rather than only in one direction.

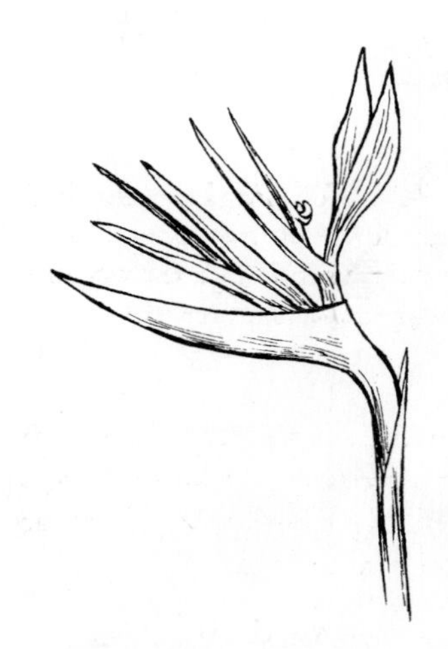

Figure 9-12. Monochasial cymes have one division that arises below the terminal flower. The florets of bird of paradise (Strelitzia reginae) *appear in a scorpioid cyme or cincinnus pattern.*

Umbel

An ***umbel*** (UM-bul) is a flower cluster that is easily recognized (see figure 9-13). A ***simple umbel*** has single pedicelled flowers all arising from the top of a main stem. The shape of an umbel is flat-topped, rounded, or globular. A ***compound umbel*** has secondary pedicelled umbels arising from the tips of the main branches, in this case called rays (see figure 9-14).

Spadix

A ***spadix*** (SPAY-diks) is a thick or fleshy flower spike usually surrounded by a conspicuous or colorful bract called a ***spathe.*** The unbranching spadix flowers

Figure 9-13. Umbels, such as Agapanthus *and* Nerine, *have pedicelled flowers all arising from a common point at the top of the stem. A compound umbel is a branched umbel, such as Queen Anne's lace* (Ammi majus).

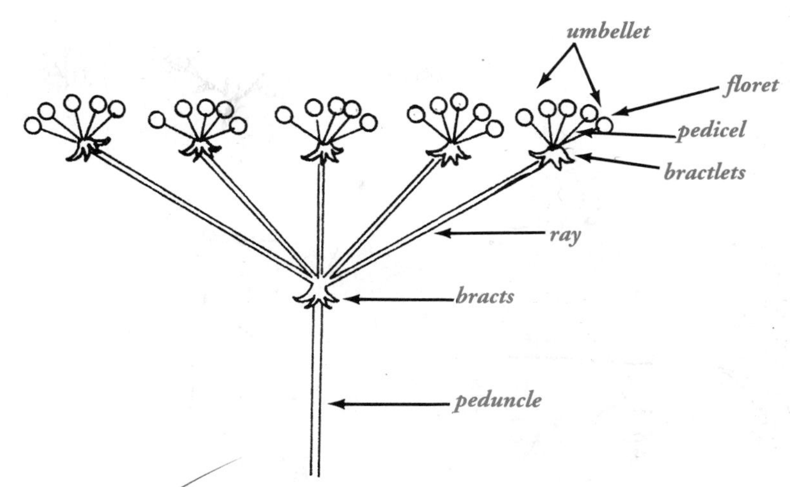

Figure 9-14. In a compound umbel, secondary pedicelled umbels arise from the tips of the main branches, or rays.

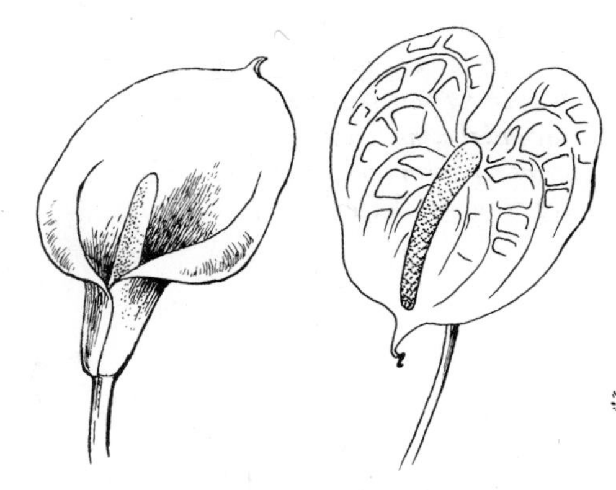

Figure 9-15. Calla lily (Zantedeschia) *and* Anthurium *are two examples of flowers with a spadix inflorescence surrounded by a showy spathe.*

Figure 9-16. Catkins are slender spikelike, often drooping, sessile flower clusters found on woody plants, such as willow, alder, and birch. Pussywillow (Salix discolor) *is an example of an upright catkin.*

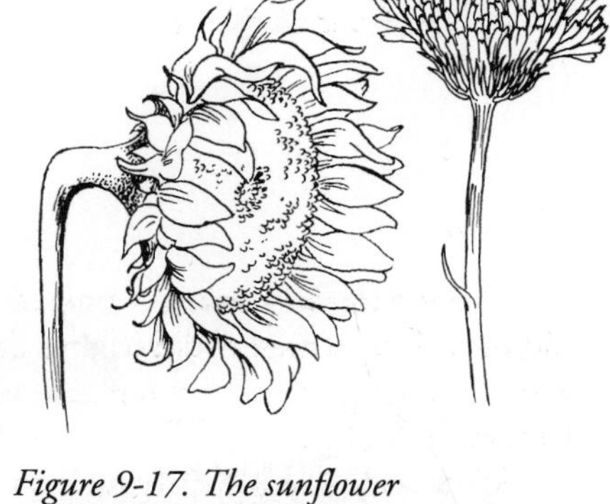

Figure 9-17. The sunflower (Helianthus annuus) *is an example of a head inflorescence with a crowded group of sessile or nearly sessile disc and ray flowers on a compound receptacle. China aster* (Callistephus), *safflower* (Carthamus), *and* Chrysanthemum *are other examples of head flowers.*

are tiny and, as a result, the colorful spathe is mistaken as a flower petal, as with the bright red spathes of anthuriums (see figure 9-15).

Catkin

A ***catkin*** (KAT-kin), also called ***ament,*** is a slender scaly-bracted, usually drooping spike or spikelike inflorescence, found on woody plants such as willow, alder, birch, and poplar (see figure 9-16). The flowers are usually tiny and without petals.

Head

A ***head,*** or ***capitulum,*** is a short, dense cluster of flowers. The flowers are generally sessile in a rounded or flat pattern (see figure 9-17).

Illustrated in figure 9-18 are the tubular flowers that compose the central, often yellow, part of the head flower, called ***disc flowers.*** In contrast, ***ray flowers*** radiate out from the disc flowers and in many species are present on the margin of a flower head.

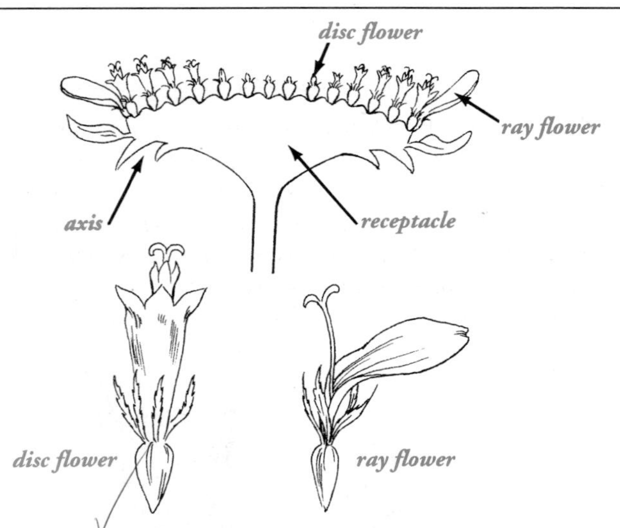

Figure 9-18. A diagram of a head flower. The individual flowers (ray and disk flowers) are subordinated to the overall display of the head, which acts as a single large flower. The disk flowers form the center area of the head, whereas the ray flowers, often mistaken for petals, form the outer portion of the head.

LEAF NOMENCLATURE

As you work with fresh flowers, pay close attention to the leaves that are present on most stems. The type, vein pattern, shape, and margin of all leaves are characteristics that can help you identify flowers correctly.

Foliage is too often an afterthought when it comes to gathering the parts to make a flower arrangement. However, it is often the foliage that can set your designs apart from all others, giving them distinction and beauty. The diversity of leaves and foliage available for floral work is great. Give thought to the foliage you select for your compositions, their leaf shapes, vein patterns, and margins. By increasing your design awareness, you also enhance your design skills. All the different characteristics discussed here will affect the style, shape, and texture of your compositions.

Figure 9-19. Design style, shape, and texture are all directly influenced by the types of foliage selected.

Leaf Parts

In its simplest form, as shown in figure 9-20, a leaf generally consists of three main parts—the ***blade,*** or leaf; the ***petiole,*** or leafstalk; and the ***stipules,*** which are two appendages at the base of the petiole. However, any of these parts may be lacking. For example, when there is not a petiole, the leaf is sessile (or attached directly to the stem).

Leaf Types

A leaf with a single blade is a ***simple leaf.*** In contrast, a leaf with more than one blade is a ***compound leaf.*** The smaller blades that make up a compound leaf are called ***leaflets.*** These leaflets may be arranged in a variety of ways, as shown in figure 9-21. Many compound leaves will display a coarse-appearing texture because of the fussy appearance of the combined leaflets.

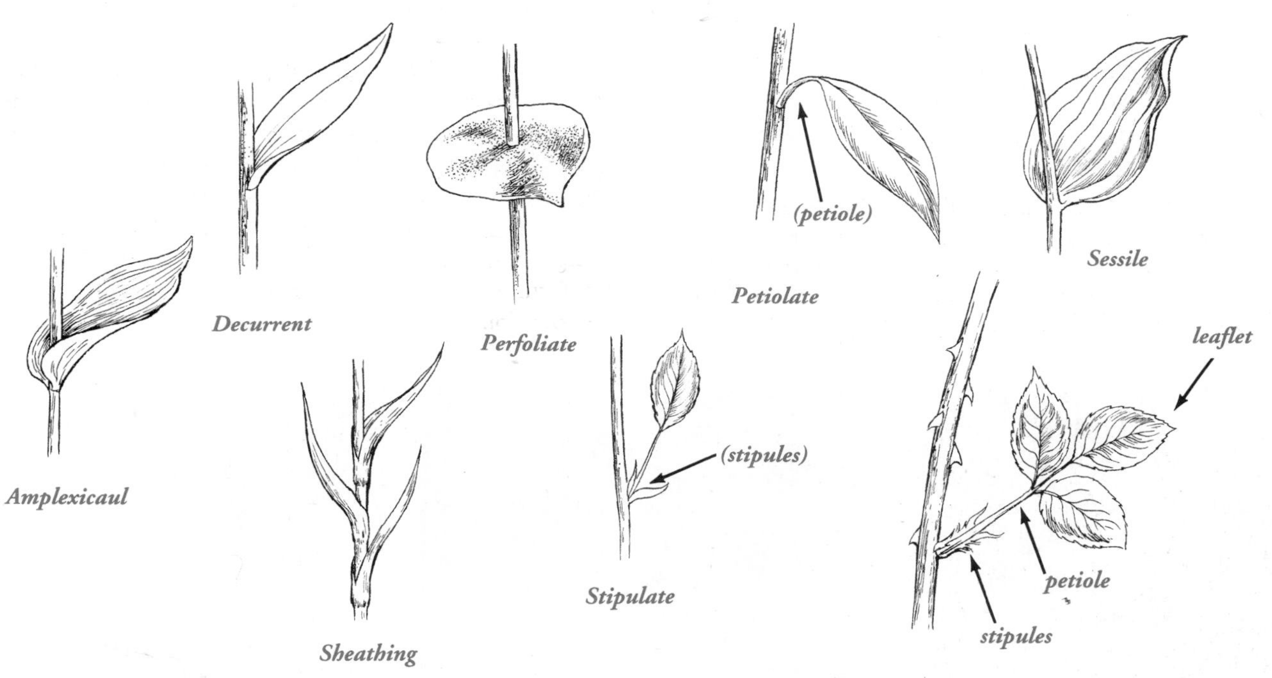

Figure 9-20. In its simplest

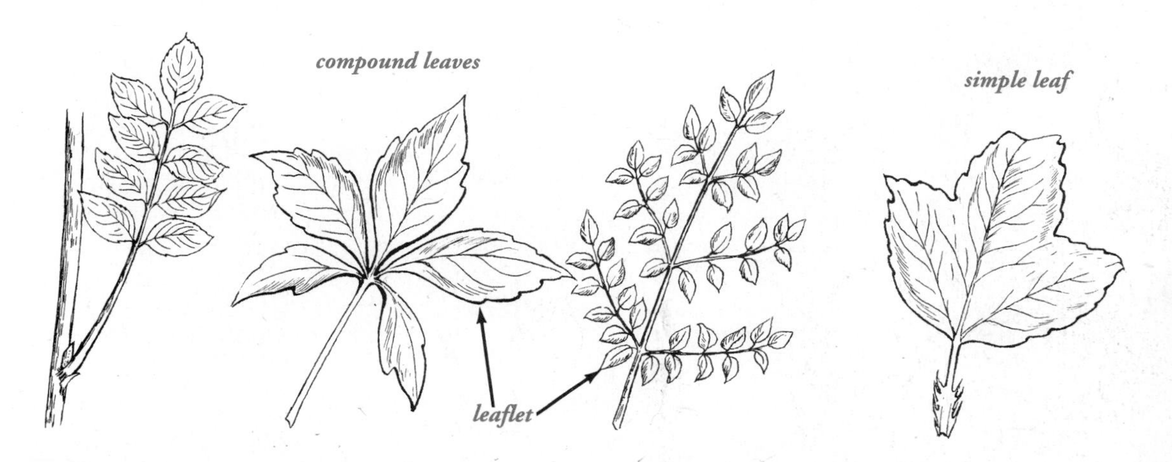

Figure 9-21. Leaves may be simple or compound. The type of leaf will effect texture, style, and form in design.

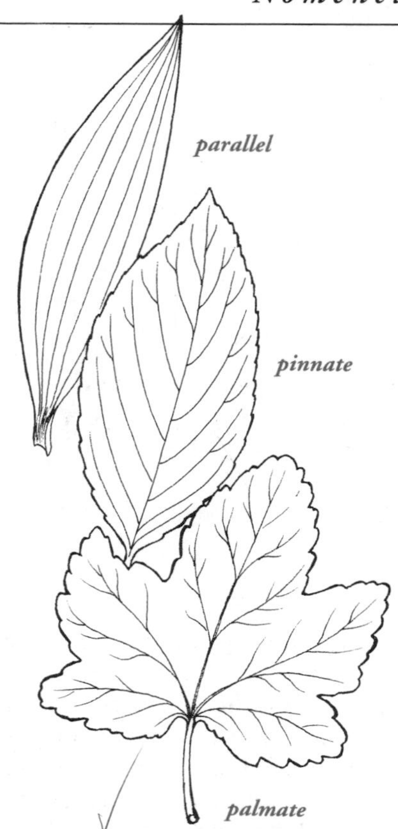

Figure 9-22. The three main types of leaf venation, or vein patterns, are parallel, pinnate, and palmate.

Figure 9-23. Venation often appears in blotched, striped, or marbled patterns, which provide visual texture and interesting color patterns to use for accent and emphasis in designs. (Top to bottom) prayer plant (Maranta), *dumb cane* (Diffenbachia), *croton* (Codiaeum), *and angel's wings* (Caladium).

Leaf Vein Patterns

The vein patterns within leaf blades are called ***venation.*** As illustrated in figure 9-22, the three main types of venation are ***parallel, pinnate,*** and ***palmate.*** Sometimes these types appear in combinations. Often venation is associated with blotched, striped, or marbled patterns (see figure 9-23). These colorful patterns add visual texture as well as emphasis and accent to floral compositions.

Leaf Shapes

The basic outline of a blade, or of all the leaflets combined in a compound leaf, make up the general ***shape*** of a leaf. The illustrations in figure 9-24 show some of the more common leaf shapes. Not every leaf fits into the shapes illustrated; often some display a combination of these shapes. A variety of leaf shapes rather than just one within a floral composition creates more visual interest and excitement.

Leaf Margins

The edge of the leaf blade is its ***margin.*** As shown in figure 9-25, margins vary greatly. Although appearing perhaps insignificant, leaf margins can directly affect the overall texture of a floral bouquet. For example, at times jagged, indented, or prickly margins cause a leaf to appear coarse in texture when actually the leaf surface may be smooth, as with holly *(Ilex).*

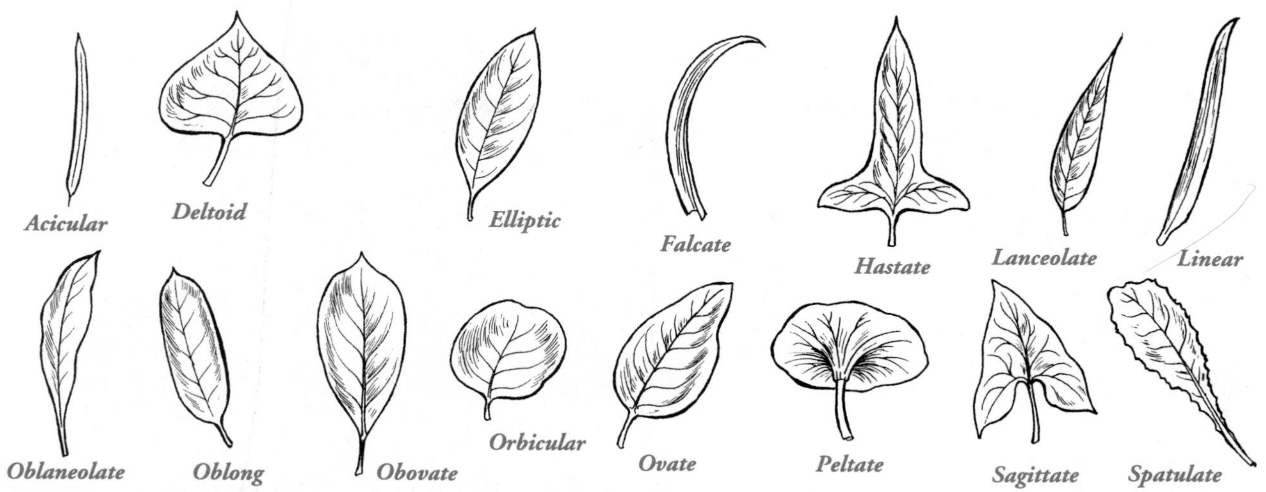

Figure 9-24. A variety of leaf shapes within a floral composition creates visual interest.

POSTHARVEST PHYSIOLOGY AND METABOLIC PROCESSES

Postharvest physiology is the division of plant physiology that deals with the ***metabolic processes*** in plant material after it has been harvested. Cut flowers and foliage are plant parts that are handled and marketed while still alive. Postharvest physiology of cut flowers and foliage spans the time from harvest to utilization by the final consumer.

It is important to remember that cut flowers are living plant parts, subject to the same basic aging process as are entire plants. Because cut flowers and foliage are still alive, they continue to function metabolically. Harvested flowers and foliage, although undergoing similar metabolic functions as their intact, parent plants (such as respiration and transpiration), undergo varying degrees of stress (such as physical damage or an undesirable gaseous environment), causing them to deteriorate quickly.

A knowledge and understanding of the metabolic processes and stresses of harvested flowers and foliage is vital in order to maintain high quality for as long as possible.

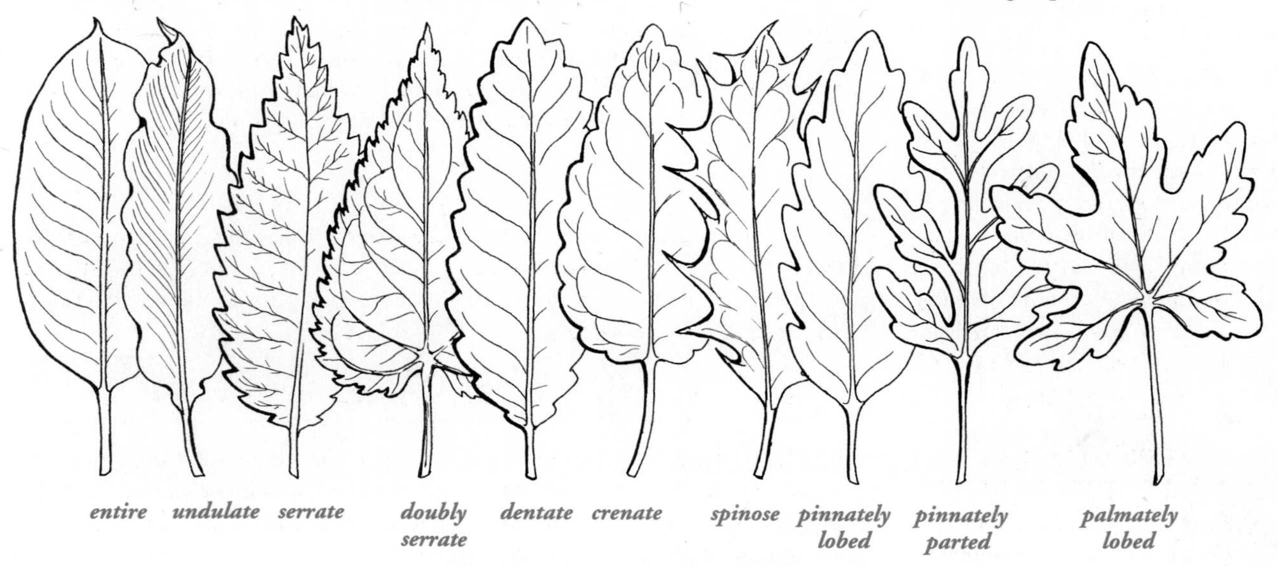

Figure 9-25. Leaf margins vary greatly, which can directly affect the overall texture and appearance of floral arrangements.

Figure 9-26. This creative and distinct composition contains a variety of foliage, all with diverse shapes, margins, and patterns. These differences are heightened by the grouping of similar elements.

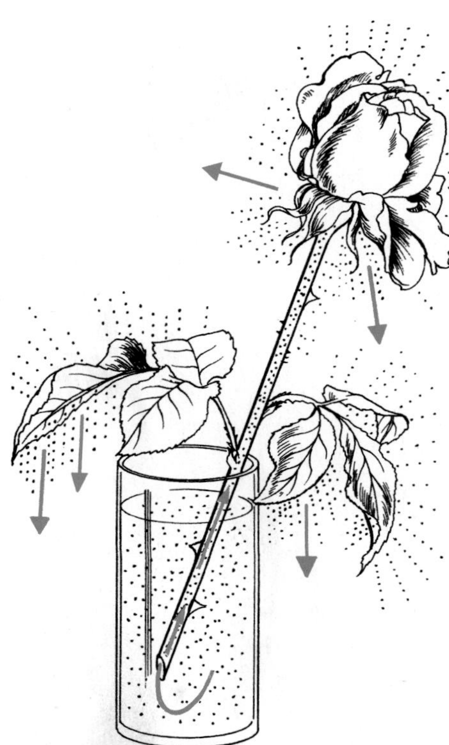

Figure 9-27. The water-conducting vessels, or xylem, drink in water which then moves throughout the stem, leaves, and flower, keeping these plant parts turgid, or firm. The loss of water vapor by plant parts, or transpiration, occurs mostly through stomata (minute pores) in the epidermis of leaves and stems.

Water Uptake and Transport

When flowers and foliage are harvested, the supply of water and mineral nutrients essential for normal metabolic activity is temporarily cut off. The leafy portions of the flowers and foliage continue to lose water. Unless this water loss is inhibited, wilting and loss of ***turgor*** (cell rigidity and firmness) will result.

Cut flowers need to drink water, which carries sugars and other compounds and helps keep stems and flower parts ***turgid*** (firm). Flower stems have a plumbing system, called the ***xylem,*** which is made up of tiny ***vessels.*** The xylem is the water-conducting tissue that carries water up the stem, to the leaves, and to the flower. The ***phloem*** is another system of plumbing; more specifically, the pholem is the food-conducting tissue. Water keeps flowers alive and fresh during their ***postharvest period.***

Transpiration

Plants are able to regulate the amount of ***transpiration,*** or water loss, by closing the ***stomata*** (pores) on their leaves. It is not possible to completely eliminate the transpiration of water from cut flowers, but water loss can be reduced by increasing the surrounding ***relative humidity*** (moisture in the air). For cut flowers the best postharvest storage conditions maintain high humidity, low temperature, and moderate air circulation (see chapter 10 for optimum conditions for cut flowers).

Reducing water loss from cut flowers will keep them firm longer. A high turgidity is necessary for flower buds to continue to develop and for plant life to continue its metabolic activities.

Respiration

Respiration is one of the main activities of living plant tissue. A process that takes place within the cells, respiration breaks down food and sugars resulting in the release of energy. Respiration is the highly complex process of utilizing stored reserves and oxygen from the surrounding environment and releasing carbon dioxide. It is essential for the maintenance of life in flowers and foliage after harvest.

In order to extend the vase life of cut flowers, it is vital to lower their respiration rate. This is accomplished by lowering surrounding temperatures. Lower temperatures slow down the respiration rate and the use of ***carbohydrates*** (such as sugars and starches) and other storage materials in plant tissues. Higher temperatures speed up floral development and ***senescence*** (the aging process). At lower temperatures, flowers produce less ***ethylene,*** a gaseous plant hormone that speeds senescence. The sensitivity to ethylene present in the atmosphere and the ability to absorb it also decreases. Lower temperatures also slow down water loss and the development of ***microorganisms.***

Photosynthesis

The process of photosynthesis is the conversion of light energy to chemical energy. It also involves the production of carbohydrates from carbon dioxide in the presence of ***chlorophyll*** (green pigment of plant cells) by using light energy. Although it is an important metabolic process, ***photosynthesis*** is not commonly considered a significant postharvest function. However, for some cut flowers and foliage, photosynthesis is an important continued metabolic activity in order to keep flowers looking fresh longer.

Tropisms

A ***tropism*** is a growth curvature caused by some external stimulus such as light or gravity. For instance, some cut flowers will curve and bend towards light or away from gravity (see figure 9-28).

When a cut flower curves or bends in the direction of light, as do anemones, this phenomenon is specifically called ***phototropism,*** from the Greek words *photos* (light) and *trope* (turning). When a cut flower such as a gladiolus or snapdragon *(Antirrhinum)* bends upward in response to the force of gravity, it is called ***geotropism,*** from the Greek words *geo* (earth) and *trope* (turning).

Curving stems due to tropisms provide exciting and unusual lines in designs. However, if curving stems will detract from the entire composition, these tropisms can be mildly reduced on some flowers by removing the top bud of a flower inflorescence as shown in figure 9-29.

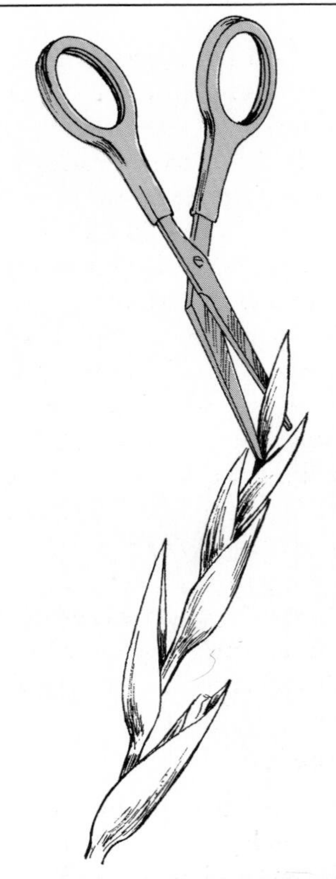

Figure 9-29. Often the removal of the top bud of a flower inflorescence will decrease tropism responses.

Plant Hormones

Natural hormones or growth regulators produced by living plants, called ***phytohormones,*** help delay or speed the process of senescence (aging). These hormones, namely ethylene and ***cytokinins,*** directly affect the longevity or vase-life of cut flowers.

Figure 9-28. Many flowers experience tropisms (growth curvatures) because of an external stimulus. Anemones and other flowers curve in response to light, called phototropism. Other flowers, such as Gladiolus and snapdragons (Antirrhinum) *curve in response to gravity, called geotropism or gravitropism. Curvature of stems, often unavoidable, produces exciting and unique lines and patterns.*

Ethylene

Ethylene is a gaseous plant hormone that stimulates deterioration and senescence of flowers and plants. Ethylene is often referred to as the ***aging hormone.*** Because it is odorless and invisible, it is easy to disregard this insidious gas. However, even in small amounts, ethylene is harmful and causes irreversible effects in cut flowers.

Ethylene gas is naturally produced by plants, fruits, vegetables, and cut flowers and foliage. Higher levels of ethylene are associated with apples and other ripening fruits and vegetables, and with damaged or disease-infected plant tissues. Considerable amounts of ethylene are produced by old, wilted flowers and foliage that have suffered from physical injury or stress. Microorganisms such as bacteria and fungi also contribute to ethylene levels. Other sources of ethylene include automobile exhaust, cigarette smoke, and air pollution. Ethylene is also a byproduct of the burning of fuels for heaters and stoves.

Not all flowers are equally sensitive to the effects of ethylene. Flowers that are highly sensitive to ethylene will oftentimes have limp and wilted stems, buds that fail to open, open flowers that close up, as well as a general wilted, faded appearance. Other symptoms include leaves that yellow and florets that easily ***shatter*** or fall off. (See chapter 10 for specific ethylene reduction and control methods.)

Figure 9-30. Ethylene is a gaseous plant hormone that quickens the senescence (aging and death) of fresh flowers. Various sources of ethylene include pollution; smoke; exhaust; ripening fruits and vegetables; and old, wilting and physically injured flowers and foliage.

Cytokinins

One type of growth hormone that helps to delay senescence is called cytokinins. Cytokinins are plant hormones that stimulate and promote cell division. As a means to extend the length and quality of cut-life for some flowers and foliage, cytokinins are occasionally an added ingredient in commercial floral preservatives.

Flower and Foliage Color

Different mixtures of various pigments, as well as cellular pH and differences in the structural and reflective properties of plant parts, all produce the characteristic pigmentation of flowers and foliage.

Color is used to ascertain quality in cut flowers and foliage. During the postharvest period, pigment characteristics of flowers and foliage change significantly. The gradual change of pigments after harvest is visible as flower colors fade.

The wide array of colors in flowers and foliage can be separated into four different groups of ***pigments***—chlorophylls, ***carotenoids, flavonoids,*** and ***betalains.*** Of these, the most important and responsible for color with cut flowers and foliage are the chlorophylls, carotenoids, and flavonoids.

Chlorophyll is the green pigment found in plant cells, prominent in leaves and stems. Chlorophyll is the primary light-accepting pigment in plant cells and is necessary for photosynthesis. The main function of chlorophyll is to absorb light energy and convert it to chemical energy.

Many red, orange, or yellow flowers owe their color to the presence of carotenoids. These pigments are associated with chlorophyll in leaves and are responsible for much of the autumn leaf colors. Carotenoids function in several metabolic processes, including photosynthesis. Carotenoids are fat soluble and are found in the plastids (the area of the cell associated with food manufacture and storage).

The most important pigments in flower coloration are the flavonoids. Three groups of flavonoids are of particular interest in plant physiology. These are the ***anthocyanins*** (from the Greek words *anthos* [flower] and *kyanos* [dark blue]), the ***flavonols,*** and the ***flavones.***

The anthocyanins are common pigments present in red, purple, and blue flowers. Anthocyanins are usually named after the particular flower from which they were first obtained. For instance, two common anthocyanins are ***cyanidin*** (named after the blue cornflower *Centaurea cyanus)* and ***delphinidin*** (named after *Delphinium).*

The flavonols and flavones are pigments in yellows and ivory. Like anthocyanins, they are contributors to flower colors.

Flavonoids are water soluble and found in the ***vacuoles*** (spaces within the cytoplasm filled with a watery fluid, the cell sap). They function in absorbing visible light, thus giving flowers their color.

Many postharvest conditions, especially light and temperature, affect the degree of change in pigmentation in flowers and foliage. Several plant hormones also have a significant effect on pigmentation during the postharvest period, especially ethylene and cytokinins. Ethylene speeds the process of pigment degradation, color fading, and senescence. The use of chemicals, such as ***silver thiosulfate,*** slows the harmful effects of ethylene on pigments. On the other hand, hormones such as cytokinins, help extend the length of the visual beauty and the color quality, especially with chlorophyll pigment.

REVIEW

A broad knowledge of the anatomy of flowers and foliage as well as botanical nomenclature will enhance your professional vocabulary and build your confidence as a designer. As flowers can be classified according to their basic shapes, this directly influences the style, shape, and texture of your designs. Also important in your design work are the leaves and foliage. The type, vein pattern, shape, and margin of individual leaves will influence design style. Often foliage can set your designs apart by giving them distinction and beauty.

Because cut flowers and foliage are living plant parts, it is vital that you have an understanding of the metabolic processes that continue during the postharvest period. Knowledge of these continuing metabolic processes will allow you to properly care for fresh flowers and foliage, thus increasing their postharvest quality as well as lengthening their otherwise ephemeral existence.

TERMS TO INCREASE YOUR UNDERSTANDING

ament
anatomy
androecium
anther
anthocyanins
axillary flower
betalains
blade
bostryx
botanically complete flower
bracts
calyx
capitulum
carbohydrates
carpel
carotenoids
catkin
chlorophyll
cincinnus
compound corymb
compound cymes
compound dichasia
compound leaf
compound umbel
corolla
corymb
cyanidin
cyme
cytokinins
delphinidin
determinate
dichasial cyme
dichasium
disc flowers
ethylene
filament
flavones
flavonoids
flavonols
floral axis
florets
gynoecium
head
helicoid
indeterminate
inflorescence
leaflets
margin
metabolic processes
microorganisms
monochasial cyme
monochasium
nomenclature
ovary
ovules
palmate venation
panicle
parallel renation
pedicels
peduncle
perianth
petals
petiole
phloem
photosynthesis
phototropism
phytohormones
pigments
pinnate venation
pistils
pollen
postharvest period
postharvest physiology
raceme
ray flowers
receptacle
relative humidity
respiration
scapose
scorpioid
senescence
sepals
sessile
shape
shatter
silver thiosulfate
simple leaf
simple umbel
solitary flower
spadix
spathe
spike

spikelike
stalks
stamens
stigma
stipules
stomata
style
tepals
terminal flower
transpiration
tropism
turgid
turgor
umbel
vacuoles
venation
vessels
xylem

TEST YOUR KNOWLEDGE

1. What advantages are there in knowing the nomenclature of flowers and leaves?
2. What are the three main types of leaf venation? Name actual examples of each type.
3. How can foliage influence the style of a floral arrangement?
4. What are the primary metabolic activities that continue during the postharvest period?
5. What are some of the effects of plant hormones on flowers during their postharvest period?

RELATED ACTIVITIES

1. Sketch a lily flower or other complete flower. Name the parts.
2. Dissect two or three different flowers and study and identify their parts.
3. Make a pictorial guide of various inflorescence types.

Chapter 10

Care and Handling

As discussed in previous chapters, the attractiveness of the flowers, their color, and their fragrance are all important. However, it is the length of time the flowers remain beautiful and usable, referred to as their ***vase life,*** that determines whether the retailer or consumer will come back for more. The appearance, quality, and longevity of flowers depend upon the conditions during cultivation, harvest, and postharvest.

Directly influenced by genetics, the longevity of flowers varies greatly among ***genera.*** Longevity even varies among ***cultivars*** of the same ***species.*** Generally, stem diameter and stiffness are factors in the length of a flower's cut life; thicker stems prevent bending and breaking and contain more stored nutrients for flowers, increasing their longevity.

Care and handling refers to all procedures that are done to help cut flowers and foliage last longer. Florists can do much to extend the life of cut flowers, thus extending satisfaction and enjoyment of flowers for the final consumer.

Figure 10-1. Impressive calla lilies are being conditioned in buckets of preservative solution for increased longevity.
(Photo taken at Evergreen Wholesale Florists Inc., Seattle, WA. Used with permission.)

CHAIN OF LIFE

The ***chain of life*** is a postharvest care program now under the direction of the Society of American Florists. The postharvest period for cut flowers, when thought

Figure 10-2. When the postharvest period of flowers is thought of as a chain, it can be separated into links, or levels of distribution, in order to analyze care and handling techniques.

of as a chain, can be separated into links, or steps of distribution. The steps of distribution, or the links of the chain, include the ***grower, transporter, wholesaler, retailer,*** and ***final consumer.*** When the postharvest period is divided into steps, each level of distribution may be better analyzed regarding care and handling techniques necessary for extending longevity.

Strengthening each link by looking at sanitation, preservatives, water quality, temperature, and humidity, as well as other factors, is the goal of this program, thereby giving the entire chain more vitality and endurance. The chain of life program, based on facts derived from scientific research rather than hearsay and myth, requires a cooperative effort by each vital link. The cliché "a chain is only as strong as its weakest link" reminds floral distributors and designers that a combination of proven techniques at each level of distribution is necessary for promoting the longevity and beauty of flowers.

CAUSES OF EARLY SENESCENCE

There are several common reasons why flowers wilt and die prematurely. They include the inability of stems to absorb water, lack of ***carbohydrates,*** excessive transpiration, bacterial growth and disease, ethylene gas, and improper surrounding conditions. In order to counteract these detrimental situations for flowers, a knowledge and understanding of care procedures necessary for extending longevity and delaying ***senescence,*** or death, is essential.

INITIAL PROCEDURES TO DELAY SENESCENCE

The initial treatment of fresh flowers at each level of distribution, or the chain of life, is referred to as ***processing.*** These first care procedures, whether at the wholesaler or retailer or in the home, are vital for ensuring the lasting

Figure 10-3. "I can let you have those for half price."

Herman © 1987 Jim Unger. Reprinted with permission of Universal Press Syndicate. All rights reserved.

Figure 10-4. Sunflowers and irises await underwater cuttings.
(Photo taken at Evergreen Wholesale Florists Inc., Seattle, WA. Used with permission.)

quality of flowers. Caring for flower stems by recutting them and removing lower leaves and any damaged parts are prerequisites to later treatments involved in the entire postharvest care program. (For specific information regarding the harvest, storage, and transportation of commercially grown cut flowers, see chapter 20.)

Recut Stems Prior to Conditioning

As mentioned previously, the inability to absorb water is typically the reason cut flowers wilt and die prematurely. A new cut at the base of the stem opens up the water-conducting vessels of the xylem, which close due to ***stem blockage.***

Stem blockage, or ***plugging,*** is the result of air, debris, healing from the original or earlier cut, and the growth of microorganisms at the base of the stem. ***Air embolism*** occurs when small bubbles of air, called ***emboli,*** are drawn into the stem at the time of cutting. These bubbles cannot move far up the stem, so the upward movement of solution to the flower is restricted. Microorganisms present in water enter the base of the stem causing plugged vessels. If flowers do not receive a fresh cut, vase life will be drastically reduced because the clogged flower stems physically cannot absorb necessary water and nutrients to keep them turgid (firm) and healthy for as long as possible.

The initial stem cuttings by the wholesaler, retailer, or final consumer, should be done underwater to prevent air from entering the xylem (see figure 10-5). Existing emboli can be removed by recutting the stems under water. As shown in figure 10-6, water droplets will adhere to the freshly cut stem ends as they are removed from the water in which cuttings are made and moved to another container of

Figure 10-5. Commercial underwater cutters can easily cut entire bunches of stems with a single motion.

water. These water droplets, in essence, seal the stem ends and prevent air from entering.

Cutting stems underwater is especially important for flowers that are severely water stressed or wilting. This will help them regain turgidity quickly. Also, if flowers have been out of water for an extended period of time, underwater cutting ensures stem blockage removal and rapid water uptake. Cutting stems underwater will accelerate the ***conditioning*** process, as well as open flower buds more quickly.

To cut flowers under water, fill a bucket, sink, or commercial ***underwater cutter*** with warm water. If you are cutting several flower stems or an entire bunch at once, make sure stem ends are all the same length so that each will receive a cut. Hold the flowers at an angle so the lower part of the stems are submerged. Cut off the lower 1/2 to 1 inch of stem.

It is helpful to recut flowers every two or three days to open the water-conducting vessels. Once flowers have been properly processed and are full of water at each step throughout the chain of life, other cuttings do not necessarily need to be made under water, unless flowers become limp and wilted.

It has long been thought that ***woody stems*** drink up water faster if they are pounded with a hammer, but this action has proven ineffective. A better practice is simply to recut woody stems more frequently than ***herbaceous*** (nonwoody) stems. Never mash, shred, or split stems in any way, as these practices can easily damage vessels that carry water and nutrients.

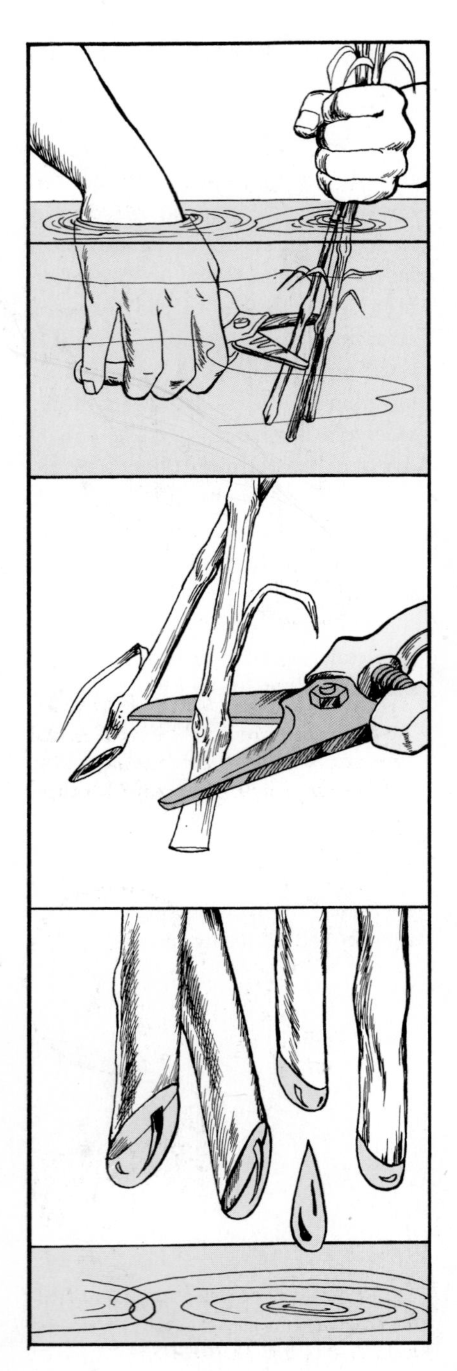

Figure 10-6 . To cut flowers underwater, hold flowers so that the lower portion of the stem is under water. When flower stems are transferred, water droplets will adhere to the stem ends, preventing air from entering the stems.

Remove Excess Foliage

Excess foliage that will be below the water level should be removed from the stems, as shown in figure 10-7. If left on the stem, the soggy foliage decomposes, increasing the growth of microorganisms and the production of ethylene. The water becomes filthy and odorous, and the stems and container become slimy with the bacterial growth. The bacteria then lodges in the stem ends, clogs the vessels, and impedes water uptake. All this leads to the early senescence of flowers.

In removing excess foliage, it is important that you do not scrape the stems. Gently pull the leaves down and off or use clippers or a stripper tool. (Removing leaves with a knife may injure tender bark.) Automatic strippers are available for larger floral operations. These machines have rotating rubber fingers that remove leaves quickly and gently.

Remove Thorns

Most roses have thorns that can damage tender petals and foliage. Once a thorn has punctured into an adjacent rose head or other flower, the puncture wound usually goes through many layers of petals. Removing sharp thorns will prevent this damage from happening and allows for easier handling of flowers. To remove thorns, it is best to just remove the thorn tip. The use of a hand stripper is more efficient than that of a knife or clippers (see figure 10-8). When using a hand stripper, be careful not to scrape or scar the stems. It is not necessary to completely remove entire thorns right down to the bark. This can damage stems, allowing bacteria to move into the newly formed scars where thorns have been. Instead, just dull the thorns by removing their sharp points.

Specialty Treatment for Stems with Latex Sap

A few flowers and foliage, when cut, bleed ***latex,*** a white milky sap. Flowers and foliage with this milky sap will wilt rapidly if not properly cared for since the latex, which is made up of resins and proteins, reseals the cut surface and blocks water uptake. It is, therefore, necessary to open up the water-conducting vessels and at the same time stop the bleeding sap. There are several treatments that may be done to open the water-conducting vessels while sealing the latex.

Many flower varieties of today do not warrant searing or scalding the stems to stop the milky latex from bleeding as shown in figure 10-10. However, some latex is still present and must be properly cared for. For these newer varieties, research has proven that simpler methods help to keep the stem clean and open for water absorption. Just cutting the stems and placing them in floral preservative for a few hours or overnight will benefit these flowers. Or use two teaspoons of ten-percent liquid chlorine bleach per gallon of water to stop latex excretion for some flowers.

Time-honored techniques used for flowers that rapidly excrete excessive latex include the use of heat (boiling water or a flame). Upon cutting a stem, if milky sap is excessive and will not stop bleeding with other treatments, to seal the latex and prevent early senescence, it may be necessary to use heat, to coagulate the latex, and allow the water-conducting vessels to remain unblocked. To seal the latex, immediately after recutting the stem, dip the stem ends in boiling or very hot water for a few seconds, and then transfer the stems to preservative solution. Or, singe individual stem ends with a flame. This is done by holding the stem at an angle and passing a flame beneath the newly cut stem, which sears the latex. These stems should also be placed in preservative solution.

Sealing the sap of stems that bleed latex—such as poppies, poinsettias, and flowering spurge—helps prevent unsightly discoloration of the water and lengthens postharvest life.

Figure 10-7. Excess foliage that would otherwise be below water level should be removed to prevent early rotting and to inhibit the growth of harmful bacteria.

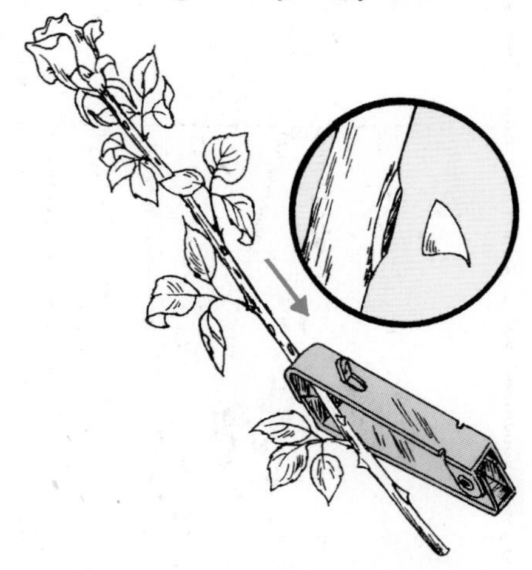

Figure 10-8. To remove sharp thorns and lower leaves, a hand stripper tool may be used. It is best not to remove entire thorns, which leaves large open wounds. Instead, dull thorns by removing their sharp tips.

Wilt-sensitive Flowers and Foliage

Some flowers and foliage are ***wilt-sensitive,*** that is, they wilt easily and transpire rapidly, resulting in lost turgidity. In dry air, or when left out of water even for a short time, these flowers and foliage have a tendency to become limp quickly. Many are prone to drying ***(dry-prone),*** causing shedding of florets or leaves. It is important for you to know which flowers and foliage are wilt-sensitive, so that you can give them priority when processing many types of flowers and foliage. Many of these flowers and foliage benefit from misting. (See the appendix for specific flower and foliage care.)

TABLE 10-1: FLOWERS AND FOLIAGE THAT ARE SENSITIVE TO DRYING AND WILTING

Flowers

Anemone
Anthurium
Antirrhinum (snapdragon)
Astilbe
Bouvardia
Campanula
Celosia
Centaurea (cornflower)
Chrysanthemum frutescens (marguerite daisy)
Dahlia
Delphinium
Erica (heather)
Euphorbia (flowering spurge, poinsettia)
Freesia
Gerbera
Gypsophila
Iris
Orchids
Ranunculus
Rose
Saponaria
Scabiosa
Tulips

Foliage

Adiantum (maiden hair)
Asparagus (ming fern, plumosa, sprengeri)
Caladium (angel's wings)
Calathea
Cordyline (ti leaf)
Nephrolepis (Boston and Oregon fern)

Figure 10-9. Special care of bleeding latex will increase longevity for poppies. (Photo taken at Evergreen Wholesale Florists Inc., Seattle, WA. Used with permission.)

Reviving Wilted Flowers and Foliage

Many flowers and foliage can be revived after being out of water for an extended period of time or if they show signs of water stress and become limp and wilted. To revive flowers and foliage, soak the entire flower (stem and head), foliage stem or single leaf, by submerging it in room temperature water for five to fifteen minutes or longer (see figure 10-11).

Figure 10-10. Generally, bleeding latex will stop with simple conditioning in a preservative solution. However, for flowers that have excessive oozing of latex or milky sap, it may be necessary to seal stem ends with a match flame. This stops the bleeding latex, while keeping the xylem vessels open for water absorption. Bunches of flower stems may be dipped in hot or boiling water for a few seconds. It is vital to protect the flower heads from hot steam.

Many tropical flowers benefit from soaking. Some examples include anthurium, ginger, bird of paradise, and heliconia. After soaking these, make sure that water does not collect on or in the open sheath; this can cause rotting of the enclosed flower parts.

Some foliage that is easily revived by soaking include Boston fern, galax, ivy, oregonia, and boxwood. Soaking brings new life to this foliage.

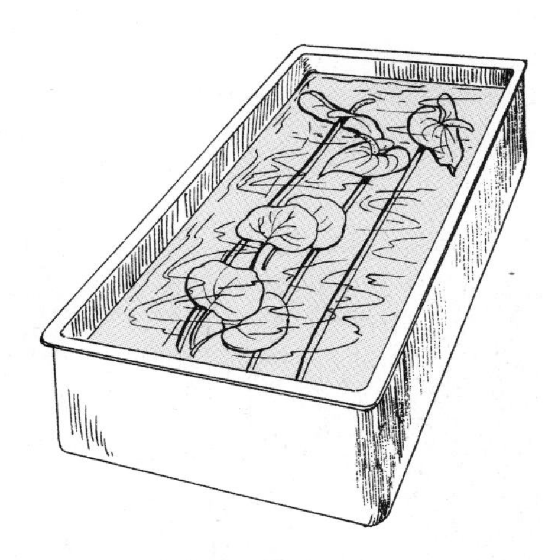

Figure 10-11. Many flowers and foliage, after being out of water for an extended period of time, or if limp and wilted, can be revived simply by soaking them in room temperature water for 5 to 15 minutes.

Figure 10-12. Cut flowers properly with sharp, quality knives and clippers. Tearing, breaking, snapping, or smashing will decrease the life of flowers by damaging water-conducting vessels.

Recut Stems Prior to Design Work

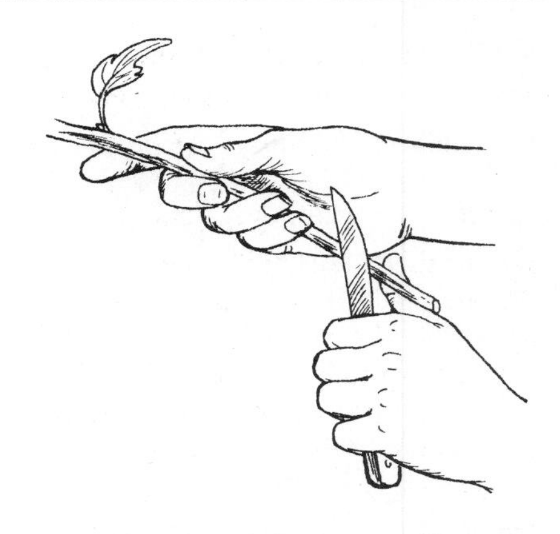

When recutting flower stems as you make arrangements, use sharp, quality tools; never use scissors or dull knives to cut flowers. Scissors tend to pinch stems and actually can constrict xylem vessels and keep them from absorbing maximum water.

Many designers prefer using knives over clippers; when using a knife, while holding the stem of the flower in one hand and the knife in the other, pull the knife quickly through the stem to form a slanted cut, keeping your thumb and fingers out of the path of the knife blade (see figure 10-13). After some practice, you will find that using a knife as you do design work will become easier and more efficient than using clippers.

Figure 10-13. When cutting flowers with a knife, hold the flower in one hand and the knife in the other. Pull the knife quickly through the stem to form a slanted cut.

WATER QUALITY

The proper temperature and quality of water are both important considerations that can increase the longevity of cut flowers. Not all water is equally suitable for flowers. The quality of water is determined by various characteristics, such as pH level and the amount of minerals and impurities present. These can drastically influence cut flower quality and the effectiveness of commercial floral preservatives.

Water Temperature

As stems absorb water, they also drink in entrapped air. It is best to use warm water (100° to 110° F), because it contains less air than cold water; warm water with less oxygen reduces the potential for air blockage. Warm water also is preferred because it flows more readily into the stem than cool or cold water. Treatment with warm water is especially beneficial for flowers that are slightly wilted. Never use hot water, which essentially cooks or scalds the stems. If you are unsure how warm 110° F water feels, use a thermometer to indicate the correct temperature.

Water Characteristics

Cut flowers need clean, pure water. Tap water often contains injurious chemical compounds and varies in pH level; it also contains organic matter and microorganisms that may prove to be detrimental to flowers. For instance, certain ions found in tap water are toxic to some cut flowers. Sodium, present in high concentrations in soft water, is toxic to carnations and roses. Another example is fluoride (which is often added to drinking water) is extremely toxic to gerbera, gladiolus, freesia, and some roses, and will cause damage to these cut flowers. The differences in tap water influence the longevity of the flowers as well as the efficiency of chemical solutions used.

Several aspects of water quality should be considered. These include ***pH level, hardness, alkalinity,*** and the amount of ***total dissolved solids (TDS).***

Figure 10-14. Warm water (about 110° F) contains less air and is more readily absorbed by flowers than cold water. Using warm water and commercial floral preservative will ensure a longer vase life for flowers.

pH Level

A pH level denotes the relative concentration of hydrogen ions in a solution. The letters pH come from the French words *pouvoir* (power) and *hydrogene* (hydrogen), literally meaning hydrogen power. On a pH value scale from 1 to 14, the lower the value, the more ***acidic*** the water, or the more hydrogen ions it contains. A pH of 7 is considered neutral, and a pH higher than 7 is alkaline, meaning that it contains less hydrogen ions.

The ideal pH solution for flowers, after water is mixed with floral preservative, is 3.0 to 4.5. Acidic solutions (containing more hydrogen ions) have better cohesive qualities and move more readily into and through stems than neutral or alkaline solutions. Microbial growth is also limited in solutions that are acidic.

Hard Water and Alkalinity

Hard water has a high mineral content, specifically magnesium and calcium. The level of hardness refers to the amounts of these dissolved salts in the water. Hard water generally leaves white, crusty deposits when it evaporates from a surface. This type of water often produces dismal results for flowers.

However, softened water should not be mixed with hard water in an attempt to counteract water hardness. Softened water does more damage to certain flowers than hard water, since water softeners add salts (usually sodium) to the water. If hard water is the only option in your city, you can increase flower longevity by using distilled or purified water.

Hard water contains minerals that make the water alkaline. Alkaline water does not move readily through stems and can substantially reduce vase life. The alkalinity of water refers to the measure of its capacity to neutralize acids and resist chemical changes; this is also its ***buffering capacity.*** Water with high alkalinity is ***highly buffered*** and resists any efforts to change pH when floral preservative is added specifically to lower pH. Alkaline water can be treated by removing the minerals using a deionizer or by adding enough acid to acidify the water.

Total Dissolved Solids (TDS)

TDS refers to the measurement of ***water salinity*** (salts) or total dissolved solids or soluble elements in the water. Some of these dissolved solids include salts from calcium, magnesium, sodium, chlorides, carbonates, and sulfates. A TDS reading is usually in parts per million (ppm). High-quality water for cut flowers should have less than 200 ppm. Higher TDS levels generally prove to be detrimental for many cut flowers.

Water Analysis

To ensure that you are doing the most for your flowers, have the water in your area tested. You can test the water frequently yourself by using a pH meter and TDS meter. These hand-held meters are available in a number of different models and sizes. They are generally available for purchase from laboratory supply companies or greenhouse supply companies.

For a ***water analysis,*** contact a reputable water treatment company. There are also many laboratories that will run the tests you need. Once you receive results of the water analysis, most floral preservative companies will evaluate it at little or no charge, revealing the overall quality of the water as it relates to cut flowers.

In most cases, all that is needed to improve ***water quality*** is the addition of the correct amount and kind of commercial floral preservative. In extreme cases, for wholesale or retail operations, for instance, it may be suggested that a water purification system be used.

CHEMICAL SOLUTIONS AND PROCEDURES

Several chemical solutions and procedures are used to improve the longevity and quality of cut flowers. Some of these procedures are done by the grower shortly after harvest and before the shipping or storage of flowers. Others may be done at the wholesale or retail level before or after shipping. Several considerations are important in deciding which treatments are necessary, including the genus or species of the flowers, the overall quality of the flowers, the length of time the flowers are at each level of distribution, and if any prior treatments have been done.

Several chemical solutions are used to extend vase life and increase the quality of cut flowers. The use of ***floral preservatives*** is a ***long-term solution,*** that is, they should be used at every level of distribution during the entire postharvest period; while the ***pretreatments*** of ***silver conditioning*** and ***citric acid,*** as well as the procedures of ***pulsing, hydrating,*** and ***bud-opening*** are ***short-term solutions,*** that is, used for a short time at one or more levels of distribution. The different solutions in which flowers may be placed after harvest have specific objectives.

Floral Preservative

You may remember a time when the vase water holding a bouquet of flowers turned murky and smelled awful; and after tossing out the wilted flowers and disgusting water, the vase and stems were slimy. The use of floral preservatives greatly inhibits unpleasant, odoriferous situations like this.

Commercial floral preservatives are chemical mixtures, added to the water, that extend the vase life of flowers. Preservatives literally ***preserve*** the quality of postharvest life by prolonging vase life, increasing flower size, and maintaining the color of petals and leaves.

Floral preservatives should be used during the entire postharvest period—by the grower, wholesaler, retailer, and final consumer—during which flowers at all levels will benefit from the action of preservatives. Whenever plain water can be absorbed by flowers, use instead ***preservative solution.*** For instance, flowers in ***holding*** or storage will benefit greatly from the use of preservatives. And when designing with flowers, use preservative solution to soak floral foam and add it to all designs including bud vases and foam arrangements. Flowers in water tubes benefit greatly from the use of preservative solution as well.

When mixing commercial floral preservatives, whether liquid or powder, always follow the directions on the label or package. Avoid the use of metal buckets and containers; metal containers react chemically with the acidifiers in the preservative, rendering them ineffective. Generally, metal buckets and containers experience some corrosion as well. The correct concentration is important, as is a thorough mixing. A lump of preservative powder can easily plug water-conducting vessels, thus inhibiting water uptake.

Ingredients

The combined ingredients in floral preservatives all work in harmony to maintain quality and extend the postharvest period. Three primary components of preservatives include ***sugar*** (carbohydrates to nourish), ***biocide*** (inhibits the growth of microorganisms), and an ***acidifier*** (lowers pH levels). Other secondary components included in some preservatives are: ***growth regulators*** (increase the vase life of some flowers) and ***wetting agents*** (aid in water aborption).

Sugars

Sugars are the main source of nutrition for flowers. Plant tissues require sugar in order to carry on vital functions, especially respiration. Normally, the process of photosynthesis supplies sugars to the plant tissues both for growth and metabolic activities. Once a flower is cut, the process of photosynthesis, through which the flower obtains most of its energy, is temporarily stopped. Although starch and sugar stored in the stem, leaves, and petals provide much of the food needed, cut flowers continue to need energy in order to open up and develop to their full potential. It is necessary to supply an alternative source of energy with the addition of sugars or carbohydrates.

Carbohydrates support the metabolic activities that continue during the postharvest period. Simply put, these sugars, such as sucrose and dextrose, provide food for flowers. Concentrations of sugars needed by cut flowers vary with species. All sugars present in water, however, create excellent conditions for the growth of detrimental fungi and bacteria, the growth of which can be enhanced by organic materials from the cut stem. Substances produced by the bacteria, and the bacteria themselves, plug the water-conducting vessels of stems. For this reason, buckets and containers should be cleaned and disinfected regularly, and the water solution should contain biocides to prevent the growth of microorganisms.

Biocides

A biocide or ***germicide*** is a chemical substance that can kill living organisms, especially microorganisms. Microorganisms that grow in the vase water include bacteria, fungi, yeasts, and molds. Through their development and the eventual blockage of the xylem, these are harmful to cut flowers (see figure 10-15). Water and carbohydrates are physically prevented from getting into the flower. These microorganisms also produce ethylene and toxins, which speed up flower senescence.

Several biocides are used in commercial preservatives to inhibit the growth of microorganisms, including 8–HQC (8–hydroxyquinoline citrate), 8–HQS (8–hydroxyquinoline sulfate), Physan–20 (an ammonium compound), silver nitrate, and household bleach. These are all used in extremely small amounts.

Acidifiers

Commercial floral preservatives also contain acidifiers for reducing the pH level of water. As mentioned previously, the pH is the measure of acidity or alkalinity of water or other solutions. The ideal pH level for preservative solutions for cut flowers is 3.0 to 4.5, which maximizes water

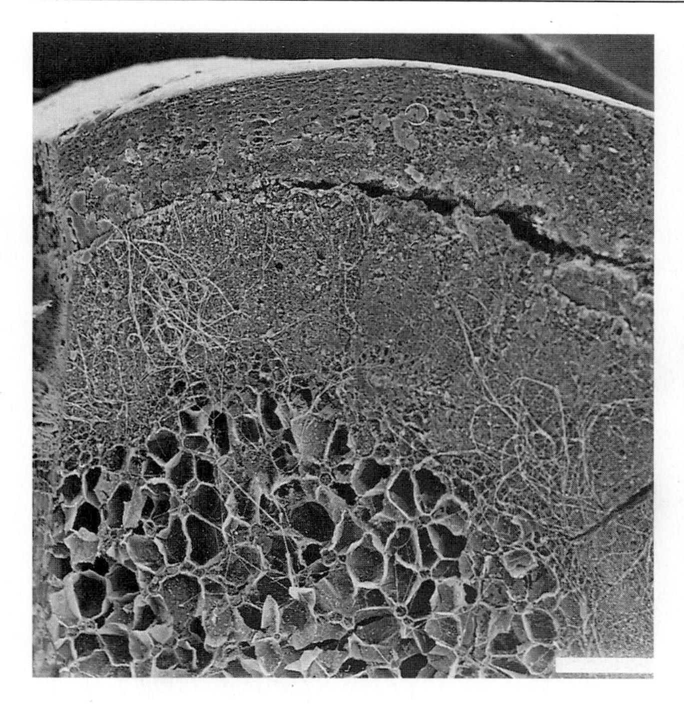

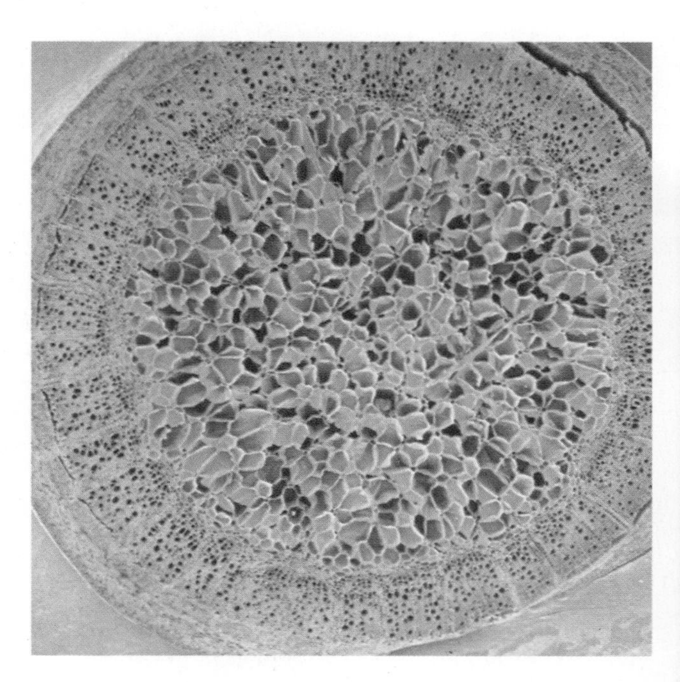

Figure 10-15. Scanning electron micro graph of a cross section of the stem of a rose flower before (top) and after (bottom) being held for 3 days in deionized water. The fine tubes conducting the water to the flower have been plugged with bacterial slime and fungal hyphae.
Photograph by Michael Reid, University of California, Davis.

uptake. This allows stems to become hydrated (full of water and preservative) more easily.

The most effective acidifier in preservatives is citric acid. Another highly effective acidifier is ***aluminum sulfate,*** which has proven to be especially helpful in water that is high in salts. Both are used in extremely small concentrations.

Growth Regulators

Some preservatives contain growth regulators, namely ***cytokinins.*** Cytokinins are plant hormones that stimulate and promote cell division. When absorbed by stems, cytokinins greatly extend the postharvest period and help maintain pigment colors, especially chlorophyll, preventing yellowing in the leaves and stems of some cut flowers and foliage. The growth regulator that is most common in commercial preservatives, increasing vase life, is 6 benzyl adenine.

Wetting Agents

Chemicals used to increase the ability of water to be absorbed by flower stems, or wetting agents, are often an ingredient in floral preservatives. These help make the water

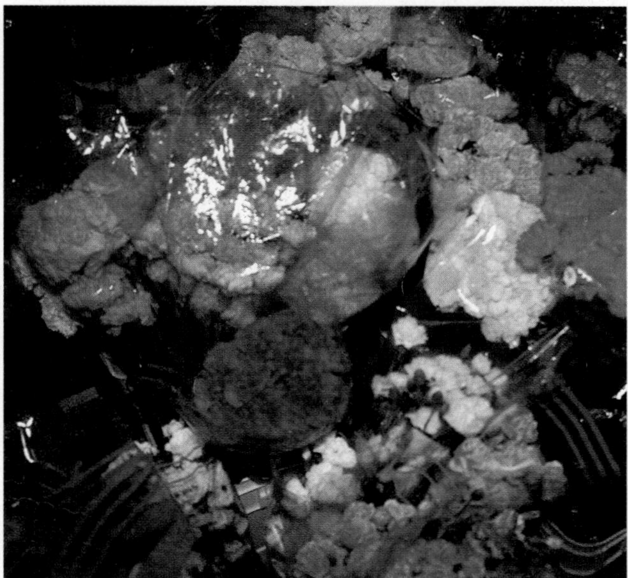

Figure 10-16. Many flowers, especially carnations, benefit from a pretreatment of silver conditioning (STS) to guard against the detrimental effects of ethylene gas.

Figure 10-17. Roses benefit from a pretreatment of citric acid. This process expedites hydration and helps to prevent bent neck, increasing flower longevity and beauty.

"wetter," accelerating the uptake of water. These chemicals, such as sodium hypochlorite, are used in extremely small concentrations.

Home Remedies

There are many who swear by the use of certain household chemicals as home remedies for extending the life of cut flowers. The addition of home remedies—table sugar, painkillers (namely aspirin), or pennies—to the water has been proven ineffective in prolonging flower life. Commercial floral preservatives have proven more reliable in extending the life of flowers.

There are, however, two products that have proven beneficial when diluted with water—lemon-lime soft drinks and medicinal-type mouthwash. However, use these products at home only when no commercial preservative is available.

Adding a lemon-lime soft drink to water can extend the longevity of flowers. Any brand of lemon-lime carbonated drink will work, with the exception of diet varieties. Be careful, however, to use a brand of lemon-lime soft drink that is clear and cannot be seen; some sodas are yellow or yellow green which, in a clear vase, prove to be visually unappealing. The soda beverage provides sugar or food for the flowers, while the citric acid lowers the pH level of the water. Both the citric acid and the carbonation help keep the water clean. Use at a ratio of 1 part carbonated drink to 2 parts water; or dilute further by using one 12-ounce can per gallon of water.

Another product that also helps extend the longevity of cut flowers is mouthwash. When added to the water, mouthwash is an effective inhibitor of microorganisms. It is best to use the medicinal types of mouthwash, rather than the mint or other flavors. Use only a small amount (about 2 tablespoons per gallon of water).

Remember, it is more beneficial to use time-tested commercial preservatives because they contain the most effective combination of ingredients for extending the life of cut flowers.

Pretreatments

Pretreatments are special care treatments used prior to floral preservatives that help to extend the life of cut flowers. Common pretreatments include silver conditioning, which greatly inhibits the action of ethylene in many flowers, and citric acid conditioning, which hydrates flowers quickly. Other names for pretreatments are ***supplements, special treatments, specific-action chemicals,*** and ***conditioners.***

Silver Conditioning

Silver conditioning is a postharvest process carried out for ethylene-sensitive flowers. Often referred to as a pretreatment (procedure used prior to floral preservative), this conditioning process helps protect flowers, such as carnations, gypsophila, lilies, alstoemeria, snapdragons, and dendrobium orchids against the irreversible effects of ethylene, therefore, increasing their postharvest period.

Silver thiosulfate (STS) is a compound of silver that is readily absorbed into flower stems. Flowers should be treated with STS only one time during postharvest life. The most effective time for silver conditioning is shortly after harvest, administered by the grower. The process of silver conditioning is similar to the use of a floral preservative; flower stems must be processed first. Next, the newly cut stems are immediately placed in STS solution and left to condition at room temperature for one to two hours, depending upon label directions. Flowers should then be placed in preservative solution (without recutting stems) and transferred to the cooler.

Citric Acid

Citric acid treatment, like silver conditioning, is a pretreatment, administered before the use of floral preservative. The procedure, also called hydrating, is a short-term technique that restores flowers to a visibly turgid condition by expediting water uptake. Generally, hydrating is done before or after dry storage and shipping. Carried out by the grower, the wholesaler, and/or the retailer, the hydrating process takes one or two hours. Hydrating solutions contain acidifiers (to obtain an optimum pH level) and wetting agents. Combined with warm water, newly cut stems readily absorb water quickly.

Roses, gerberas, and chrysanthemums benefit greatly when pretreated with a citric acid solution, which prevents the growth of microorganisms and keeps the water-conducting vessels of the xylem open for easier transport of water and nutrients. Citric acid treatment speeds the intake of water, helping to prevent ***bent neck*** in roses and gerberas (see figure 10-18).

Citric acid treatment is accomplished by placing newly (underwater) cut stems in the solution, at room temperature for one to four hours (depending on the manufacturer's directions). Then place flowers in a preservative solution (do not make new stem cuts) and transfer to the floral cooler. Other solutions are available in which the base of flower stems are quickly dipped in higher concentrations of citric acid for one or two seconds, then immediately placed in a floral preservative solution. Citric acid pretreatments can be used at all levels of the floral industry, by growers, shippers, wholesalers, and retailers.

Floral Treatments

After pretreatments are complete, flowers should be conditioned with other chemical solutions, namely floral preservatives. Depending upon the level of distribution, the type of conditioning will vary; the flowers may need to be made ready for drypacking for transportation in boxes, or they may need to be stored in a cooler, or

Figure 10-18. Bent neck in roses and gerberas can be lessened with proper hydrating and conditioning techniques.

the flower blossoms may need to open and mature quickly for use in arrangements. Common treatments include conditioning, pulsing, and bud opening.

Conditioning

The process of conditioning allows flower stems to remain in a warm preservative solution for a period of time before being placed in a floral cooler. Most flowers will absorb a maximum amount of preservative solution in one or two hours or, for extended conditioning, flowers may remain in the solution overnight. Conditioning speeds the opening of tight flower buds (for example, gladiolus, lilies, carnations, and alstromeria) helping them reach optimum beauty. Conditioning is done at all levels of the floral marketing chain.

Most flowers may be conditioned together with other flowers, with the exception of daffodils *(Narcissus),* which should always be put in separate buckets because they produce a sappy secretion that flows from the stem ends after they are cut. The sap is harmful to other flowers, especially tulips. Once daffodils have been conditioned in a preservative solution for 6 to 24 hours, they may be mixed with other flowers. However, do not cut them again after this conditioning process.

After flowers have been conditioned in room temperature and are fully turgid, these buckets, or holding containers, should be transferred to a floral cooler where stems become firm and solid. This process is called ***hardening*** and is done at the grower, wholesale, and retail levels of distribution.

Pulsing

Pulsing is a postharvest technique used to load or fill flowers with sugar and other chemicals, giving them "strength" before they are ***dry-packed*** and shipped long distances. Pulsing is usually performed at the grower level and sometimes at the wholesale level. Not all flowers respond equally to pulsing solutions. Sucrose is the main ingredient of pulsing solutions, generally 2 to 20 percent or more sugar (depending on the flower). A biocide to help keep flowers healthy by inhibiting the growth of microorganisms is also used. Some cut flowers are pulsed with STS to reduce the effects of ethylene. Flowers can be pulsed for short periods at warm temperatures, such as 10 minutes at 70° F, or long periods at cool temperatures, such as 20 hours at 36° F. Some crops, such as China aster *(Callistephus*), *Gerbera,* and maiden hair fern *(Adiatum*) respond well to short 10-second pulses in solutions of silver nitrate.

Bud Opening

Bud opening procedures are generally intended for flowers that are harvested at an immature stage and would otherwise not open up to their full potential and optimum beauty. These solutions contain a biocide and sugar. Buds should be opened in an environment of high relative humidity (60 to 80 percent), warm temperatures (70° to 80° F), and with high light intensity. Tight flowers may be placed in bud-opening solutions until buds begin to open on their own or until flowers are at the desired maturity stage. This procedure may be done at the grower, wholesaler, or retailer.

REFRIGERATION

After flowers are cut, and before they are enjoyed by the final customer, they are kept refrigerated for much of the time. Proper ***refrigeration*** is essential for increasing the overall storage and vase life of cut flowers and foliage.

Three main functions of refrigerated storage for fresh flowers are (1) to reduce the rate of respiration, (2) to reduce water loss or transpiration, and (3) to reduce the rate of bacterial growth and ethylene production and action.

Refrigeration removes heat that flows naturally from flowers and slows respiration, or the breakdown of food. Refrigeration keeps flowers at their peak longer by delaying further development and bud opening.

Refrigeration decreases the rate of transpiration or water loss in flowers. Because water loss is reduced, flowers remain fresh and turgid, or full of water.

The growth of bacteria is slowed and the production and effects of ethylene are reduced as the temperature in refrigerated storage is lowered. By reducing bacteria and ethylene levels, the deterioration of flowers is dramatically decreased.

Although these functions of refrigerated storage all help extend quality and longevity of fresh-cut flowers, a combination of both correct ***temperature*** and ***humidity*** is required to maintain flowers at optimum conditions. Other factors such as ***ventilation*** and ***lighting*** are also important considerations for keeping flowers in refrigerated conditions.

Temperature

The goal of refrigeration is to store flowers at the lowest temperature possible without causing the flowers to suffer

Figure 10-19. It has long been known that the cooling of flowers increases their vase life. Left: Early floral coolers in the 1920s were actually fancy iceboxes. Middle: Walk-in coolers became popular in the 1930s. Right: Today's coolers are engineered for increased length and quality of flower storage, as well as enhancing the display of flowers and arrangements.

cold damage or, worse, freeze. The ideal storage temperature for most cut flowers is 32° to 36° F. The species of flowers as well as the length of time the flowers will be in refrigerated storage will dictate the correct temperatures. The optimum temperature for storage of most common cut flowers is 32° F (when stored for more than two or three days). Flowers kept in refrigerated storage for a period of less than two days can be maintained satisfactorily at 34° to 40° F.

Small changes in temperature greatly alter the quality of flowers by influencing the metabolic reactions within the flowers. It has been shown that flower deterioration increases two to five times faster for every 15°- to 18°-F increase in temperature above optimum refrigerated storage conditions. Low temperatures must be combined with humidity for increased flower longevity.

Freezing of flowers may result if the temperature in a cooler drops below 31° F. Freezing injures cut flowers by causing ice crystals to form in stems, leaves, and flowers; these ice crystals puncture the cell walls of these various plant parts. When flowers thaw, tissues appear water-soaked, translucent or discolored, and soft.

Many tropical and subtropical flowers are sensitive to low temperatures and should not be stored in a regular floral cooler with other flowers. These flowers are cold sensitive and subject to ***chilling injury*** at temperatures below 50° F. Symptoms of chilling injury include darkening (browning, blackening, and lesions) of the petals, water-soaking of the petals, and sometimes collapse and drying of the leaves and petals.

Table 10–2 lists common tropical and subtropical florist flowers that are subject to damage if temperatures are too low. These flowers store best in a warmer environment at 50° to 60° F.

TABLE 10-2: TROPICAL AND SUBTROPICAL FRESH-CUT FLOWERS THAT ARE SENSITIVE TO CHILLING INJURY

Alpinia (ginger)
Anthurium
Banksia
Camellia
Eucharis
Euphorbia (poinsettia and flowering spurge)
Godetia
Heliconia
Hippeastrum (amaryllis)
Leucadendron
Leucospermum
Orchids
Protea
Strelitzia (bird of paradise)
Tapeinochilus (wax ginger)
Telopea
Zingiber (shampoo ginger)

Humidity

Humidity refers to the amount or degree of moisture in the air. ***Relative humidity*** is the amount of moisture in the air as compared to the maximum amount that the air could contain at a certain temperature. Cold air holds less moisture than warm air. Humidity during refrigeration is essential for cut flowers and foliage, and without it at low temperatures, flowers and foliage dehydrate rapidly. After proper hydration, flowers remain turgid in cold, humid conditions; as the result of less transpiration, less water is lost from the flowers.

The recommended percent of relative humidity for a refrigerated storage unit is a minimum of 80 percent. Many flowers last longer, however, with a 90 to 95 percent humidity level. If the humidity level is too high (especially if temperatures are above 40° F) or if temperatures fluctuate, condensation may form on the petals and this moisture may lead to the consequent growth of ***botrytis*** and other pathogens.

Ventilation and Lighting

Other considerations such as ventilation and lighting are important for maintaining the quality of flowers. Ventilation, or air flow, is essential while flowers are refrigerated. Some ventilation is necessary for removing excess heat that may have built up in flowers. The circulation of fresh, cool air at a low speed in a gentle stream will allow the flowers to remain cool and avoid drying up.

Lighting is essential for flowers while in refrigeration. High-intensity lamps that simulate natural sunlight help flowers maintain the photosynthetic processes and their colors will remain more accurate.

ETHYLENE

Ethylene is naturally produced by cut flowers and foliage, ripening fruits and vegetables, and decaying and wilting plant tissues. Other sources contributing to high ethylene levels include exhaust, smoke, pollution, and bacteria. Ethylene-induced disorders account for a high percentage of flowers dying prematurely.

Effects on Flowers

The general symptoms of ethylene shown by flowers include buds and flowers failing to open, as well as the closing of open flowers, and overall ***sleepiness*** or a wilted appearance. Some flowers, especially carnations and some roses, perish rapidly if exposed to minute concentrations of ethylene and become wilted and limp. In many compound inflorescence types, such as snapdragon, delphinium, and larkspur, ethylene causes flower ***abscission,*** or shattering, and florets fall off easily. Levels of ethylene in the air above one-tenth part per million (0.1 ppm) surrounding most cut flowers can cause damage. Color is dramatically affected and many flowers quickly fade when ethylene is present.

The effects of ethylene cannot be reversed, but the amount of ethylene present in the atmosphere surrounding fresh flowers and foliage can be minimized. This is accomplished by avoiding sources of ethylene, minimizing ethylene production, and inhibiting its action.

Reduction and Control

It is impossible to eliminate all sources of ethylene, but some sources are relatively simple to avoid. When certain flowers begin to deteriorate sooner than others, it is important to remove them from the floral cooler or from floral arrangements (see figure 10-20). Their increased production of ethylene will accelerate the aging process in surrounding flowers. Remember also that almost any exhaust or smoke is detrimental to fresh flowers. Keep apples and other fruits, vegetables, and food away from fresh flowers.

Ethylene production can be minimized best by keeping floral areas and equipment clean. It is essential that buckets, containers, tools, water, and design and storage areas are kept clean and ***sanitized*** with a disinfecting agent on a regular basis. Adding fresh flowers to slimy buckets or containers will further the growth of bacteria and aid in the production of ethylene. Storage and handling areas should allow ventilation to remove ethylene that does occur.

Inhibiting the action of ethylene is all important. Some flowers are much more sensitive to ethylene. Chemical treatment with STS to inhibit the action of ethylene and reduce the effects of ethylene (both exogenous and endogenous) is extremely beneficial for some flowers. Silver treatment dramatically increases the vase life of ethylene-sensitive flowers. The treatment actually provides protection against the effects of ethylene. Refrigeration greatly reduces ethylene production and the sensitivity of flowers to ethylene. Table 10–3 lists flowers that are ethylene sensitive.

Figure 10-20. Ethylene can be controlled and minimized in several ways. Removing old or wilting flowers from arrangements will lessen the production of ethylene. Keeping floral work and storage areas, equipment, and tools clean and sanitized will reduce the growth of bacteria and the production of ethylene.

TABLE 10-3: ETHYLENE-SENSITIVE FLOWERS

Aconitum (monkshood)
Agapanthus
Alstroemeria
Anemone
Antirrhinum (snapdragon)
Astilbe
Bouvardia
Campanula
Centaurea (cornflower)
Consolida (larkspur)
Dianthus (carnation, spray carnation, and sweet William)
Delphinium
Dendrobium
Eremurus (foxtail lily)
Freesia
Gypsophila (baby's breath)
Iris
Kniphofia (red hot poker)
Lathyrus (sweet pea)
Lilium (lily)
Matthiola (stock)
Narcissus
Ornithogalum
Phlox
Physostegia
Scabiosa (pincushion flower)
Solidago
Solidaster

Figure 10-21. Many garden flower varieties are grown commercially. Summertime favorites (such as zinnias, calendulas, and cornflowers shown here) have brilliant colors. Garden flowers offer unusual forms, textures, and fragrances that are not available year-round.

(Photo taken at Boston Flower Exchange, Boston, MA. Used with permission.)

CARE AND HANDLING OF GARDEN FLOWERS

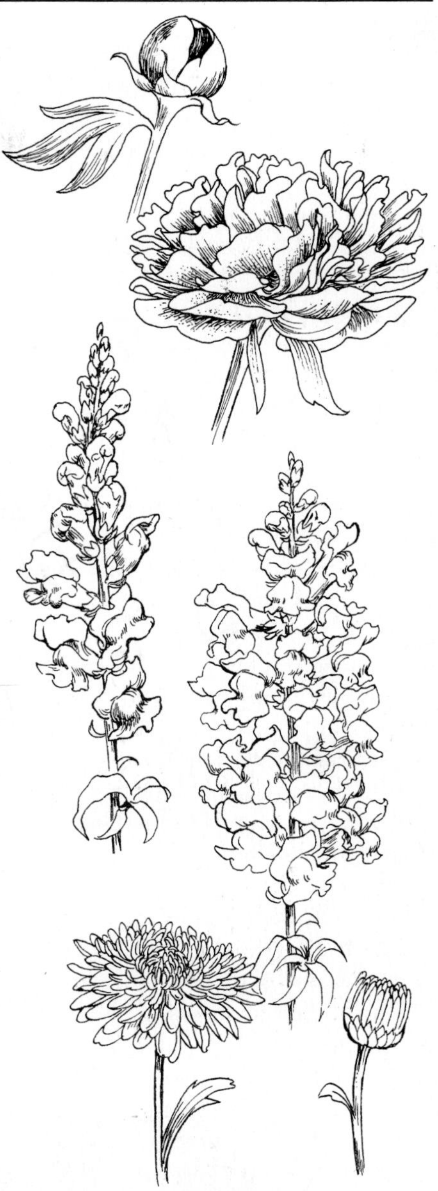

Figure 10-22. Garden flowers should be cut at certain stages of development for maximum vase life. Many flowers, such as peonies, should be cut at the bud stage because once cut, they continue to open. Other flowers, for example snapdragons, are best cut at the half-open stage, with some florets open and others buds. Still other flowers, like chrysanthemums, should be cut at the fully-open stage, because if they are cut too early, they rarely continue to open to their full potential.

Many flowers grown in a garden make excellent cut flowers, giving floral bouquets a casual, country look. If you have the space and time, growing a flower garden can provide a wonderful array of bright and unique flowers that can be used alone or with commercially grown flowers. As with flowers that are commercially grown, it is important to follow proper harvesting, care, and handling techniques in order that maximum vase life can be enjoyed.

Harvesting Garden Flowers

Gather garden flowers only in the early morning or in the evening. This is because flowers in the morning or evening are more fully turgid and contain higher carbohydrate food reserves. Never cut garden flowers during midday or in the late afternoon. Weakened by the heat of the day, flowers are less turgid and contain less carbohydrates. Flowers that are cut in the afternoon have a much shorter vase life than those cut in the morning or evening. They never gain full turgidity and usually appear limp and wilted, even after following postharvest care and handling techniques.

Use sharp, bladed clippers or floral shears to cut flowers from plants that are robust and healthy. Choose flowers that are at the correct stage of bud development. If flowers are harvested too early, they usually will not open fully and will soon wither and die with the buds still closed. However, other flowers when cut fully opened, start to fade and drop petals quickly. With some practice, you will know when to cut garden flowers for maximum vase life.

Stage of Flower Maturity

Garden flowers can be grouped into three stages of bud development for cutting: ***bud stage, half-open stage,*** and ***fully open stage*** (see figure 10-22).

Flowers that should be cut when they are still ***buds*** are generally those that are ***solitary.*** Examples of these flowers include roses, peonies, daffodils, irises, and tulips. These single flowers usually continue to open and develop after cutting. Therefore, the use of floral preservative is essential to provide these flowers with more carbohydrates, giving them energy for buds to open up.

The half-open stage is a general description for spikelike flowers and racemes that have a number of florets per stem. Other compound inflorescence types like umbels, corymbs, and many panicles are also best cut at the half-open stage. These flowers are best cut with about half of the florets on the stem open and the other half still in the bud stage. Since florets continue to open, these flowers will extend the life of a floral arrangement. Examples include gladiolus, delphinium, larkspur, eremurus, and snapdragon. Examples of umbels and other inflorescence types with more than one floret per stem include agapanthus, nerine, alstroemeria, lilies, and phlox.

Some flowers are best cut when fully open before they shed their pollen. Once cut, they do not continue to open to their full potential. Many of the

head inflorescence types are best cut when fully open. Examples include sunflower, zinnia, aster, chrysanthemum, dahlia, marigold, and calendula.

Care and Conditioning

When you gather flowers, take a bucket of warm, preservative solution with you out to the garden. As you cut your flowers, immediately place stems into the preservative solution. Never place harvested flowers on the ground where they may become contaminated with disease organisms.

Take the cut garden flowers indoors, recut stem ends underwater, and allow them time to drink in the preservative solution for at least two hours or, better still, overnight before using them in designs. This time is needed for sufficient conditioning.

Garden flowers, like commercially grown flowers, should have lower foliage removed from stems that will be in water. Remember specialty treatments as well, such as conditioning daffodils by themselves and sealing the latex of poppies.

Garden flowers should be used as soon as possible after a conditioning period rather than storing them in a floral cooler. With planning and proper postharvest care, most cut, garden flowers will last five to ten days. Table 10–4 lists easily grown garden flowers that are medium- to long-lasting cut flowers.

TABLE 10-4: TOP-RATED CUT GARDEN FLOWERS (COLORFUL, LONG-LASTING, AND FAIRLY EASY TO GROW)

Achillea (yarrow)
Agapanthus
Anemone
Antirrhinum (snapdragon)
Aquilegia (columbine)
Aster
Astilbe
Calendula
Callistephus (China aster)
Celosia
Centaurea (cornflower)
Chysanthemum
Consolida (larkspur)
Coreopsis
Cosmos
Dahlia
Delphinium
Dianthus (pinks, sweet William)
Digitalis (foxglove)
Gaillardia
Gladiolus
Gomphrena
Gypsophila (baby's breath)
Helianthus (sunflower)
Iris
Lilium (lily)
Limonium (statice)
Molucella (bells of Ireland)
Narcissus (daffodil)
Phlox
Paeonia (peony)
Ranunculus
Rosa (rose)
Rudbeckia (black-eyed Susan)
Scabiosa (pincushion flower)
Sedum
Tagetes (marigold)
Tulipa (tulip)
Zinnia

Forcing Flowering Branches

Branches from deciduous shrubs and trees may be forced to bloom beginning in January, after they have experienced cold winter temperatures. Attempts earlier than January 1 usually fail because dormant buds have not had enough cold temperature. After this cold requirement has been met, budded branches may be cut from outdoor plants and forced to flower indoors (see figure 10-23).

After cutting woody branches, take indoors. Recut again under water and place stem ends in warm preservative solution. The length of time from cutting to flowering depends upon the variety of the flowering branch being forced, the temperature, and when the branches are cut. Warmer temperatures of air and water will hasten flowering. Colder temperatures delay blossoms. The later in the season the branches are cut (that is to say, the closer to natural flowering patterns), the less time is required to force branches into bloom. Generally, forcing will take place in several days to two to eight weeks. It is essential during the forcing period to replenish and change preservative water as well as recut the stems underwater every two or three days. Table 10–5 lists proven favorites for bringing forth an early spring.

Figure 10-23. Branches from deciduous shrubs and trees can be cut from trees (after a period of cold winter temperatures), placed in warm preservative water, and forced to blossom early indoors.

TABLE 10-5: DECIDUOUS FLOWERING BRANCHES THAT ARE EASILY FORCED

Cercis canadensis (eastern redbud)
Chaenomeles speciosa (flowering quince)
Cornus florida (flowering dogwood)
Crataegus phaenopyrum (Washington hawthorn)
Daphne
Forsythia
Magnolia
Malus (apple and crabapple—edible and ornamental types)
Philadelphus (mock orange)
Prunus (almond, apricot, cherry, nectarine, peach, and plum—edible and ornamental types)
Pyrus (pear—edible and ornamental types)
Salix discolor (pussy willow)
Spiraea
Wisteria

CARE AND HANDLING OF FLORAL ARRANGEMENTS

Flowers in arrangements will last longer, as a whole, if they are given continued care in the home. ***Care tags*** should be attached to bouquets to educate the final consumer on how to care for arrangements to increase longevity and enjoyment. The water level in the container should be checked once or twice a day and replenished with warm preservative solution. Periodic misting of floral arrangements is beneficial. Misting helps maintain turgidity of flowers and foliage. For best results, spray gently with a spray bottle. If one or two of the flowers become wilted or limp within the arrangement, remove them from the bouquet, recut their stems underwater and immerse them in warm preservative solution. If these flowers revive, replace them in the bouquet. If they don't, discard them.

Keep floral arrangements away from ethylene sources including cigarette smoke, exhaust, and ripe fruits and vegetables. Also, remove flower heads as they die to keep a floral arrangement looking attractive for a longer period of time.

Figure 10-24. To extend the life of floral arrangements, it is essential to check water level daily and replenish with warm preservative water. Misting helps restore vigor to many flowers, especially tropical flowers.

Floral arrangements will last much longer if kept out of direct sunlight and off of warm surfaces, such as the top of a television set. Warm temperatures cause flowers to wilt quickly. Keep floral arrangements out of cool or warm drafts, which speed transpiration and cause flowers to wilt quickly. With a few simple care techniques, floral arrangements may be enjoyed for an extended length of time.

REVIEW

Much can be done to extend the longevity of cut flowers and foliage. The methods of treatment after harvest are referred to as care and handling. For flowers to have lasting qualities, proper care and handling techniques are necessary at each level of distribution (often referred to as the chain of life). The grower, transporter, wholesaler, retailer, and final consumer each play an active part in determining the longevity and quality of cut flowers.

An understanding of the causes of premature aging in flowers gives insight and reason to the many procedures that help delay senescence. The stage of flowers at harvest, processing, water quality, chemical solutions, refrigeration, ethylene, and sanitation all are important considerations in postharvest care of commercial and garden-grown flowers.

TERMS TO INCREASE YOUR UNDERSTANDING

abscission
acidic
acidifier
air embolism
alkaline
alkalinity
aluminum sulfate
bent neck
biocide
botrytis
bud opening
bud stage
buds
buffering capacity
carbohydrates
care and handling
care tags
chain of life
chilling injury
citric acid
conditioners
conditioning
cultivars
dry-packed
dry-prone
emboli
final consumer
floral preservatives
fully-open stage
freezing
genera
germicide
grower
growth regulators
half-open stage
hardening
hardness
head
herbaceous stem
highly buffered
holding
humidity
hydrating
latex
lighting
long-term solution
pH level
plugging
preservative solution
preserve
pretreatments
processing
pulsing
refrigeration
relative humidity
retailer
sanitized
senescences
short-term solutions
silver conditioning
sleepiness
solitary
special treatments
species
specific-action chemicals
stem blockage
sugar
supplements
silver thiosulphate (STS)
temperature
total dissolved solids (TDS)
transporter
underwater cutter
vase life
ventilation
water analysis
water quality
water salinity
wetting agents
wholesaler
wilt sensitive
woody stems

TEST YOUR KNOWLEDGE

1. Define the chain of life.
2. Describe basic care and handling procedures for cut flowers.
3. What are the primary ingredients in commerical floral preservatives? What is the function of each in increasing longevity?
4. When during the day should flowers be cut from the garden for use in fresh arrangements? Why?

RELATED ACTIVITIES

1. Experiment with the longevity and vase life of various cut flowers. Processflowers differently and place stems in various water types, with and without commercial floral preservative. Record your results day by day.
2. Visit a wholesale or retail operation. Observe and learn how flowers are processed.

Chapter 11

Flower and Foliage Forms

Flowers and foliage have different shapes or forms, which influence the style of a floral arrangement. These shapes are classified accordingly for design and arranging purposes. Using a variety of forms within one design can provide greater visual interest.

Flowers are classified in four design shapes: line, form, mass, and filler. Each group is named after its visual characteristics, as shown in figure 11-2. Linear in shape, ***line flowers*** add eyeflow and establish the framework of a design. ***Form flowers*** have unique forms which easily create

Figure 11-1. The various shapes, or forms, of flowers can influence to a great degree the shape, style, and impact of a floral arrangement.
Randi Seiler, De Cicco Floral Corporation, Miami, FL.

Figure 11-2. It is important to recognize flowers according to their basic shapes. These flowers represent the four basic forms: line (liatris), form (iris), mass (tulip), and filler (statice). Various forms influence the style, texture, and shape of floral designs.

emphasis. ***Mass flowers*** are rounded, adding mass and weight to a composition. ***Filler flowers,*** smaller in scale than other arrangement flowers, act to fill in and complete designs.

Some flowers are not limited to one group and may instead be classified several ways depending upon any of the following: the degree of openness of a blossom, the color intensity or value of a flower, interesting characteristics of texture or pattern, adjacent flowers, and the size or style of a completed floral arrangement (see figure 11-3).

A floral composition need not contain all four shapes. One group can be used alone successfully as well as in combination with the others. In order for you to become a skilled designer, it is important for you to recognize the shapes of flowers, learn about the variety of materials available, and know how flowers function according to their shape.

LINE MATERIAL

Line Flowers

Line flowers are named simply for their linear shape. Since their shape is generally tall and long with several blossoms, they can be used effectively to create height, width, and depth. Line flowers will easily set the framework, shape, and size of an arrangement. As shown in figure 11-5, line flowers are generally placed first when used in combination with other groups because they function in setting the skeleton of a design.

Many of the individual line flowers have a variety of blossom sizes on each stem. Often the larger, more open blossoms are at the base of a stem, gradually decreasing in size towards the top, which ends in several small buds. An example is the gladiolus, shown in figure 11-6, where ***gradation*** of size helps to create a focal point.

Figure 11-3. Flowers are not limited to one shape group, but may shift according to their function within a floral composition, as shown by the lilies in both of these designs. From top to bottom: *Lilies serving as mass flowers; a single lily serving as a form flower, establishing the focal point; and several lilies parallel to one another, function in emphasizing line in a design.*

Figure 11-4. Delphinium, pussywillow, and bells of Ireland are examples of line flowers.

Figure 11-5. Line flowers function in setting the framework of an entire design and are generally placed first.

Whether these flowers are straight or curved, they provide a line that moves the eye from the focal point out to the perimeters of an arrangement then back to the focal point. This eye movement helps to create a pleasing design.

Examples of line flowers are gladiolus, stock, delphinium, larkspur, snapdragon, liatris, tall heather, cattail, and spring flowering branches, such as forsythia, quince, leptospermum, and pussywillow.

Line Foliage

Foliage having a linear shape are often used with line flowers to repeat the framework already created. A linear repetition of flowers and foliage sets a strong, unifying pattern within a design. Adding curved line foliage (such as bear grass) to a composition creates motion, which helps the eye move speedily throughout the arrangement (see figure 11-8).

Some examples of line foliage include scotch broom, spiral eucalyptus, gorse, gladiolus and iris leaves, ferns, myrtle, and flax.

Containers and Accessories

Take care to choose a container that will not only mechanically hold the weight of the material, but will also visually enhance line flowers and foliage. As shown

Figure 11-6. Many line flowers (such as gladiolus) have an inflorescence pattern with the larger, more open blooms at the base of the stem. These blooms gradually decrease in size towards the tip, helping to create a traditional focal area at the base.

Figure 11-8. Curved foliage, such as bear grass, speeds eye movement.

in figure 11-9, tall containers repeat the shape of vertically placed line flowers and foliage.

Long and low containers repeat the shape of line flowers and foliage that are placed horizontally. This emphasis on line from various elements produces harmonious design.

Examples of accessories providing line include candles, grasses, tree branches, wheat, feathers, and ribbon streamers. These can be used to repeat and strengthen the visual impact of line.

Figure 11-7. Spiral eucalyptus, Scotch broom, and sansevieria are examples of line foliage.

FORM MATERIAL

Form Flowers

These flowers are so named because they have distinctive shapes that are interesting and captivating. Form flowers are often used at or near the rim of the container to provide a traditional focal point. Some form flowers, however, are more successfully placed in the perimeter or beyond the framework of a composition, so their silhouettes may be fully appreciated (see figure 11-11).

Unique form flowers, often called ***exotics,*** demand visual attention. Leave plenty of space around form flowers in order that their bold forms may be seen. Since form flowers speak for themselves, other flower types are often not necessary within the same composition.

Form flowers are generally expensive, but because of their visual importance in a design, fewer are needed. Often, a single form flower will have greater impact than a dozen.

You can create a pattern and a line for the eye to follow throughout an arrangement through ***directional facing,*** or the direction that each flower faces as shown in figure 11–12.

Figure 11-9. Tall containers repeat the shape of vertically placed line flowers and foliage.

Examples of form flowers include anthurium, bird of paradise, Easter lily, calla lily, iris, protea, orchids, heliconia, and ginger.

Form Foliage

Form foliage also have distinctive and interesting shapes. Oftentimes they will also possess other characteristics that make them stand out and attract the eye, such as an interesting texture, color, or pattern.

Figure 11-10. Bird of paradise, lilies, and anthurium are examples of form flowers.

Figure 11-11. Form flowers are often placed near the rim of the container to provide a traditional focal point. They may also be placed in the perimeter of a design to emphasize each flower's individual silhouette.

Figure 11-12. Directional facing creates a path for the eye to follow.

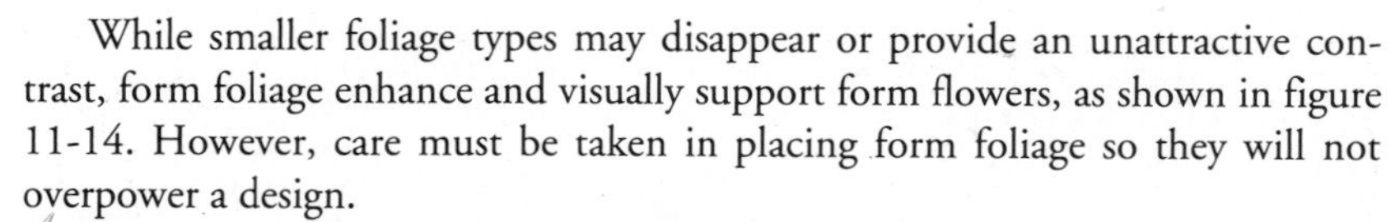

While smaller foliage types may disappear or provide an unattractive contrast, form foliage enhance and visually support form flowers, as shown in figure 11-14. However, care must be taken in placing form foliage so they will not overpower a design.

Examples of form foliage include caladium, croton leaves, ti leaves, monstera, calathea, papyrus, and cyperus.

Containers and Accessories

Choose containers carefully. Simple plastic, utility-type containers will generally detract from the flowers. It is better to avoid their use with expensive form flowers. Often, containers used with form flowers are uniquely shaped and the repetition of interesting forms helps to unify the entire composition. Those with distinctive shapes and styles are effective with form flowers.

Accessories often used with form flowers include sticks, pods, and other plant materials. Choose accessories so they will harmonize, enhance, and unify the entire design.

Figure 11-13. Form foliage varieties often have other unique pattern and texture characteristics. The monstera, calathea, and cyperus leaves harmonize with tropical and exotic form flowers.

Figure 11-14. Tropical Emphasis. Form foliage (dieffenbachia, philodendron, and ti leaves) enhance and visually support form flowers (heliconia, anthurium, red ginger, and canna).

Figure 11-15. *Carnations, roses, and chrysanthemums with rounded forms are common mass flowers.*

MASS MATERIAL

Mass Flowers

Mass flowers are so named because their purpose is to add mass to an arrangement. They are solitary flowers, consisting of a single, rounded flower head at the top of a stem. Because of this shape, they add bulk and weight to an arrangement quickly. When mass flowers are used alone in a floral design, it is important to vary flower color, size, spacing, or depth to avoid monotony (see figure 11-16). Roses, carnations, chrysanthemums, and tulips are only a few of the mass flowers available.

Mass Foliage

Mass foliage, like mass flowers, add weight and bulk to an arrangement quickly. As shown in figure 11-17, they are efficient in covering up the floral foam and other mechanics of a design. Often it is necessary to use more than one type of mass foliage to avoid monotony. Examples include leatherleaf, salal, pittosporum, huckleberry, and camellia foliage.

Figure 11-16. When mass flowers are used alone in floral designs, monotony can be relieved by varying flower color, size, spacing, or depth.

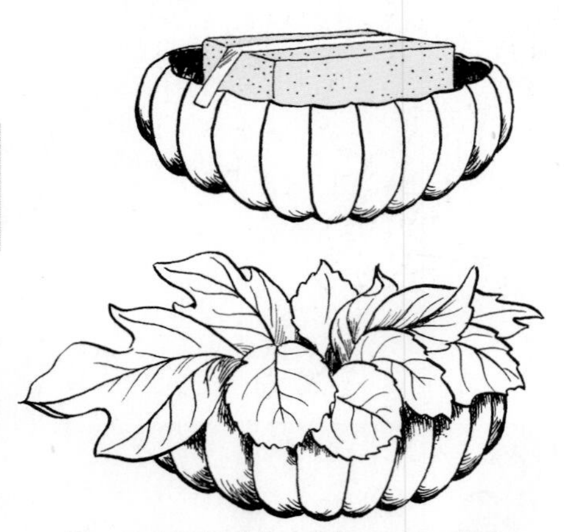

Figure 11-17. Mass foliage are efficient in covering up floral foam and other mechanics of construction.

Figure 11-18. Accessories, like this plush bear, are often used with mass flowers to help break monotony and add interest.

Figure 11-19. Examples of filler flowers shown here are bouvardia, gypsophila, and button chrysanthemums.

Containers and Accessories

A variety of containers can be used with mass flowers and should be chosen with the shape, style, and purpose of the completed arrangement in mind. Often, accessories help break up the monotony that may occur by using only mass flowers. Novelty items such as toy animals, candy, and seasonal or holiday decorations are commonly used with mass flowers and foliage (see figure 11-18).

FILLER MATERIAL

Filler Flowers

Filler flowers generally have a complex branched system of stems and flowers. They should be smaller in size, and much smaller in scale, to the other flowers next to them in the arrangement. They are used to fill in empty spaces, add accent, and complete a design. Filler flowers are not always needed; but when they are incorporated into an arrangement, they should be placed within the framework of a design below or behind other flower types. These flowers should visually support all other flowers present in a composition. It is important not to crowd or detract from the main flowers in a design by using too much filler, which also unnecessarily increases the cost of a bouquet (see figure 11-20).

Figure 11-20. Overdone. Excess filler flowers compete with main flowers, detract from the design, and increase cost unnecessarily.

Figure 11-21. Filler foliage add accent and interest to designs without introducing another color. The mahonia and string smilax give distinction and increased visual appeal to this pink monochromatic centerpiece.

Filler flowers are generally the last flower type placed in a bouquet. Oftimes, using a filler flower will change the entire design because of the new color or texture added. Baby's breath, for example, generally adds an airy, delicate appearance, whereas bright yellow button mums may add a bold and lively look. Other examples of filler flowers include statice, bouvardia, short heather, Queen Anne's lace, aster, and waxflower.

Filler Foliage

As shown in figure 11-21, ***filler foliage*** fill in an arrangement and complete a design without introducing a new color that might detract. These foliage are smaller in scale to other foliage in the bouquet. Foliage that are wispy, such as plumosa and tree fern, can soften a design, while the textures of huckleberry and ming fern can perk up a bouquet. Other examples of filler foliage include sprengeri, boxwood, and ivy.

Accessories

Material other than flowers and foliage used to fill in and complete a design can be effective in helping with the overall message or theme of an arrangement (see figure 11-22). These materials are generally smaller in scale and serve the same function as filler flowers and foliage. Examples of filler accessories that may be used throughout a bouquet

Figure 11-22. Berries and other small accessories throughout an arrangement help to fill in and complete a design.

include tiny pine cones, berry clusters, valentine hearts, birds, or ribbon loops.

REVIEW

The size and style of your designs will be greatly influenced by the various shapes of the flowers and foliage you use. The size of an opened bloom, the size of flowers in proportion to the whole composition, and interesting colors, textures, or patterns can all account for flowers being classified in several groups. So although classified in one shape group, some flowers and foliage may shift into another. For example, closed tulips are mass flowers; but after they open up, many reveal dynamic color patterns that demand attention and are thus then classified with form because they provide emphasis. In addition, the overall size and style of an arrangement will help to determine how flowers and foliage will be classified.

As shown in figure 11-23, when you are planning to use all four flower groups together, place line flowers

Figure 11-23. When all flower shapes are combined together, place line flowers first to establish the framework; second, place form flowers to achieve emphasis; third, arrange mass flowers to support the focal area; and last, complete the design with appropriate filler flowers.

Figure 11-24. A variety of forms within one bouquet creates greater visual interest. The various shapes influence the style and shape of a design.

first to set the pattern, framework, shape, and size of a design. Then arrange the form flowers to achieve emphasis. Place the form flowers near the container rim to provide a traditional focal point or set these flowers in the perimeter of your design to emphasize individual silhouettes. Next, arrange the mass flowers throughout the design to support the focal area. These flowers add bulk and weight to the arrangement. To add accent and complete your design, place filler flowers last, behind and below other flowers. Choose filler flowers that will harmonize with other flower types and unify the entire composition. This variety of forms within one composition will provide greater visual interest.

TERMS TO INCREASE YOUR UNDERSTANDING

accessories
directional facing
exotics
filler flowers
filler foliage
form flowers
form foliage
gradation
line flowers
line foliage
mass flowers
mass foliage

TEST YOUR KNOWLEDGE

1. After defining each shape classification, sketch the typical flower forms.
2. When combining all four shapes into one bouquet, what is the proper order of arrangement and the placement of flowers?
3. Name several examples of flowers and foliage from each shape group.
4. What factors allow flowers and foliage to fall into several shape classifications?
5. What shapes of containers enhance line flowers and form flowers?

RELATED ACTIVITIES

1. Visit a retail flower shop. Notice which line, form, mass, filler flowers, and filler foliage are readily available.
2. While at a retail floral shop, observe the fresh arrangements on display. Notice how the different flower and foliage shapes are combined together in each design.
3. Prepare an information guide using pictures to illustrate the different shape classifications of flowers and foliage.

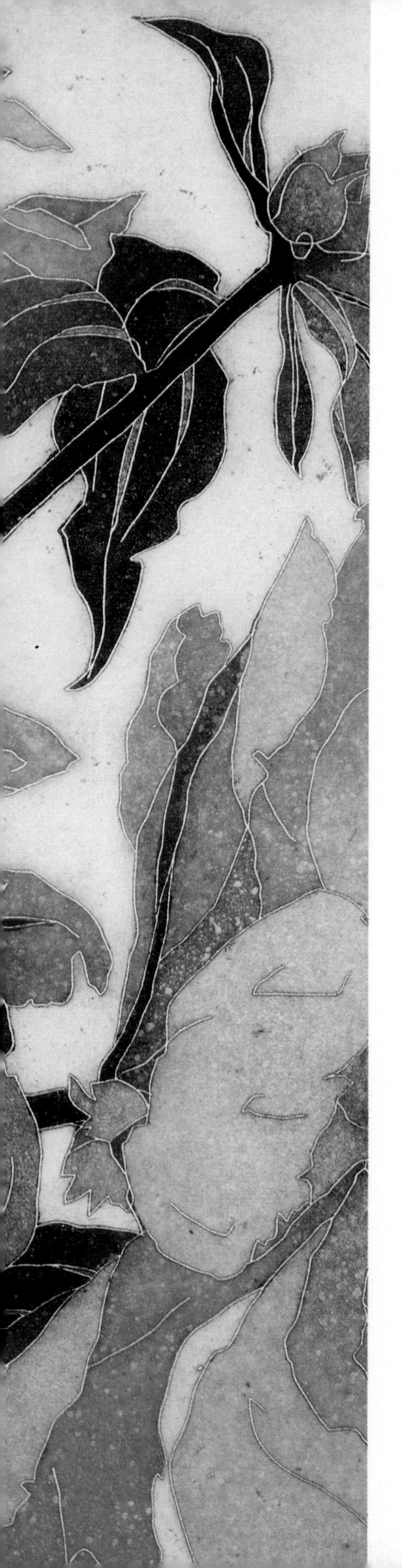

Section 3

BASIC TECHNIQUES AND STYLES

Chapter 12

Shapes of Floral Arrangements

Figure 12-1. Many of today's floral arrangements are a blending of two diverse styles. The oriental style is simple and elegant with emphasis on line, while the European style is generally round, full, and overflowing.

Floral arrangements of today are a blending of two diverse styles—the ***oriental style*** or ***line design*** and the ***European style*** or mass design. Often this line-mass blending is referred to as ***geometric design*** and is commonly known as ***Western style.*** Some arrangement shapes are dominantly mass and full, while others are more clearly linear with emphasis placed on line and negative space. Either way, floral arrangements, with the exception of some oriental and abstract designs, can be classified into certain geometric shape groups such as triangular, circular, or vertical.

The silhouette of an arrangement is its basic shape. It will be helpful for you to become more aware of the basic shapes of arrangements, which will allow you to expand your design options.

Generally, you will find it simpler to achieve a successful floral arrangement by starting with a definite design shape in mind rather than having no plan. This will enable you to choose a proper container, as well as the type and quantity of flowers and foliages appropriate for a particular shape and style. Imagining the silhouette of your design before you begin construction will also expedite the process of arrangement, allowing you to become more efficient with supplies as well as time.

Figure 12-2. A successful designer begins with a definite design shape in mind.

FACTORS INFLUENCING ARRANGEMENT SHAPE

Many factors influence the choice of a design shape. First and foremost is the placement or location for the arrangement. A specific location and viewing angle will help dictate what shape and size your arrangement will take and whether the arrangement needs to be ***all-sided*** (designed to be seen from all sides) or one-sided (designed with a front, to be viewed from one side). If an arrangement is designed for a coffee table, for instance, it is often long, low, and all-sided. A floral design intended for a fireplace mantle can take on a number of different shapes, all of which are generally one-sided (see figure 12-3).

The table size and shape on which the floral design will be sitting is also an important factor in determining an arrangement's form. A round table is enhanced with a rounded arrangement. Likewise a rectangular dining room table is an ideal setting for an all-sided, horizontal design. A small bedside table against a wall dictates a smaller arrangement, and often one that is one-sided. However, a bouquet made for the center of a spacious hotel lobby needs to be all-sided and larger in scale, often circular in form (see figure 12-4).

Figure 12-3. A specific location and viewing angle will help determine whether a floral arrangement needs to be all-sided or one-sided.

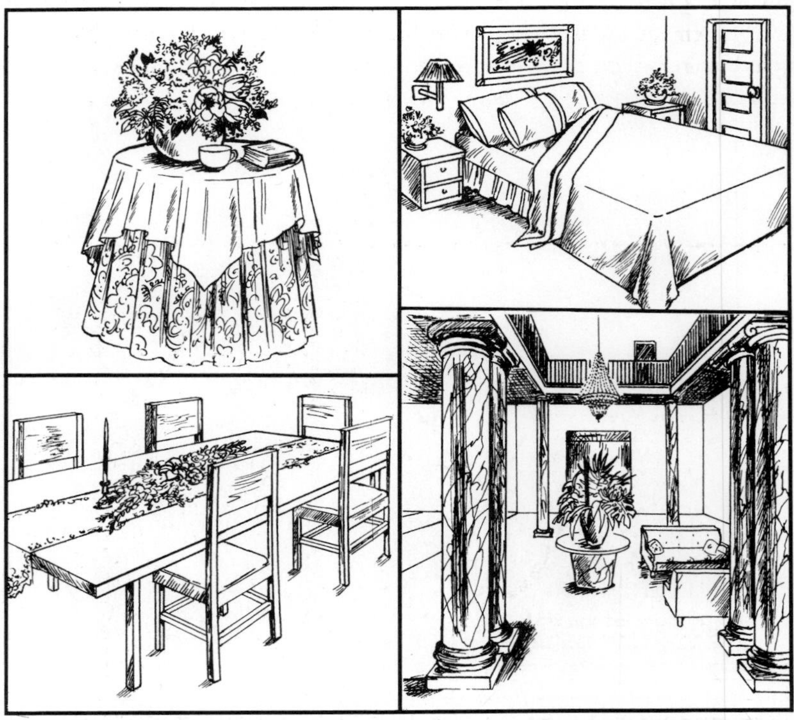

Figure 12-4. The form of a floral arrangement is determined in part by the table size and shape on which it will be placed.

The level at which the floral arrangement is viewed can also be a factor in determining the shape of an arrangement. For example, a bouquet displayed in a tall container that rests on the floor will certainly have a different shape compared with an arrangement that sits on a ledge at the top of an arched doorway, as shown in figure 12-5.

The kinds of flowers and foliages you select will also influence your choice of design. Some flowers lend themselves to tall arrangements, while others do not (see chapter 10). Your choice of container can also affect the overall design form. A tall vertical container implies a tall vertical design, while a round, basket generally suggests a different form, such as a circular floral design.

The occasion and purpose for a bouquet are also factors that often influence the shape a design must take. The formality of an occasion can also be a governing factor for choosing a shape. ***Symmetry*** is often associated with formal settings, as found in churches, funeral homes, auditoriums, and hotel lobbies (see figure 12-6). ***Asymmetry,*** on the other hand, can lend a more casual feeling and is often associated with less formal occasions and settings. Remember that the overall shape of a floral arrangement must be best suited for the occasion.

Figure 12-6. A formal setting frequently calls for arrangements that are symmetrical in form.

BASIC SHAPES OF ARRANGEMENTS

The basic shapes of arrangements illustrated in figure 12-7 are the most common geometric shapes used in florists' designs. After considering the requirements of a bouquet, choose one of these forms that will best suit the needs for your arrangement and use it as a guideline for setting up the framework for your design. It is important that you visualize a completed floral design before actually arranging, so that you will know how to begin construction.

Figure 12-5. The level at which the arrangement will be placed also influences the shape of an arrangement.

Most floral arrangements can be named ***triangular, circular, vertical,*** or ***horizontal*** based on their geometric form. However, some arrangement shapes, rather than fitting into one shape category, are really a blending of several combinations (see figure 12-8).

Once the basic outline of the arrangement is determined, keep your chosen shape in mind, especially as you set up the ***framework*** or basic structure of the design. The framework for a design can be constructed with foliage or flowers, as shown in figure 12-9. Generally, the first few flowers or stems of foliage that establish the framework are referred to as ***skeleton flowers*** or foliage, because they create the main and essential outline for forming a specified design shape. These first few stems set the geometric limits and establish outer boundaries for a design, including height, width, and depth.

To increase the feeling of depth in one-sided designs, it is important to add a skeleton flower or foliage stem

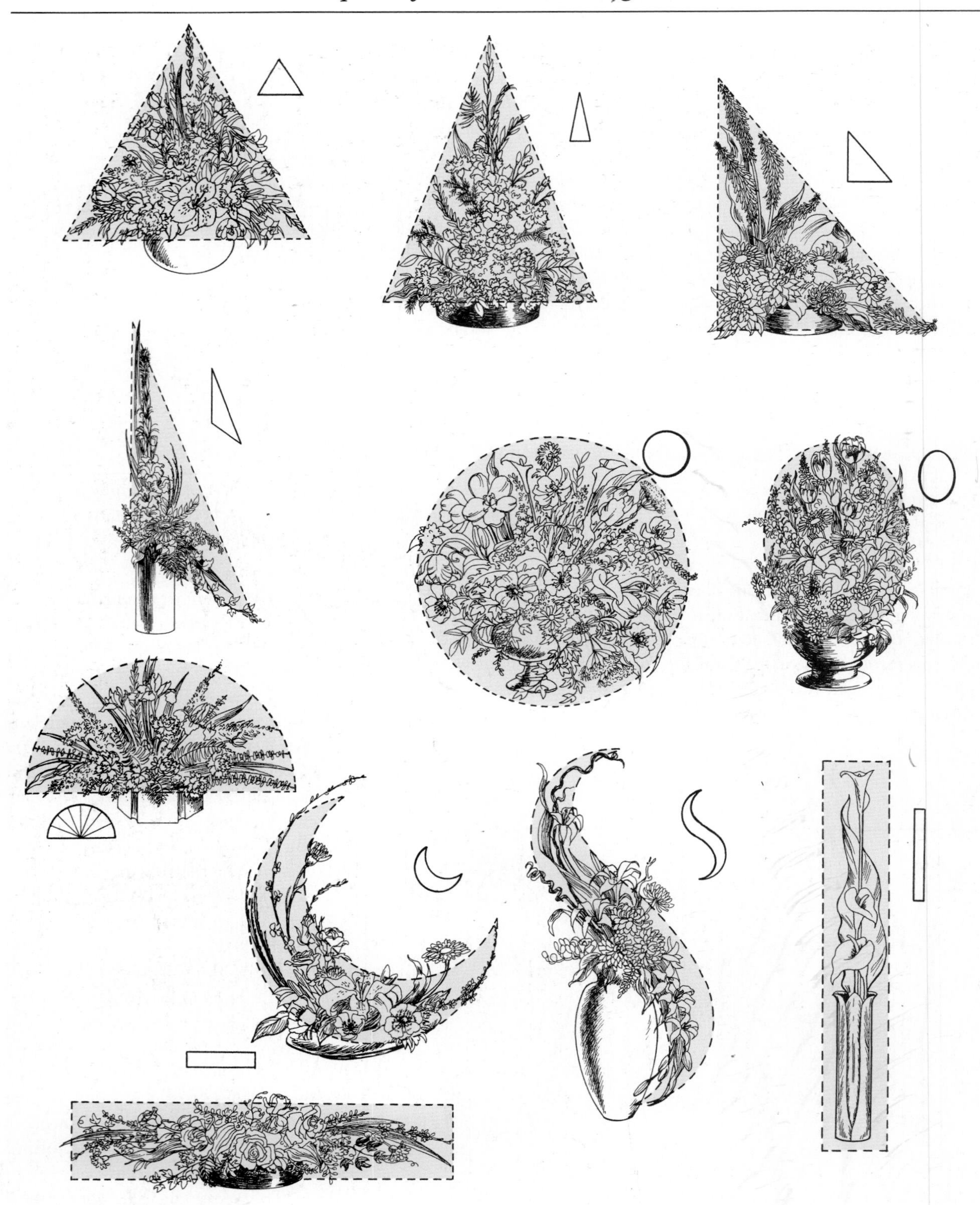

Figure 12-7. Floral arrangements often follow geometric shapes. The basic shapes of floral arrangements include: a. equilateral triangle, b. isosceles triangle, c. right triangle, d. scalene triangle, e. circular, f. oval, g., fan, h. crescent, i. hogarth, j. vertical, k. horizontal.

Figure 12-8. Some arrangements are a blending of several shapes, rather than falling into a single shape category. The shape of this arrangement of gerbera, irises, delphinium, lisianthus, Queen Anne's lace, wax flower, heather, and statice is based on a circle as well as a triangle.

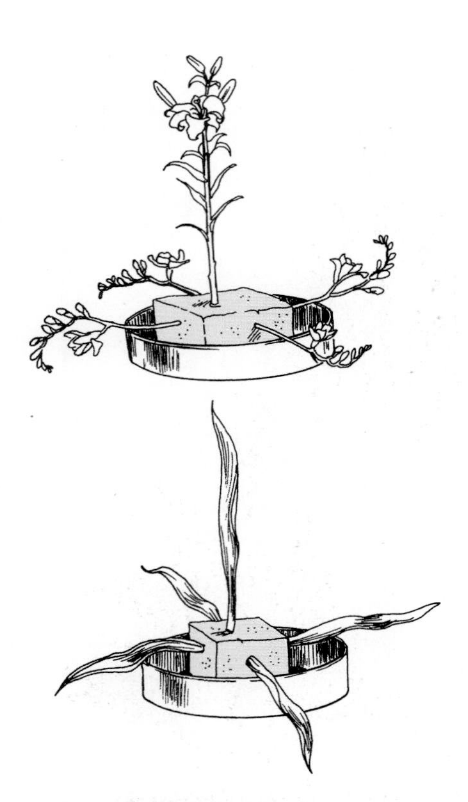

Figure 12-9. The framework, or skeleton, may be constructed with flowers, foliages, or a combination of both.

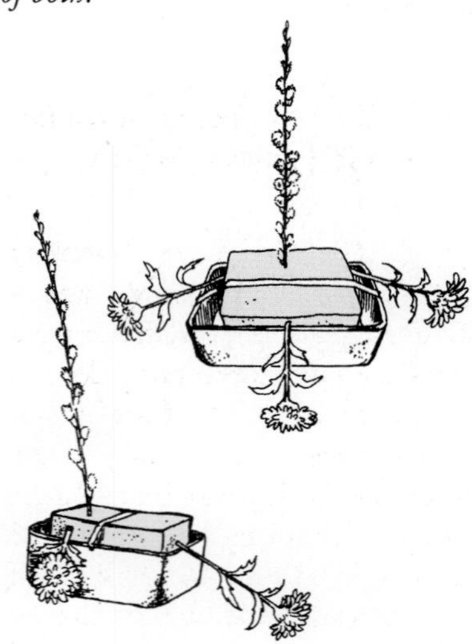

Figure 12-10. Angling the tallest stem slightly backward and placing a skeleton flower or foliage stem in front of the arrangement that angles downward will add depth to one-sided designs.

in the front of the arrangement near the rim of the container that angles downward. In order to visually balance this downward positioning, angle the tallest stem slightly backward, as shown in figure 12-10.

Flowers and foliages may then extend slightly beyond the framework edges or they may stay within. Establishing a simple framework will give a foundation for construction and ease the process of arrangement.

TRIANGULAR DESIGNS

The triangle shape is a popular choice for floral arrangements. This form has three distinct sides and three corners, angles, or tips. Triangular floral arrangements are generally referred to as ***one-sided designs,*** normally viewed on just one side. However, some of these designs can be arranged to be viewed from all sides and still maintain their triangular shape. Many variations of height, width, and size exist, all of which are either symmetrical ***(formal)*** or asymmetrical ***(informal).***

Symmetrical Designs

The symmetrical triangle design is formal, and when divided in half vertically, each side of the shape is a mirror image to the other. Remember that when creating symmetrical triangle designs, it is important to keep the framework a mirror image. The flowers within the triangle, however, do not need to be repeated exactly on both sides. This would make an already formal and rigid bouquet much too stiff and unnatural. The basic symmetrical designs offering a traditional formal look include the ***equilateral triangle, isosceles triangle,*** and ***cone design.***

Equilateral Triangle

A triangle that is called equilateral simply means the three sides forming the triangle are all equal in length. These designs are usually one-sided and placed against a wall or in an area where most people will see only one side, such as a buffet table or at the front of a church.

The equilateral design will generally have three skeleton flowers or foliage stems that create the three points of the triangle. It is important to remember that the three sides of the triangle are equal, not the three stems that form the corners of the triangle. The materials used in forming the base line of the triangle need a combined length similar to the length of the stem forming the top center point of the triangle. This will result in the three sides remaining equal to one another (see figure 12-11).

Isosceles Triangle

An isosceles triangle is also symmetrical, having two sides equal in length with the third or base side unequal (see figure 12-12). The most common isosceles floral arrangement has two equal sides that come together to form the height of the arrangement.

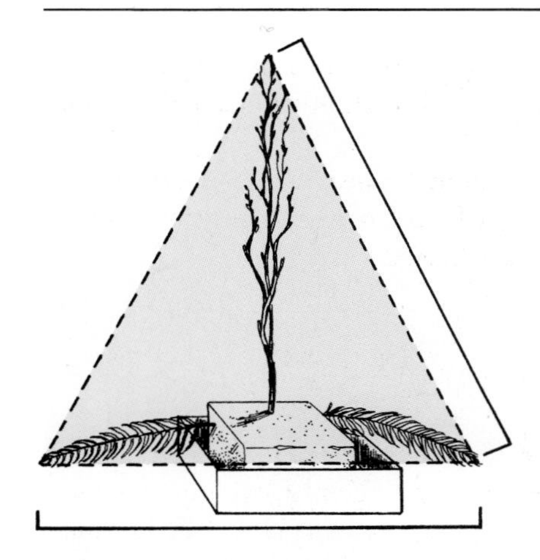

Figure 12-11. Remember, for equilateral triangle designs, the three sides of the triangle are equal (not the three stems forming the corners of the triangle).

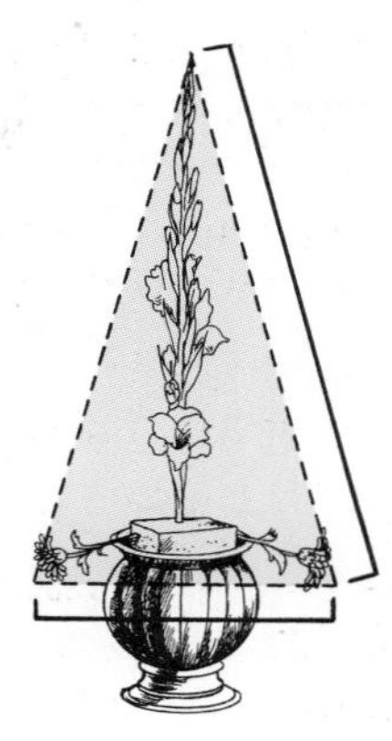

Figure 12-12. An isosceles triangle is symmetrical, with two of the three sides being equal in length. A floral arrangement based on an isosceles triangle is generally a vertical design with two equal sides forming the height.

Figure 12-14. After establishing the framework of symmetrical triangle designs, add mass flowers and strengthen the desired shape with more foliage and fillers.

These one-sided isosceles designs are a popular choice for flower arrangements because they can be easily created from just a few flowers. Whether small or large in scale, these designs generally appear highly stylized.

One-sided Symmetrical Triangle—Steps of Construction

The equilateral and isosceles triangles share similar steps of construction (see figure 12-13).

Figure 12-13. For symmetrical triangle designs, secure a piece of soaked floral foam that extends above the lip of the container. To establish the framework of height and width for an equilateral or isosceles triangle, insert three flowers or stems of foliage.

Step 1. Secure a piece of soaked floral foam that extends above the lip of the container.

Step 2. Then add three flowers or stems of foliage to establish the framework of height and width. Since these designs are most often one-sided, to make better use of the floral foam, insert the tallest stem centrally in the back portion of the foam block. Insert the flowers or foliage that create the side triangle points into the sides of the foam, toward the back portion of the foam, so their tips angle forward slightly.

Step 3. Next, add mass flowers to fill the body of the design, as illustrated in figure 12-14. Complete the

Figure 12-15. As the last step to complete your arrangement, add more flowers and foliages within the framework you have set. As shown in this design, all stems—leptospermum, heather, and scotch broom—seem to radiate out from a central location. The roses face out toward the perimeters, enhancing the feeling of rhythm.

triangle shape by adding filler flowers and foliages within the framework you have set (see figure 12-15).

Remember to sufficiently cover the floral foam not only in front but also in back. Even though these one-sided designs will be viewed mostly from the front, the designer may not know if these arrangements will be placed against a mirror or if some people will view the arrangement only from the back side, as with stage pieces (see figure 12-16). To achieve a more professional and aesthetic appearance, remember to cover foam, other mechanical aids, and large unsightly stems, with foliage.

Cone Designs

The cone-shaped design is typical of the Byzantine period style. It actually is a three-dimensional vertical isosceles triangle. These designs require some type of foam base for mass flower insertion. Cone arrangements are formal and rigid and generally appear the same on all sides.

Often pieces of floral foam are gathered together within chicken wire to form these tight designs or a block of floral foam can be trimmed to make a cone. Manufactured floral foam cones are also available.

Figure 12-16. Although one-sided designs are intended to be viewed from just one side, often, as with stage pieces, they will also be viewed from the back. Always conceal unsightly floral foam, other mechanics, and stems with foliage.

Asymmetrical Triangle Designs

The asymmetrical triangle design is informal; when it is divided vertically in half, one side is visually heavier than the other. In order to achieve true balance, something must compensate on the side with less visual weight, such as color, texture, form, or the length or angle of a stem.

Asymmetrical designs offer a unique arrangement form. They appear less contrived and more natural. However, more planning is necessary in order that these designs achieve proper balance. Asymmetrical (like symmetrical) triangle arrangements are generally one-sided but can be made into all-sided designs with initial framework planning.

These design shapes often rely on the use of a focal point for visual success (see chapter 5). Asymmetrical triangle designs have the height of the bouquet

Cone Designs—Steps of Construction

The most time consuming part of making a cone design is establishing the foam base foundation (see figure 12-17). The foam must be thoroughly soaked for fresh arrangements.

Step 1. Secure the foam and chicken wire in the container.

Step 2. Add a thin layer of moss with green pins. The moss will easily hide mechanics, and less foliages and flowers will be needed to fill in the cone stucture.

Step 3. Add fruits, vegetables, or pine cones if desired. Secure with green pins, wire, or wood picks, depending upon their individual shape and weight.

Step 4. Insert short-stemmed flowers and foliage pieces through the moss and into the foam to form a colorful cone design as shown in figure 12-18.

Figure 12-17. A foam cone may be used, a block of foam may be sculpted in the shape of a cone, or several pieces of foam may be gathered together with chicken wire. Chicken wire may also be wrapped around a foam cone, giving the entire arrangement increased support. A thin layer of moss can then be attached with green pins. Start with the biggest elements, such as fruits and vegetables, which can be added to the design by inserting the sharp end of a wooden pick into the fruit and then the other end into the cone. Heavy wires can also be used for added security. Short-stemmed flowers and foliages are then inserted into the wet foam.

Figure 12-18. Once the foundation of a conical design, or an all-sided isosceles triangle, has been established, the design possibilities are endless.

shifted to one side, and the width of the triangle extending out on the opposite side. The basic asymmetrical designs are ***right triangle*** and ***scalene. triangle***

Right Triangle

A right triangle arrangement is named for the prominant ***right angle*** that forms the design. These triangles are generally set in a vertical format with a tall vertical line perpendicular to the base line of the arrangement. The vertical emphasis or height is usually seen on the left side (explaining why these designs are often called ***L-shape),*** but it can just as easily be established on the right side of a design.

Right-triangle designs can point to the right or left, according to the location selected for displaying them.

Figure 12-19. A right-triangle pair will emphasize, frame, and give importance to a painting or other object.

Figure 12-20. Right-triangle pairs can point toward the center or away; either will emphasize the center object.

These designs can also be made in pairs and placed on a mantle or table to frame and give importance to a picture or object in between the two arrangements as shown in figure 12-19.

The right-triangle pair can also be used in an opposite manner (see figure 12-20). This placement will still give importance to a central object, but at the same time provide direction and length with the base horizontal line extending outward.

Scalene Triangle

A scalene triangle is one that has unequal sides and angles. A floral design that is made in a scalene form generally has vertical emphasis. This visually impressive design style has an apparent ***obtuse angle,*** with height on one side. To create the open angle, the base line extends downward on the other side opposite to the height. Because flower and foliage stems angle downward, it is often necessary to use a compote or taller container; if a low container is used, the arrangement should be placed on a mantle or other location where the flowers are allowed to fall freely preventing blossoms from getting damaged (see figure 12-21).

Figure 12-21. Scalene triangle arrangements are generally made in vertical containers to allow for downward stems. However, they may be made in shallow containers, as long as they are located where the downward stems may fall freely.

One-sided Asymmetrical Triangle—Steps of Construction

(Refer to figures 12-22 and 12-23.)

Both the right-triangle and scalene-triangle designs have similar steps of construction.

Step 1. Extend the floral foam above the container rim to allow for horizontal or downward positioning of stems.

Step 2. Once the floral foam is secure in the container, the framework may be established with flowers or foliages.

Step 3. Insert the tallest flower or foliage stem to one side of the block and towards the back of the foam. This allows more foam space for the flowers and foliages that will later be placed within the framework

Step 4. The base of the triangle is formed by extending a flower or foliage stem into the opposite side of the design. Insert this stem into the side of the foam block, generally towards the back of the container.

Step 5. To soften the tall vertical line and to help balance the horizontal tip of the triangle, insert a short flower or piece of foliage next to the vertical height of the arrangement.

Step 6. Establish a ***focal point*** near the rim of the container where the tall vertical stem and the extended horizontal stem meet together. This helps give further balance and stability to the asymmetrical triangular form.

Step 7. Fill in the triangle form with mass flowers and foliages, this gives emphasis to the focal area and conceals foam and other mechanics.

Figure 12-22. Right triangle—Steps of construction from left to right: a. Secure in the container a piece of floral foam that extends above the container rim. Insert stems to establish the framework for the right triangle, inserting the tallest stem in the back corner of the foam, and opposite to it, insert horizontally a short stem. b. Next, add foliages near the container rim to conceal portions of the foam and blend the container with the arrangement. c. Establish a focal area near the rim of the container where the vertical and horizontal lines of the arrangement meet. Add other flowers within the triangle shape. d. Add fillers and foliages to strengthen the shape, making sure any mechanics of construction are hidden.

Figure 12-23. Scalene triangle—Steps of construction from left to right: a. Generally, a vertical container is required for scalene triangle designs. Secure a piece of foam that extends above the container rim to allow for the downward angling of stems. Establish the framework in the desired scalene triangle shape; the stem establishing the height of the design should be inserted towards the back on one side or the other. The front stem opposite to the height of the design is inserted horizontally into the foam and allowed to angle downward to the side. b. Add foliages near the rim of the container. Other foliages may be added to strengthen the framework. c. Establish a focal point and add mass flowers within the framework. d. Continue adding fillers and foliages to strengthen the shape, making sure any mechanics of construction are hidden.

CIRCULAR DESIGN STYLES

Even though the following arrangements are simply grouped as circular, many diverse styles exist, including symmetrical, asymmetrical, all-sided, and one-sided. Although some circular design styles are more popular than others, it is helpful to know about each and how each may be constructed.

Mass Styles

Designs emphasizing mass with a circular or rounded silhouette are symmetrical and can be made in a variety of sizes and arrangement styles. Most mass designs can be categorized as round, oval, ***fan,*** or a ***topiary ball.***

Round Arrangements

The round arrangement shape appears the same on all sides and from all viewing angles. This design form is also called circular, ***round mound,*** nosegay, and tussie-mussie. Because there is no front or back side to this arrangement, these designs offer more versatility and can be placed in a number of locations (see figure 12-24).

Figure 12-24. Circular arrangements may be placed in a variety of settings.

Arranging flowers and foliages in a circular manner adds an interesting element of repetition that is pleasing and harmonious. However, in order to avoid the feeling of monotony, use foliage that offers a pleasant contrast to the dominant round flower forms.

If a round or sphere-shaped arrangement is made only one-sided, it usually lacks balance and appears unfinished, as shown in figure 12-25. When an arrangement must be one-sided, it might be better to select and plan an arrangement shape other than round before beginning construction.

Figure 12-25. One-sided circular designs lack balance. Even after foliage is added to the back to conceal foam and other mechanics, they still will look unbalanced and unfinished.

Oval Arrangements

Oval arrangements are symmetrical, yet unlike round arrangements, they are commonly made in both all-sided and one-sided styles. These designs offer an extention of the round form with the circular shape being generally elongated vertically. All of the flowers and foliages are placed as if to radiate out from a central location similar to round designs, the stem lengths being more varied to form the oval.

Fan-shaped Designs

A fan-shaped design is sometimes called a ***radiating arrangement.*** These designs are one-sided and similar in shape to a fan or half circle. Often line flowers and foliages are used to set up the framework of the design. Stems should appear to radiate from a central location in the design. These designs are symmetrically balanced with a formal, planned appearance. Voids or negative spaces between the linear framework material give importance to the radiating lines in this mass design.

One-sided Rounded Designs—Steps of Construction

(Refer to figures 12-26 and 12-27.)

Fan-shaped arrangements and one-sided oval designs, though different in shape, share similar steps of construction.

Step 1. Secure foam in an appropriate contain. Foam should extend above the lip of the container.

Step 2. Establish the framework. (Remember to tilt the tallest stems slightly away from the front of the design. Also, angle the front, lower stems downward for an increased appearance of depth.)

Step 3. Add line flowers and foliages to emphasize the radiating appearance of the fan-shaped form. Use a variety of flower and foliage forms to avoid monotony in these circular, mass designs.

Figure 12-26. Fan design—Steps of construction from top to bottom: a. Secure a piece of floral foam that extends above the container rim. Establish the foundation of heighth and width, as well as the intermediate "spokes" of the fan. Since this is generally a one-sided design, angle the top stems slightly backward for better balance. b. Add foliage near the container rim and between the framework. c. Add mass and line flowers in a radiating fashion. d. Continue adding fillers and foliages, keeping them within the established framework. Conceal any mechanics, especially on the back of this one-sided design.

Figure 12-27. Oval design—Steps of construction from left to right: a. Oval designs are commonly made both one-sided and all-sided. Extend the foam above the lip of the container. Establish the oval framework all around. b. Add foliages to conceal the mechanics as well as to give a fuller look to the design. c. Add mass flowers in a radiating fashion within the framework of the design. d. Continue adding fillers and foliages to the desired thickness and shape. Remember to conceal any mechanics.

Step 4. Continue adding flowers and foliages to form the final shape. Add fillers if desired to soften these concise shapes.

All-sided Circular Designs—Steps of Construction

Before beginning a circular arrangement, it is important to consider both size and mood needed. Choose a container that will offer the appropriate size and proportion relationship with the finished rounded design.

Step 1. Cut floral foam to fit container, leaving 1/2 to 1 inch above the rim. Secure foam in the container with an anchor pin or waterproof tape.

Step 2. Often with rounded designs, it is helpful to ***pre-green*** the arrangement with some foliage before flowers are inserted, as shown in figure 12-28. This will help define the shape all around and will cover large portions of the foam and other mechanics of construction. Use only enough foliage to establish the width on all sides, placing stems near the rim of the container.

Step 3. Add a stem of foliage to the center of the block to help establish height. Be sure to allow open spaces in the foam for flower insertion. If too much foliage is used to pre-green an arrangement, it is often difficult to find any space left in the foam for flowers (see figure 12-29).

Step 4. Establish the height of the arrangement with a small flower or bud.

Step 5. Next, insert flowers near the rim of the container that extend out horizontally or slightly downward to establish the width on all sides.

Step 6. Insert additional mass flowers to fill out the circular format on all sides. All flowers should radiate from the center of the design. It may be helpful to constantly turn your arrangement or place it on a rotating Lazy Susan in order that no one side is neglected. That way your design will end up symmetrically circular on all sides.

Step 7. Leave some voids or negative space for visual rest. Voids also allow room for insertion of further foliage to soften the design and to conceal any mechanics still visable.

Step 8. Add filler flowers last to help complete the design shape and style.

Topiary Ball Designs

The topiary ball design is a perfectly round sphere. The shape appears the same from any side or viewing angle. Round topiaries are symmetrical and offer a formal decorating idea in a variety of design styles, sizes, and applications.

Topiary Ball—Steps of Construction

The topiary itself is made by inserting flowers and foliage closely together into floral foam.

Step 1. Make the foundation from pieces of foam or sculpture foam and wrap in chicken wire (see figure 12-31). Foam spheres and foam holders are manufactured specially for topiary ball construction.

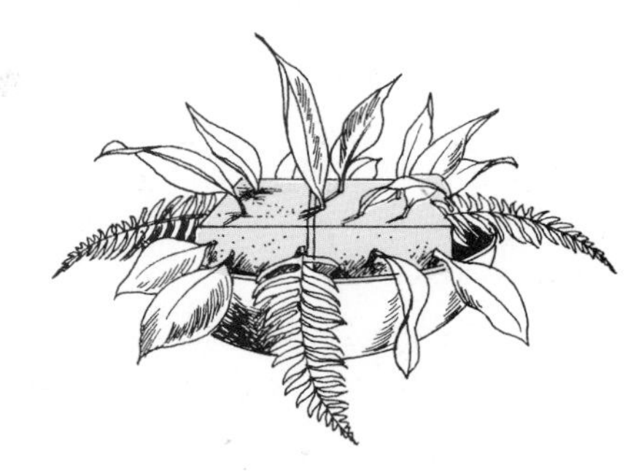

Figure 12-28. Round arrangements are commonly pre-greened around the container rim (often referred to as collaring) before flowers are inserted. This helps to define the shape and cover the floral foam.

Figure 12-29. When too much foliage is used to pre-green an arrangement, there is little space left for flowers. If pre-greening, or collaring, is done before flowers are inserted, the stems must be angled downward in the foam, rather than horizontally. Too many horizontal stems use up foam space and make it difficult to insert the flower stems.

Figure 12-30. Round arrangement—Steps of construction from top to bottom: a. Secure a piece of floral foam that extends above the container rim. Pre-green, or collar, the arrangement with foliage all around to establish the rounded form. Establish the framework of height and width on all sides. b. Add flowers and foliage in a radiating pattern, keeping the outside edges rounded on all sides. c. Continue adding fillers and foliages in the skeleton, leaving some negative spaces at the edges of the design.

Step 2. Attach the foam ball to a sturdy branch, piece of piping, dowel, or other narrow linear material that is anchored into a heavy pot for use as a centerpiece.

Step 3. After the foundation is secure, cover the ball with a thin layer of moss.

Step 4. Add flowers and foliage stems through the moss, into the foam (see figure 12-32).

Step 5. Remember to cover any mechanics, especially in the pot that holds the plaster of Paris or other anchoring material. Mechanics at the base can easily be covered with a thick layer of moss or with an accenting floral arrangement as shown in figure 12-33.

Topiary balls have other floral design applications. For instance, they can easily be hung as a room decoration. A small topiary, often called a ***floral pomander,*** may be carried. Before flowers are inserted, the ribbon or string that the topiary hangs from must be securely anchored into the ball (see figure 12-34).

Line Styles

Some circular designs are constructed so that the line of the design, rather than the mass, is emphasized. These arrangements are generally highly stylized one-sided designs but with initial framework consideration they may be made all-sided.

The curved lines of these designs create a unique and sophisticated appearance and display graceful rhythm, resulting in arrangements that are a pleasure to view. Circular line designs that are named for their curvilinear form are the ***crescent*** or ***c-shape*** and the ***Hogarth*** or ***s-shape.***

Crescent Design

The crescent floral design is a portion of a circle with one concave edge and one convex edge, similar to the visable shape of the moon just before the first quarter or after the last quarter phases. An arrangement requiring skill and experience, the crescent shape is also referred to as c-shape

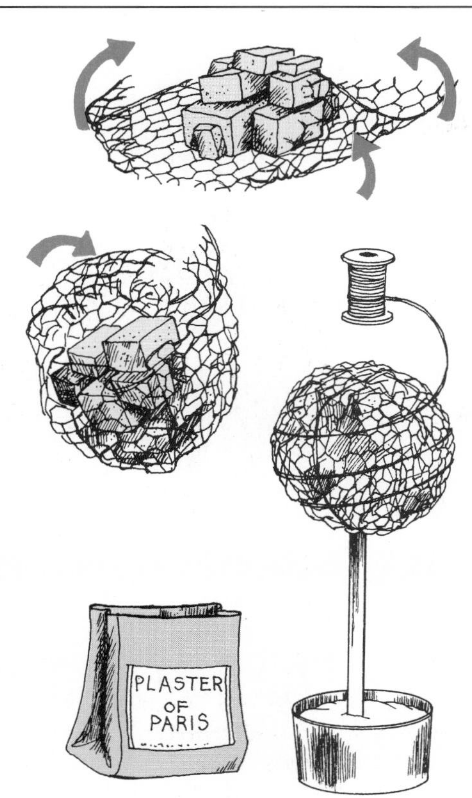

Figure 12-31. The foundation of a topiary ball may be made from pieces of wet foam wrapped in chicken wire. However, foam spheres and foam cage holders which are manufactured for the purpose of making topiary designs, will save design time. The topiary ball must be elevated by a dowel or branch that is secured in a container. Plaster of Paris is commonly used to anchor the design. Follow the label directions.

Figure 12-32. After foam is secured to a sturdy branch or dowel, the ball is covered with a thin layer of moss before inserting flowers and foliages.

Figure 12-33. Cover mechanics at the base of a topiary arrangement with a thick layer of moss or a floral design that harmonizes with the topiary ball.

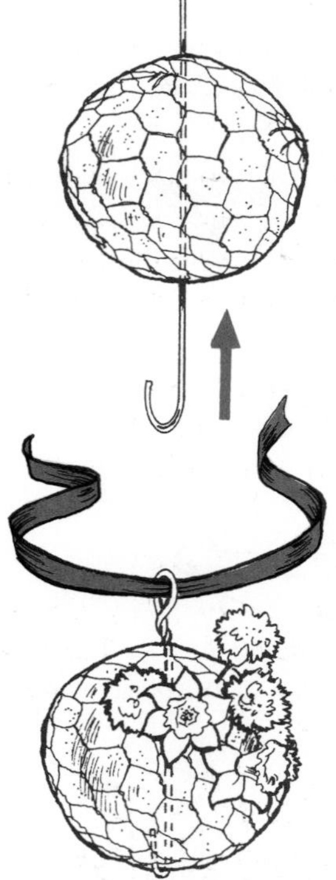

Figure 12-34. To make a floral pomander, anchor wire and ribbon or sting before inserting flowers. Insert a heavy wire through the center of the foam ball wrapped in chicken wire. Form a hook on one end and pull the hook back into the foam, catching the chicken wire so that the wire does not cut through the ball. Ribbon or string can be attached through the top of the wire piece as shown.

Figure 12-35. The crescent shape can be constructed in a variety of sizes, heights, widths, and styles.

and can be constructed in a variety of sizes, heights, and widths (see figure 12-35). Because of its asymmetry, the crescent requires a great deal of negative space with a focal point generally near the container rim for visual balance.

Like other asymmetrical designs, crescent arrangements may be used in pairs accenting and giving importance to a picture or object placed between them.

A crescent design is also beautiful when inverted and symmetrical (or asymmetrical) as illustrated in figure 12-36. In creating crescent designs, select an upright container and flowing stems to create the downward curves at the sides. All-sided, inverted crescents form a graceful, rhythmic centerpiece, although these designs may also be made as one-sided arrangements.

Crescent Designs—Steps of Construction

Choose a container that will be suitable for the type of c-shape arrangement you will be making. An inverted crescent design generally requires a compote or other tall container, allowing the side stems to hang downward gracefully, whereas an upright crescent most often is made in a low container.

Step 1. Secure a piece of floral foam into the container, allowing the foam to extend above the container rim to allow for side stem insertions.

Step 2. Select curving stems to form the c-shape. Some stems are naturally curved, while others are not. With hidden mechanics, however, many stems will form the desired tight or loose curve. As shown in figure 12-37, scotch broom will easily curve with some gentle persuasion. Hold the stems and blow hot air directly onto the foliage a number of times. The stems will begin to take on the desired curvilinear form. You may also use wires for curving flower stems, but remember to conceal any wiring mechanics you use.

Step 3. Insert first the line foliage to form the curves that form the "c" (see figures 12-38 and 12-39). Because of the length and curve of these skeleton stems, insert these stems deep into the foam.

Figure 12-36. An inverted crescent design is graceful and rhythmic.

Step 4. Next add some mass foliage that will blend the curvilinear foliage with the massed area of the design. This will help establish the inner framework as well as conceal portions of the foam.

Step 5. Insert mass and line flowers next. Taper the flower sizes, placing buds and smaller flowers at the outer perimeters of the design. The flower sizes should gradually get bigger or increase in visual weight towards the center of the design. As stems radiate out from a central area, so should the flower heads. This will enhance the feeling of rhythm.

Step 6. Establish a focal area where the two side portions meet. This may be done simply by massing flowers closer together. A single form flower will also create a focal point as well as give stability to the design. (See chapter 5 for more ideas on creating a successful area of emphasis.)

Step 7. Add mass flowers and foliages to form the crescent shape.

Step 8. Complete the arrangement by adding filler flowers, foliages, or accessories that will unify the design and complete the desired shape. The crescent form relies on negative space for visual success. Do not be tempted to add more material than is needed (see figure 12-40).

Figure 12-37. Many foliages and flowers will provide natural curves to create the crescent line. Others may be made to curve. Scotch broom, for example, can easily be curved by blowing hot air directly onto it while shaping it with the heat of your hands. Flowers and foliages may also be wired to form curves.

Figure 12-38. Inverted crescent design—Steps of construction. fom top to bottom : a. Select a raised container. Secure floral foam that extends above the container rim. Establish the curvilinear framework. These designs may be made one-sided or all-sided. b.Add foliage near the rim to conceal mechanics and blend the arrangement with the container. Add other stems of foliage to blend the massed area with the curving linear portions. c. Add flowers within the established framework. d. Establish a focal area and continue adding fillers within the framework, concealing mechanics. Do not fill in important negative spaces.

Figure 12-39. Upright crescent design—Steps of construction from left to right: a. Secure floral foam making sure that it extends above the container rim. Establish the framework with curving stems. For one-sided designs, place the stem forming the height of the bouquet towards the back of the foam. b. Next add foliages near the rim of the container to conceal mechanics and blend the massed area with the curvilinear portions of the design. c. Add mass flowers within the framework. Establish a focal point at the rim of the container, where the height of the bouquet meets with the width. d. Continue adding fillers and foliages within the established framework. Do not be tempted to fill in important negative spaces.

Figure 12-40. A crescent form relies on negative space for visual success. A design, such as this that is too full will lose its crescent shape and look more like a round design with curved lines added.

Figure 12-41. Self-portrait of William Hogarth dated 1745. The palette in the foreground shows the famous serpentine line, Hogarth described as The Line of Beauty.

Tate Gallery, London/Art Resource, New York.

Figure 12-42. These two plates from The Analysis of Beauty, *published in 1753 are attempts by Hogarth, both serious and comical, to explain and justify his theories and beliefs*

The Analysis of Beauty, Plate One and Plate Two, *by William Hogarth, published 1753. Used by permission of the British Library, London.*

Hogarth Curve Designs

Of all the arrangement forms, this design style is the only one that is named for a person rather than a geometric shape. These designs are named for the English artist William Hogarth (1697–1764). Their shape comes from Hogarth's self-portrait titled Portrait of the Painter and his Pug, dated 1745, in which the artist drew a serpentine line on a painter's pallet with the words under it, "The Line of Beauty." Many were puzzled with this phrase and curious about its meaning. In response to questioning, Hogarth theorized that all beauty was based on the serpentine s-line, documenting his beliefs later in *The Analysis of Beauty,* published in 1753. A two-dimensional s-line he called *The Line of Beauty,* whereas a three-dimensional s-line he called *The Line of Grace* (see two of the plates from *The Analysis of Beauty,* figure 12-42). Floral arrangements made in this serpentine line truly display the graceful rhythm for which Hogarth is credited with discovering.

Often called the s-curve style, these floral arrangements display a sophisticated asymmetrical appearance. Because these arrangements have a downward sweeping curve that extends below the container rim, they must be constructed in a ***compote,*** or tall vase. The rhythmic line is easiest to achieve with vines, pliable branches, and naturally curving flower stems. These arrangements offer an elegant design, and they are often the choice of design shape for formal gatherings.

Although the s-shaped arrangement is generally not as popular as other design shapes, the graceful, serpentine line has application in many floral designs and floral decorations (see figure 12-43). Learning the steps for making a Hogarth design will allow you to choose this design more readily to serve as an option for floral arrangements.

Figure 12-43. Although less popular than other shapes, the Hogarth s-line is lovely and versatile in many decorating purposes, for example, weddings.

Hogarth Curve—Steps of Construction

(Refer to figure 12-44.)

Step 1. Select a raised container for the Hogarth curve design. Secure a block of floral foam that extends above the rim for horizontal stem insertions.

Step 2. Establish the framework with curving linear material. Insert the stem that forms the height into the back portion of the foam on one side. Insert the stem that forms the lower curve of the "s" in the front (opposite to the height of the "s"). Secure these stems deeply into the foam.

Step 3. Similar to the steps of construction for the crescent, add some mass foliages next in order to soften the curvilinear lines. This blends lines into the central, massed area of the design and conceals some of the foam and tape.

Step 4. Strengthen the "s" line by adding more linear material.

Step 5. Next, insert mass and line flowers. If a variety of flower sizes are used, place the buds and smaller flowers at the perimeters. Flowers should gradually increase in size so they are largest at the container rim. The visual weight of materials should also become heavier in the central, massed area. All stems should appear to radiate from the center of the design.

Step 6. Create a focal point. For example, mass flowers closer together to create an area of emphasis.

Figure 12-44. Hogarth design—Steps of construction. from left to right: a. Select a compote or other raised container. Secure floral foam above the container rim to allow for downward positioning of stems. Establish the framework with curving linear stems. Since these designs are generally one-sided, insert the stem forming the height of the bouquet in the back portion of the foam. The stem forming the downward curve may be inserted in the front, side portion of the foam. b. Add foliage all around at the rim of the container, by blending the curvilinear lines with the center massed area for a gradual transition. c. Establish a focal area near the rim of the container where the height and the width seem to intersect. Add flowers and foliages within the established framework. d. Continue adding fillers and foliages to form the desired shape and style of design. Remember to conceal foam and other mechanics in the back.

Step 7. Add filler flowers and foliages to finish the design. If desired, add fruits, vegetables, or other accessories to the central area of the arrangement. A cluster of grapes that cascades over the container rim, combined with other fruits anchored in the foam, offers a classic design typical of the baroque period.

Vertical Design Styles

Standard proportion rules may be easily overlooked when constructing vertical designs. Often height will be emphasized by exaggerating the vertical emphasis, thereby creating a dynamic appearance of strength. A vertical container enhances and strengthens the vertical line in a floral arrangement. These arrangements, with their perpendicular line, are often the choice of design shape when display space is limited or when a strong vertical line is to be emphasized in a room decoration (see figure 12-45).

Bud Vase Arrangements

The simplest vertical design is a ***bud vase*** with a single flower or a limited grouping of flowers and foliage. The flowers repeat the shape of the tall slender vase in which they are placed. A bud vase design is generally one-sided, but with initial planning can be made all-sided for use as a simple centerpiece.

Because a bud vase is small in scale to other designs, it is usually placed on a small table or in a small room or area. This way the delicate arrangement is not overwhelmed. Placed in a large room or on a large table, a single bud vase is out of scale and tends to disappear. However, bud vase arrangements may be used in groupings for increased visual impact.

Figure 12-45. ertical designs are useful when display space is limited or when an emphasis for a strong vertical line is desired.

Bud Vase Arrangements—Steps of Construction

Containers used as bud vases are generally tall and narrow. Because containers are narrow, flowers are easily held in place. Some vases, however, might be too narrow for the number and thickness of the stems intended for a bud vase design. A good "rule of thumb" for selecting a bud vase is this: if you can not fit your thumb down the neck of the vase, it probably will not be suitable for more than one or two stems of plant material (see figure 12-46). Forcing flowers into a narrow-necked vase is difficult and frustrating and may crush tender flower stems.

Step 1. Select a suitable vase and fill with tepid water and preservative.

Step 2. nsert the primary flowers to establish height and width. Vary the height and positioning of flowers, as illustrated in figure 12-47. Keep the number of flowers and the size of the design in scale with the visual weight of the vase.

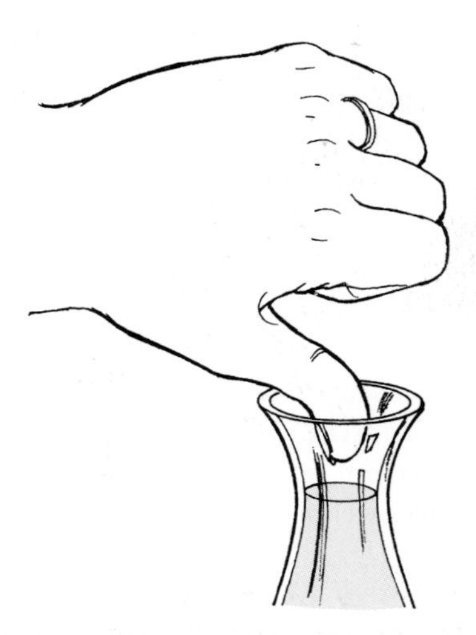

Figure 12-46. When selecting a bud vase, choose one that has a wide enough neck to suit your design needs. If your thumb will not fit down the neck, it might also be a challenge to fit in flower and foliage stems.

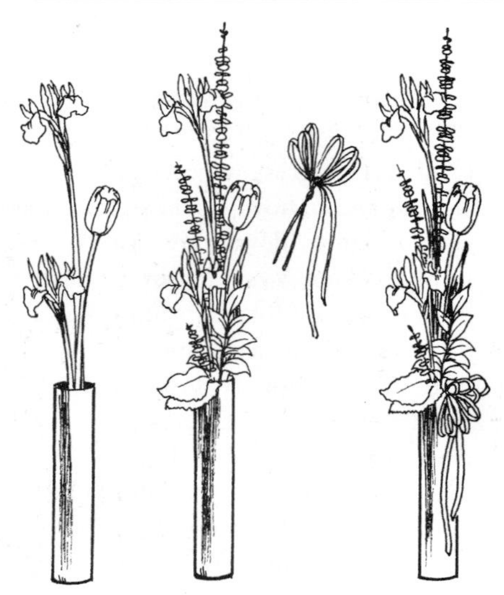

Figure 12-47. Bud vase—Steps of construction from left to right: a. Choose a suitable bud vase. Fill the vase with preservative treated water (floral foam generally is not used). Vary the height and position of flowers, keeping the size of the design in scale with the vase. b. Add foliage near the rim of the vase, to help blend arrangement with container. Add line foliage or flowers behind mass flowers to strengthen the vertical line and to help keep flowers securely in place. c. Filler flowers and foliages may be added to complete and give accent to the design. A bow may be added in front at the container rim to serve as an accent to the flowers.

At times a vertical arrangement may simply have a vertical line of plant material with no emphasis on width. In order to achieve visual interest and balance, these designs often require a focal point either near the rim of the container or at the upper edges, an accent may be placed near the container rim (see figure 12-48).

Stylized Vertical Designs—Steps of Construction

Although these arrangements may be made in low containers, tall containers emphasize the vertical importance of these designs.

Step 1. Select an appropriate container and secure the floral foam if necessary. The style of design, as well as the type of flowers and foliages needed for a particular vertical design, will dictate whether the foam should extend above the container rim or stay below.

Step 2. Establish the height of your arrangement with a line flower or foliage, as shown in figure 12-49. Since these designs are most often one-sided, insert

Figure 12-48. Often vertical designs rely on a focal point near the container rim (as shown by the lilies at left), or at the perimeter of the design with an accent near the container rim (as shown by the heliconia at the top and the leaves and curly branch inserted at the rim in the vase on the right). Areas of accent or emphasis are needed to give strictly vertical designs visual interest and balance.

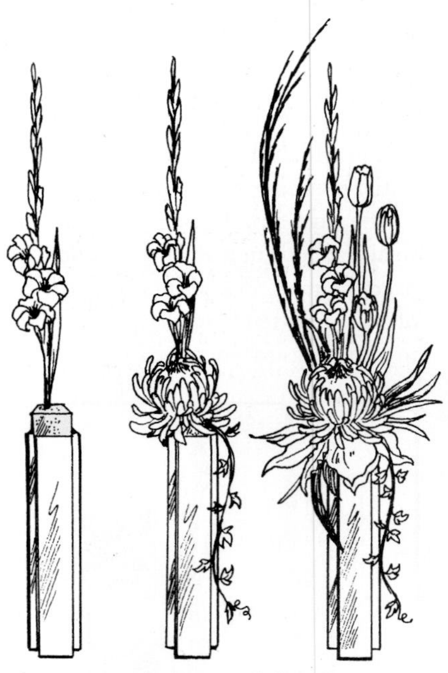

Figure 12-49. Vertical designs—Steps of construction from left to right: a. Choose an appropriate container. If floral foam is used, secure a piece that extends above the container rim. Establish the height of the design with linear material. Space will be tight, so position the stem that forms the height towards the back of the foam. b. Establish a focal point and add foliage to blend the container with the arrangement. c. Continue adding flowers and foliage, keeping the design prominantly vertical.

Step 3. Next add some linear and massed foliages to enhance the vertical line as well as blend the arrangement to the container.

Step 4. If desired, add a bow of ribbon, twine, bear grass, or other material to the arrangement at the container rim, to provide accent and visual balance.

Stylized Vertical Designs

Stylized vertical designs are often constructed in vertical containers with the use of floral foam. These designs are often one-sided but may be arranged to be viewed on all sides.

the first stem towards the back of the block of foam, allowing more foam space for other stem insertions.

Step 3. Add primary flowers first (before mass foliage is added) because foam space is extremely limited in these designs.

Step 4. To enhance the vertical line of the design, add a stem of ivy or other foliage or flower, that extends down in front. This downward line helps visually blend the container with the arrangement.

Step 5. Next, add other flowers and foliages, keeping the line of the design as vertical as possible.

Step 6. Add foliage last to accent the focal point and conceal foam and other mechanics.

Figure 12-51. Table centerpieces should be kept low so they won't interfere with conversation at the table.

Horizontal Design Styles

Horizontal designs, as the name implies, have a strong horizontal line emphasis, one that is parallel with the table top or other plane surface. The horizontal line provides a restful, peaceful feeling. These designs are generally symmetrical (but can also be asymmetrical) and can be made all-sided or one-sided. The horizontal line is combined with the triangular or the circular shape, resulting in horizontal style designs as shown in figure 12-50.

Horizontal arrangements are generally made in low and sometimes long containers. Often, the sides of the arrangement will extend past normal width proportion, enhancing the feeling of horizontal line.

This design shape is especially effective when designing flowers for the center of a dining table and is attractive when viewed from any side or angle. Candles are commonly used with this style of arrangement for a formal dinner party. Also, because it is kept low, it will not interfere with conversation across the table.

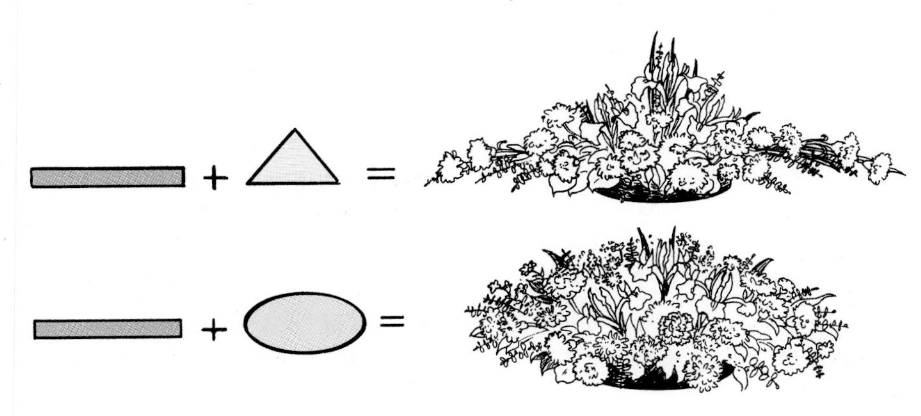

Figure 12-50. Horizontal arrangements are generally based on a horizontal line, plus a triangular or circular shape.

Horizontal Design Styles—Steps of Construction

Step 1. Select a container that is low for your horizontal design, generally one that is oval or rectangular. A longer container will hold enough foam and provide a larger area for water.

Step 2. Place the foam so that it extends above the container rim for insertion of side stems.

Step 3. If candles will be used, insert them directly into the foam, or use candle holders. Leave the candle wrappers on the candles to help protect them from scratches during construction. After the arrangement is completed, the wrappers may be removed.

Step 4. Next, establish the framework of the design. Insert flower or foliage stems to establish the length of the arrangement. Insert these stems deeply into the foam, angling them downward, inside the foam if possible, into the water supply.

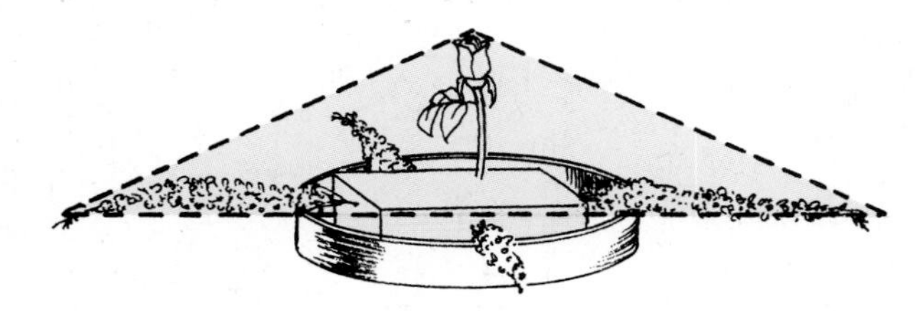

Figure 12-52. Horizontal designs—Steps of construction: a. Choose an appropriate low container. Secure a piece of foam that extends above the container rim, allowing for horizontal stems. Establish the framework of height and width.

Figure 12-54. c. After setting the initial framework, continue to add flowers and foliages to fill in the body, avoiding the temptation to add too many flowers.

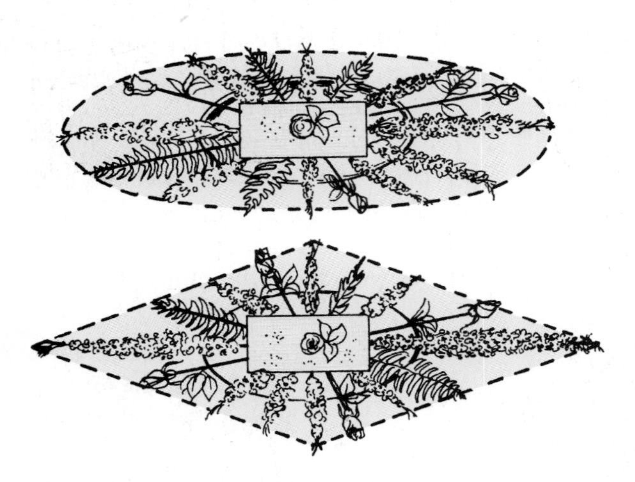

Figure 12-53. b. For symmetrical arrangements, extend flowers and foliages far enough to form an oval or diamond shape (top view).

Step 5. Insert two more stems to establish the narrow width of the framework. Keep these stems shorter, because space is generally limited on the tables where these arrangements are often placed (see figure 12-52).

Step 6. Place a flower or foliage stem in the center to set the height limit for your bouquet.

Step 7. Add more flowers and foliages near the rim of the container. When viewed from above, these flower heads and foliage tips should extend out far enough to form a pattern of an oval or diamond, as shown in figure 12-53. This will help blend the length of the arrangement with the massed area.

Step 8. Add flowers and foliages to fill in the body of the arrangement and conceal any mechanics (see figure 12-54). As with your s-curve designs, do not be tempted to add too many flowers. Keep the profile of the entire design low.

Step 9. Finally, add filler flowers and foliages if desired to complete the design (see figure 12-55).

Figure 12-55. Horizontal arrangements rely on a low profile. This arrangement of carnations, roses, and star of Bethlehem becomes unified as stems of heather, leptospermum, and statice radiate out from a central location.

REVIEW

Most floral arrangements can be classified as triangular, circular, vertical, or horizontal. Learning about these various shapes and the steps of construction for each allows you more design options.

With continued practice, you will become proficient in constructing the basic shapes of arrangements described in this chapter. As you gain more and more experience, your skills and confidence will increase and your desire for further self-expression in the art of floral design can be more fully realized.

TERMS TO INCREASE YOUR UNDERSTANDING

all-sided design
Analysis of Beauty, The
assymetry
bud vase
circular form
compote
cone design
crescent
c-shape
equilateral triangle
European style
fan design
focal point
floral pomander
formal
framework
geometric design
Hogarth
horizontal form
informal
isosceles triangle
line design
L-shape
mass design
obtuse angle
one-sided designs
oriental style
pre-green
radiating arrangment
right angle
right triangle
round mound
scalene triangle
skeleton flowers
s-shape
stylized vertical designs
symmetry
topiary ball
triangular form
Western style
vertical form

TEST YOUR KNOWLEDGE

1. What are factors that influence the choice of shape for floral arrangements?
2. How can each of the triangular designs be used in decorating to the best advantage for display purposes?
3. Which arrangement shapes generally require a focal point for maintaining visual balance and interest?
4. Why is the s-shaped arrangement known as the Hogarth curve?
5. When can the serpentine line be incorporated into floral decorating work?
6. Describe the steps of construction for a one-sided triangular design. How do these steps differ from other arrangement styles?

RELATED ACTIVITIES

1. Examine several empty containers and vases. What shapes of arrangements would be most appropriate for each? (This can also be a class activity. Everyone bring several container types to class and discuss, as a group, the most appropriate shapes and styles for each.)
2. Visit a floral shop or look through floral magazines. Notice the shapes of arrangements. Which design forms are most common?
3. Put together a "shapes of arrangements" poster or notebook with pictures from magazines of the various design forms.
4. Practice setting up the framework of different design shapes by using styrofoam and sticks of various lengths.
5. Follow the steps of construction listed in this chapter and make a one-sided bud vase, a one-sided triangle design, and an all-sided circular arrangement.
6. Put together a topiary cone, topiary ball, or floral pomander by following the steps of construction listed in this chapter.

Chapter 13

Seasonal, Holiday, and Special Occasion Designs

Rarely does a week of the year pass by without some sort of special occasion taking place. Perhaps it is the birth of a baby, a dinner party, an anniversary, or a birthday. Or maybe it is a holiday such as Grandparents Day, Secretaries' Day, or Thanksgiving. Whatever the occasion, flowers can be an important part of a celebration.

Much of the floral industry's business is seasonal, revolving around holidays and other special occasions. Many holidays, such as Valentine's Day and Mother's Day, are associated particularly with the giving and receiv-

Figure 13-1. Floral arrangements are often made to portray a particular season. Each season, whether spring, summer, fall, or winter, is associated with seasonal flowers and colors.

ing of flowers; these are sometimes referred to as ***floral* holidays** because of the significant increase in sales just prior to and during these days. Since flowers convey messages of love, friendship, and respect, they are an important part of these sentimental holidays.

SEASONAL THEMES

Often a floral bouquet may be designed to portray a particular season. Spring, summer, autumn, and winter are each associated with various colors and types of flowers as shown in figure 13-1.

Although most commercially grown flowers are available year-round, some flowers are unique or abundant during one season or another. ***Seasonal flowers*** give variety and distinction to everyday designs.

Spring

Floral arrangements representing spring are made in a rainbow of colorful flowers. Pastel and intense colors, as well as the complementary combination of purple and yellow, are all common to spring designs. Colorful bulb flowers begin to show themselves in the garden—crocus, daffodils, hyacinths, and tulips. Unique to spring are flowering branches of forsythia, quince, and decorative fruit tree and shrub blossoms. Fluffy masses of acacia, lilac, snowball, and catkins of willow, birch, and alder add unique textures to spring designs.

Spring flowers often dictate informal treatment for bouquets. Often a simple glass container or a basket are natural choices to hold the masses of flowers.

Summer

Summer is generally the longest of the floral seasons; its flowers are also the most vibrant in color. Bright blue delphiniums and cornflowers, vivid red peonies and roses, radiant orange calendulas and zinnias, and glowing golden lilies and sunflowers are just a few of the hundreds of richly colored flowers available during the summer.

Wildflowers and tall grasses of varying colors, textures, and shapes are natural fillers in summer arrangements. Vases featuring bright colors or bold textures and forms harmonize well with the array of bright color.

Full massed and overflowing bouquets of mixed garden flowers are common for typical summer designs. Take advantage of all the lovely flowers growing outdoors that are yours for the asking, and mix them with commercial flowers to make full bouquets.

Autumn

The brightly colored flowers of summer give way to the warm deep hues of reds, oranges, golds, and browns in the autumn, predominate colors in typical autumn bouquets. Chrysanthemums, dahlias, and gladiolus are all long lasting flowers abounding in the fall months. Since autumn is the time of the harvest, colorful vegetables and fruit are aften combined with flowers. Wheat, barley, and other grains, as well as thistles and branches of autumn-colored leaves, work well in harvest bouquets.

Autumn is the season for natural-looking vases: baskets, earthenware pots, terracotta bowls, and wooden trays. Their warm tones help to harmonize with the flowers in warm colors.

Winter

Winter affords fewer flowers in the garden. However, evergreen and conifer trees and shrubs retain their foliage throughout the year. Junipers, pines, cedars, yews, and mahonia are just a few of the greens available. Foliage designs can add beauty with their extreme colors and textures. The differences in foliages emerge when various types are juxtaposed with one another. Some appear blue, gray, or silver, while others appear golden or reddish-brown.

Cones and berries are the flowers of winter. Berries abound in just about every color. The bright red berries of cotoneaster, skimmia, and holly contrast strikingly with a variety of foliages. People who live in frosty climates, may wish to supplement the foliage, cones, and berries, with the wide range of year-round commercially grown flowers available from around the world. Containers that are simple and muted in color reinforce the colors and textures of the foliage, berries, and flowers.

Take advantage of the various seasonal flowers and foliages when they are plentiful. As a designer you can make a name for yourself by experimenting with the seasonal plant materials available. In combination with florist flowers, the possibilities in design are endless.

FLORAL HOLIDAYS

Table 13–1 is a calendar of the civil holidays and various other popular days observed in the United States of America. Many of these days are associated with the giving and receiving of flowers; however, some days are more geared to simply using flowers as decorations or tributes. Flowers that are a form of celebration in themselves can play a large part in these special occasions.

TABLE 13-1: FLORAL AND CIVIC CALENDAR

Month	Special Occasion	Date of Celebration
January	New Year's Day	January 15
	Martin Luther King, Jr. Day	Third Monday in January
	Spouse's Day	Fourth Friday in January
February	Groundhog Day	February 2
	Lincoln's Birthday	February 12
	Valentine's Day	February 14
	Washington's Birthday observed, (Presidents' Day)	Third Monday in February
March	International Women's Day	March 8
	St. Patrick's Day	March 17
	Easter	First Sunday following the first full moon on or after the vernal equinox (March 20), always occurring between March 22 and April 25
April	Professional Secretaries' Week	Last full week from Sunday to Saturday in April (Professional Secretaries' Day is the Wednesday of Secretaries' Week)
May	May Day	May 1
	PTA Teacher Appreciation Week	First full week in May
	Mother's Day	Second Sunday in May
	Armed Forces Day	Third Saturday in May
	Memorial Day	Last Monday in May
June	Children's Sunday	Second Sunday in June
	Flag Day	June 14
	Father's Day	Third Sunday in June
July	Independence Day (Fourth of July)	July 4
August	Friendship Day	First Sunday in August
September	Labor Day	First Monday in September
	National Grandparents Day	First Sunday in September following Labor Day
	National Good Neighbor Day	Fourth Sunday in September
October	Columbus Day	Second Monday in October
	National Boss Day	October 16
	Sweetest Day	Third Saturday in October
	Mother-in-law Day	Fourth Sunday in October
	Halloween	October 31
November	Veterans Day	November 11
	Thanksgiving	Fourth Thursday in November
December	Christmas	December 25
	New Year's Eve	December 31

Major and Regional Floral Holidays

Christmas is actually the largest floral holiday of the year since flowers, plants, and decorations are needed during the five weeks prior to Christmas Day. Most of the other major floral holidays involve only a single day. Two of the days that florists consider the busiest are Valentine's Day and Mother's Day. Other major floral occasions include Easter and Thanksgiving.

The use of flowers for gifts or decorations is often regional, meaning that flower shops in one area will be swamped with business for a particular holiday, whereas in another area that same holiday is not associated with the use of flowers.

Often flower giving and floral decorating is determined according to the historical background and religious or cultural make up of a geographical area. For instance, in several areas, Memorial Day is extremely busy for florists, especially where national military cemeteries are located. Similarly, florists located in areas with large populations of a certain religion or culture in which decorating graves is traditionally an important custom will also experience greater sales.

Another example of the regional sales of flowers is St. Patrick's Day. Floral shops located in cities with a large number of people of Irish descent, such as New York and Boston, generally find St. Patrick's Day to be an important sales period.

Along with cultural and religious special occasions, many holidays recognize and honor individuals such as secretaries or teachers for diligent service. Other holidays focus on relationships with people: grandparents, parents, friends, and neighbors.

In order for you to understand the popular flowers and types of arrangements that are associated with many floral holidays, it will be beneficial for you to know the history, background, and dates of celebration for each. This, in many cases, will help give you a better understanding as to why certain colors, accessories, and design styles are popular. In addition to discussing the well-known floral holidays, other less popular flower-giving days such as Groundhog Day, Columbus Day, and Mother-in-law Day are also included. These and many other lesser-known days can become fun days for people to express feelings and emotion through the gift of flowers. Through their promotional efforts, the floral industry can help people enjoy the beauty of arrangements more frequently (see chapter 20).

New Year's Day

January 1, the first day of the calendar year, is a nationwide holiday in the United States known as New Year's Day. However, people in almost every country celebrate this day as a public holiday.

New Year's celebrations begin on December 31 with festive parties in anticipation for the beginning of the new year. These noisy, happy celebrations have their roots in the early beliefs of pagan people who believed evil spirits were all around. To prevent these spirits from entering the new year, it was common to ring the church bells, causing a ruckus. From this activity comes the popular expression "ring in the New Year." Toward the end of the nineteenth century, a carnival-like celebration prevailed with drinking and loud noises from not only bells, but whistles and horns as well.

The noisemakers and fireworks of today are reminiscent of those early customs of causing noise to turn away evil spirits and keep them from entering the new year. Today both formal and informal parties and dinner dances, including masquerade balls, are also traditional activities during the evening. This wide variety of activities invites an equally broad range of floral arrangements. Therefore, in addition to the more elegant floral arrangements for formal gatherings, New Year's Eve bouquets can also reflect fun-loving celebrations with the addition of many accessories, including glittery party hats and horns, balloons, confetti, and streamers.

In contrast to the pagan's fearful celebration of the new year, the Romans in 153 B.C. were the first to celebrate the New Year's Day. They believed that commencing on the first day of January, each person's life would begin with a clean slate; the previous year's events would be erased, signaling a fresh beginning. To symbolize this new opportunity for growth, the ancient Romans gave each other New Year's gifts of branches from sacred trees. In later years, gold-covered coins were exchanged, imprinted with pictures of Janus, the god of gates, doors, and beginnings. Janus had two faces, one looking forward and the other looking backward.

Toward the end of the nineteenth century, the well-known symbols of the New Year's baby and the gray-bearded Father Time bearing a scythe, became popular during the late nineteenth century. Both of these symbols are often incorporated into New Year's Day party floral arrangements.

Traditionally, New Year's Day is a time for personal stocktaking, for making New Year's resolutions to break bad habits and setting goals for the coming year. Celebrations are both festive and serious. Today many Americans devote part of New Year's Day visiting friends and relatives, giving gifts, attending religious services, and watching parades in person or on television, such as the Mummers' Parade in Philadelphia and the Tournament of Roses Parade in Pasadena, California (see figure 13-2). For many people, a good portion of the day is devoted to watching football games and attending parties to watchfootball games.

Floral arrangements designed for New Year's Day vary according to the purpose for which they are intended. Whether they are decorations designed for a religious service or for a football-watching party, or a simple hand bouquet to give to a friend, fresh flower arrangements are always appropriate as a message of renewal, happiness, joy, and hope for a new beginning. Cheery bright colors, football team colors, or even all white are common color schemes for floral designs.

Figure 13-2. From the most popular parade, this float is made totally from plants and flowers.
Pasadena Tournament of Roses

In areas that are populated with people of Chinese origin, the Chinese New Year lasts for four days. It begins at sunset on the day of the second New Moon following the winter solstice, which always occurs sometime between January 21 and February 19. Celebrations include elaborate meals, fireworks, and a parade of colorful dragons.

The color red is one of good omens in the Chinese culture believed to ward off evil spirits. Red flowers are used in floral bouquets for centerpieces and other decorations to celebrate the New Year to complement other nonfloral decorations.

Martin Luther King, Jr. Day

Martin Luther King, Jr. (1929–1968), an African-American Baptist minister, was the main leader of the civil rights movement in the United States during the 1950s and 1960s. The winner of the Nobel Peace Prize in 1964 for leading nonviolent civil rights demonstations, he was nevertheless the target of violence and was finally shot and killed at the age of 39. King became the second American whose birthday is observed as a national holiday, the first being President George Washington. Also commonly called Civil Rights Day and Human Rights Day, this holiday is celebrated on the third Monday in January in observance of King's birthday (January 15).

Floral tributes, wreaths, and decorations are used in celebrations and festivities honoring King and other civil rights leaders in Arlington, Virginia, and many other cities across the nation.

Spouse's Day

The fourth Friday in January is annually set aside in celebration of one's spouse. Husbands and wives are encouraged to share jobs, roles, and responsibilities to better understand and appreciate each other. Although not a well-known holiday, it is still an opportune time to convey love and appreciation with a gift of flowers.

Groundhog Day

Groundhog Day is based on a custom brought to America by people of Germany and Great Britain. They believed that February 2 was a time for forecasting the weather for the next six weeks. According to legend, the groundhog or woodchuck comes out of hibernation on this day. If he sees his shadow, he supposedly returns to his hole for six more weeks of winter weather. If the weather is cloudy when the groundhog comes out of his hole, according to legend, spring weather will soon come.

Groundhog Day is an ideal time to celebrate an oncoming spring with daffodils, tulips, and hyacinths. Dormant branches of forsythia, quince, and other flowering varieties once cut from trees and shrubs can be encouraged to blossom in warm preservative solution (see Chapter 9). Giving a friend a bouquet of "sticks" on Groundhog Day that will soon become colorful and fragrant sends a message of happy anticipation for an early spring.

Lincoln's Birthday

About 30 states celebrate the birthday of Abraham Lincoln (1809–1865). Born on February 12, the sixteenth President of the United States is remembered and honored at many formal dinners and celebrations throughout the country. Red, white, and blue floral arrangements, along with bunting and flags, are traditional and familiar means of commemorating this holiday.

Valentine's Day

Several theories attempt to explain the origin of Valentine's Day. Some authorities trace the origin to Lupercalia, an ancient Roman festival and feast connected with fertility rites held on February 15. Others connect the event with one or more saints of the early Christian church, one of whom was named Valentine. According to legend, he was imprisoned because he refused to worship the Gods of the Romans. He had made friends with many children, who missed their friend and expressed their love by tossing notes through the bars of his cell window. Another theory believed by many is linked to an old English legend that says birds choose their mates on February 14.

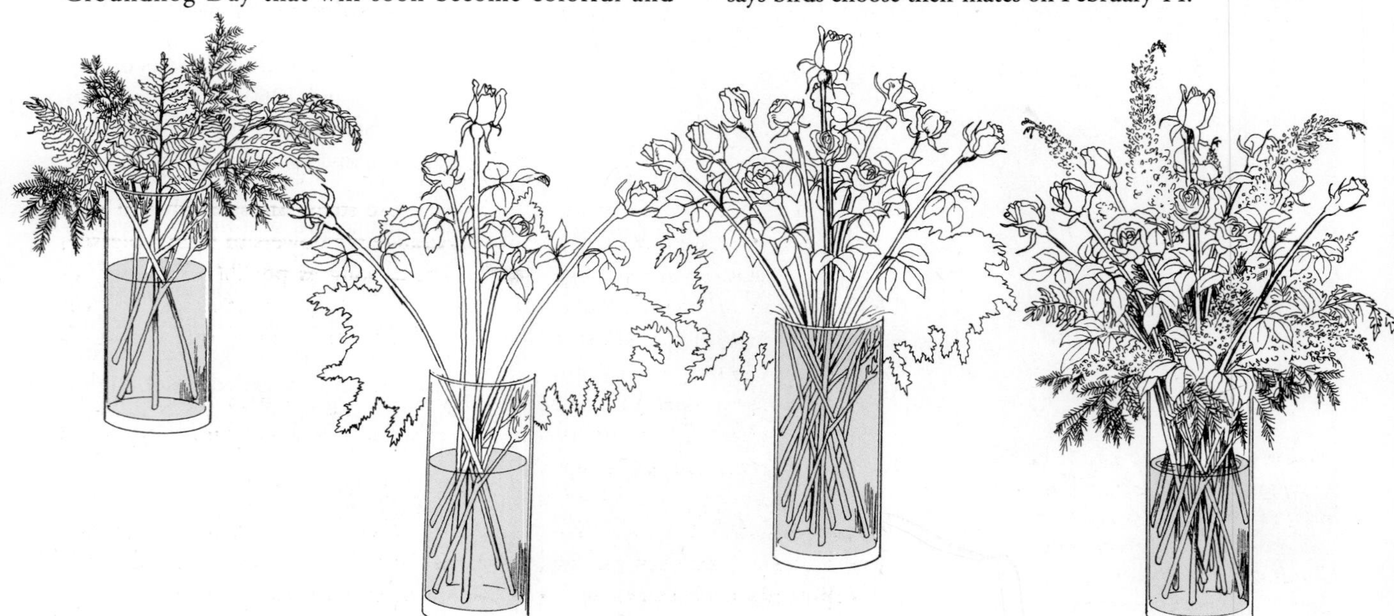

Figure 13-3. When arranging a dozen roses, or other flowers, use foliage grids to eliminate unsightly mechanics of chicken wire or tape. Once a foliage grid is formed, arrange flowers to form a pleasing outer shape. Add more foliage and filler flowers to complete the design.

The celebration of Valentine's Day on February 14 is one of the most widely observed unofficial holidays, which is one of the busiest for the floral industry. Expressions of love and admiration are often exchanged on this day through flowers, particularly through roses, the most requested flower for Valentine floral bouquets (especially red ones, which have long symbolized "true love"). Long-stemmed roses are a traditional favorite, in bouquets or boxes of one dozen.

Figure 13-5. Bouquets for Valentine's Day can incorporate imaginative Valentine accessories.
Florist's Review Magazine

Arranging a dozen roses in a vase, as shown in figure 13-3, is relatively simple to do if you begin with the proper foundation, allowing all the long stems to stay in place. When using a clear vase, remember to keep the mechanics to a minimum. Foliage grids eliminate the worry of hiding unsightly tape or chicken wire (see chapter 8). Once the foundation is ready and weak rose stems are strengthened with green wire, the roses can be put in place. Arrange the flowers to form a pleasing outer arrangement shape, keeping all stems as long as possible. A variety of foliages can then be added as well as a touch of flower fillers to help accent the roses. Bows and other accessories can be added to help convey the Valentine's Day theme of love.

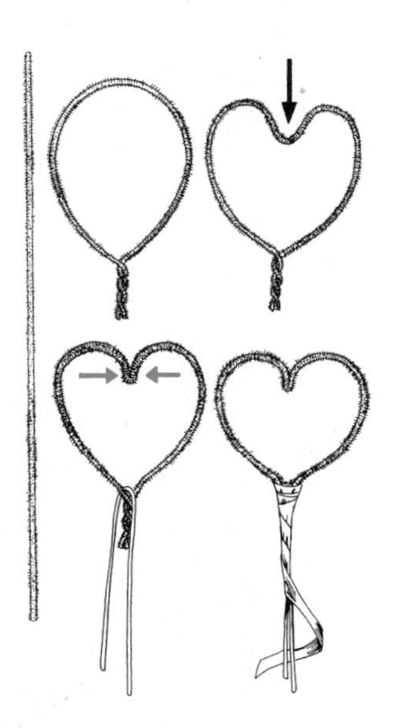

Figure 13-4. Hearts may be made inexpensively with chennile stems to adorn bouquets. Step 1: Select a coordinating color of chennile. Step 2: Twist the chennile stem into a circle. Step 3: Pinch the top-center of the chenille downward into the form of a heart. Step 4: Add a wire as shown to form a stem and floral tape, to conceal mechanics.

Because red roses on Valentine's Day are in highest demand, this causes prices to increase. Therefore, other colors of roses, as well as all other types and colors of flowers, are also in demand.

The most popular color scheme on Valentine's Day is monochromatic with red and various value and intensity levels of red. Pastel and hot florescent pinks, dusty mauves, and deep maroons in various combinations with each other and in combination with white and cream for pleasing contrasts are traditional.

Common accessories include hearts and cupids. Hearts can be inexpensively made with chennile stems to help trim bouquets (see figure 13-4). An imaginative floral designer can also devise more accessories to decorate bouquets and at the same time convey a clever message of love as shown in figure 13-5.

Washington's Birthday

George Washington, the first president of the United States, was born on February 22, 1732. His birthday is celebrated as a federal holiday on the third Monday in February. This holiday is often called Presidents' Day in honor of both Washington and Abraham Lincoln as well as other presidents of the United States. Floral decorations and tributes are generally in the patriotic colors of red, white, and blue.

International Women's Day

Also called International Working Women's Day, this day is said to commemorate a New York City march and demonstration by female garment and textile workers in 1857. It is believed to have been first proclaimed for this date at an international conference of women in 1910, held in Helsinki, Finland.

This holiday has recently become more widely recognized in the United States, and although originating in the United States, it is well known and observed internationally. Many nations, including the People's Republic of China (where it is a national holiday), honor working women with flowers and other gifts.

St. Patrick's Day

Although St. Patrick's Day is celebrated on March 17 as both a holiday and holy day, it is not known exactly how or when this holiday began, only that it is an occasion for the Irish to enthusiastically pay tribute to their homeland. St. Patrick was a missionary to Ireland in the fifth century who brought Christianity to the Irish people. According to legend, St. Patrick explained the idea of the trinity to the Irish people by using a shamrock.

This day is a national holiday in Ireland, and is also widely celebrated in other countries in areas having large concentrations of people of Irish descent.

In the United States, St. Patrick's Day, a festive day encompassing parties, parades, and the wearing of the green, was first celebrated in Boston in 1737. The biggest St. Patrick's Day parade is held in New York City.

Flowers for boutonnieres, corsages, centerpieces, and party decorations are traditionally green. The flowers can be naturally green (Bells of Ireland, gladiolus, and cymbidium orchids) as well as dyed (carnations and gypsophila); white flowers can be spray painted green for harmonizing with the theme. The use of varieties of foliage as well as all-foliage arrangements make distinctive and unique designs. Many accessories are featured in these theme arrangements, including green hats, pipes, shamrocks, and leprechauns.

Easter

Easter, one of the most important days of the Christian calendar, commemorates the resurrection of Jesus Christ. Because this religious day in the spring celebrates new life, many churches are decorated with arrangements of Easter lilies and spring bouquets.

The date of Easter varies from year to year, but it is always on the first Sunday after the first full moon following the first day of spring (the ***vernal equinox*** marks the beginning of spring on March 20 or 21). This means that Easter can occur on any Sunday between March 22 and April 25.

Easter floral arrangements are designed in a variety of formal or informal styles as shown in figure 13-6. Many display a more religious theme using the traditional Easter lilies, while others show more secular themes with emphasis placed on the spring season or commercial Easter activities and symbols. Accessories include bunnies, chicks, and colored eggs, and arrangements may be placed in novelty containers as well as traditional baskets. Daffodils, tulips, and hyacinths with flowering branches all convey a spring, as well as an Easter, theme.

Professional Secretaries' Week

Professional Secretaries' Week is set aside to honor, show appreciation for, and thank both male and female secretaries. This week was inspired by the president of the National Secretaries Association as a way of recognizing and increasing the integrity and image of secretaries everywhere. In 1952 the U.S. Secretary of Commerce, Charles Sawyer declared the first National Secretaries' Day and week in June. The dates changed in 1955, so that today this holiday is observed annually in the United States the last full week in April from Sunday to Saturday. The Wednesday during the week is designated as Professional Secretaries' Day.

Since most secretaries have limited space on their desks in which to place floral bouquets, many floral arrangements are designed in coffee mugs or other novelty containers. Small baskets and narrow vases are also common. Whatever container is used, it should be stable and hold an adequate supply of water. The arrangement must not interfere with the work space on the desk. Clever accessories

Figure 13-6. Easter bouquets can be both formal and informal. © *1993 FTD.*

help make these bouquets even more enjoyable. Pens, pencils, and other office supplies, as well as candy, small picture frames, gift certificates, and balloons, are all often included in designs for secretaries.

May Day

May Day was originally established in a Roman festival called Floralia, in honor of Flora, the mythological Roman goddess of flowers and spring. This day signified the changing of the seasons.

In medieval times, May Day was the favorite day of celebration for many English villages. Flowers were gathered and people decorated their churches and homes. On May 1, villagers danced around a Maypole, holding the ends of ribbons that were attached to the top of the pole. As they danced around the pole, the ribbons became beautifully intertwined. A May queen was selected and crowned with a floral wreath.

In the nineteenth century, May baskets filled with fresh flowers and candy were secretly hung on the doors of friends and neighbors expressing friendship and romance.

In the United States, the Puritans frowned on May Day and, as a result, it has never been celebrated with the same vigor as it is in areas of Great Britain. However, in many American cities May Day parties with festive Maypoles and dancing are traditional.

In making May baskets, arrangements should reflect the gathering of flowers with a casual, unstructured look. Colorful floral wreath centerpieces as well as miniature maypoles with colorful ribbons are appropriate for party decorations.

PTA Teacher Appreciation Week

The National Congress of Parents and Teachers, commonly called the National PTA, is a volunteer organization helping to unite the forces of home, school, and commu-

Figure 13-7. This bouquet for Secretaries' Day incorporates appropriate accessories. Florist's Review Magazine.

Figure 13-8. A simple floral arrangement made in a mug, expressing thanks or appreciation, is a welcome gift.

nity to promote the welfare of children. Local PTA units serve in individual elementary schools, middle, junior, and high schools, with approximately 27,000 local PTAs across the United States and about six and one-half million members.

PTA Teacher Appreciation Week is celebrated the first full week (from Sunday to Saturday) in May. This week is devoted to honoring and supporting teachers and the teaching profession. Many local PTA units honor teachers by giving them flowers, including corsages, boutonnieres, and arrangements. Accessories such as chalk, erasers, pencils, apples, and balloons are often added to these designs.

Mother's Day

Mother's Day, an important floral holiday in the United States as well as many other countries, is a day set aside to honor mothers and motherhood.

The history of Mother's Day is actually a series of events. The first known suggestion for a Mother's Day in the United States was made by Julia Ward Howe in 1872. Several other people later rallied and launched campaigns and celebrations for the observance of Mother's Day. However, it was Anna Jarvis of Philadelphia, who in 1907 began a campaign for a nationwide observance of Mother's Day; she is, therefore, credited as being the founder of Mother's Day. In 1914 Congress designated the second Sunday in May as Mother's Day.

In some regions it is a custom to wear a carnation on Mother's Day (a red carnation symbolizes appreciation and love for a living mother; a white carnation, symbolizes love and respect for a mother who has passed away).

In many regions popular colors for floral bouquets are those in the tints and shades of pink and lavendar (see figure 13-9).

Whether a floral gift is a single wrapped rose, a colorful corsage, a basket of mixed flowers, a flowering plant, or a dozen roses in a cystal vase, a gift of flowers will convey a pure message of love and appreciation on this special occasion.

Figure 13-9. This design in pink is a welcome gift for Mother's Day.
From Flowers & Magazine, Teleflora.

Armed Forces Day

Established in 1950 by President Harry S. Truman, Armed Forces Day honors all branches of the armed forces in the United States. It is celebrated on the third Saturday of May with military exercises on land, at sea, and in the air. Floral decorations, wreaths, and centerpieces help celebrate this day, and are generally made with red, white, and blue flowers.

Memorial Day

A legal public holiday in the United States, Memorial Day is observed the last Monday in May. This patriotic day was established to honor all Americans who have given their lives for their country, although originally this day honored those who died in the Civil War. Many southern states also have their own days for honoring their Confederate dead. Many countries also designate a day each year for decorating graves with flowers and flags (hence the name Decoration Day).

Military exercises and programs are held at national cemeteries. Red, white, and blue floral tributes, wreaths, and other decorations repeat the pattern set by the national flags that are placed on thousands of graves.

In some areas of the country, decorating graves of loved ones on Memorial Day (see figure 13-10) is a popular activity for the day, and for many, a great deal of planning goes into the selection of colors and styles of floral tributes. While floral tributes vary, the most common designs are hand-tied bouquets, flat sprays, large radiating bouquets, easel pieces in various shapes, and potted chrysanthemums.

Flag Day

Another patriotic holiday, Flag Day, is celebrated on June 14 in commemoration of the day when the Stars and Stripes were adopted as the official flag of the United States. Although Flag Day is not an official national holiday, it is proclaimed every year by the President of the United States and widely observed. Flag Week is

Figure 13-10. In some areas of the country decorating graves of loved ones on Memorial Day is a popular activity. For many, a great deal of planning goes into the selection of floral tributes.

designated the week from Sunday to Saturday that includes June 14.

Floral tributes are often made to look like the national flag. Red, white, and blue hand-tied bouquets and other floral decorations are also popular.

Father's Day

One of the first known requests for a set day to honor fathers was made by Sonora Louise Smart Dodd of Spokane, Washington. After listening to a sermon on Mother's Day in 1909, she wanted to honor her own father, William Jackson Smart. He raised six children on his own after the death of his wife in 1898. Sonora Dodd recommended the adoption of a national father's day. Through her efforts the first Father's Day was celebrated in Spokane on June 19, 1910 .

Many people tried over the years to make this day an official national holiday, but it wasn't until 1972, when President Richard M. Nixon proclaimed the third Sunday in June that it became a national day of recognition. The United States and Canada celebrate Father's Day on this day.

In recognizing and honoring fathers, many people often give floral gifts to their fathers. For example, potted plants and dish gardens are popular florist gift items for Father's Day. Favorite floral arrangements for fathers include the bold, daring tropical flowers or other arrangements in the warm colors.

Accessories that suggest a favorite hobby, are commonly placed in the floral design adding areas of emphasis and accent. The addition of creative accessories can set arrangements and dish gardens you design apart from all others. For instance, the addition of golf balls, golf tees, fishing lures, gardening tools, slippers, or even a tie, can make a clever, more interesting design.

Independence Day

The United States celebrates its own birthday every year on July 4. This day commemorates the anniversary of the day on which the Declaration of Independence was adopted by the Continental Congress on July 4, 1776. Independence Day was first celebrated on July 8, 1776, in Philadephia. Since that time, Independence Day has been widely celebrated with parades, pageants, picnics, and parties. The day concludes in most cities with colorful fireworks. Congress declared July 4 to be a federal legal holiday in 1941.

Red, white, and blue flowers are popular for Independence Day festivities and decorations. Hand-tied bouquets as well as centerpieces often incorporate the use of tiny American flags. Remember, as you mix flowers in the red, blue, and white hues, use red in lesser amounts, because of its intense, dramatic qualities. Too much red may overwhelm the blues and whites.

Friendship Day

A regional holiday, Friendship Day, is an unofficial day set aside to honor friendships and show appreciation to others by expressing thanks, respect, and honor, often through giving flowers and other gifts. Friendship Day is a day to follow the advice of Samuel Johnson (1709–1784), a great English writer and philosopher who advised, "A man should keep his friendships in constant repair." This day was approved by Congress in 1935 for the first Sunday in August of each year.

Colorful, summer flowers in hand-tied bouquets as well as wrapped flowers are popular gifts among friends.

Figure 13-11. Friendship Day is an ideal time to honor friendships and show appreciation to others with a gift of flowers.

Labor Day

A day set aside to honor working people, Labor Day is a legal holiday in the United States and Canada, and is observed on the first Monday in September. First celebrated in New York City in September 1882, Labor Day was signed into law in 1894 by President Grover Cleveland, making Labor Day a national holiday.

Labor Day for most people is a day of rest and recreation, symbolizing the end of summer. Although flowers are generally not associated with Labor Day, colorful floral centerpieces add a festive touch for end-of-the-summer parties.

National Grandparents' Day

National Grandparents' Day is observed annually on the first Sunday in September following Labor Day. This day, first observed in 1978 and proclaimed as an annual event in 1979, recognizes and honors grandparents and older people.

This day, like many others, is fairly regional; that is, it is more popular and successful in certains areas than in others. Floral arrangements in various styles and colors are given as a token of love, honor, and respect for grandparents.

National Good Neighbor Day

Although not a well-known day, National Good Neighbor Day, unofficially observed on the fourth Sunday in September, is a day set aside to show appreciation for the good people who live nearby. A day of promoting respect and understanding for our fellowman, National Good Neighbor Day, is an ideal time to bridge communication and promote friendship with a memorable gift of flowers.

Figure 13-12. Many creative touches can be added to bouquets for Grandparents' Day. © 1993 FTD.

Columbus Day

Although not thought of as a floral holiday, floral decorations and centerpieces can help celebrate festivities on Columbus Day, which is a public holiday in most countries in the Americas. First observed in the United States on October 12, 1934, Columbus Day is the anniversary of Christopher Columbus's (1451–1506) arrival in the New World on October 12, 1492. Since 1971, Columbus Day has been observed on the second Monday in October.

National Boss Day

A day set aside to honor employers and supervisors, National Boss Day is recognized throughout the United States and Canada annually on October 16. This day was originated by Patricia Bays Haroski, a secretary, who was impressed with how well her boss treated her and the other employees in the company.

Boutonnieres and corsages are popular floral gifts for bosses, as well as arrangements, plants, and dish garden plants.

Sweetest Day

Observed annually on the third Saturday in October, Sweetest Day is a day named for recognizing and honoring those who have been "sweet" or kind and helpful throughout the year. Sweetest Day is said to have its beginnings in the 1930s when a man from Cleveland, Ohio, wanted to do something special for forgotten orphan children and lonely shut-in people.

Sweetest Day is enthusiastically celebrated in various regions throughout the United States, while in some areas it is virtually unknown. Floral gifts of all types, similar to those given on Valentine's Day, are given to friends and sweethearts. Sugar cubes or candy incorporated into a floral arrangement help to convey the "sweet" message of friendship and love.

Mother-in-Law Day

A special day set aside to honor and thank the mother of one's spouse is held annually on the fourth Sunday in October. This day, like many others, is regional and more celebrated in certain parts of the country. A gift of flowers is a thoughtful gesture that is always welcome and may serve to strengthen family relationships.

Halloween

Halloween, held annually on October 31, developed from ancient new year festivals and festivals of the dead, combining Druid (Celtic religious order of priests and soothsayers in ancient Britain, Ireland, and France) autumn festivals and Christian customs. The celebration historically marked the beginning of the season of cold, darkness, and decay, a time associated with death. Later when pagan customs became part of this Christian holy day, the Christian church established All Saints' Day, held on November 1. The church then began to honor the dead on All Souls' Day, held on November 2.

Many early American settlers brought their various Halloween customs with them. However, it wasn't until the 1800s that Halloween celebrations became popular. Customs such as carving pumpkins into jack-o'-lanterns, bobbing for apples, dressing in costumes, and trick-or-treating are still popular activities to help celebrate Halloween.

Floral arrangements for decorations and gifts are often made in novelty containers with the addition of miniature scary witches, black cats, ghosts, spiders, and cobwebs. Popular color schemes are orange and black or autumn colors. A unique container you may choose to use during the Halloween and autumn season is a real pumpkin (see figure 13-13).

Figure 13-13. A pumpkin is a novel and festive container for floral arrangements at Halloween. Step 1: Cut out top of pumpkin to form a lid, which may be incorporated into the bouquet later. Insert sharp end of a wooden stick into the inside of the lid, as shown). Cut low and wide enough to allow for floral foam. Remove the inside pumpkin and seeds. Step 2: Add floral preservative to the inside walls of the pumpkin. Next, a layer of cellophane may be added before adding a block of floral foam. The floral foam must extend above the top of the pumpkin to allow for side and downward stems. Step 3: Arrange autumn-colored flowers and Halloween novelties to form desired shape and style. Insert pumpkin lid into focal area if desired. Add preservative water to pumpkin.

Veterans Day

Veterans Day, a legal federal holiday in the United States, is held annually on November 11. President Woodrow Wilson in 1919 proclaimed this day (formerly called Armistice Day) to remind Americans of the tragedies of war. This day is set aside to honor and give recognition to men and women who have served in the United States armed services. Other countries, such as Canada and Great Britian, have a similar day of honor.

Celebrations in the United States include speeches and parades, and special services are held at the Tomb of the Unknown Soldier in Arlington National Cemetery.

Floral wreaths and other tributes often in the patriotic colors of the national flag, are used at special activities on Veterans Day.

Thanksgiving

Thanksgiving is a family day set aside each year as a time for feasting and giving thanks. Thanksgiving Day is commemorative of the Pilgrims' celebration of the good harvest of 1621 shared with Indians in Plymouth. Similar harvest Thanksgivings were held in Plymouth for several years.

The custom of celebrating the harvest spread from Plymouth to other New England colonies. Eventually, many states had their own Thanksgiving days. Several people, including Sarah Josepha Hale, the editor of Godey's Lady's Book, promoted the idea of a national Thanksgiving Day. In 1863, President Abraham Lincoln proclaimed the last Thursday in November as a day of thanksgiving.

In 1939, President Franklin D. Roosevelt proclaimed Thanksgiving Day one week earlier to promote business by lengthening the shopping period before Christmas. After 1941, Thanksgiving Day, a national holiday, was observed on the fourth Thursday of November.

Canada also celebrates Thanksgiving Day in a manner similar to the United States. In 1957, the Canadian

Figure 13-14. A cornucopia may be filled with fruits, vegetables, grains, grasses, or autumn flowers, foliages, and berries and used as a decoration or centerpiece. Step 1: Secure with tape, glue, or wire, a low container or liner with soaked floral foam into the bottom of a cornucopia basket. Step 2: Establish the framework of the design with linear and mass materials. The shape of the design should seem to "spill out" of the basket. Step 3: Anchor heavier fruits, vegetables, and novelties into the focal area of the design. Step 4: Add flowers and smaller novelties into the design. Complete the design with fillers to conceal all mechanics. Add preservative water.

government proclaimed the second Monday in October as Thanksgiving Day.

Popular floral decorations for Thanksgiving are designs that represent a full and abundant harvest. The ***cornucopia,*** or horn of plenty, is a symbol of abundance and nature's plenty. Often filled with fruits, vegetables, grains, grasses, and autumn colored flowers and foliages, it has become a popular decoration and centerpiece for the Thanksgiving Day table (see figure 13-14).

Because most flower arrangements will be used as centerpieces, the height and width need to be kept within reasonable and functional bounds. Candles offer a warm glow at the table and may be added to these floral designs without obstructing the view of guests.

Novelty accessories are often included in these harvest designs to suggest a Thanksgiving or autumn theme. Birds, nests, and novelty turkeys, pilgrims, and Indians, nuts, berries, Indian corn, gourds, and a colorful array of fruits and vegetables can be used individually or in combination to create and enhance the style and theme of a design.

Christmas

Christmas is the most popular Christian holiday. The date, December 25, was probably influenced by early pagan festivals and adapted from the Roman calendar in the fourth century. The word Christmas comes from the early English phrase "Cristes maesse" meaning "mass of Christ." The Christmas celebration centers on the events surrounding the birth of Jesus Christ as told in the New Testament.

Many celebrations depict a nativity scene, also called a ***creche,*** with the infant Christ, Mary, and Joseph, with shepherds, wise men, and various animals surrounding the Holy Family. Many Christians participate in religious services on Christmas Eve or Day.

The Christmas season typically begins following Thanksgiving. For many Christians, the Christmas season begins on the Sunday nearest November 30, marking the first day of Advent, which refers to the coming of Jesus on Christmas Day. Many Christians decorate their homes with an Advent wreath, Advent calendar, or Advent candles.

Many customs, decorations, and activities from non-Christian festivals have been adopted as part of the Christmas celebration, such as belief in Santa Claus; decorations of lights, evergreen trees, wreaths, holly, and ivy; and participation in activities such as feasting, caroling, and gift-giving.

No other holiday season inspires the use of flowers and foliages as does the Christmas season. More so at Christmas than at any other season,

Figure 13-15. A gift basket is an attractive way to say thank you or other sentiments at Christmas or at any time. Step 1: Secure with pan-melt glue, a container, vase, or liner into one side of a basket. Step 2: Add a filler material, such as shredded grass, tissue, or cellophane, into the bottom of the basket around the vase. Step 3: Next, add a single layer of larger items, such as fruits and vegetables, cheese and crackers, etc. Add a bottle of non-alcoholic cider or juice to one side of the basket, leaning it against the basket edge for security. Small pieces of tissue or cellophane may be used to cushion each item into place. Step 4: Continue adding smaller items, such as candy and holiday nuts, and delicate fruits, such as peaches and bananas for the top layer. Spill grape and banana clusters over the basket edges. Anchor them with tape if necessary. Step 5: Next, arrange flowers in the liner or vase that is secured into the basket. Provide height with linear materials. (An arrangement may also be designed separately into a container that is placed into the liner already glued into the basket.)

florists are key suppliers to many of the decorations associated with Christmas, especially fresh arrangments, wreaths, and garlands. Decorated poinsettia plants and gift baskets designed by a floral designer are also in high demand during the season (see figure 13-15).

Green and red are the traditional colors of Christmas, both highly symbolic to Christians. Green signifies everlasting or eternal life through Jesus Christ. Red represents the blood that Christ shed while being crucified. Green and red are often mixed with the metallic silver, gold, brass, and copper. In addition, deep shades of red, blue, and green represent a rich, winter theme.

The Christmas season can be expressed through a variety of themes and styles of design. Whether the theme is religious, wintery, or youthful and fun, accessories can be added to designs to enhance the message or style. Popular accessories include such items as shiny glass balls, candles, ribbons, cones, berries, fruit, birds, candy canes, figurines, and novelties (such as Santa, reindeer, shiny packages, and elves).

Christmas wreaths can easily be made with a wire frame, a variety of foliages, and spool and straight wires (see figures 13-17, 18, 19, and 20).

Figure 13-16. A traditional Christmas wreath is easily assembled with a wire frame, various foliages, and accessories, and spool and straight wires.

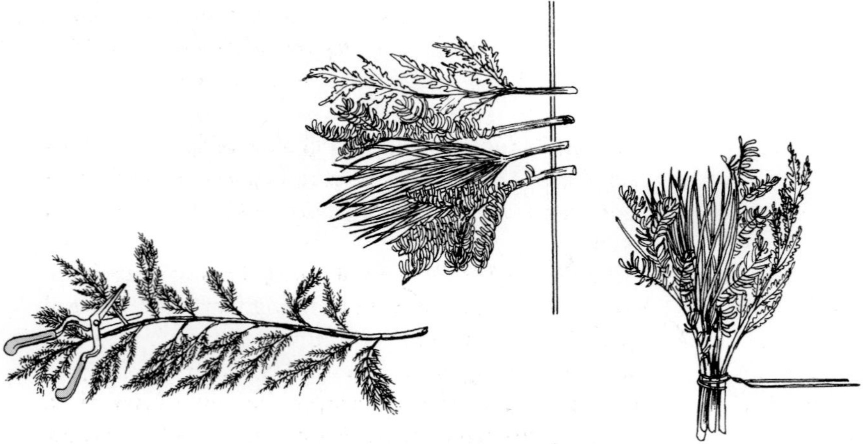

Figure 13-17. To make a fresh wreath, cut apart larger foliage branches and group them into smaller bunches with wire.

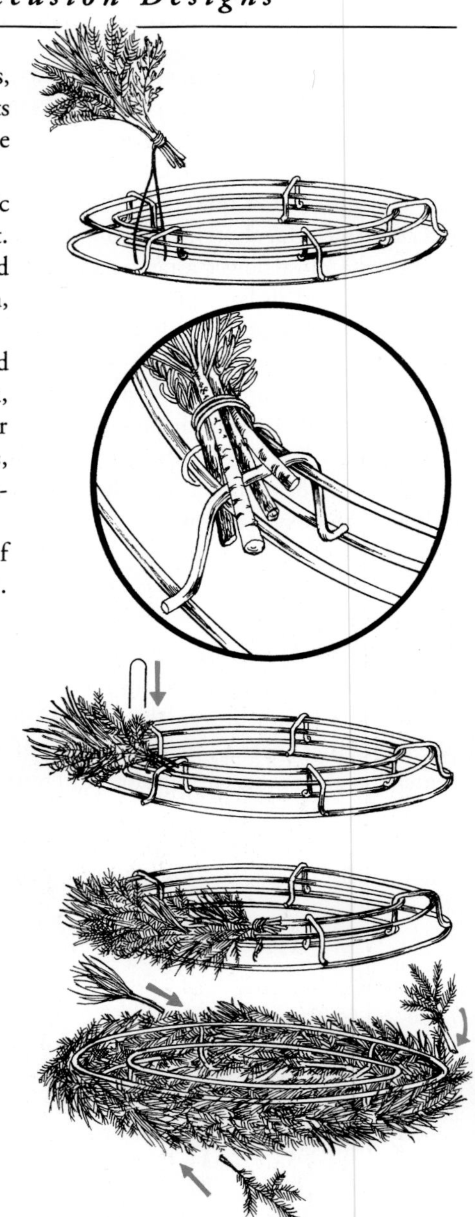

Figure 13-18. Step 1: Attach foliage groups onto wire wreath frame. Wrap wires around foliage groups. Remember to tuck loose wire ends back into the foliage (so the back of the wreath will not scratch a door or table. Step 2: Use small pieces of wire, formed into a U, to anchor the tops of the foliage groups onto the frame. Step 3: Continue adding foliage groups, in a pattern that conceals the previous group's wire and stem ends. Step 4: Add more groups or single foliage stems into the wreath to create a fuller wreath if necessary.

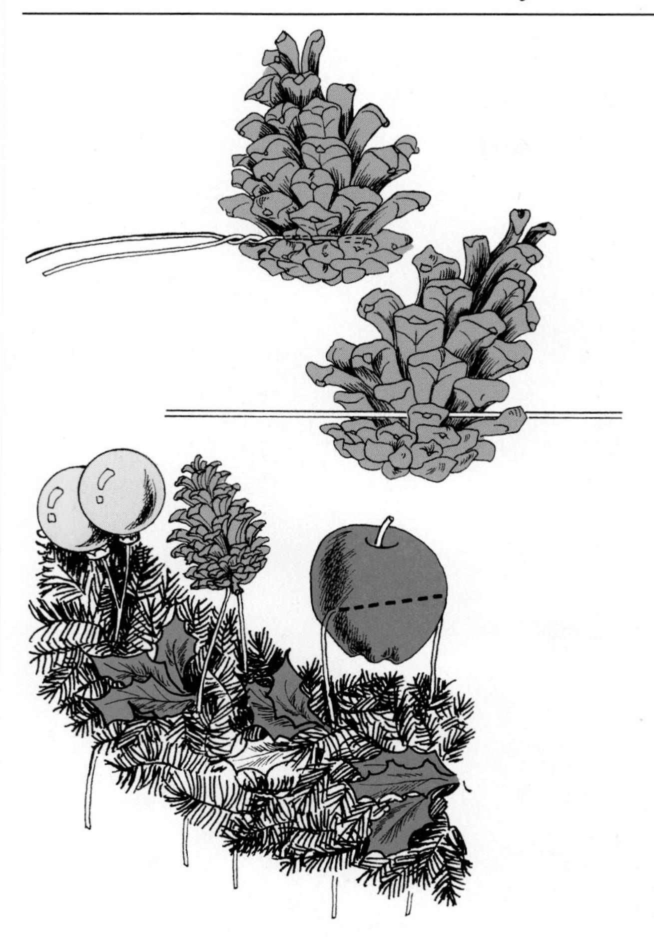

Figure 13-19. Accessories may be easily added (and then later removed) by first attaching a wire. Lay a piece of wire across the base of the cone. Wrap the wire around the cone, inside the bottom layer of bracts and twist the wires together. The wire ends will provide a way to anchor the cone into the wreath. Other accessories, such as fruit and novelties may also be secured with heavy wire and then may be inserted into the wreath.

Figure 13-20. Trim excess foliage and complete wreath with ribbons, bows, and accessories into desired pattern and style.

Wreaths for centerpieces can be made in the same manner, or they may be designed in wet foam to keep the foliages from drying out throughout the season.

Religious Feasts and Festivals

Along with the popular Christian special occasions such as Easter and Christmas and the Jewish holidays such as Passover and Hanukkah, there are many other special days celebrated annually in religions across the United States and the world. Important festivals are observed by Muslims, Buddhists, Hindus, and other religious faiths. As a floral designer, it is important to be aware of special religious days and traditions that may require floral gifts and decorations.

Feasts and festivals are important times of celebration. Most of them take place annually and may last for several days. Many feasts celebrate a harvest, the beginning of a season or a new year, or honor great leaders, saints, gods, or spirits. Other feasts celebrate the anniversary of an important historical event. Most feasts and festivals are happy and joyous occasions, many of which require floral decorations; others are more solomn.

Figure 13-21. A floral bouquet celebrates the joy of new life.
Used with permission by McGinley Mills Inc.

Special Occasions of States and Nations of the World

Most states and nations set aside one or more days each year as public holidays, often recognizing the anniversary of an important historical event or the birthday of a significant person in the past or present. These dates may vary from year to year or may be the same date each year. These special days often are marked with parties and parades where the use of floral gifts and decorations are important.

THEME DESIGNS FOR SPECIAL OCCASIONS

In addition to traditional holidays, other special occasions occur throughout the year. Many of these occasions are associated with the giving of flowers, for instance the birth of a baby, a birthday, graduation ceremony, or wedding anniversary. Flowers are a symbolic sharing of another's happiness.

New Baby

Floral designs to celebrate the birth of a child are commonly made in novelty containers. Color schemes in pastel pink often symbolize the birth of a baby girl while blue,the birth of a baby boy. Smaller flowers such as spray carnations, sweetheart roses, daisies, and button mums, combined with delicate filler flowers, such as baby's breath (Gypsophila) and heather is popular for these smaller designs. Rose arrangements, as shown in figure 13-21 are also popular for new baby and mother.

Accessories common to new baby arrangements include rattles, bottles, bibs, balloons, and plush toy animals. However, floral designers will limit themselves if they use only teddy bears and rattles. New baby arrangements can vary from the usual "darling and cute" all the way to tropical and high style.

Figure 13-22. A birthday bouquet, often festively decorated, is a cheery message that tells someone to have an especially nice day.
Used with permission by Carik Services Inc.

Birthday

To many, there is no greater gift on a birthday than receiving fresh flowers. Mixed floral arrangements, rose designs, boxed flowers, hand-tied bouquets, and even corsages and boutonnieres can all express good wishes and cheerful sentiments for someone on this special day.

A "birthday bouquet" often incorporates brightly colored flowers, party horns and hats, and streamers as shown in figure 13-22. Balloon bouquets are also popular with flowers. Receiving fresh flowers on a birthday is a memorable gift, not easily forgotten.

Graduation

Graduation is an important sentimental time, signifying accomplishment and new beginnings. From elementary school to high school and college, flowers are often given to a graduate to express congratulations (see figure 13-23). Flowers have long been powerful expressions of best wishes for another's success, and may be given in a variety of styles, for instance,

Figure 13-23. Flowers have long been used to help celebrate graduation services.
Graduation from Prep School, Salt Lake City, circa 1890s. Photo by Charles Ellis Johnson, Wadsworth Photo Archives.

arrangements, boxed flowers, wrapped flowers, and leis.

Anniversary

An anniversary may be any date on which a memorable event has occurred, such as an engagement, wedding, office opening, or other occasion or activity. Floral gifts and centerpieces can help to celebrate any type of anniversary party celebration.

Many couples celebrate their wedding anniversaries with festive parties and floral gifts. Many women, and men as well, find nothing more romantic than a gift of beautifully fragrant, fresh flowers from a spouse. Some couples carry on a tradition of giving each other the same number of

Figure 13-24. Floral arrangements for the 25th silver anniversary and the 50th golden anniversary, are often made in corresponding colors and novelty accents.

Figure 13-25. To assemble a presentation bouquet, first gather materials, including, long-stemmed flowers, foliages, spool wire, and ribbon. Step 1: Bind together with spool wire several tall foliages to form the back of the bouquet. Step 2: Next, lay several tall flowers on top of the foliage base, and bind their stems with the foliage. Step 3: If desired, add linear foliages or flower accents. Bind with spool wire. Step 4: Make a full, rounded shape by adding more flowers and fillers. Bind securely with spool wire. Step 5: Add a full bow and ribbon streamers where all the stems come together.

flowers as the number of years they have been married.

Anniversary parties and dinners are often held for couples who have been married for a certain number of years, such as 25 or 50. The 25th anniversary is often called the ***silver anniversary*** and the 50th anniversary is referred to as the ***golden anniversary.*** Accessories and containers used to make floral decorations for these special anniversary parties may be selected either in silver or golden colors to help enhance the 25th or 50th theme (see figure 13-24).

Recognition and Honor

Floral arrangements and gifts of flowers are often used to honor an individual for achievements such as reaching a certain age or retirement. Individuals may also be distinguished with gifts of flowers for acts of heroism or praiseworthy service.

Presentation bouquets, sometimes called ***queen's bouquets*** or ***arm***

Figure 13-26. Athletes and other performers are often recognized with presentation or hand-tied bouquets.

Figure 13-27. Hand-tied bouquets, also call European hana-tied bouquets and Dutch spiral bouquets, differ slightly from presentation bouquets because they generally appear rounded and massed on all sides. Step 1: First, prepare all flower and foliage stems by removing the lower foliage from the stems. This will ensure that no foliage will fall beneath the binding point of the stems. Step 2: From start to finish, all the materials will be held in one hand. Begin by holding a stem of foliage between your thumb and index finger, as shown. Step 3: Place a flower stem at a slight angle, against the foliage stem. Step 4: Continue adding stems of flowers and foliages in a spiral fashion, allowing the tops to angle outward to create a full, round bouquet. The stems will also spiral outward at the base. Step 5: While still holding the bouquet, add ribbon, string, raffia, or other binding material at the binding point (where the stems all cross each other). Wrap the ribbon tightly around the stems several times for security. Tie the ribbon ends together. Step 6: Cut the stem ends straight across, which will allow the bouquet to stand on its own. Step 7: The bouquet can be then easily inserted into a vase. Clear glass will allow the spiral pattern of the stems to be seen.

bouquets, are popular gifts to help recognize people for the performance of certain duties or activities, such as athletics, beauty contests, or cultural performances. Presentation bouquets are generally one-sided and tall, with the individual flower heads spread apart (see figure 13-25).

Hand-tied bouquets, also called European hand-tied bouquets and Dutch spiral bouquets, generally appear the same on all sides, and are most often massed and rounded. The rounded tight cluster of the entire bouquet is the emphasis and individual flowers often lose their focus (see figure 13-27).

HOSPITAL DESIGNS

A hospital stay is often brightened with a gift of flowers. Flowers may communicate a happy and joyous "Congratulations" for a new baby, or a more solomn and serious message. Flowers may express a tender "Hang in there," "Thinking of you," or "Wishing you a speedy recovery."

Whatever the message for a hospital arrangement, a few practical guidelines of construction and design will assist you as you select containers, flowers, foliages, accessories, and decide upon a shape and style of arrangement.

"Would you believe it? I've been all over town trying to get you some flowers."

Figure 13-28

An important scale consideration is the size of the floral design, since hospital rooms are generally small with limited space for displaying a floral bouquet. When floral arrangements are too large, they consume too much space and can be a nuisance to hospital personnel. But just because the flower arrangements must generally be kept small doesn't mean they need to look cheap and boring. As you consider the parts of the bouquet allow individual blossoms to be seen; use fewer, showy flowers, because you are in essence, "charging for space." Keep hospital designs compact, but not tight and massed.

Hospital arrangements must be "low maintenance," that is, the flowers must be arranged in a container or vase and should not be "loose." For example, a dozen roses in a box or a hand-tied bouquet without a vase are inappropriate as the patient
will not have access to a proper sized vase nor should the patient have to worry about arranging flowers. Containers and vases should allow an adequate water supply for the flowers (especially for the duration of the patient's stay) or one that can have water easily added, if necessary, for longer periods of time. Generally, hospitals lack the personnel to care for flowers properly, and the patient should not have to worry about adding water to the design.

It is also important to select flowers and foliages that are maintenance free and long lasting. Flowers and foliages that droop and shed are not suitable in the hospital setting. If lilies are used in designs, remove the pollen (which will stain). Avoid flowers with a strong fragrance (which may cause nausea to the patient). Fragrance is generally an added pleasure of flowers, but in a small, warm hospital room it can be overpowering to a weakened patient.

Most hospitals welcome floral arrangements, and hospital employees realize the benefit of flowers to the morale of the patients. However, it is helpful to understand hospital rules and regulations. For example, flowers are generally not permitted in Intensive Care Units.

EXPRESSING SENTIMENTS WITH FLOWERS

Flowers are a versatile gift that can touch the human heart and transcend speech; many thoughts and feelings can be expressed more easily with flowers than they can with any other gift. It has been said that flowers speak an international language, understood by all peoples of the world.

Flowers have a unique ability to express sincere, honest emotion without the use of words. Flowers have power to

Figure 13-29. There are many ways to depict romance in a design.
Provided by the American Floral Marketing Council (AFMC) © 1993.

shout "I love you" or "Have a happy day." They can say "Thank you" or "I appreciate your friendship." Flowers can express tender emotions by whispering "Please forgive me," or "My thoughts are with you" or simply "You're not alone."

When designing a floral arrangement, consider the message that the bouquet will be "speaking" to the recipient. Designers have a special kind of stewardship over the flowers. This will help you select more carefully not only the flowers and foliages, but the container, accessories, and the total style and look of the bouquet.

REVIEW

Floral decorations and arrangements can be designed to go with a season, color scheme, or special occasion. Flowers can help celebrate religious, civic, and cultural events. As a floral designer, you can create an environment at parties, celebrations, and religious ceremonies. Special occasions take place daily. Flowers help make these occasions more unique and memorable.

TERMS TO INCREASE YOUR UNDERSTANDING

arm bouquets
creche
cornucopia
floral holiday
golden anniversary
hand-tied bouquets
presentation bouquets
queen's bouquets
regional
seasonal flowers
shamrock
silver anniversary
vernal equinox

TEST YOUR KNOWLEDGE

1. Name popular flowers, foliages, accessories, containers, and color schemes that depict the seasons of spring, summer, autumn, and winter.
2. What are the major floral holidays?
3. What are some regional holidays?
4. What is the rule that sets the date for Easter each year?
5. What are some important guidelines of design for hospital arrangements?

RELATED ACTIVITIES

1. Visit a floral shop before a floral holiday. Notice the variety of styles and shapes of bouquets on display.
2. Assemble a design scrapbook of pictures of seasonal and holiday floral designs.
3. Make a hand-tied bouquet with a special theme or recipient in mind.
4. Design and assemble a presentation bouquet for a particular occasion.
5. Plan decorations for a religious, cultural, or civic event.

Chapter 14

Flowers to Wear

Figure 14-1. Spring *by Giuseppe Arcimboldo.*
Cliche des Musees Nationaux, Paris

Flowers have long been used for personal adornment and decoration. Modern floral designs meant to be worn or held come from a number of floral traditions. A few examples show clearly that the idea of wearing flowers is not a new one. The ancient Grecian garlands and chaplets, the Polynesian floral leis, the Georgian period formal gown accents, as well as the Victorian Era tussie-mussies all testify of the popularity of using flowers in one's dress.

As illustrated in figure 14-2, flowers can be arranged in an infinite number of design patterns and worn literally from head to toe. This chapter will discuss guidelines of design, wiring and construction techniques, and various design styles.

GUIDELINES OF DESIGN

Many of the principles of design that you follow in making flower arrangements in vases can be directly applied to the art of making corsages and other floral pieces. However, the following principles are additional guidelines for the construction of

Figure 14-2. Floral arrangements for wearing may be made in a variety of shapes, sizes, and styles.

floral arrangements meant to be worn, often called "body flowers."

Theme and Style

The color and style of the gown, suit, hat, purse, or even hair and hair style to which the flowers will be attached is of utmost importance in determining the color and style of the floral piece. For example, flowers designed to be attached to the decollatage of a sequinned black velvet gown will have a certain theme and style totally different to a floral ***chaplet*** for a young flower girl.

The occasion or event for which these designs are made is also an important consideration, and knowledge of the environment in which the flowers will be worn will help you select the parts. A formal black tie dinner-dance will most surely dictate a style different from a luncheon honoring the volunteer candy-stripers at a hospital. The black-tie event suggests glitz and glamour while a hospital luncheon does not.

Figure 14-3. Flowers have long been worn by men and women on special occasions.
James McGaw Slade and Agnus Burnham on their wedding day, 1897. Photo files of Jarad and Debbie Slade. Used with permission.

Figure 14-4. The parts of a corsage (or other floral piece) must be harmonious to each other as well as for the event for which the design is made.

The flowers, foliage, ribbon, and other accessory fillers, often called the "parts," used to make a floral piece, must be harmonious to one another in color, texture, and style. What kind of flowers you use—whether they are carnations, roses, orchids, or zinnias—is not as important as how well they blend with other flowers and foliages. Ribbon and other fabric materials added to the design should complement the flowers; the textures, colors, and patterns of ribbon are all elements to consider when adding loops or bows to corsages. Delicate lace ribbon will suggest a different texture and style to that of shimmering metallic gold or silver ribbon. ***Filler accessories*** such as hearts, pearls, butterflies, and other tiny novelties, when used, should also be in harmony with the flowers and fit the style.

Proportion and Scale

The size and amount of flowers, foliages, and accessories of the corsage must be proportionate to each another. A corsage with too much ribbon, netting, or other accessories does not allow the flowers and foliages to be fully seen and appreciated.

The size of the completed design should also be in proportion to the person who will be wearing the corsage, boutonniere, or hair piece. This is especially true in the case of small children or petite women who can easily be smothered and frustrated by a large floral piece, as shown in figure 14-5.

Sizes of corsages will vary with trends and styles. As shown in figure 14-6, corsages have not always been small and compact. For some occasions, such as homecoming, and in some regions, the philosophy is "the larger, the better." Corsages and boutonnieres are designed purposely large and out of proportion to the wearer.

Shape

The basic shape of the floral piece should take into consideration where the design will be worn. For example, the hairstyle will help determine the actual shape of a hair piece. Whether worn on the shoulder, wrist, or neck, the shape of a corsage must be in harmony with its placement. If the shape is wrong for the way the floral piece is intended to be worn, as shown in figure 14-7, problems will probably occur. Flowers become a nuisance when they are the wrong shape.

Figure 14-5. Proportion and scale are important considerations, especially when designs are made for children. Unlike this corsage, designs should "fit" the wearer.

Figure 14-6. Corsages have not always been petite as shown by the corsages worn by these women on graduation day. Salt Lake City, circa 1880.
Symons Family. Wadsworth Photo Archives. Used with permission.

Always remember to blend the shape of a floral piece to be an added accessory to the total "look."

Mechanics

Just as you must carefully arrange flowers in a container for proper balance and stability, you must also design corsages, boutonnieres, and other floral pieces securely. The design should be well constructed to retain its original shape, letting nothing fall out of the design (see figure 14-8). When flower petals shed or parts drop off all together, a curious and embarrassing situation arises for not only the person wearing the floral piece, but for the designer and gift giver as well.

Whether attached with pins, a wristlet, a barrette, or a clip, a floral piece must be lightweight and easy to wear. Heavy corsages put a strain on clothing and a bulky corsage results in discomfort and self-consciousness (see figure 14-9). Heavy wrist corsages are undesirable as well, often making the design a burden to wear. A minimum of stems, wires, and tape will keep the design less weighty. Several guidelines and techniques are listed later in this chapter to help you construct lightweight floral pieces.

Balance

Both visual and physical types of balance are important when making floral pieces. Choosing asymmetrical, symmetrical, or radial balance before beginning construction of a floral piece will help promote mechanical balance throughout construction (see figure 14-10). Corsages will be more apt to lay flat on the shoulder or wrist when visually balanced.

Figure 14-7. Problems may arise if a corsage is the wrong size or shape.

Figure 14-8. Corsages and other designs to be worn must be secure and well constructed so nothing can fall out and their original shape can be retained.

Figure 14-9. Corsages must be lightweight! Heavy designs cause discomfort and put a strain on clothing.

Figure 14-10. For visual balance, as well as mechanical balance, choose an asymmetrical, symmetrical, or radial design before constructing a corsage or other floral piece.

For stability, the heaviest portion of the design should be located at the point where all the stems are physically bound together. This central area should also be the point of attachment to a wristlet, barrette, or wherever the pins hold the floral piece to clothing.

Focal Area

A focal area or center of interest in a corsage or hairpiece draws attention and provides a visual, as well as physical area where all lines converge to and from, just as it does in floral arrangements. A focal point can be created in various ways. A larger or more unique flower will easily create a focal point, as will a bright color, dark shade, or any contrasting color to the rest of the design. Sometimes, for special events, accessories, used in greater proportion to the flowers, will provide an instant focal point (see figure 14-11). Place the focal point at the center of gravity; never place a focal point at the design edges (for example, flowers at the top and a bow, to serve as a focal point, at the bottom). This results in a lopsided design, both visually and physically.

PREPARATION OF MATERIALS

Before construction of corsages, boutonnieres, and other floral pieces begins, the materials going into the design must be prepared. Fresh flowers and foliages must be conditioned, wired, and taped, occasionally sprayed with paint or an anti-transpirant, and accessories such as ribbon loops and tulle fans must be made. Having once done all the initial design work, you will be more efficient during the construction and formation of a floral piece.

Conditioning

Conditioning is a technique that allows flowers and foliages to fully hydrate with water and preservative before using them in designs. Fully turgid or firm flowers and leaves will hold up better after losing their stems to wire and floral tape.

It is especially important to condition blossoms and leaves that are harvested from blooming and green potted plants. Do not wait until the last minute to select various blossoms for corsage work. Give these flowers and leaves time to drink in water and preservative and fully hydrate before using them in corsages and boutonnieres.

Wiring and Taping

Wiring and taping flowers and leaves replaces natural stems; if left on, they would be too bulky and heavy. Wire allows more freedom in design, making it easier to manuever stems and keep flowers in position while being worn. Wire strengthens and for some flowers lengthens their stems.

Several wiring techniques are used in corsage work. The wiring method used will depend mostly on the actual shape of the flower head or cluster of flowers. The thickness or gauge of the wire used is determined by the weight of the flower head and where in the design it will be placed. Large or heavy flowers closer to the binding area will require thicker wires. Small, delicate flowers positioned on the perimeters of the design will not require the same thickness of wire.

The most useful wire gauges are medium to fine in thickness. Gauges #24, #26, and #28 are the most common for corsage work. Use the lightest gauge wire possible that will do the job. When making corsages and boutonnieres, trim out excess wire to keep design pieces lightweight. As you prepare your floral piece be aware that not every addition to a corsage (or other floral piece) must be wired and taped in. Many types of low-temperature glues and floral adhesives are available to add in lightweight accents of ribbon and blossoms.

Roses and Carnations

Roses and carnations have rounded heads and a visable calyx. The most common method of wiring for these and other similarly shaped flowers is often referred to as the pierce method (see figure 14-12).

Figure 14-11. Flowers that are large, brightly colored, or uniquely shaped will create a focal area, as will accessories in a theme floral piece.

Step 1. Remove all but about 1/4 inch of stem. Do not remove the sepals from roses; left on, they provide color and accent, and look more natural.

Step 2. Insert a medium-gauge wire (#24) into the calyx. (It may also be inserted across the base of the petals for roses.)

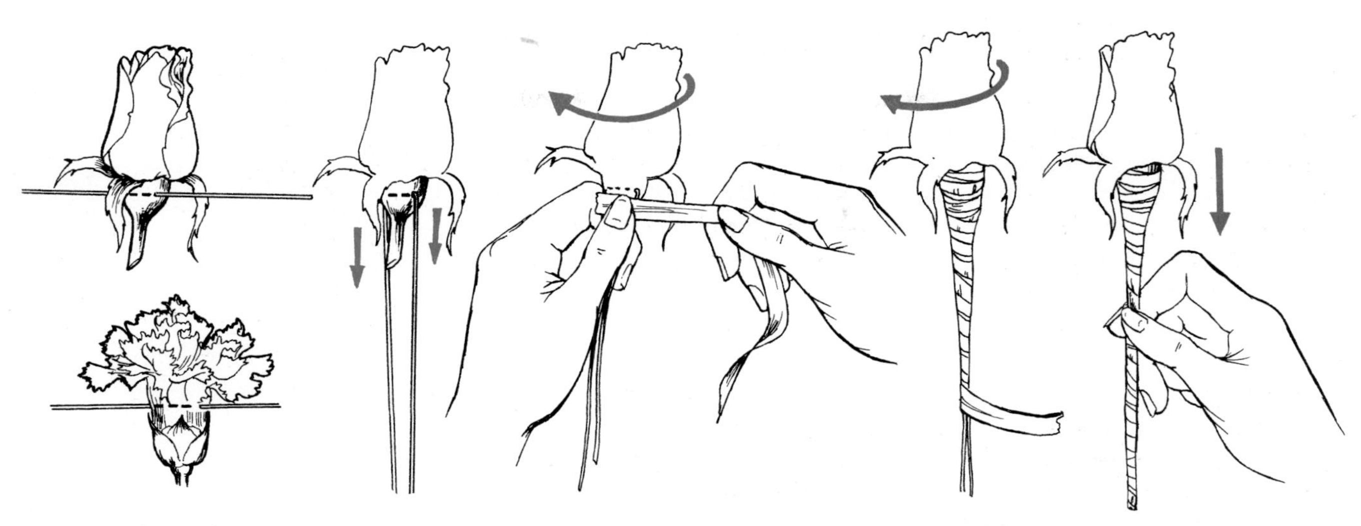

Figure 14-12. The most common method of wiring roses, carnations, and other similarly shaped flowers is the pierce method.

Step 3. Bend the two wire ends downward, keeping them parallel to each other. Never twist wires. This causes too much bulk in the stem.

Step 4. Floral tape the wires by stretching a piece of floral tape around the calyx. Wrap around in one spont until the tape begins to stick to itself. Continue stretching the tape downward onto the wire in a spiral pattern, covering all the wire. After floral taping, the tape can be further tightened by applying pressure with fingertips.

Rose Petals

Individual rose petals may be used to form tiny rose buds for accents and contrast in floral pieces (see figure 14-13).

Step 1. To make a rose bud, trim a tiny half circle at the base of the petal in the center.

Step 2. Roll a single petal from side to side.

Step 3. Add some floral tape or low-temperature glue to keep the petal from unrolling.

Step 4. Add a second or third rose petal around the bud to give it more fullness if you wish. Attach these other petals with tape or glue.

Step 5. Insert a wire above the floral tape and bend wire ends downward.

Step 6. Floral tape to form a natural-looking stem.

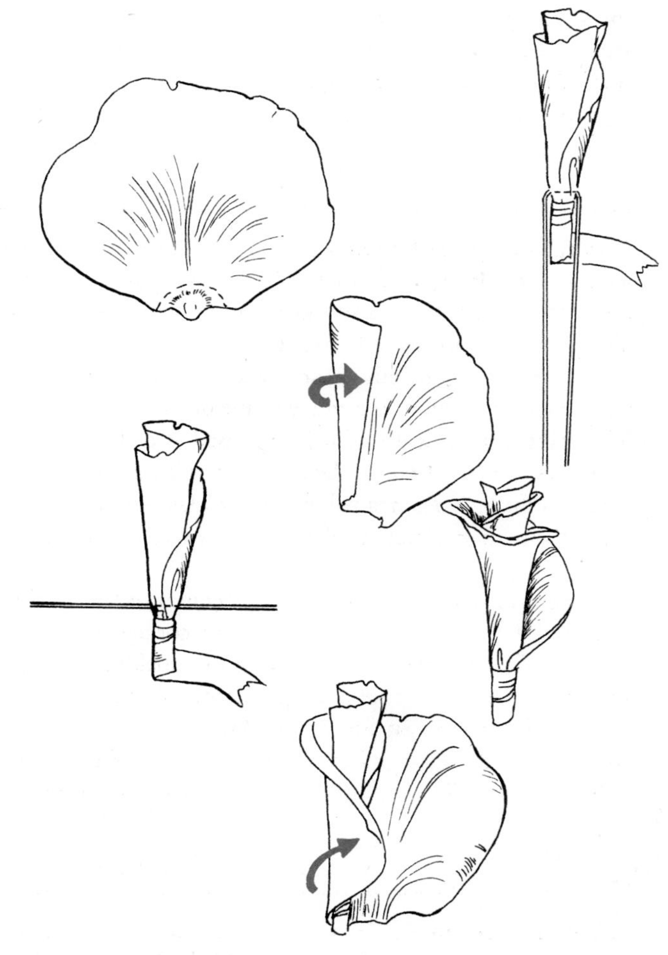

Figure 14-13. A designer may roll individual rose petals to form tiny rose buds for accent or contrast.

Feathering Carnations

Petals from carnations may also be used to make smaller flowers. The process of splitting a carnation apart is often called ***feathering.*** Feathering large carnations takes a little more time and effort but allows for more versatility in design. Knowledge of this technique is especially beneficial if small carnations are needed for a design and miniature (pixie) carnations are not available. Several flowers (two to six) may be made from a standard-size carnation.

Step 1. To feather a carnation, remove its stem. While pressing firmly at the base of the calyx, roll from side to side until the ovary with its attached center part (the pistil) slips out and can be removed (see figure 14-14).

Step 2. Next, peel the calyx sections downward, away from the petals. This will open up the flower head, yet still keep the petals attached to the base.

Step 3. Remove a section of petals to form the new smaller flower.

Step 4. While holding the petals tightly in one hand, wrap the base of the cluster together with a piece of floral tape. Once the floral tape is tight around the petals (forming a new calyx), the section can be wired.

Step 5. Insert a thin wire just above the floral tape, as shown in figure 14-15, and bring the wires down together, keeping them parallel to each other.

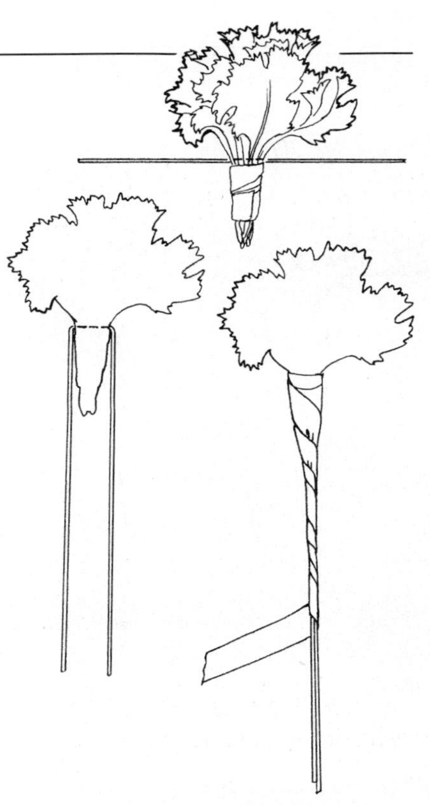

Figure 14-15. As the final steps for feathering a carnation: 5. pierce-wire above the floral tape; 6. bend the wire down; and 7. floral tape the wire stem.

Figure 14-14. To feather a carnation to make smaller florets: 1. press and twist (roll) the base of the calyx until the center comes out; 2. peel down the calyx sections (like peeling a banana) keeping petals intact; 3. divide (or section) pieces of petals; and 4. floral tape base of section together.

Step 6. Floral tape to form a new stem.

Chysanthemums

Chrysanthemums, asters, daisies, and other flowers with flattened heads lacking a visable calyx, are wired for security using hook-wiring (see figure 14-16).

Step 1. Remove all but 1/4 to 1/2 inch of stem.

Step 2. Insert a hooked or U-shaped wire through the top of the flower head, and pull the wire ends down until they can not be seen from the top.

Step 3. Floral tape the wire to form a new stem.

Figure 14-16. The hook-wire technique is used for daisies and other flowers with flattened heads and no visable calyx.

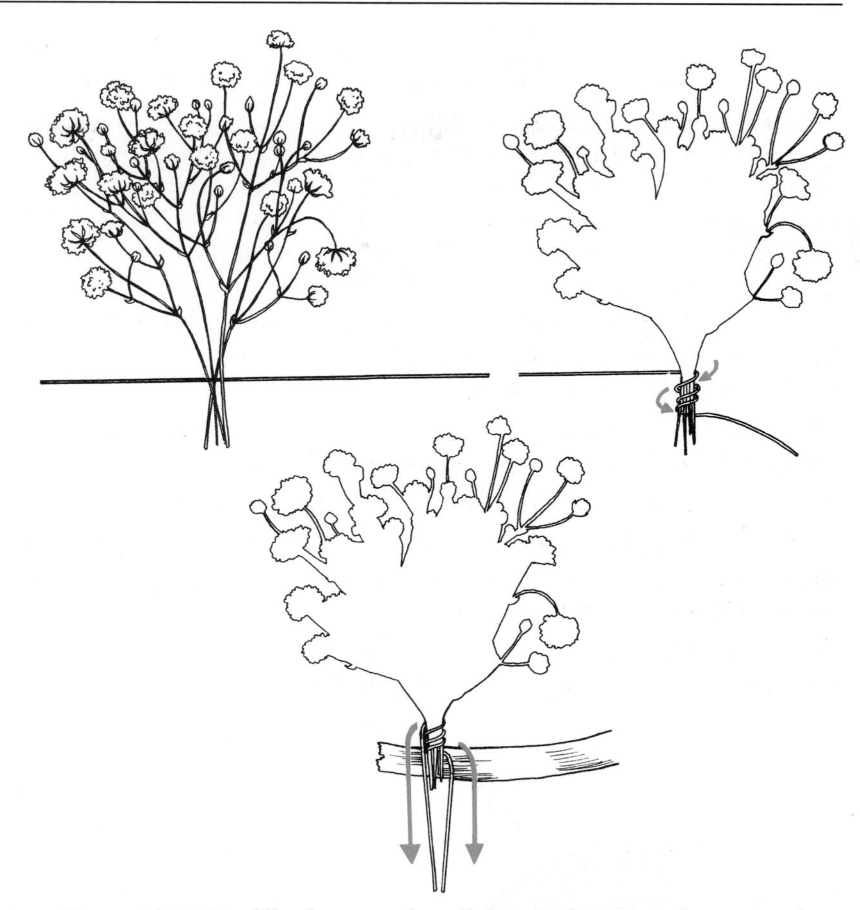

Figure 14-17. For filler flowers and small clusters of tiny mass flowers, use the "clutch" or "wrap-around" method.

Filler Flowers

Filler flowers and small clusters of tiny mass flowers are often wired by using ***clutch*** or ***wrap-around wiring*** (see figure 14-17).

Step 1. Gather the stems of baby's breath, statice, waxflower, or other tiny fillers into a small cluster.

Step 2. Wrap a lightweight wire tightly around the stems.

Step 3. Bend the wire ends downward, keeping them parallel to one another.

Step 4. Floral tape the stems and wires together.

Stephanotis

Stephanotis flowers can be wired in a number of different ways; however, the most efficient method is to use a ***stephanotis stem.*** These stems are manufactured specifically to provide a stem and keep stephanotis flowers from wilting. Before using stephanotis, it is best to condition them in cool water to firm them up.

Step 1. Moisten the cotton portion of the stem in water for a few minutes.

Step 2. Remove the tiny green sepals and tiny stem from the blossom.

Step 3. As shown in figure 14-18, use the wired stem end to push out the tiny ovary through the top of the flower.

Step 4. Insert the cotton stem into the flower. (It can be poked down through the top or can be pushed up through the base of the flower.) Generally, the flower and stem do not need any further wiring or taping and are ready to use in designs.

Cymbidium, Cattleya, and Japhet Orchids

Orchids are wired differently than other flowers because of their unique shapes. Orchid stems can be left in a tiny plastic tube for a constant water supply or the stem can be wrapped with wet cotton or tissue (see figure 14-19). Generally,

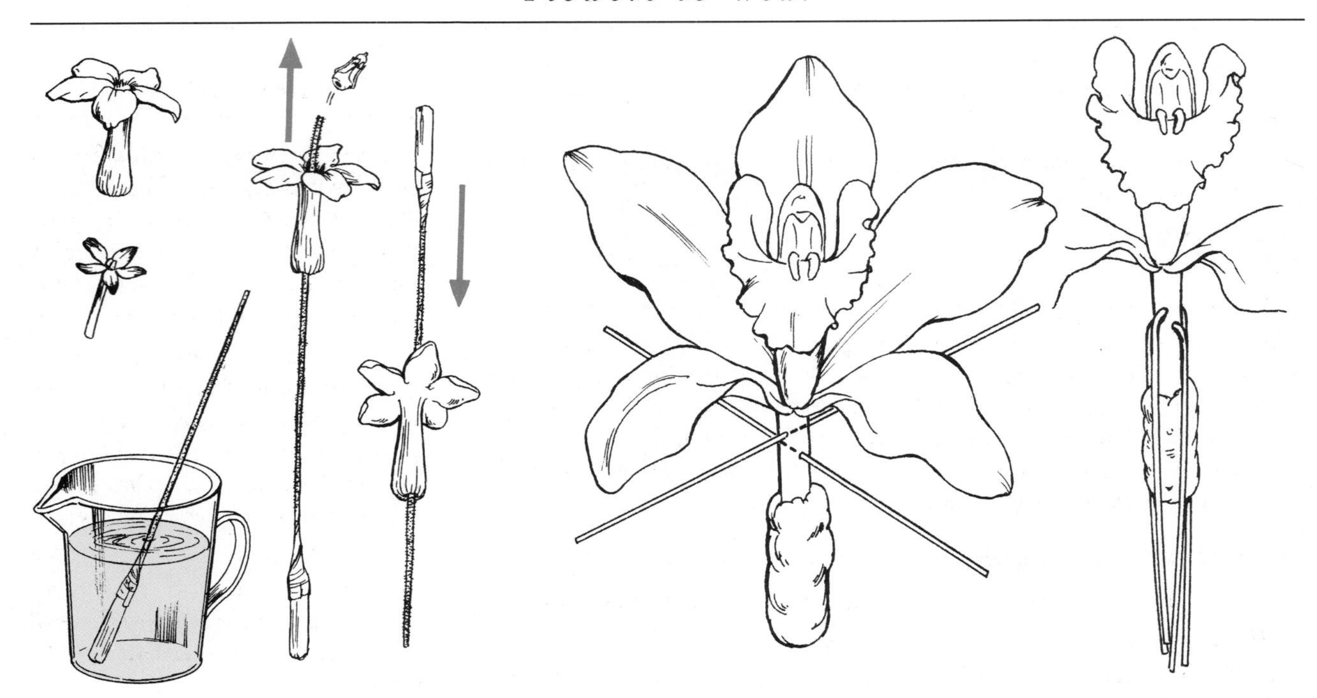

Figure 14-18. Manufactured stephanotis stems may be used to wire delicate stephanotis and keep them fresh while in designs.

Figure 14-19. A wet tissue or cotton held in place by cross-wiring, keeps larger orchid flowers fresh.

two wires are needed when wiring cymbidium, cattleya, and japhet orchids.

Step 1. Recut stem and place in corsage water tube, or wrap stem end in a piece of moistened cotton or issue.

Step 2. Insert a heavy wire (gauge #22 or #24) through the orchid stem just below the flower. This wire will provide strength at the same time holding the flower in the proper position.

Step 3. Insert a finer wire (gauge #26) through the stem above the previous wire at a right angle to the first wire. This thinner wire will help keep the moistened cotton, tissue, or water tube in place.

Step 4. Bend all wires downward, keeping them parallel with one another.

Step 5. Floral tape the source of moisture inside the wires, covering all the mechanics.

Dendrobium and Paphiopedilum (Lady's Slipper) Orchids

There are several ways these orchids may be wired, depending upon how they will be used in the design.

Step 1. Insert a medium wire (#24 or #26) vertically up the short stem.

Step 2. Push the wire out through the throat of the orchid.

Step 3. Curve the wire end into a small hook. Then pull the wire hook gently back into the orchid until it reaches the flower base. The end of the hook will protrude slightly through the back of the orchid.

Step 4. Floral tape the wires to provide a new, natural-looking stem.

Phalaenopsis and Vanda Orchids

Phalaenopsis and vanda orchids are wired differently than other orchids because they are quite fragile and require extra support within the flower head itself (see figure 14-20).

Step 1. Floral tape a wire (gauge #24 or #26) with white floral tape.

Step 2. Curve the wire into a hook or U-shape.

Step 3. Insert the wire carefully from above the orchid over the middle section and through the visable spaces. The taped wire does not puncture any part of the flower; instead, it provides a source of support for the orchid.

Step 4. Wrap a tiny piece of moistened cotton or tissue around the orchid's stem if you wish. This may be held in place with another wire.

Step 5. Tape the stem, wires, and cotton together to provide a longer, more natural-looking stem.

Gardenias

Highly fragrant gardenias are extremely fragile and easily bruised. Browning of petals can be lessened by keeping your hands wet while working with gardenias. Most gardenia flowers are mounted on special ***collarettes*** with background leaves and packaged in protective boxes that keep them fresh.

To wire a gardenia, it is best to leave the protective collar on the flower. The collar keeps the flower positioned and protects the petals.

Step 1. Insert a wire (#24) through the stem, beneath the collar.

Step 2. Trim the end of the stem and add a moistened piece of cotton or tissue.

Step 3. Insert another wire (#26) through the stem, higher than the first wire and perpendicular to it. This wire will help hold the cotton in place.

Step 4. Bend the wires parallel to the gardenia stem and floral tape.

Step 5. If you wish, trim the leaves on the gardenia collar with scissors.

Lilies and Alstroemeria

Lilies, alstroemeria, and other flowers can be wired several different ways. The flower stem thickness and position of the flower (in a corsage or other floral piece) will determine the most efficient wiring technique.

Step 1. Remove most of the stem except for about 1/2 to 1 inch. Large lilies will benefit by wrapping a small moistened piece of cotton or tissue against the stem end. Remove the anthers (pollen) to prevent staining.

Step 2. Insert a wire (#24 or #26) through the top of the stem. Sometimes a second wire will need to be inserted perpendicular to the first wire.

Step 3. Bend the wire ends downward, keeping them parallel to one another.

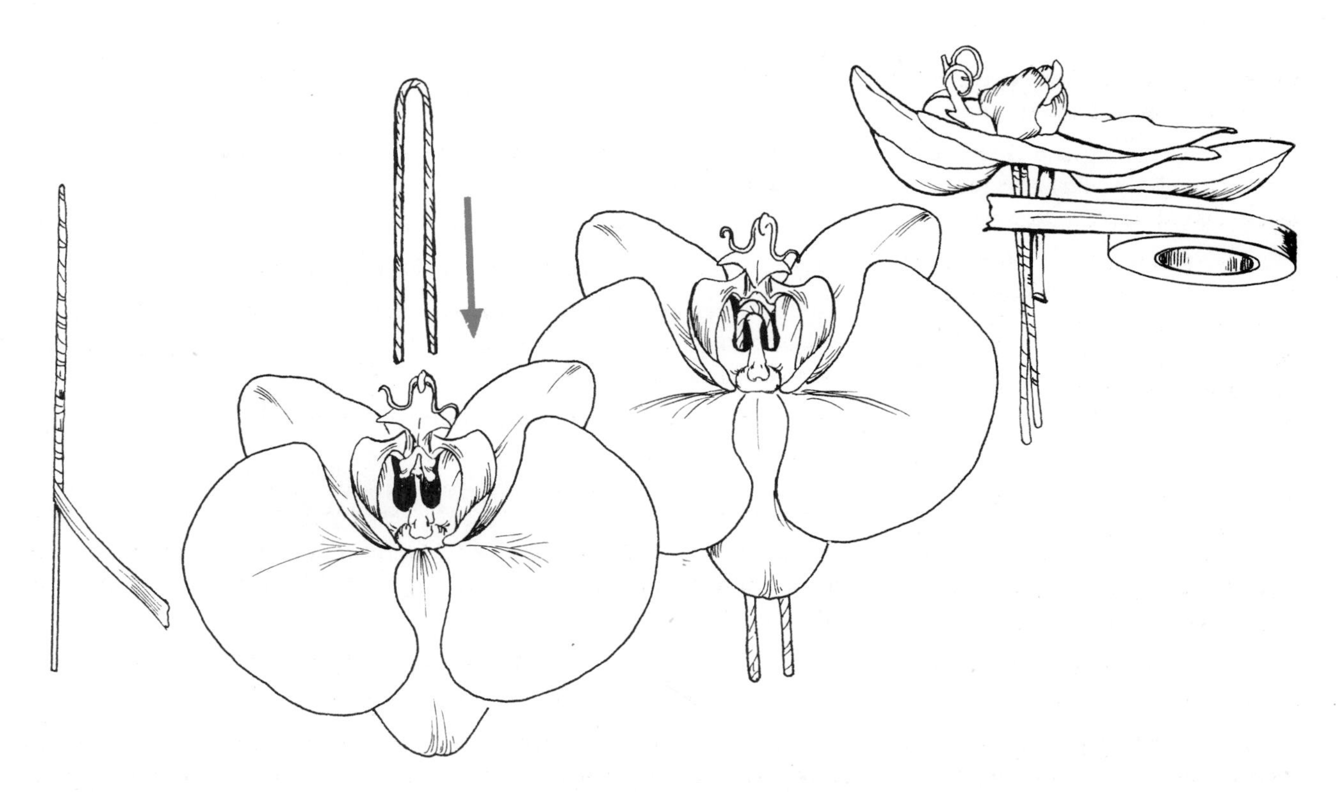

Figure 14-20. Phalaenopsis and vanda orchids require extra support within the flower head itself.

Step 4. Floral tape the stem and wires together, being careful not to damage the flower petals.

Broad-leaf Foliages

Individual leaves of camellia, ivy, salal, and other ***broad-leaf foliage*** can be wired using ***stitch wiring*** (see figure 14-21).

Step 1. Pierce a wire (#26) through the back of the leaf near the center vein and make a tiny "stitch." Make this stitch high enough on the leaf to gain control and support the leaf, but not so high that the stitch will be visable in a design.

Step 2. Move the wire until there are two equal ends, and bend them downwards.

Step 3. Keep one of the wire ends parallel with the mid-rib of the leaf. Twist the other wire end around the stem and the straight wire, holding it in place.

Step 4. Tape the leaf stem and wires to form a new, natural-looking stem.

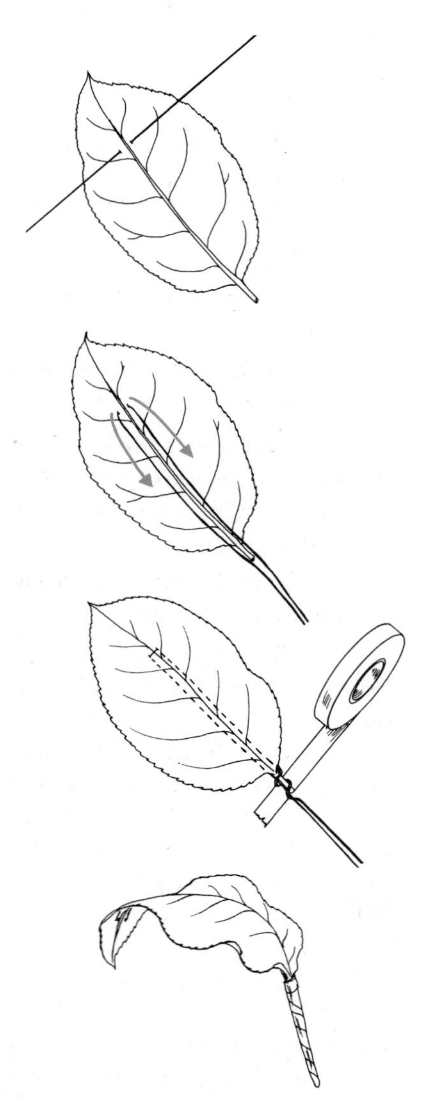

Figure 14-21. The stitch method is used to wire camellia, ivy, salal, and other broad-leaf foliages.

SPRAY PAINTING

Carnations and a few other flower types can be lightly spray painted with special floral tints. It is important that just the petal edges are painted. This technique is called ***tipping*** (see figure 14-22).

Step 1. Wire and floral tape carnation.

Step 2. Pierce wired stem through the center of a paper towel (to protect hands from spray paint).

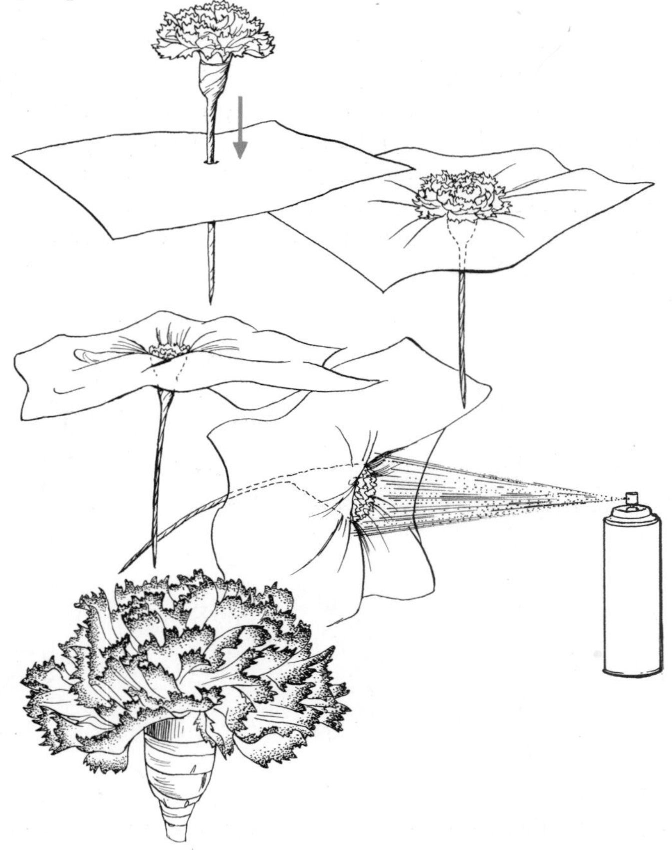

Figure 14-22. When spray painting, or tipping, carnations, be careful to paint only the petal edges.

Step 3. Gather the paper towel up around the flower head.

Step 4. Squeeze the flower head together with the paper towel tight against the petals. Only the petal tips will be exposed to the paint.

Step 5. Apply an even coat of paint. If the tipped carnation is not dark enough or another paint color will be added, the process can be repeated until the desired effect and color is achieved.

ACCESSORIES

Accessories, such as ribbon loops, bows, tulle netting fans, and other novelties, can be added to corsages to enhance a theme and create a unified design. Not all corsages rely on accessories for visual success—some will look better without any extras.

It is important to give thought to the selection of these design extras. Their main purpose is to accent and give importance to the flowers in the design. To keep the entire design lightweight, be sure to choose light extras that are not heavy.

Figure 14-24. Double and triple loops, as well as loops with flags serve to add color and texture and strengthen unity in a design.

Ribbon Loops and Flags

Loops and flags of ribbon are more suited to corsages than multi-looped bows. A large bow in a corsage can be overpowering and unfitting to the flowers and the entire design in which it is placed. Ribbon loops and flags accent the flowers and add color and texture throughout the entire corsage, helping to unify the whole design. (See figure 14-23). Loops can be made in a number of different ways. The following steps explain how to make a simple single ribbon loop.

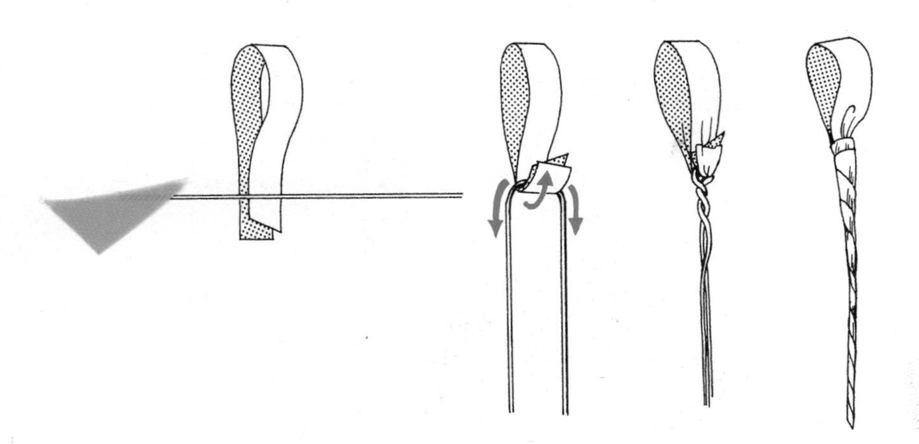

Figure 14-23. To make a single ribbon loop, start with a 3- to 4-inch piece of ribbon and: 1. fold the ribbon in half with the right side out; 2. Place a wire on top of both ribbon ends; 3. fold the ribbon ends up, and bend the wire ends downward. 4. Finish by floral taping the base of the ribbon and wire to hide the mechanics.

Step 1. Cut a piece of ribbon 3 to 4 inches long.

Step 2. Fold the ribbon in half with the right side out.

Step 3. Place a wire (#26, #28, or #30) across and on top of both ribbon ends. Fold the ribbon ends up toward the top of the loop.

Step 4. Bend the wire ends downward and twist them together.

Step 5. Floral tape the base of the ribbon and onto the wire, hiding any mechanics.

As shown in figure 14-24, double and triple loops can be made, as well as loops with tails, called ***flags.*** A variety of ribbons can add shimmering, glitzy, velvety, or lacy textures and patterns throughout a corsage.

Net and Lace Fans

Net and lace may be added to designs to provide a background and also to create fullness in corsages without adding weight. Net, also known as ***netting*** and ***tulle,*** is sold on small bolts, usually six inches wide. Netting is available in a wide variety of colors, patterns, textures, and styles. When using netting in corsages, choose materials that are not scratchy and stiff. These can be annoying and sometimes irritating to the skin when placed in shoulder or wrist corsages.

Net and lace can be cut into sections, forming fans, butterflies, or tufts (see figure 14-25). Several methods can be used to make net background for corsages.

Step 1. Cut a long piece of net (12 to 18 inches) from the bolt.

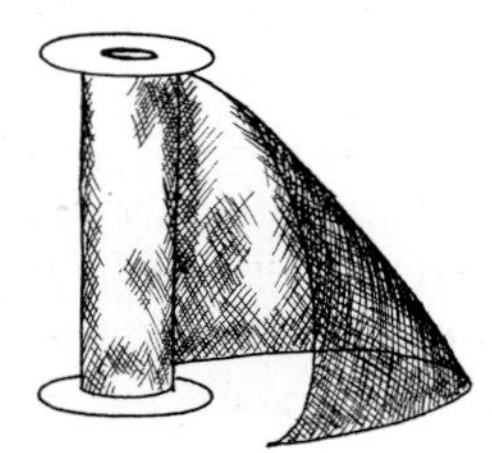

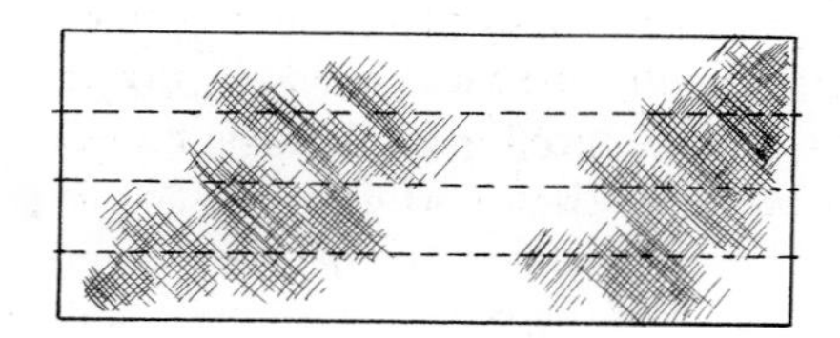

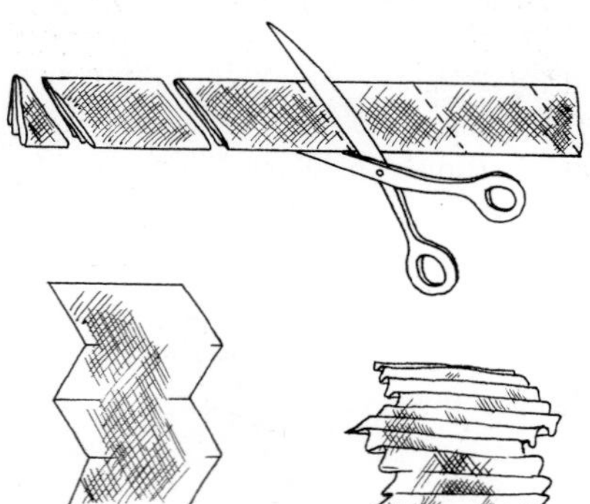

Figure 14-25. To make a simple tulle fan-tuft for a corsage: 1. fold the piece of tulle in half, then half again; 2. cut diagonally into pieces 3 to 4 inches wide; and 3. lay the piece flat gather up with 1/4- to 1/2-inch pleats.

Figure 14-26. As the final steps for making tulle fan-tufts: 4. lay fine wire across all layers of tulle; 5. fold tulle sides up, while bending wire ends down; and 6. twist the wire several times for security. Tulle fans can be made in many other ways as well. The tulle tuft on the lower right is made by folding a piece of tulle and then wiring and taping. This results in a softer, fluffier tuft of netting.

Figure 14-27. Tulle works well as a background accessory in corsages.

Step 2. Fold the piece of netting in half (from side to side). Fold in half again so the piece is folded into fourths.

Step 3. Cut the net, while still folded, diagonally every 2 1/2 to 3 inches. Trim the ends diagonally as well.

Step 4. Open up an individual piece of net and lay flat.

Step 5. Starting at a straight edge, fold up in 1/4 to 1/2 inch pleats (like an accordian).

Step 6. Place a thin wire (#28 or #30) across the top and center of the piece of net (see figure 14-26).

Step 7. Fold the wire ends down together, while folding the two net sides Twist the wires together. The net will form the shape of a butterfly or fan. These sections may be floral taped to hide wires and mechanics.

For a smoother appearance, pieces of netting may also be folded to form a tuft without cut edges. Net pieces are placed to form a background to help visually as well as physically support flowers in a design (see figure 14-27).

Figure 14-28. Corsage accessories range from butterflies and bees to pearls and faux jewels as well as holiday and seasonal novelties.

Novelties

Many novelties are manufactured for use in corsages (see figure 14-28). They range in style from cute, youthful bees and butterflies, to more elegant-looking pearls, rhinestones, and other faux jewels. Lightweight holiday and seasonal novelties are available for use in corsages as well. Novelties generally have a wire attached for convenience in design. However, do not assume that everything must be wired and taped. Low-temperature glues and liquid or spray adhesives may be used to secure some novelties into designs.

Artificial Leaves

Artificial leaves, sometimes called ***glamour leaves,*** can be added to corsages and boutonnieres in addition to fresh foliages. Artificial leaves are available in a wide range of colors, sizes, textures, and cluster groupings. These leaves add color and texture accents throughout a design and generally add a touch of elegance.

BOUTONNIERES

Many events such as weddings, proms, banquets, and other formal events are special occasions for men to wear flowers. However, the occasion need not be formal. Special holidays such as Father's Day, Valentine's Day, Boss Day, or sentimental occasions, such as an anniversary or birthday are also times when flowers are worn by men. A single flower or small cluster of flowers is a classy addition to a suit coat for any day of the year.

A floral piece worn by a man is called a ***boutonniere*** and is generally worn on the lapel of a formal jacket or less formal suit coat. Flowers are most often worn on the left lapel near the buttonhole (hence the name "boutonniere"). Smaller pins with black heads (boutonniere pins) or the larger pins with pearl heads (corsage pins) may be used to attach the boutonniere to the lapel. Pins may be inserted from the back side of the lapel if desired.

Whether the boutonniere is a single rose, carnation, or dendrobium orchid, or a cluster of stephanotis, statice, or a sprig of holly, the success and security of any design will depend upon proper construction techniques. A single flower with some filler and foliage is a popular boutonniere choice (see figure 14-29).

Figure 14-29. After the main flower for a boutonniere is wired and floral taped, filler flowers and foliage may be added. The stem may be curled easily with the help of a pen or pencil.

Figure 14-30. Pins should be inserted through the calyx before the boutonniere is worn. Remember to bend carnation flowers slightly forward to allow the boutonniere to lie flat on the lapel.

Single-Flower Boutonniere

Carnations and roses are the most requested flowers for single flower boutonnieres. There are, however, many other flowers that work well in boutonnieres, such as alstroemeria blossoms and dendrobium orchids.

Step 1. Wire and floral tape a rose (or other flower).

Step 2. Add a sprig or two of baby's breath (or other filler flower, such as statice, heather, Queen Anne's lace, etc.). Use just enough filler flower to accent the main flower, not overpower it. Next, position baby's breath blossoms above and behind the rose head or to the side, or lower in the front of the rose. Floral tape the stems of the baby's breath onto the main flower stem.

Step 3. Add an ivy leaf to the back or side of the design (or other small broad leaf such as camellia, lemon leaf, or pittosporum). Lacy foliages can be used as well, such as leather leaf, ming fern, cedar, tree fern, or plumosa fern. The foliage may have to be wired for additional support before taping it into the boutonniere (generally a stitch wire or wrap-around wire).

Step 4. Floral tape the boutonniere stem tightly and evenly. A tight wrapping will secure the design and look nicer, without wires showing or poking through the tape.

Step 5. Curl the stem of the boutonniere forward around a pencil, if you wish, for an interesting, less blunt look.

Step 6. Attach the pin by inserting it through the thick portion of the wired stem. As shown in figure 14-30, after making a carnation boutonniere, bend the carnation head forward. The flower will look nicer and fit more closely to the lapel.

Multiple-Flower Boutonniere

Once you have learned how to make a single-flower boutonniere, the design possibilities are endless. Two or three little flowers may be clustered together to form a unique boutonniere. It is important to remember to keep these designs fairly small. Avoid making boutonnieres that are so large and excessive that they look like corsages.

Step 1. Wire and tape two or three small flowers.

Step 2. As shown in figure 14-31, select the smallest of the flowers for the height of the boutonniere. Add the other flowers onto the stem of the first in a slight zigzag pattern. Angle the lower flower heads facing out.

Figure 14-31. Two or three small flowers may be clustered together to form a boutonniere.

Step 3. Add filler flowers for accent if desired. Add a few small broad leaves or tiny sprigs of foliage behind the main flowers.

Step 4. Floral tape and complete the stems as desired. Add a pin.

Nestled Boutonniere

Smaller flowers (such as sweetheart roses, pixie buds, or a tiny cluster of statice, or other fillers) can be inserted into the center of a carnation. Knowledge of construction techniques for these unique, novelty boutonnieres is important, for they are often requested for proms and other special occasions (see figure 14-32).

Step 1. Wire a small or sweetheart rose with a #22 wire. Do not tape.

Step 2. Select a carnation with a different color than the rose. Remove the pistil from the center of the carnation. This will allow you to nestle the rose further down into the carnation.

Step 3. Insert the ends of the wire through to the top of the carnation head, down through the center, and out the base of the calyx. Position the rose down into the center of the carnation.

Step 4. Insert a wire (#24) through the carnation. Keep all wires parallel to each other. Tape the wires together.

Step 5. Add filler flowers and foliage to the design if you desire.

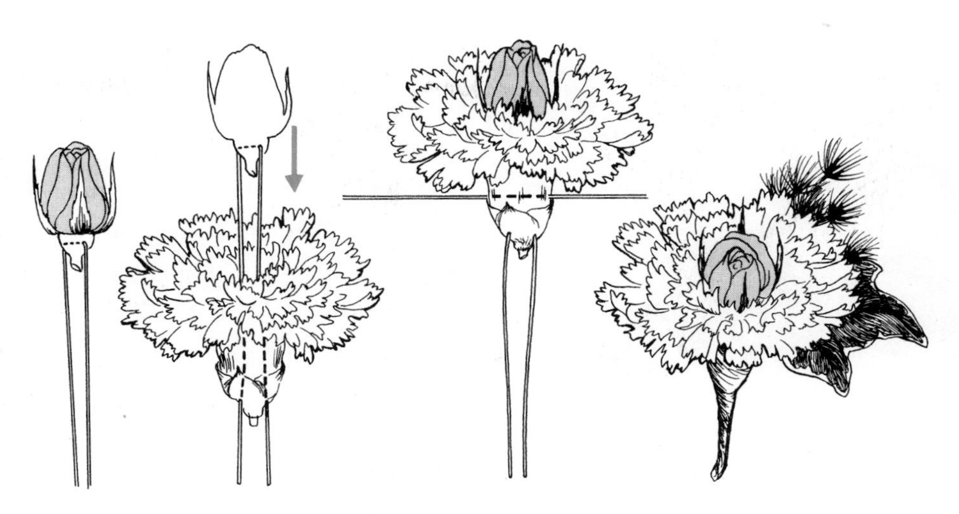

Figure 14-32. For nestled boutonnieres: 1. wire a rose; 2. remove the pistil from the center of a carnation; 3. insert the ends of the wire through the top of the carnation down through the base of the calyx; 4. position the rose in the center of the carnation and, keeping the wires parallel, tape the wires together. 5. For the final step, add filler flowers and foliage.

Figure 14-33. The simplest method to finish stems is to leave them straight; however, curled and clustered stems provide interesting options.

Boutonierre and Corsage Stem Ends

The wired stems at the base of boutonnieres and corsages are generally visable. Because of this, it is important to cover all wires with floral tape. Often a distinctive corsage and especially a boutonniere may be set apart from all others by a simple twist or turn of the taped stems.

One of several simple techniques may be used for putting the "final touch" on the design piece itself. As shown in figure 14-33, the simplest method is cutting the stem and leaving it straight. However, curled stems offer more unique options than does a straight stem. The stems may be easily curled by spiraling the stem around a pencil. The curl can be pulled to the side or stretched out in a number of ways.

Figure 14-34. Corsages and boutonnieres may be created in various flower combinations and styles for a wide variety of occasions.

Stems may also be given a more natural look by keeping individual wired and taped stems separated from one another, taping them together only at the top. In order for the design to appear as if the flowers have been plucked from the garden, vary the stem lengths.

CORSAGES

Corsages are worn by women on such special occasions as weddings, proms, and other formal events. Corsages are popular floral gifts during holidays such as Mother's Day, Easter, and Secretaries' Day, as well as many other times. Whether a corsage is worn by the mayor at a town fair, the woman giving the main address at a conference, the volunteer candy stripers at the hospital, a woman on an anniversary dinner date, or a young girl graduating from elementary school, the corsage sets the wearer apart. Corsages and boutonnieres as floral gifts show appreciation and distinguish the wearer.

Corsages are commonly worn on the shoulder or the wrist. Smaller corsages may be worn in the hair, at the waist, or pinned to an evening purse. The style and fabric of the dress, current fashion trends, the occasion, and personal preference all dictate what type of corsage is preferred. And remember, the success of any design is dependent upon proper and secure mechanics.

Figure 14-35. Single flower corsages are made more elaborate with the addition of filler flowers and foliage as well as a variety of accessory items such as ribbon loops, bows, and netting.

Single-Flower Corsages

Corsages with one main flower share similar steps of construction to a single flower boutonniere. However, corsages differ from boutonnieres in the additions of bows or ribbon loops, and accessories such as tulle, and decorative novelties. The single flower may be a simple carnation or a more exotic gardenia, cymbidium orchid, or cattleya orchid (see figure 14-35).

Step 1. Wire and tape an orchid (or other flower for a single-flower corsage).

Step 2. Prepare accessory items that are to be added into the corsage, such as 2 or 3 net (tulle) fans, 1 to 3 ribbon loops, or a tiny, unobtrusive bow. Lightly spray paint net fans, if desired, to provide color and accent.

Step 3. Prepare several leaves or sprigs of foliage and a few tiny sprigs of filler flowers.

Step 4. Place filler flowers behind and around the main flower as desired, and tape in position.

Step 5. Position foliages at the top and sides of the flower, and tape into place.

Step 6. Add net fans as background material, if desired, around the perimeters of the corsage. Add ribbon loops for color and accent.

Step 7. Trim excess wire from the corsage as you add additional materials. Floral tape securely and curl up the stem end. Add one or two corsage pins to the back of the corsage.

Multiple-Flowers Corsages

Corsages often have several small flowers grouped together accented with foliages, ribbons, net tufts, and filler flowers. Constructing this type of corsage is similar to making several boutonnieres or single-flower corsages and then putting them all together into a secure, lightweight, visually lovely design. A variety of flowers may be used, such as roses, feathered carnations, miniature carnations, chrysanthemums, stephanotis, and alstroemeria. It is important to establish a theme and style for the corsage, so that harmony and unity are achieved. (Refer to figure 14-36 to make a multiple-flower corsage.)

Step 1. Wire and floral tape three to seven small flowers. Choose a variety of flower types and sizes. Select some leaves or sprigs of foliage, and filler flowers, and wire and tape as needed.

Step 2. Prepare accessory materials, such as net tufts, ribbon loops, and a bow.

Step 3. Begin the corsage with a small flower bud. Position a ribbon loop behind the flower and one in front. Tape these together. This flower will serve as the "backbone" or ***spine*** of the corsage.

Step 4. Tape filler flowers, leaves, and accessory materials around and behind each of the flowers. Position individual flower sections (similar in appearance to boutonnieres and single flower corsages) onto the main spine of the corsage in a zigzag pattern. Flower heads will begin to face out.

Step 5. If you wish to, add a bow in and among the flower sections. Make the integral to the design of the corsage, so that it repeats the color and texture of the ribbon loops already taped into the design.

Step 6. To complete the visual flow of the design, position remaining flower sections below the bow so the flower heads are facing out and downward.

Figure 14-36. Multiple flower corsages are made by combining flower groupings and accessories into a harmonious and unified design.

Step 7. Trim excess wire from the corsage as you tape in materials. Floral tape securely and roll the stem end forward or to the side. The spine of the corsage should remain flat in order for the design to lie closely against the shoulder or the wrist.

Step 8. The corsage may be maneuvered into a variety of styles or shapes, such as a crescent, round, vertical, or triangle, simply by bending and curving wired stems into desired positions. Add two pins to the spine of the corsage.

Over-the-shoulder Corsage

Over-the-shoulder corsages, sometimes called ***epaulet designs,*** are made to be worn on top of the shoulder and cascade down, both in front and back. Although these elegant, graceful designs are constructed similar to other multiple-flower corsages, the smallest flowers on both ends must be wired with a fine gauge wire (#28 or #30 or finer) allowing them to cascade freely. These small flowers at the ends should be more widely spaced than the rest of the corsage, giving the corsage a more natural, graceful appearance.

Glamellia Corsage

A ***glamellia corsage*** is made from various sizes of gladiolus florets arranged in a way to resemble a camellia flower (hence the name for this corsage). The florets are removed from the stem, and the petals of the flowers are cut and wrapped around a gladiolus bud. These novelty corsages, though time consuming, are distinctively glamorous for special occasions (see figure 14-37).

Step 1. Choose a slightly opened gladiolus bud. Remove the green calyx portions. Wire and floral tape.

Figure 14-37. A glamellia corsage made from gladiolus petals offers a distinctive and glamorous design option for a special occasion.

Step 2. Cut the base of a floret, as shown in the illustration. The center portions of the flower will fall out, but the petals should remain intact. Position a section of petals against and around the center bud. Wrap a fine gauge wire around several times to secure the petals and then floral tape. (Floral adhesive may also be used to attach the petals to the center bud).

Step 3. Add another section of petals circling around the other side of the bud. The glamellia will begin to take on a rounded shape.

Step 4. Continue adding sections of petals around the perimeters of the glamellia until the desired size is achieved. Wire and floral tape the base of the glamellia.

Step 5. Add several broad leaves underneath the glamellia to help support the newly formed flower and give it a more natural appearance. The glamellia flower may then be used in a corsage design.

Wrist Corsage

Corsages may also be designed to be worn on the wrist rather than the shoulder. However, wrist corsages must remain lightweight, fairly small, and unobtrusive. Although made in the same way as shoulder corsages, these corsages must have a ***wristlet*** of some kind securely attached.

Several commercial wristlets or wristbands are available. Some have a band that is elasticized while others are plastic latch-type bands. The wristlet is attached to the back of the corsage with metal clamps (see figure 14-38). For security, after pressing the clamps in place around the spine of the corsage, add floral tape around the clamps to keep them tightly in place.

Figure 14-38. To attach a wristlet to a corsage, clamp down the metal brackets onto the back of the design and secure them with floral tape.

Football Mum Corsage

Football mum corsages are traditional for high school and college homecoming football games, dances, parades, and other festivities. These corsages are regional; that is, in some areas they are considered the ultimate status symbol and in other areas they are virtually unknown.

Corsages are designed with one, two, or three standard incurve chrysanthemums (also called football mums) and a variety of long, cascading ribbons in the school colors. Accessories such as miniature footballs, cowbells, megaphones, and tiny plush animals are popular additions to these designs. Often the first initial of the school name is represented on top of one of the mums with the name of the individual wearing the corsage spelled out with glitter or stick-on letters on one of the flowing ribbons.

Figure 14-39. Traditional for many high school and college homecoming events, football mums are considered by some to be the ultimate status symbol.
Photography courtesy Florist's Review Enterprises, Stephen Smith, photographer.

Because of the size and weight of these corsages, mechanics are all important and everything must be wired and glued securely. To keep petals from ***shattering***, mums must be sprayed on the backside with a clear glue. The following steps of construction may be altered for making single and simple, or large and lavish designs.

Step 1. Choose a standard incurve chrysanthemum (football mum) free of blemishes. Remove all but one inch of stem below the flower head. Hook-wire the flower with a gauge #22 wire. (Alternate wiring techniques such as pierce or insertion may be used.) Floral tape the wired stem. If desired, wire and floral tape a second or a third football mum, depending upon the style and size of design.

Step 2. Spray the flower head with a spray sealer, which will prevent petals from shattering.

Step 3. Stitch-wire camillia or salal leaves with gauge #24. These wired leaves will help to support the otherwise fragile mum. Floral tape the stems of the wired leaves. Place these leaves around the mum to add background while giving support to the flower. Leave all wires parallel to each other. Tape the wires tightly together.

Step 4. Long ribbon streamers and a large bow may be added at the base of the design. Streamers may be braided, knotted, or twisted.

Step 5. Accessories such as dangling miniature footballs or cowbells may be added to the design. Letters and numbers may be made out of chenille stems; these may then be glued onto the top of the flower head or attached with a wire.

Step 6. Securely floral tape the stem. Make sure all accessories are secure. The entire design may be sprayed with a light coat of glitter or sealer spray.

Figure 14-40. When attaching a floral piece to a barrette, secure it with wire or glue.

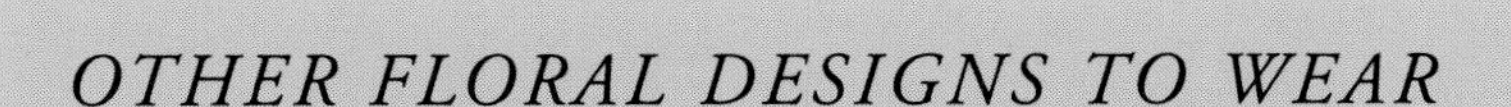

OTHER FLORAL DESIGNS TO WEAR

Floral pieces are often worn in the hair, attached to a hat, pinned to a purse, or worn as a lei. As in other designs, the occasion, clothing, individual preference, and fashion trends all help determine the desired design. As with wrist corsages, keep designs lightweight, with an appropriate shape, and size for the setting.

Flowers for the Hair

Small floral pieces are often designed for individuals to wear in their hair for special occasions such as proms and weddings. Tiny flowers or filler flower clusters may be secured in the hair with hairpins. Small floral designs (similar to boutonnieres) may be attached to a barrette, comb, or hair clip with wire or glue (see figure 14-40).

Chaplet

A chaplet is a floral wreath or garland for the head. These designs are popular for special occasions such as proms and weddings. Since floral wreaths are often worn by young flower girls at weddings, it is important that head measurements be taken before making these designs to ensure a proper fit.

Step 1. Measure the head. Cut a #22 gauge wire to the length needed, adding on an extra inch to allow for hooking the wire ends together to form a wreath (two straight wires may be floral taped end to end in order to form a longer piece, or spool wire may be used).

Step 2. Floral tape the wire, as shown in figure 14-41.

Step 3. Wrap strands of ivy around the wire. Secure the ivy periodically with floral tape. Or, another common procedure for adding foliage is to bind tiny clusters of foliage, such as ivy or leather leaf with fine wire (#28); each additional cluster of foliage should cover the stems of the previous cluster. Floral tape will secure foliage clusters in place.

Figure 14-41. Chaplets, circlets, or floral wreaths for the head are popular designs for proms, weddings, and other special occasions.

Step 4. Add tiny flowers and fillers in a random or orderly pattern with wire and floral tape.

Step 5. Form the wire into a ring and insert the straight wire end into the opposite hooked end, as shown.

Step 6. If you wish, add delicate ribbon streamers and a tiny bow in the back where the garland comes together, concealing the final mechanics of the chaplet.

Lei

The ***lei*** originates in Hawaii and is a garland or wreath of flowers and leaves, generally worn around the shoulders about the neck. Leis vary greatly according to the flowers, foliages, and the manner of assembly. For example, a simple lei may be made with carnations (see figure 14-42).

Step 1. Gather 25 to 35 carnations and remove their stems.

Step 2. Using a special medium-weight lei needle (or long darning needle), thread fishing line, dental floss, or lei string (about 7' long, doubled to 42" long) into the needle hook.

Step 3. In order to make each carnation fuller, gently brush across the petal tips of each flower, as shown.

Step 4. Begin threading a carnation onto the string by first inserting the needle into the calyx. Continue pushing the needle through the calyx until it goes through the seed pod and comes out through the center of the flower. Thread another carnation onto the string in the same fashion. The carnation floewers should fit snugly next to each other, and the green calyx should be visible between each carnation head.

Step 5. Thread enough carnations to make the lei of the desired length. Remove the lei needle and tie the string ends togher. The connecting point may be concealed with a bow.

A "double" carnation pattern is made in much the same way as a "single carnation lei.

Step 1. Gather 45 to 55 carnations and remove their stems.

Step 2. Using a special medium weight lei needle (or long darning needle), thread fishing line, dental floss, or lei string (about 7' long, doubled to 42" long) into the needle hook.

Step 3. Split carnation sepals (calyx) as shown, keeping the flower intact. This will allow the carnation flower to spread out flat, forming a wider, thicker, lei.

Step 4. Begin threading a carnation onto the string by first inserting the needle into the calyx. Continue pushing the needle through the calyx until it goes through the seed pod and comes out through the center of the flower. Feed carnations onto lei string with the green calyx tucked into the center of the previous carnation. No calyx should be visible. This will produce a thicker, fuller pattern.

Step 5. Thread enough carnations to make the lei of the desired length. Remove the lei needle and tie the string ends together.

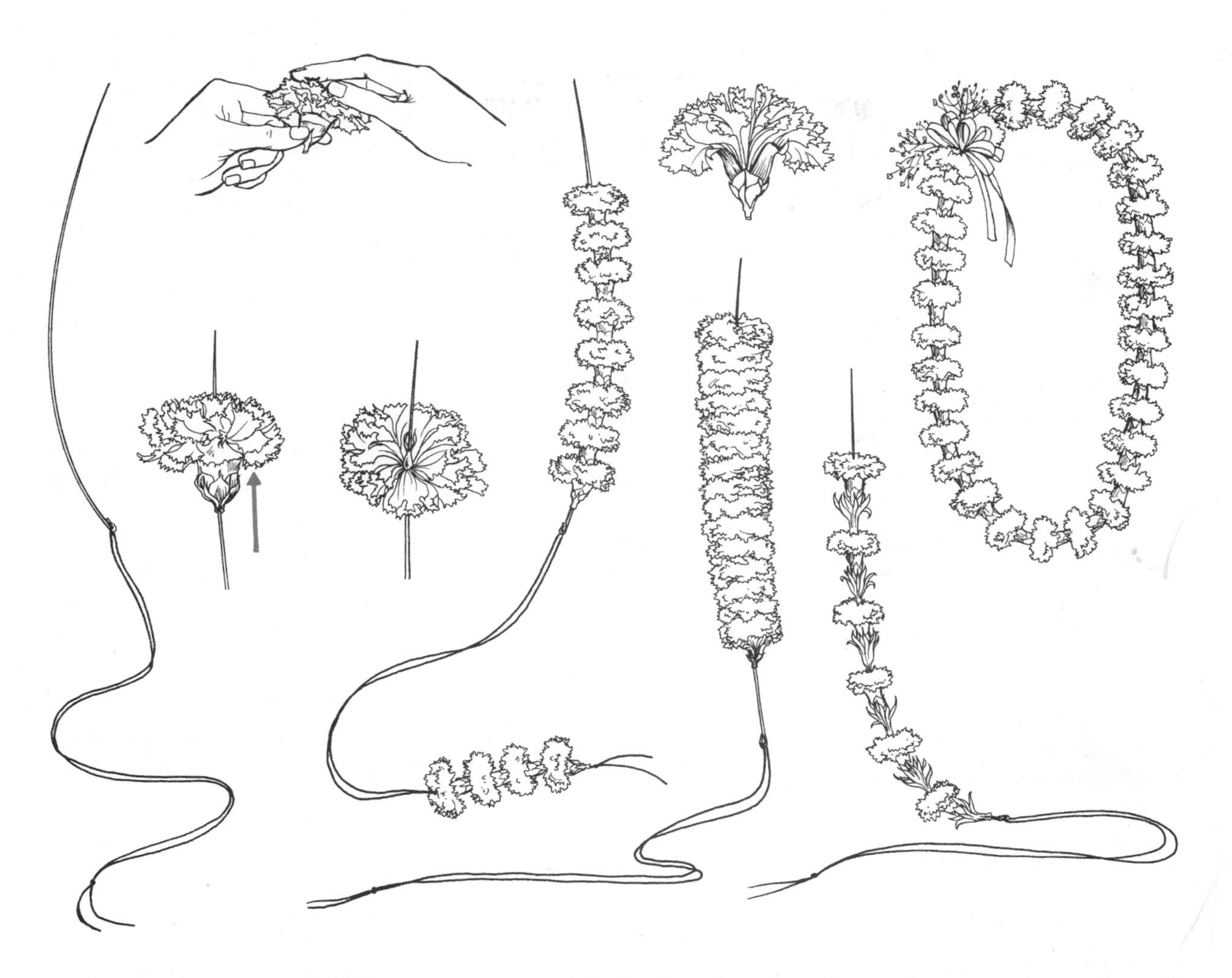

Figure 14-42. A lei, or wreath of flowers, is worn around the shoulders about the neck. Lei styles vary greatly according to flowers, colors, and alternating patterns (shown here are a "single" carnation lei, a "double" carnation lei, and a lei designed with carnations and tuberose blossoms).

SEALERS

Aerosol and liquid ***sealers,*** often referred to as ***finishing sprays*** and ***dips,*** may be used to seal the porous surfaces of flowers and foliages. Sealers inhibit water loss, helping flowers and leaves in corsages, boutonnieres, and other floral pieces to remain firm for a long period of time. Sealers should only be applied to firm, healthy flowers and foliages; they will not help flowers and foliages that are already wilted. After spraying a floral piece, allow the sealer time to dry before packaging the design.

PACKAGING

Corsages, boutonnieres, and other floral pieces for special occasions should be packaged carefully to prevent moisture loss, protect the floral design, and provide an attractive presentation for the receiver. Several types and sizes of bags and boxes are available for packaging flowers to wear.

Step 1. Place the floral piece on a layer of shredded wax paper (often called orchid grass). The orchid grass will cushion and help protect the design while it is packaged.

Step 2. Gently place the floral design and ***orchid grass*** into an appropriately sized bag. Fold the bag and close with boutonniere or corsage pins (or if pins are already attached to the design, staple the bag shut). It is not always necessary to use a bag within clear plastic boxes or boxes with windows; simply place the orchid grass in the bottom of the see-through box and place the floral piece on top.

Step 3. Place the floral design (within the bag) into an appropriately sized box. A ribbon in a corresponding color may be tied in a pretty bow around the box.

Figure 14-43.
The Far Side © 1992 Far Works Inc. Distributed by Universal Press Syndicate. Reprinted with permission. All rights reserved.

REVIEW

Flowers to wear are special accessory items that may be designed in an infinite number of ways. Many special occasions and events are traditionally associated with the wearing of flowers such as weddings, proms, homecomings, Mother's Day, and many other formal or sentimental times. Floral pieces to wear must be designed in the appropriate shape, size, and style to fit the purpose and place for which they are intended to be worn. Proper wiring and construction techniques are essential to keep these fresh designs secure and long lasting. Knowing and practicing basic techniques of construction for boutonnieres, corsages, and other floral pieces to wear will give you confidence in your own ability to create truly distinctive designs.

TERMS TO INCREASE YOUR UNDERSTANDING

accessories
boutonniere
broad-leaf foliage
chaplet
clutch wiring
collarettes
conditioning
corsage
dips
epaulet designs
finishing sprays
flags
football mum corsages
glamellia corsage
glamour leaves
feathering
filler accessories
hook-wiring
lei
netting
orchid grass
over-the-shoulder coursages
pierce wiring
sealers
shattering
stephanotis stem
stitch wiring
tipping
tulle
wrap-around wiring
wristlet

TEST YOUR KNOWLEDGE

1. Name some special occasions in which the wearing of flowers is traditional.
2. What are the basic visual as well as physical guidelines of design for making "body flowers"?
3. Explain the reasons why natural stems are removed and replaced with wire and floral tape when making body flower designs.
4. Why is the time-consuming process of feathering often done on carnations when there are mini (pixie) carnations?
5. What are the common wiring methods called?
6. Why and how is conditioning done?

RELATED ACTIVITIES

1. Make a boutonniere by incorporating all of the following wiring techniques: pierce, stitch, and clutch.
2. Feather a standard-size carnation, making a variety of new flower sizes. Wire and tape each for use in a corsage.
3. Make several rose buds from rolling rose petals. Wire and floral tape each for use in a corsage.
4. Make a nestled boutonniere.
5. Make several ribbon loops and netting tufts for use in a corsage.
6. Assemble a multiple flower corsage.
7. Make a small floral design for a particular hair style and attach the design to a barrette, comb, or clip.
8. Assemble a floral chaplet for a little girl.

Chapter 15

Everlasting Flowers

Figure 15-1. A lovely garden bouquet of pressed wildflowers.

Although there is a magical quality about the fleeting aspect of fresh flowers, the pleasure and enjoyment of flowers can be prolonged. With advancements in preservation, the abundant variety of real flowers and foliage now available is greater than ever. Today, ***preserved flowers*** are more durable, last longer, and are more colorful and beautiful than the preserved flowers of the past. The practice of decorating with preserved flowers, herbs, and foliage is not new. Dried plant material has been used throughout time to decorate, scent, and beautify interiors.

The popularity of both dried and ***artificial flowers,*** or ***everlasting flowers,*** is widespread. Today artificial flowers, or ***silk flowers,*** are made with realistic-looking fabrics, formed in natural shapes, and given colors and textures that create believeable facsimilies of the real thing (see figure 15-2).

Design possibilities are unlimited with everlasting flowers. Requiring little maintenance, everlasting designs are ideal for home and commerical interiors where fresh designs are not practical, especially from day to day; such as, in a dark corner, in a warm or sunny spot, on the wall, high on top of a china hutch or on a piano (where water in fresh designs could cause disaster).

Figure 15-2. These everlasting floral designs complement the interior design with a blend of colors and textures. Flowers & Magazine, Teleflora.

Several advantages of everlasting designs over fresh include:

- designs may be made prior to their sale or use eliminating the usual rush associated with fresh flower arrangements
- seasonal, holiday, and special occasion designs may be easily designed in advance and stored
- the worry of water in designs is eliminated
- a wide assortment of containers may be used, or a design may be constructed without a container
- great versatility exists when designing with everlasting flowers because more time may be taken to construct designs
- stems may be easily lengthened and manipulated
- flowers and foliage will not wilt
- designs may be dismantled and designed again
- exact or unusual colors may be found in everlasting flowers to match a theme or interior color scheme (see figure 15-3)

Figure 15-3. This loose spray of garden blossoms showcasing snapdragons, roses, lilac, zinnias and gerbera daisies is a good accent to this contemporary setting.
Florist's Review Magazine, Photograph by Kate Wootton.

- artificial flowers do not irritate allergies (although some preserved flower and foliage fragrances might)
- when compared with fresh designs and the amount of time they are usable and enjoyed, everlasting designs are relatively inexpensive

While a few disadvantages do exist with everlasting designs, these problems may be easily overcome. Some disadvantages include:

- designs enjoyed day to day may begin to collect dust, fade, or look tired or outdated
- storage of seasonal and holiday designs may present a problem at home and commercially
- artificial flowers lack the fragrance of their fresh counterparts (although this is an advantage for some people)
- although preserved and artificial designs are longlasting, they are not permanent. They must be cleaned, maintained, refurbished, and updated.

PERMANENT FLOWERS AND FOLIAGE

Permanent flowers and foliage (also called artificial and silks) have evolved into sophisticated, and realistic-looking materials. Many flowers are botanically correct replicas of their fresh counterparts (appropriately named ***botanicals***).

There are literally thousands of different types of permanent flowers, foliage, berries, and fruit available at various prices. Their popularity is greater than ever because of the development of fine, realistic fabrics. Although many are initially expensive, over time they are extremely cost-effective. Whether silks are used in combination with fresh flowers, dried flowers, or used exclusively, it is important

Figure 15-4. Modern and contemporary in style, this design capitalizes on simplicity forming a colorful crescent of dendrobium and cymbidium sprays, Pink Mink Protea and Torch ginger.
Florist's Review Magazine.

for you to gain knowledge of the various types of permanent flowers and how to design with them.

Types

The most common permanent flowers are made from polyester by machine. Generally, they are not botanically correct but resemble certain flowers. Quality varies greatly with these basic silk flowers.

Some artificial flowers are described as being "handmade," which usually means the flower parts are manufactured individually and then assembled by hand. These generally are more expensive and look more realistic than other types of artificial flowers. Many imitate natural flowers in some ways, such as color or shape, but do not pass for the real thing like many new botanicals. Other flowers, often called art or fantasy flowers, give the illusion of specific flowers. Product designers will often pattern these flowers after several real flowers, using the shape of one, the color of another, and the texture of another.

Another variation of silk flowers is the paper or parchment flowers. These often are made to look like dried material. They hold up well and do not attract dust like other types of permanent flowers.

Botanicals are patterned after real flowers and product designers spend much effort to produce fabric flowers that simulate all the real characteristics—color, texture, shape, pattern, and size—as they exist in nature.

Like fresh flowers, permanent flowers may be classified according to their shape, which is useful in design work. Line flowers are called spike flowers and are tall, long, and narrow. Form flowers are referred to as focal flowers. These have large, interesting heads and are central to the design. Mass flowers may also be referred to as focal flowers. These flowers have medium- to large-sized heads and may be single flowers on a stem or in sprays of flower heads. Filler flowers are those with tiny, delicate blossoms in sprays. They function within a design to complete, and give accent to a composition. Having a variety of permanent flowers on hand will allow a wide selection and provide you with increased inspiration to help you work more creatively.

DRIED PLANT MATERIAL

The abundance of natural dried plant materials available is increasing in both selection and popularity. Dried flowers are used to provide warmth, color, and accent to interiors. Age-old methods of drying, such as air-drying, drying with a desiccant, and pressing, as well as the modern advances in preservation with glycerin and freeze-drying methods have greatly increased the selection of flowers, foliage, seedpods, grasses, berries, and other plant materials.

While preserved flowers and foliage are available in stores, you may wish to experiment drying plant material for use in your floral designs. A variety of unusual flowers, foliage, seed pods, and grasses that may not be available commercially, may be successfully preserved. A retail florist should be familiar with basic preservation techniques.

Often, floral shops sell products and chemicals used for drying and preserving flowers. Retail florists are often asked how to dry and preserve individual flowers or whole designs. A basic knowledge of the various drying techniques and how products (see figure 15-6) are used will generally lead to increased sales at the floral shop. Often flowers are dried for their sentimental value. They may hold a certain memory or provide special meaning. It is somewhat rewarding to experiment with flowers and other plant materials and be able to dry and preserve them successfully. Drying your own flowers, foliage, grasses, and seed pods gives you a large selection of materials from which to choose for your designs. Learning various drying techniques and finding success also keeps you from buying commercially preserved plant materials, which are often expensive.

Gathering Plant Material

Gather flowers and other plant materials for preservation on a dry day in the afternoon when the dew has evaporated. Excess moisture in and on plant material encourages mold and delays proper drying.

In order to achieve the best results for most drying techniques, cut flowers with sharp pruning shears. Immediately after cutting flowers, place the bases of their stems in a bucket of tepid water, to prevent them from wilting.

When gathering fresh flowers, whether from the florist, garden, or the wilds, choose the very best specimens, those without imperfections. Once dried, any blemishes—bruises, holes, and spots—become magnified.

Many flowers may be harvested at various stages of development, to provide a more natural variety for your designs. However, avoid gathering flowers and other plant materials that are overmature and past their prime. Generally, older flowers may lose their colors, shatter, and deteriorate quickly in the drying process. For drying success, most flowers, pods, and grasses should be harvested before they shed their pollen or release their seeds. It is

Figure 15-5. Plant materials that have been dried and preserved create a stunning look. Larkspur, suveroni, cockscomb, and lavender make this charming arrangement of blues, pinks, and purples.
Florist's Review Magazine.

Figure 15-6. Depending upon the method of drying, it is helpful to have a variety of products and tools on hand to preserve flowers and foliage.

Ideally, the room conditions where flowers are air-dried should be dry, cool, dark, and well-ventilated. Damp conditions will produce dismal results, with flowers often becoming infected with mold or mildew. Although materials can dry in warmer temperatures, cool conditions help flowers maintain truer colors and prevent flowers from drying too quickly and becoming extremely brittle. Sunlight or constant indoor lighting will cause colors to fade.

Variations of air-drying for some flowers, grasses, and other plant material, is to dry them standing upright

important when harvesting flowers and other materials from the wilds that you enquire which flowers may be protected, endangered, or toxic.

Overview of Preservation Techniques

There are several methods of drying flowers and other plant materials. The method you choose will depend upon the type of flower or foliage you want to preserve and how it will be used in a composition. A variety of techniques may be used on the same material to achieve different effects. It will be helpful for you to set aside an area for drying and storing the plant materials you collect. The various methods of preservation include air-drying, drying with desiccants, treating with glycerin, pressing, and freeze-drying.

Air-Drying

Air-drying is the simplest of all drying techniques. This method takes a minimum of time without requiring special equipment or supplies to achieve excellent results. With this drying method, the moisture is removed from the plant materials simply by the circulation of air, without using any drying agent.

Many flower varieties will dry more successfully if thorns and lower or all foliage is removed from their stems (see figure 15-7). Gather similar flowers into small bunches with flower heads fanned apart and at different levels, to allow for better circulation. Gather the bunches together with a rubberband, which will continue to grip stems as they shrink during the drying process. Hang the bunches upside-down with wire or string. Hanging flowers this way will keep stems straight.

You may want to dry some plant materials standing upright, rather than upside-down. Many flowers, seedheads, and grasses that are naturally pendulous or droopy will take a curving shape as they dry. These arching or twisting stems will add interesting lines to your designs.

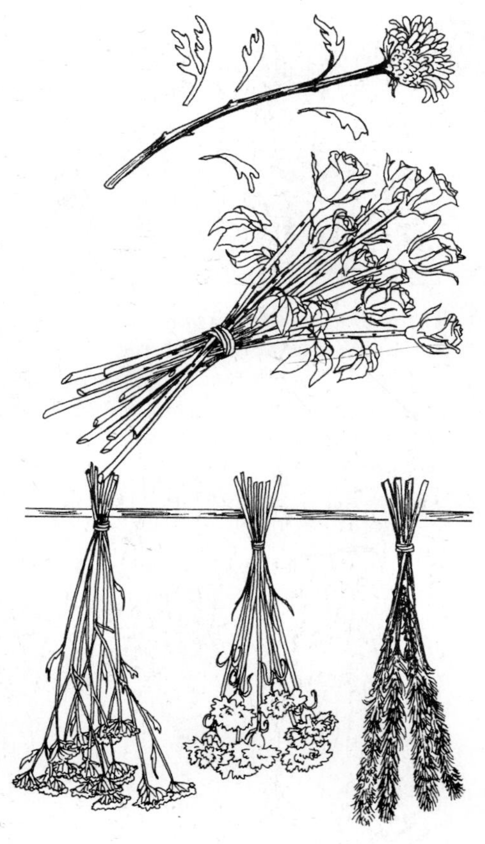

Figure 15-7. To air-dry most flowers, remove some or all of the leaves and bunch similar flower types together with a rubberband. Hang bunches upside-down to promote straight stems.

in one or two inches of warm water or to lay them flat on newspaper, as shown in figure 15-8. Room conditions should be the same as when drying flowers that hang upside-down.

Standing flowers such as hydrangea, heather, bells of Ireland, thistle-type flowers, and grasses in water will take slightly longer to dry; but the flowers and grasses dry better and retain more of their natural shape this way. It is important to recut stems in order for the material to take up some water. Do not replenish the water, simply let it evaporate.

Another method is to lay material flat in a single layer on top of newspaper. Some plant material such as heavy grasses and deciduous and evergreen plant stems, will dry more successfully when allowed to dry flat in this way allowing for natural curves.

With ideal conditions, air-drying plant materials will take one to three weeks. Delicate flowers will dry in about a week, while heavy or thick-stemmed materials may take two or three weeks to fully dry.

Once dried, many flowers, grasses, and seedpods will benefit from a light coating of commercial sealer or hair spray as shown in figure 15-9. The sealer will help keep the dried material intact, keeping seeds and plumes from shedding.

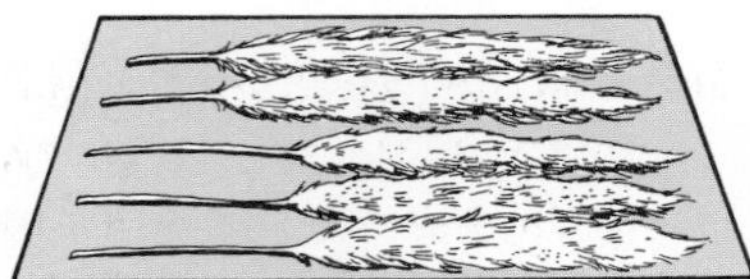

Figure 15-8. Air-drying may also be accomplished for some flower types by the evaporation method. Simply place flowers upright in a few inches of water. Larger grasses and plumes may be dried successfully by laying them flat.

Figure 15-9. Commercial sealers or hair spray may be applied to plumes, grasses, and flowers before or after drying to prevent shattering.

Figure 15-10. Flowers may be dried successfully by being buried in sand, silica gel, and other desiccants. Flowers dried in a desiccant often retain truer colors.
© Piper Floral International Inc. Used by permission.

Drying with Desiccants

Flowers may be dried using a ***desiccant,*** which is a substance that absorbs moisture. Silica gel, fine sand, borax powder, alum, detergent powder, and cornmeal are some examples of desiccants that may be used to draw moisture from plant tissues.

Most flowers dried in a desiccant retain superior color and retain their original shapes. For best results, remove stems from flowers. Flower heads may be wired for added support before being buried in desiccant. Fill a container with at least one inch of desiccant. As shown in figure 15-11, place flowers onto the desiccant, either upside-down, upright, or on their sides, depending upon their shape. Ideally, the same flower types should be dried together, rather than mixing flowers. This will ensure similar drying time. Be careful that flowers do not overlap or touch one another. Gently sprinkle them with desiccant until they are completely covered, then cover the container.

Depending upon the desiccant, drying time will vary from several days to several weeks. To determine if the flowers are dry, gently tilt the container, to reveal a flower; if the petals are soft, they need more time to fully dry; if they are papery or crisp, the flowers have dried and need to be removed in order not to become overly brittle. Slowly tilt the container until the desiccant slides away and gently remove the flowers.

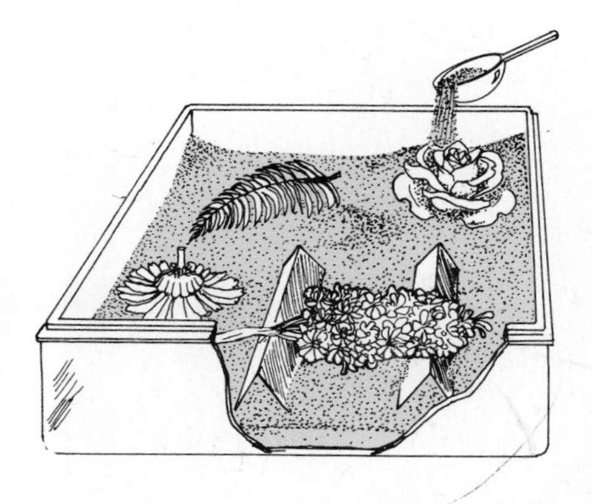

Figure 15-11. To dry flowers and leaves with a desiccant, place them on top of a layer of desiccant. Gently sprinkle them with more desiccant until they are completely covered.

Figure 15-12. For quicker drying, silica gel may be used to dry flowers in a microwave. First, warm a layer of silica gel in a microwaveable dish. Next, add a layer of flowers and sift more silica gel on top of them until they are completely covered. Place a small bowl of water in the microwave to prevent flowers from becoming too brittle. Then, microwave on medium or high for several minutes.

Silica Gel

Silica gel is similar in appearance to granulated sugar. It is by far the most effective desiccant, because it dries many flowers in two to four days and retains more of the natural colors. Silica gel may be purchased for drying flowers at hobby and craft shops, chemical supply companies, and some florist supply shops. There are several types of silica gel with various trade names. Silica gel is available as white crystals or as color-indicator crystals. When the color-indicator crystals dry they are blue; while still absorbing moisture they are pink. Although silica gel is initially expensive, it may be used over and over again.

To speed the drying process, flowers may be dried with silica gel and warmed in a microwave oven. For best results, preheat the silica gel in the microwave on high for about one minute. Then pour about an inch of silica gel into a glass or other microwaveable dish. Place fresh flower heads (without any wires) on the warm layer and gently cover with more prewarmed silica gel. Place a small dish of water into the oven to prevent flowers from becoming overly dry and brittle (see figure 15-12). Set the microwave on medium for one to three minutes. Because drying is so rapid, experiment with various types of flowers and with different microwave settings in order to dry successfully. After removing the container from the microwave, let the silica gel cool for 15 to 30 minutes before tilting the container to reveal the flowers.

Fine Sand

The oldest desiccant used for drying flowers, fine sand is still useful as a drying agent. Use only the finest grades of white beach sand, which may be purchased at hardware and building suppliers. Sift the sand to remove organic materials, then wash it. To dry, bake the sand in shallow pans in an oven set at 250° F for 30 to 60 minutes. Some silica gel with color-indicators may be added to the sand which will turn blue when the sand is dry.

Sand is heavy and, therefore, not suited for drying delicate flowers. Flowers buried in sand take one to two weeks and often up to four weeks to dry.

Figure 15-13. While not considered glycerinated, these foliage have been preserved by special methods and some are painted. Knud Nielsen Company Inc., Evergreen, Alabama.

Borax

Borax may be used alone to dry flowers or may be mixed with cornmeal, alum, or fine sand. A borax and cornmeal mixture is lightweight and suited for dainty and delicate flowers. Drying time will vary usually taking between 3 and 10 days. Once dry, remove all desiccant from flowers to prevent blotches.

Treating with Glycerin

Glycerin (GLISH-er-ihn) is the non-technical term for glycerol. It is an odorless, colorless, syrupy liquid that may be purchased at chemical supply companies. When used to preserve evergreen and deciduous foliage and some flowers and berried branches, glycerin draws water from its surroundings. Rather than removing water from the plant, it replaces the water in the plant's tissues with glycerin, keeping the material supple and smooth. Glycerin often alters foliage and flower colors, turning them to deep shades of the original plant color or dramatically changing colors to deep browns, reds, greens, tans, and yellows.

Use only mature foliage and firmly attached berried branches and keep them fresh in water until ready to treat with glycerin. Young, immature leaves, tender branches, or wilted material do not preserve well and collapse when treated with glycerin.

For best results, mix one part glycerin to two parts boiling water in a non-metal container. Allow the diluted glycerin to cool. Recut stem ends at an angle and place in several inches of the glycerin solution. Commercial automobile antifreeze (undiluted) preserves some plant material and in some cases, may be substituted for the glycerin mixture. For tall, heavy branches, mix the solution and put in a smaller jar, then stand the smaller jar in a taller bucket to give the heavy material support. (See figure 15-14.)

The mixture is taken up more quickly in a warm room than in a cool one. Leaves at the top of tall stems are the last to be preserved and have a tendency to wilt. To speed the glycerin infusion process to the tops of stems, brush on or wipe the upper leaves of tall branches generously with the glycerin solution. This will help prevent curling and wilting. Replenish the glycerin solution as needed. The preservation process generally takes 3 or 4 days to several weeks, with woody branches taking longer.

Remove the stems from the liquid mixture when beads of glycerin form on the leaf surfaces and the entire stem changes color out to the leaf tips. Wipe off excess glycerin with a warm, damp sponge. Hang stems upside-down to dry.

Individual leaves may also be preserved in a glycerin solution (see figure 15-15). Submerge them into a non-metal dish. Leaves absorb the solution in 2 to 6 days.

Figure 15-14. Plant material must be kept fresh and never allowed to wilt before preserving with glycerin. To successfully preserve some flowers and foliage, prepare the glycerin solution by mixing one part glycerin to two parts boiling or hot water in a non-metal container. Allow the diluted glycerin to cool. Next, recut stems at an angle and place in several inches of the glycerin solution. The mixture is taken up by the stems and replaces the water in the plant's tissues. To speed the glycerin infusion process, brush the tops of the stems generously with the glycerin solution.

Remove leaves after they have turned color and wash them in warm water with a mild soap and rinse. Lay them flat on top of newspaper or another type of absorbant paper.

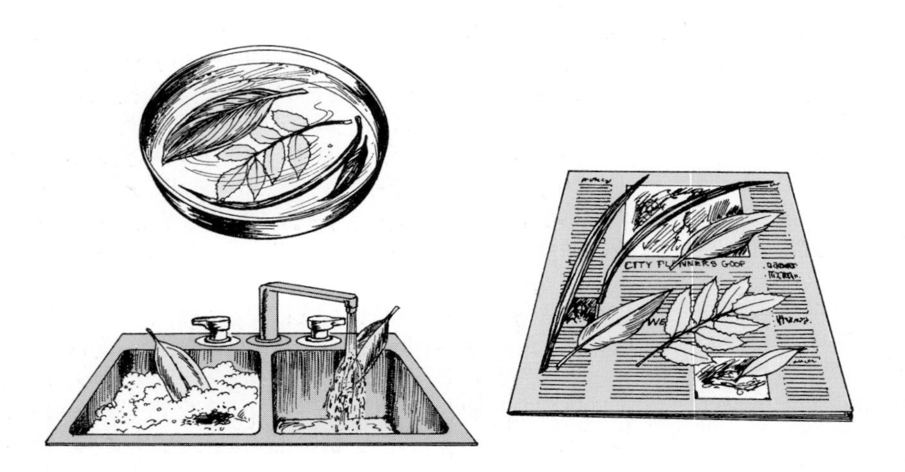

Figure 15-15. Individual leaves may also be preserved in the glycerin solution. Simply submerge leaves in a non-metal dish. They will absorb the solution in several days. Remove them after they have turned color. Wash them in warm sudsy water and then rinse. Lay them flat on absorbant paper to dry.

Pressing

A time-honored method of preserving leaves and flowers is by ***pressing*** (see figure 15-16). This method is used mostly for green and autumn-colored leaves and dainty flowers. Although pressed material is flat, naturally flat foliage such as gladiolus and iris leaves, ferns, and juniper press well. They may be used in a variety of ways in both arrangements and with pressed flowers in glass-covered frames.

Flowers and leaves may be flattened in a traditional flower press, as

Figure 15-16.
© Old-fashion Hollyhock and Delphinium Bouquet by Lynne Oliver, Flower Farm, Provo, Utah. All rights reserved. Used with permission.

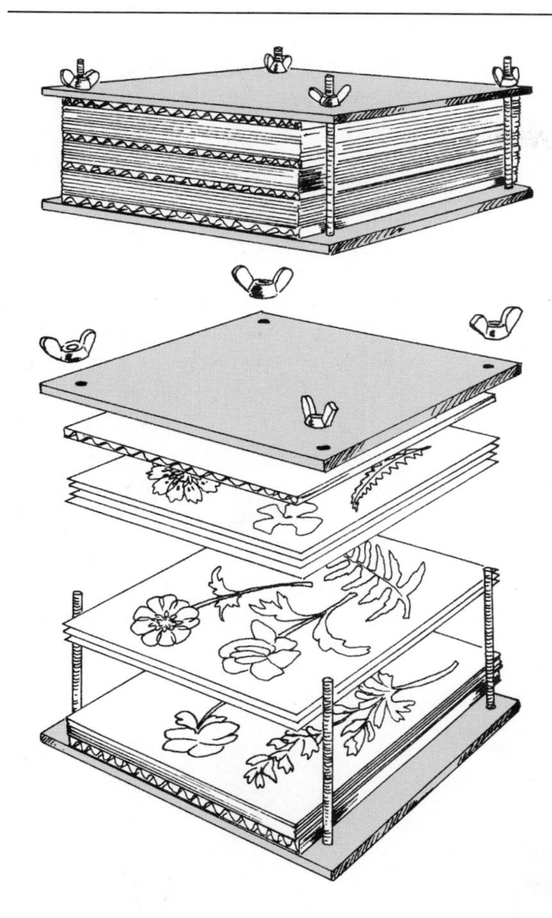

Figure 15-17. Flowers and leaves may be flattened and dried in a traditional flower press. Layers of wood and absorbant paper are alternated. Pressure is applied by tightening the nuts or clamps. The pressing process generally requires two or three weeks to accomplish complete drying.

Figure 15-18. A simple method of pressing is accomplished by laying flowers and leaves within the pages of an old telephone book.

shown in figure 15-17. Allow adequate space between materials and never overlap them. Layers of wood and absorbant, blotting paper are alternated and pressure is applied by tightening the nuts or clamps. As materials dry, they shrink; and the press must be tightened. A simpler method of pressing is to lay flowers between absorbant papers between heavy books or within the pages of old telephone books, as shown in figure 15-18. The pressing process requires two to three weeks to accomplish complete drying.

To design with pressed flowers, handle them carefully with tweezers. Place them in the desired pattern (see figure 15-20). Next, remove flowers and leaves with the tweezers, and brush the underside with diluted glues (purchased from a hobby or craft store). Return the flower or leaf to the desired position. Simple projects include cards and bookmarks (see figure 15-21). With experience and persistence, more elaborate compositions may be designed with your pressed flowers (see figures 15-22 and 15-23).

As Harriet turned the page, a scream escaped her lips: There was Donald—his strange disappearance no longer a mystery.

Figure 15-19.
The Farside © 1985. FarWorks Inc. Distributed by Universal Press Syndicate.

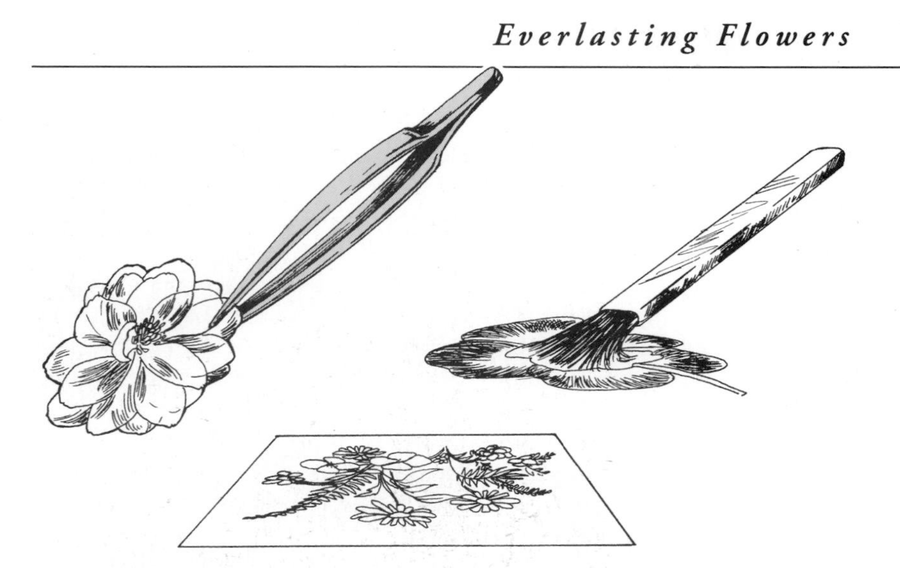

Figure 15-20. To design with pressed flowers once they are dried and flat, handle carefully with tweezers and place them in the desired pattern. Once the design or pattern is decided, remove each flower or leaf. Brush the underside gently with glue and replace it in the desired position in the design.

Figure 15-21. Simple projects include designs of pressed flowers and leaves on cards and bookmarks.

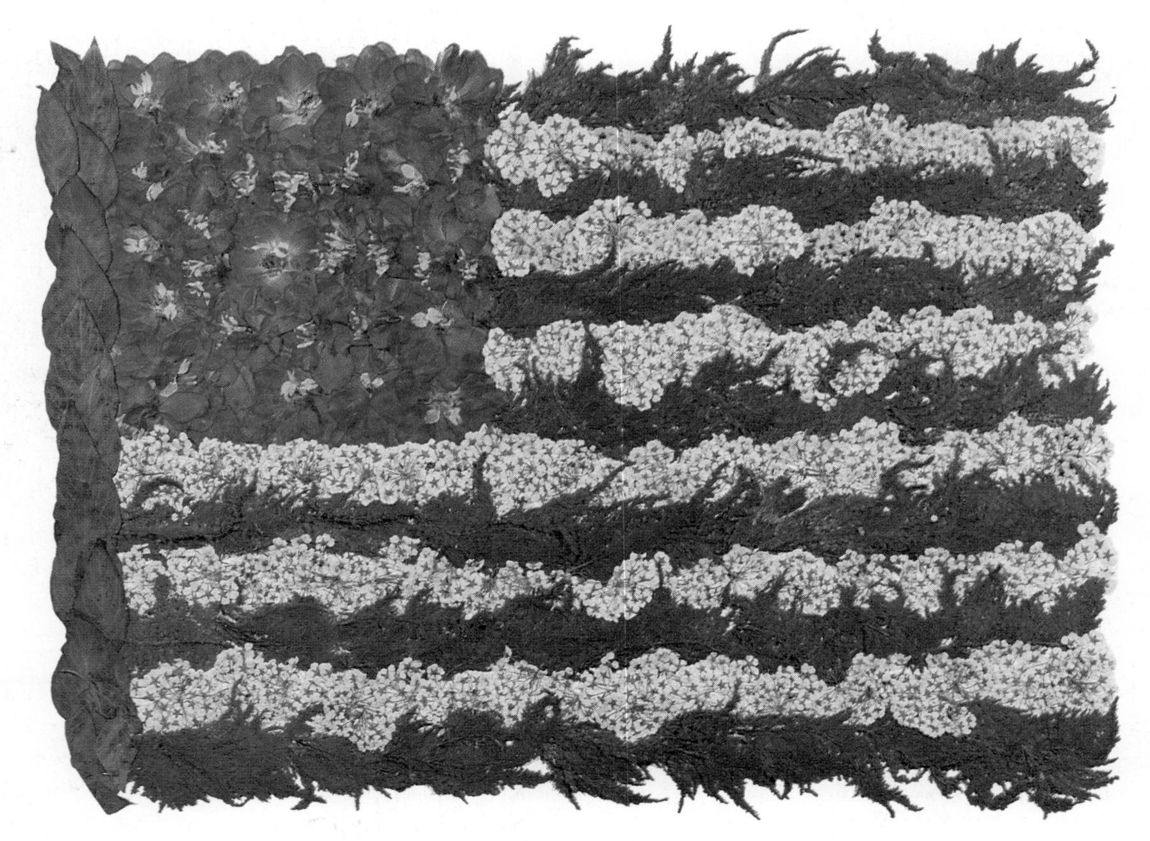

Figure 15-22. With practice, more elaborate compositions may be designed with pressed flowers and leaves. This design of the American flag is made from delphinium, celosia, and bridalwreath florets.

Figure 15-23. Wildflowers and commercially grown flowers arranged together in a pressed design.

Figure 15-24. Producing a dried flower with the beautiful color and shape of fresh is a phenomenon that has taken the industry by storm.
Northstar Freeze Dry Equipment.

Freeze-Drying

The process of ***freeze-drying,*** which is generally done commercially, differs from other drying methods because flowers are frozen before being dried. Flowers are placed in a refrigerated vacuum chamber. While in the chamber, any water in the flowers changes from ice to water vapor. Most freeze-dried flowers, herbs, foliage, vegetables, and fruits retain their original characteristics of color, shape, and texture (see figure 15-24).

Storing Preserved Plant Material

Preserved plant material does not have to be used in designs right after the drying process. It can be stored for several weeks or months without deteriorating. Flowers that are hung dry in bunches, can stay where they are for their decorative value, as long as there is plenty of space. Air-dried material may be layered between tissue paper. Take special care to support delicate or large flower heads. Place layers of flowers in a cardboard box. Plant material that has been treated with glycerin must be stored alone (not in the same box with dried materials) in a well-ventilated box, between layers of newspaper.

While an enormous range of plant material can be dried, table 15-1 lists some of the more common flowers that may be easily dried and preserved. The botanical and common name, the part of the plant that dries well, and the most appropriate drying method are listed.

TABLE 15-1: FLOWERS SUITABLE FOR DRYING

Botanical Name	Common Name	Air-Drying	Desiccant	Pressing	Glycerin
Acacia	mimosa, wattle	flowers and leaves			
Achillea	yarrow	flowers			
Aconitum	monkshood	flowers			
Alchemilla	lady's mantle	flowers and leaves		flowers and leaves	
Allium	onion flower	flowers and seedheads			
Amaranthus	love lies bleeding	flowers and seedheads			
Ammi	Queen Anne's lace		flowers	flowers	
Ananas	ornamental pineapple	flowerhead			
Anemone	anemone		flowers	flowers	
Anethum	dill	flowers		flowers	
Anigozanthos	kangaroo paw	flowers			
Antirrhinum	snapdragon		flowers		
Astilbe	astilbe	flowers		flowers	
Banksia	banksia-protea	flowers and leaves			
Calendula	pot marigold	flowers		flowers	
Callistephus	aster		flowers		
Campanula	bellflower		flowers	flowers	
Carthamus	safflower	flowers			
Celosia	cockscomb	flowers		flowers	
Centaurea	cornflower	flowers	flowers		
Chrysanthemum	chrysanthemum (small heads)	flowers	flowers		
Consolida	larkspur	flowers	flowers	flowers	
Convallaria	lily of the valley		flowers	flowers and leaves	
Cosmos	cosmos		flowers	flowers	
Cynara	globe artichoke	flowers and seedheads			
Dahlia	dahlia		flowers		
Delphinium	delphinium	flowers	flowers	flowers	
Dianthus	carnation	flowers	flowers		
Digitalis	foxglove		flowers		
Dryandra	dryandra-protea	flowers			
Echinops	globe thistle	flowers			
Erica	heather	flowers			

Botanical Name	Common Name	Air-Drying	Desiccant	Pressing	Glycerin
Eryngium	eryngium	flowers and seedheads	flowers and seedheads		
Freesia	freesia		flowers	flowers	
Gaillardia	blanket flower	flowers			
Gerbera	gerbera		flowers		
Gladiolus	gladiolus		flowers		
Gomphrena	globe amaranth	flowers			
Gypsophila	baby's breath	flowers			
Helianthus	sunflower	flowers	flowers		
Helichysum	strawflower	flowers			
Hydrangea	hydrangea	flowers			flowers
Iberis	candytuft		flowers	flowers	
Iris	iris			flowers and leaves	
Kniphofia	red hot poker		flowers		
Leptospermum	lepto	flowers			
Leucodendron	leucodendron	flowers and leaves			
Liatris	gayfeather	flowers			
Limonium	statice	flowers			flowers
Lupinus	lupine		flowers		
Matthiola	stock		flowers		
Moluccella	bells of Ireland	flower spray			flower spray
Muscari	grape hyacinth		flowers		
Narcissus	daffodil		flowers	flowers	
Nigella	love in a mist	flowers and seedheads			
Ornithogalum	star of Bethlehem		flowers		
Paeonia	peony	flowers		flowers	
Papaver	poppy	flowers and seedheads			
Protea	protea	flowers			
Ranunculus	buttercup	flowers	flowers		
Rosa	rose	flowers	flowers		
Rudbeckia	gloriosa daisy		flowers		
Sarracenia	pitcher plant	leaves			
Scabiosa	pincushion flower		flowers	flowers	
Solidago	goldenrod	flower spray		flower spray	
Tagetes	marigold		flowers		
Trachelium	throatwort	flowers			
Tulipa	tulip		flowers	flower petals	
Zinnia	zinna		flowers		

Figure 15-25. A variety of supplies and tools help make the design process with everlasting flowers and materials more efficient.

DESIGNING WITH EVERLASTINGS

With everlasting flowers, there is large diversity in design. Stems may be easily manipulated and lengthened and designs may be placed in a greater variety of places in a greater variety of containers than can fresh designs. There is no wilting, limited mess in design, no worry of water or changing water, and no dead flowers to constantly pick out of a design. There is greater freedom in design with both dried and permanent materials than with fresh. Once a design is created, it will last for a long time without much maintenance.

To ease design work with dried materials, a pick machine (see figure 15-26) may be used. Pick machines add a metal pick to single or grouped stem ends, allowing for easier insertion into floral foam.

As shown in figure 15-27, other possibilities exist for lengthening, strengthening, or creating new stems for dried materials. Depending upon the flower shape and stem thickness, wires, floral tape, glue, and sticks may be used for this purpose.

To secure some silk flowers, their heads and leaves can be removed and glued to the main stem (see figure 15-28). When you work with silk flowers, because most types have wire inside the stem and many have wires within the leaves and petals, as shown in figure 15-29, manipulate the flower parts (stem,

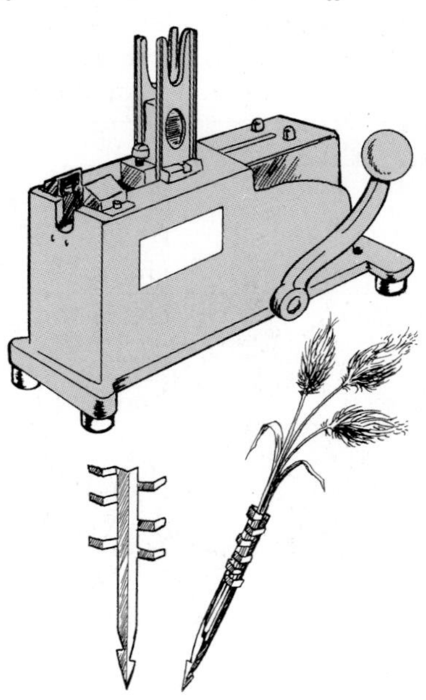

Figure 15-26. Metal picks can be attached to single or grouped dry or silk flowers with a pick machine for easier insertion into floral foam.

Figure 15-27. Many possibilities exist for lengthening, strengthening, and creating new stems for dried materials. Depending upon the flower shape and stem thickness, a variety of techniques may be done. Shown here top row, left to right: A new stem is created for a flower head by using the hook wiring technique; another stem is lengthened with a wired-wooden stick and floral tape; and a third flower's stem is created with heavy wire and hot glue. Bottom row, left to right: A hollow stick can be used to create a new stem; a combination of wires and floral tape can create a cluster of flowers; and new stems for leaves can be easily made with wired wooden sticks.

leaves, and petals) into more realistic and exciting shapes.

Use the type of dry foam that is specially designed for use with fragile drieds or heavier, thick stems of dried or silk material. For overall sturdiness, it may be necessary to glue the foam into the container and for even greater security, add a drop of glue to stems as they are inserted into the foam. The mechanics of foam and the base of the stems should be camouflaged with some type of moss (see figure 15-30).

To secure a foam foundation onto a wreath or other wall piece, a block of foam may be cut and glued onto the frame (see figure 15-31). The foam must be further secured with wire. To prevent the wire from cutting through the foam, however, glue wooden sticks or stem ends to the edges of the foam. As you add wire to secure the foam in place, the wire will not sever through the foam. Once the foundation is secure, add moss with green pins to camouflage the mechanics of construction. Flowers and foliage may then be inserted into the foam to produce a three-dimensional look (see figure 15-32).

Figure 15-28. For added security with silk flowers and leaves, remove flower heads and leaves. Add a drop of glue and reattach parts to the main stem.

Figure 15-29. Most silk flowers have wire inside the stems, leaves, and petals and may be manipulated into more realistic curves and shapes.

Figure 15-30. The mechanics of foam and the base of stems should be camouflaged with some type of moss. This design uses American Moss, which is a clean and dust free, natural excelsior alternative to Spanish moss.
"American Moss" is a registered trademark of FIBEREX INC. Florence, AL. (800) 243-3455. Photo used with permission.

Maintenance and Cleaning

Although somewhat maintenance free, everlasting flowers and designs must be maintained, dusted, and often cleaned with water and soap. If designs are maintained on a regular basis, perhaps once a week, every other week, or even once a month, dust will not become a big problem. A variety of silk flower and foliage cleaners and conditioners are available in both aerosol and pump sprays that dissolve dust. (Test first to make sure fabric will not stain.)

Another way to clean some flowers is to pour 1/2 cup of table salt into a small paper sack. Insert the flower head into the sack. Hold the sack tightly around the stem and shake for several minutes. This method removes dust and grease.

Depending upon the fabric of the silk or the drying technique that was used to preserve fresh material, the cleaning technique will vary from flower to flower (see figure 15-33). Often a hair blow-dryer will do the job to clean away dust and debris from everlastings. Other times, the commercial sprays are needed to spruce up tired and dusty-looking materials. Feather dusters, soft paintbrushes, or cloths may work for larger flower petals and leaves. Some silk flowers may be swished in warm, mild, soapy water and then rinsed, allowed to dry, and

Figure 15-31. To secure a foam foundation onto a wreath or other wall piece, cut a block of foam and glue wooden sticks to the corner edges. Next, glue the foam piece to the focal area of the design. Then, secure the block further with wire. The wooden sticks will prevent the wire from cutting through the foam. Finally, add moss to conceal the mechanics before adding flowers.

Figure 15-32. Once the mechanics are secure on a wreath, flowers and foliage may be inserted to provide a three-dimensional design.

Figure 15-33. A variety of cleaning techniques may be used to keep everlasting flowers clean and fresh looking. Often a hair blow-dryer will clean away dust and debris. Commercial sprays may be used to spruce up tired and dusty-looking materials. Feather dusters and soft brushes and cloths may be used for larger flowers and leaves.

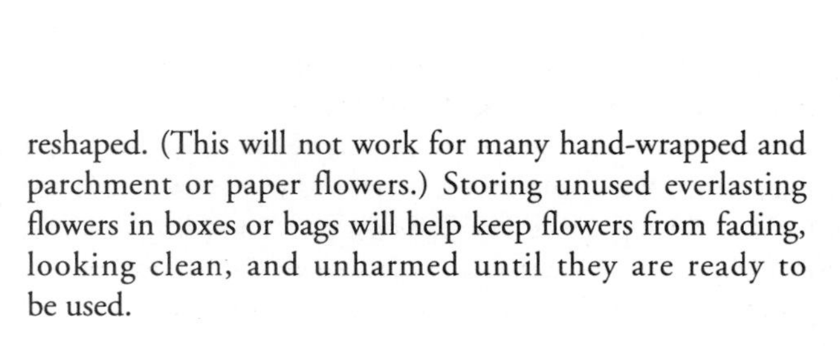

reshaped. (This will not work for many hand-wrapped and parchment or paper flowers.) Storing unused everlasting flowers in boxes or bags will help keep flowers from fading, looking clean, and unharmed until they are ready to be used.

REVIEW

Because the popularity of both dried and artificial flowers is widespread, it will be helpful for you to become knowledgeable of the various types of materials available and vitally important that you become proficient and comfortable in designing with dried and permanent materials. Their versatility and long-lasting qualities are unmatched by fresh materials.

TERMS TO INCREASE YOUR UNDERSTANDING

air-drying
artificial flowers
botanicals
desiccant
everlasting flowers
freeze-drying
glycerin
preserved flowers
pressing
silica gel
silk flowers

TEST YOUR KNOWLEDGE

1. Name five advantages of everlasting designs over fresh designs.
2. Describe the basic methods used for drying and preserving plant material. What are the characteristics of flowers and foliage after being dried and preserved with each drying technique?
3. Why take the extra time and effort to preserve and dry flowers and foliage when they are available commercially?
4. For best drying results, when should flowers be gathered and at what stage of bud development?
5. Name the ideal surrounding room requirements for air-drying plant material.
6. What is a suitable substitute for the glycerin solution when preserving foliage?
7. Name various types of permanent flowers.

RELATED ACTIVITIES

1. Individually or as a group, experiment with the various drying techniques for several types of flowers and foliage. Keep a data sheet for this activity.
2. Create a floral design from flowers and foliage that you have preserved.
3. Visit a florist, craft shop, or wholesale florist where artificial flowers are sold and notice the wide array of colors, types, and quality. Report on your visit.
4. Press a variety of fresh flowers and make a flat composition.
5. Construct an everlasting design by combining artificial and dried materials.

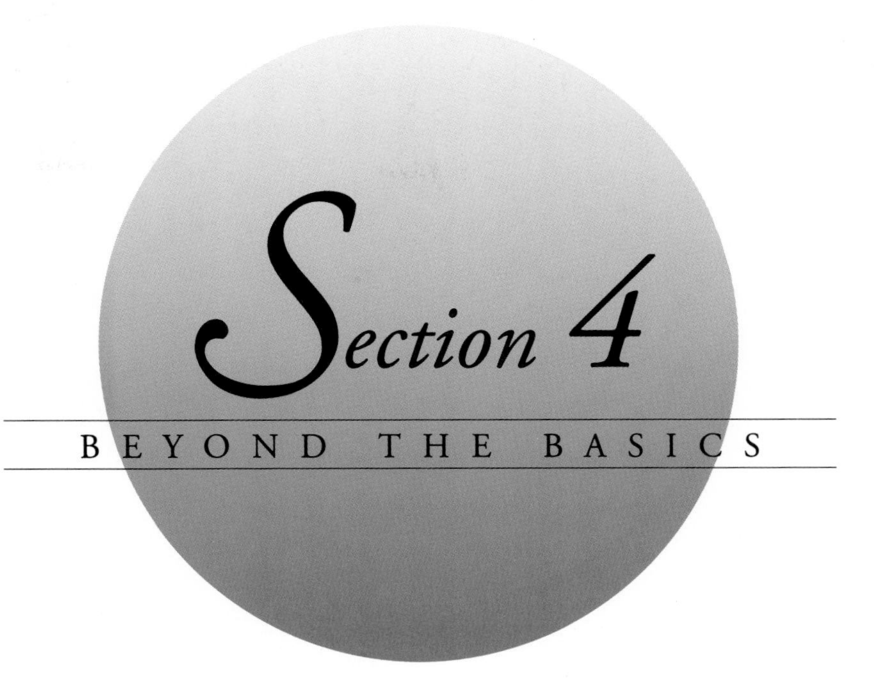

Section 4

BEYOND THE BASICS

Chapter 16

Oriental Style of Design

Figure 16-1. This distinctive, freestyle design with strong, dynamic lines of agapanthus, curly willow, and bear grass has an oriental flair. Chinese and Japanese floral arranging styles have greatly influenced the American or Western style of floral arrangement.

Throughout the various periods of history in the Western world, floral art has generally emphasized massed arrangements. In contrast, oriental art has focused on line, simplicity, symbolism, and the careful placement of plant material. The ***oriental style*** of arrangement originated the use of linear, asymmetrical balance in floral designs. Although contrasting considerably to the European style of arrangement, the oriental style of floral design has greatly influenced the development of the art of floral arrangement. The floral arrangement styles of both China and Japan share similar philosophy of nature. Because oriental design has influenced the American or Western style of arrangement, study of these important styles and design techniques can help you become more confident in your creativity as a professional floral artist.

CHINESE INFLUENCE

Glimpses of the ancient floral art are preserved in paintings, prints, and scrolls. Many of these art pieces show how flowers were used and arranged in vases. Although the ***Chinese style*** of floral arrangement is less stylized and less recognized and studied than the ***Japanese style,*** this style of design is a respected art form that has greatly influenced the Japanese style. The ancient floral style of China is symbolic of nature through the use of a few flowers and interesting branches. This quiet, refined style of design was passed on to Japan, along with Buddhism, and developed into a highly symbolic art.

Seasonal plant material is characteristic of Chinese arrangements. As shown in the four prints in figure 16-2, each season of the year has its special branches of flowers and foliage signifying the time of year.

Chinese arrangements appear less carefully planned than those of the Japanese (see figure 16-3). However, these unstructured, naturalistic designs require thought

Figure 16-2. Four silk tapestry panels depict the seasons. The various plant materials are of interest, as well as the vases and accessories. From left to right: Spring is depicted with peach blossoms, peonies, and a spray of orchids. Summer shows branches of magnolia blossoms and a large peony blossom. Autumn is evident with sprays of chrysanthemums and fragrant osmanthus branches. Winter has blossoming branches of prunus, bamboo, berries, and paperwhite narcissus (representing the new year). Kakemono; Spring, Summer, Autumn, Winter *(1662–1722), China—Kang Hsi Period. Philadelphia Museum of Art: Given by Mr. and Mrs. John S. Jenks.*

Figure 16-3. Characteristic of many Chinese floral designs include dominant, vertically placed branches combined with delicate, horizontally placed flowers in a decorative porcelain vase.

Figure 16-4. Chinese-style vases, in a variety of shapes and patterns, were made of bronze, pewter, pottery, and porcelain.

and planning on the part of the designer. Anciently, arrangements included the use of a dominant vertical element and the use of a more delicate, horizontally placed element. Usually large, these arrangements were and are made with a limited variety of plant material. The emphasis is on naturalism, not stylized design. Branches and flowers never appear tight. The voids or negative spaces within the design emphasize form, color, and texture. The overall appearance of a floral design made in the Chinese style reflects nature through luxurious abundance and elegance in plant material and container. Ornate porcelain vases, jars, and dishes are characteristic of Chinese designs. Other popular container materials include bronze, pewter, and pottery (see figure 16-4).

The design style of ancient China can be easily incorporated into professional floral arrangements of today in a variety of inspiring ways. The use of a few seasonal branches and flowers arranged together in a decorative container will portray important elegance.

TABLE 16-1: FLORAL DESIGN CHARACTERISTICS OF THE CHINESE STYLE

Flowers Typical to Chinese Floral Arrangements

Acacia
Aster
Camellia
Chrysanthemum
Dianthus
Iris
Jasmine
Lily
Lotus
Magnolia
Narcissus
Orchid
Peony
Rose

Other Plant Materials Commonly Used

spring flowering branches
berried branches
willow branches
pine branches
bamboo
fruits
mosses and fungus
gourds

Common Containers

- bowls, jars, flasks, tall vases, and low basins made from bronze, pewter, pottery, and porcelain in decorative patterns
- spouted vases made from pottery and porcelain

Accessories

- flat or footed bases of wood, porcelain, and stoneware
- rolled scroll of a painting
- incense burner
- painting hanging behind the flowers
- fresh fruits
- Chinese porcelain or jade sculptures
- single flower blossoms and flowering branches adjacent to design

Figure 16-5. Traditional oriental designs, both Chinese and Japanese, require thought and planning in the selection of appropriate vases and mechanical aids.

JAPANESE INFLUENCE

Adapted from the ancient Chinese style in the 6th century, Japanese floral design is steeped in tradition and symbolism. However, the Japanese style of floral arrangement, in contrast to the Chinese style, is highly stylized and adheres to strict rules of construction. For centuries, arranging flowers has been a recognized art form. Still today, floral designers often incorporate the Japanese style of design into contemporary arrangements. The Japanese influence has notably contributed to the development of the American or Western style of floral design.

Although many schools, styles, techniques, and literature has developed throughout time, all share the same philosophy—to depict and symbolize nature. Generally, cut flowers and accessory materials are placed in arrangements to represent nature. Natural growth patterns are reflected and plant material is pruned and groomed to perfection.

The use of flowers and plant materials in arrangements was first practiced as a part of sacred religious ceremonies (decorating alters and given as symbolic offerings). Later, flower arranging became a popular and important art, practiced in the home. Traditionally, Japanese designs are placed in what is called the ***tokonoma,*** an alcove in a home in which a flower arrangement, a hanging scroll, and other art is displayed (see figure 16-8).

The word ***Ikebana*** ("giving life to flowers") is generally interpreted as the art of Japanese flower arrangement. Several schools in Japan have greatly influenced the floral style of Ikebana. For example, the ***Ikenobo*** school is noted for the development of the formal ***rikka*** or upright style; the ***Ohara*** school is known for the ***moribana*** style, a more practical and naturalistic design; and the ***Sogetsu*** school is known for free style and abstract designs. All Japanese styles place great importance on simplicity and line (see figure 16-9). Generally, designs display three main lines or elements. The varying heights and placements of the three lines are set apart from each other. Visually, the three tips form an asymmetrical triangle (see figure 16-10).

Emphasis is placed on depth. Plant material is placed three dimen-

Figure 16-6. The Japanese method of grooming, bending, and arranging plant materials in the hand before placing them in a container is shown in this early wood block print.

Japanese wood block print, Domestic Scene, 18th Century, 1788 by Hosada Yeishi. The Metropolitan Museum of Art, Samuel Isham Gift, 1914. (JP947)

Figure 16-7. This 19th century Japanese New Year's print by the artist Itchinsai Utagawa from a Japanese surimono or greeting card features an informal Ikenobo arrangement of curled aspidistra leaves, which form the three important, classic lines. The Metropolitan Museum of Art, Bequest of Mrs. H. O. Havemeyer, The H. O. Havemeyer Collection, 1929. (JP2366)

Figure 16-8. Traditionally, Japanese floral designs are placed in the tokonoma or alcove of a home, along with other works of art.

sionally in compositions to portray the spaciousness of nature. An entire landscape can be created by properly spacing a few materials. Evident in Japanese designs is a feeling of growth through the use of dynamic, rhythmic, and unconfused line.

Generally, the line of most Japanese styles of arrangements is emphasized without regard to color or mass. Simplicity and negative space are important in order to recognize the three main elements. Uncluttered design is regarded as fundamental. To make this possible, each flower, leaf, or branch is groomed for its best appearance and each flower, leaf, and stem tip is angled upward to appear still growing. Flowers are arranged to face the viewer. Filler and helper flowers and other plant materials are placed in designs to help strengthen the three main elements. They are generally placed sparingly along the axis of the stem elements they are to strengthen and are shorter than the tips of these three main elements.

Figure 16-9. Japanese arrangements in contrast to Chinese designs emphasize simplicity, line, and uncluttered line.

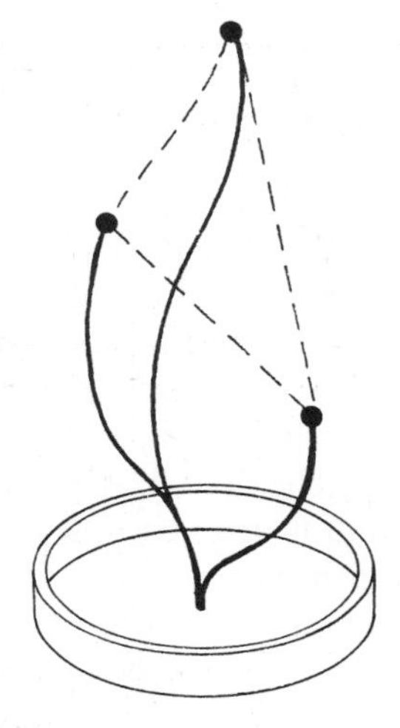

Figure 16-10. Most Japanese designs display three main lines. The heights and placements of these main elements are set apart from each other. Imaginary lines connecting the tips of the three elements form an asymmetrical triangle.

Figure 16-11 illustrates the many types of containers and bases that are typically used for Japanese designs. Most are simple in design. Container shape and type are selected carefully to complement and enhance a particular style of design.

Basically, Japanese flower arrangements may be divided into three groups: the ***classic*** or ***formal style,*** the ***naturalistic*** or ***informal style,*** and the ***abstract*** or ***freestyle.*** These three groups are general types. Variations of each exist that have been promoted by many teachers and schools.

TABLE 16-2: FLORAL DESIGN CHARACTERISTICS OF THE JAPANESE STYLE

Flowers Typical to Japanese Floral Arrangements

Aster
Azalea
Camellia
Chrysanthemum
Hydrangea
Iris
Lily
Magnolia
Narcissus
Orchid
Peony
Rose

Other Plant Materials Commonly Used

aspidistra
bamboo
Scotch broom
cedar
flowering branches
berried branches
willow branches
pine branches
galax
grasses
gourds
hosta
juniper
maple
reeds
mosses

Common Containers

There is a great variety of shapes and materials of containers, depending upon the style of design; including bronze, pottery, bamboo, baskets, wood, lacquered wares, and gourds. Most containers are simple and elegant in shape and design.

Accessories

- generally Japanese containers stand on bases
- stands with short legs, claw feet, or brackets
- lacquered or polished, round or rectangular panels
- mats
- rocks, stones, glass marbles
- driftwood

Figure 16-11. There is a tremendous variety of shapes and materials for Japanese containers. Japanese containers generally stand on bases. Various Japanese styles will often dictate the type of container required.

Figure 16-12. Similar to many 19th century Japanese prints, this floral arranger is shown gently coaxing curves into plant material, which is groomed to perfection prior to placement in a design.

Classic or Formal Style

Early Japanese designs, called classic or formal, are linear and focus on nature and symbolism. The classic or formal style includes rikka and ***shoka*** or ***seika.***

Rikka or Rikkwa ("standing flowers") is the ancient temple style of Japanese flower arrangements. These designs were introduced as religious offerings in the 11th century by a flower master of the Ikenobo school. These highly stylized designs adapted from the Chinese floral art, depict scenes of nature by using a great variety of standing plant material. They are generally massive, pyramidal, and symmetrical, as shown in figure 16-13. A typical container for these early temple designs was a bronze ceremonial vase. Anciently, these designs required several days to complete and would reach a height of six or more feet.

There are a number of Japanese floral schools, all of which use the art principles of the three main lines. The first rules of construction utilized three main structural elements. In most schools, these three elements are named ***shin, soe,*** and ***tai.*** *Shin,* the tallest background element, represents the distant view of nature and is depicted by the use of large tree branches in the background of the design. *Soe,* the secondary line of the design, represents the middle view of nature. It is represented by placements of low shrubs placed in front of the trees. The close view, or *tai,* is the tertiary line of the design and is depicted with small flowers and other plant materials in the foreground of the arrangement. In the Ohara school, the three main lines are named the ***subject line,*** the ***secondary line,*** and the ***object line.***

By the middle of the 15th century, floral styles had become less formal and rigid in design. In contrast to the rikka style, the *shoka* or *seika* ("quiet flowers") style is an

Figure 16-13. The ancient temple art of Japanese flower arrangement, the Rikka *style, is made with a great variety of plant materials. These highly stylized designs depicting scenes of nature were adapted from the Chinese.*

asymmetrical triangle design, based on three main elements adapted from the three main structural elements used in rikka styles of arrangement. The three elements of *shin, soe,* and *tai* were changed to "heaven, man, and earth," symbolizing that man is found between the sphere of heaven and the soil of earth.

Two characteristics of the classic shoka design are asymmetry and line. The subject line is the shin or heaven element and is at least one and a half to three times the height of the container. The secondary line, soe or man, is two-thirds the length of the heaven line. The object line, tai or earth element is two-thirds the length of the secondary or man line. The tips of these three elements, when connected with imaginary lines, form an asymmetrical triangle. Today, the shoka style generally takes the same triangular form that has continued for centuries (see figure 16-14).

This less formal style requires skill in bending branches into curved lines and creating mechanics to keep them in place, as shown in figure 16-15. Forced bending must be done gently. It is best accomplished with thumb pressure. Sometimes hot water will make stems more pliable. Because few materials are used, flowers and branches were traditionally, and often still today, held in place with a forked stick fitted across the mouth of a container (see figure 16-16). These forked twigs, called ***kubari,*** wedged across vase openings have long been used in classical designs. Other sticks wedged against forked twigs can further secure the flowers and branches in the vase. For low, flat, and open containers, decorative metal holders have long been used. Often pebbles or stones are used to conceal any mechanics of construction. The pin holder or ***kenzan*** is also used to construct designs. Today, however, floral foam, once secured and concealed in a vase or low container is generally the favored foundation for most Japanese design styles.

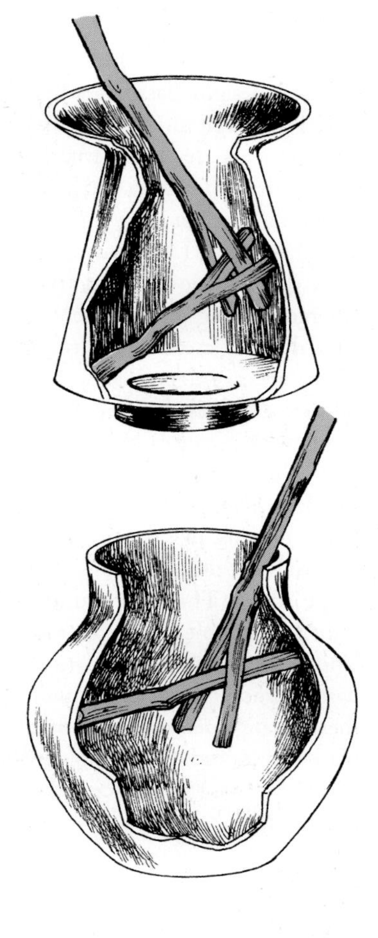

Figure 16-14. The asymmetrical triangular form has continued for centuries. Evident in these classic designs is a feeling of growth through the use of dynamic, rhythmic, and unconfused line.

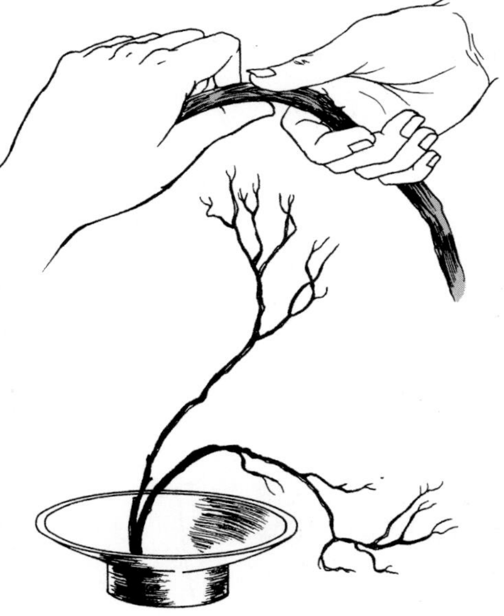

Figure 16-15. Bending and curling branches or flower stems is done by gently applying pressure with both thumbs, which keeps the branch from snapping.

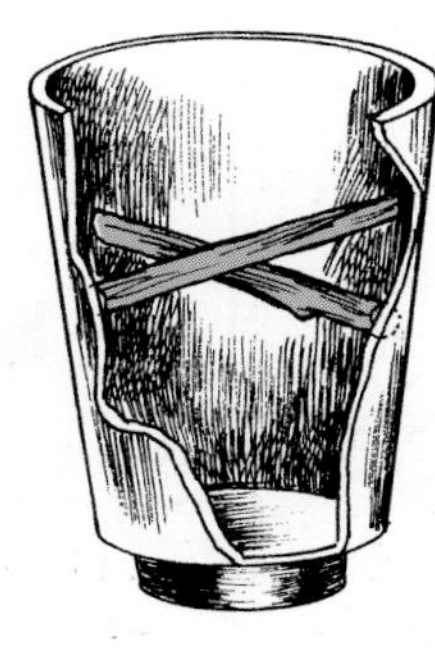

Figure 16-16. Traditional mechanics can be created by using branches within the vase to support the arrangement. Kubari *is a Y-shaped piece of wood, which when fitted inside the neck of a vase serves as a flower holder. Depending upon the shape of the vase, a variety of cross pieces can be used to serve the purpose of holding plant material securely.*

Naturalistic or Informal Style

Japanese designs classified as naturalistic or informal include ***nageire*** and *moribana.* Although these designs generally follow basic rules of construction, arrangements are characterized as more natural and casual when compared with the traditional classic styles. Importance is placed on the appearance of the design rather than its technical form.

The *nageire* ("thrown in") style of design incorporates curving lines, rather than the traditional rigid triangular shapes. This prominent style of design emerged during the 16th century and is known for using tall upright vases with curved stems creating the traditional asymmetrical triangle (see figure 16-17). These arrangements imply a more natural and less confined presentation of flowers. ***Heika*** is also an arrangement style in a tall vase and the word is synonymous with *nageire.* Although known for using tall vases, nageire designs are often created in hanging containers and other styles of vases.

Moribana ("piled up flowers") is a contemporary style of arrangement. These designs, like nageire, are informal, casual, and free of exacting rules. Plant materials are placed in low, shallow containers in order to create a natural landscape scene. The foreground, middle ground, and distance or subject, secondary, and object lines are established with seasonal plant materials. Often the landscape or garden scene made of colorful, short-stemmed flowers are grouped by type and are often surrounded by a pool of water (see figure 16-18). Because these arrangements are constructed in low containers, the plant material is often placed in a needle point holder or a piece of foam anchored into the container. Mechanics may be concealed with such materials as moss, driftwood, stones, and marbles.

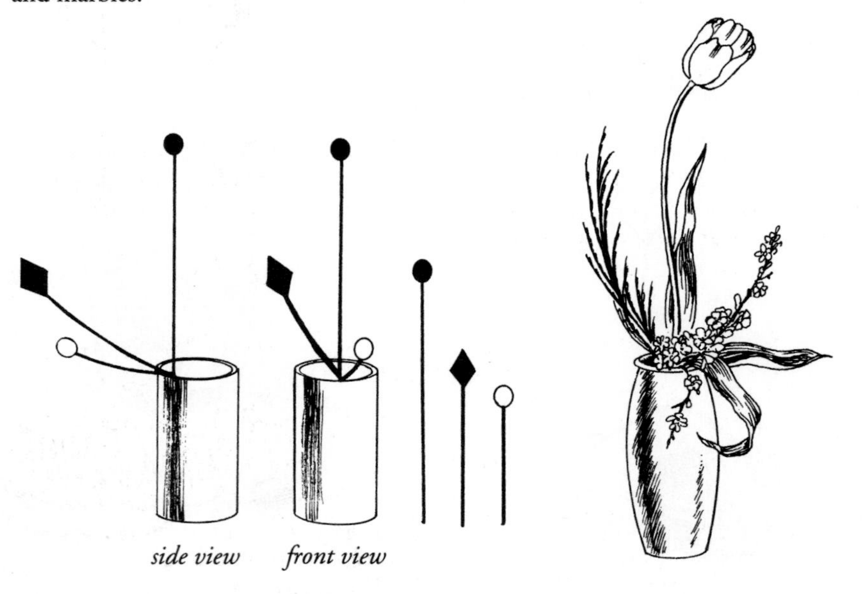

Figure 16-17. The Nageire *or* Heika *style uses curving lines to form the traditional asymmetrical triangle. This natural, less confined style is made in tall, upright containers. This nageire design shows from left to right, a side view, front view, and the relative lengths of the three main stems in proportion to each other for the upright style variation.*

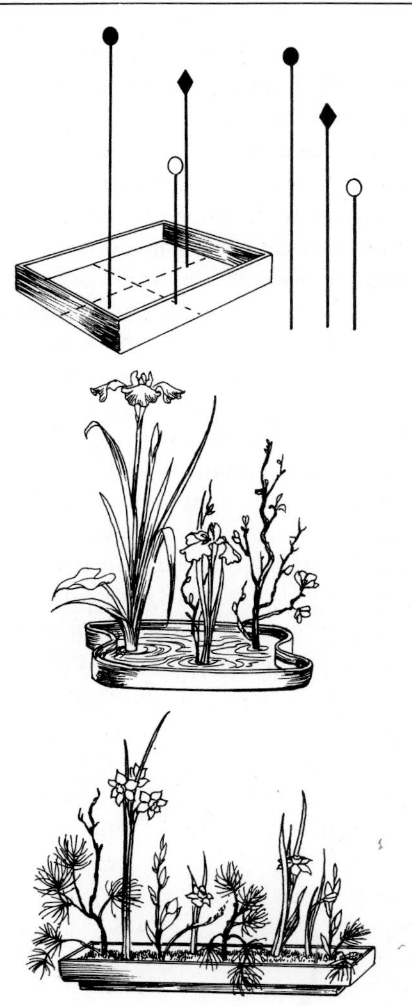

Figure 16-18. Moribana *designs may be made in a variety of design patterns. Often, similar flower types are grouped and surrounded by a pool of water. A more contemporary application is the one-row moribana design, often made in a long, rectangular container.*

Variations in Style

Several variations in line arrangements determine different styles of design. These variations may be easily adapted to both the tall-vase heika or nageire and the low-vase moribana styles of arrangement. The five basic variations include the upright, slanting, cascading, contrasting, and vertical or heavenly styles.

In the upright style, the placements of the three main stems form an asymmetrical triangle (see figure 16-19). The subject or shin line is the tallest stem and is placed in a vertical position above the container. The secondary or soe line is generally two-thirds the length of the subject line and is located to either the left or right of the subject line. The object stem is generally one-half the length of the subject line and is placed on the opposite side of the arrangement to the secondary stem.

The slanting style places the tallest element, or the subject stem, extending outward, dramatically to the side, instead of upward, as shown in the nageire design in figure 16-20. The subject or shin line dominates the entire design, as well as the lengths of the other two elements. In the true-form slanting variation, the secondary stem is generally one-half the length of the subject stem and the object or tai stem is equal in length to the secondary stem, yet they differ in the angle in which they are placed in the design. A slight height difference in the secondary and object stems may also be used in the design, as shown in the moribana design in figure 16-21.

In the cascading style, the main or tallest stem cascades from the vase and extends below the rim of the container (see figures 16-22 and 16-23). The other two stems creating the asymmetrical triangle may vary in length and differ in the angles in which they are placed.

The subject and secondary stems extend outward in opposite directions in the contrasting style. They are unequal in length and at slightly different levels. The object or tai element is the shortest in length and is located in the center of the design (see figures 16-24 and 16-25).

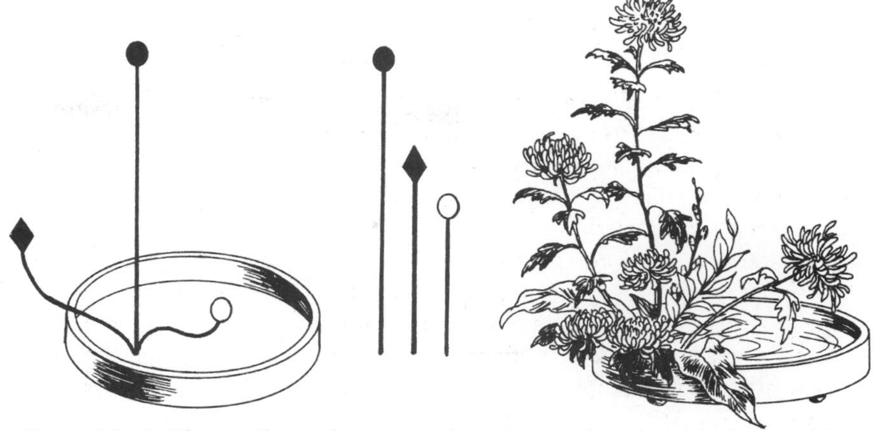

Figure 16-19. This moribana *design is made in the upright style, with the tips of the three elements forming an asymmetrical triangle. The subject line is tallest and generally is vertically set above the container. The secondary and object lines are two-thirds and one-half the length of the subject line and angle forward at various levels.*

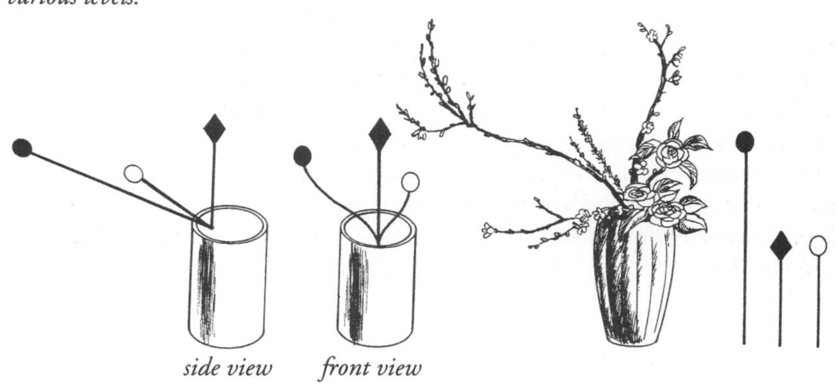

Figure 16-20. This tall container nageire *design is made in the slanting style. The tallest stem extends outward, dramatically to the side, instead of upward. This stem dominates the entire design. The secondary and object lines are generally each one-half the length of the subject line.*

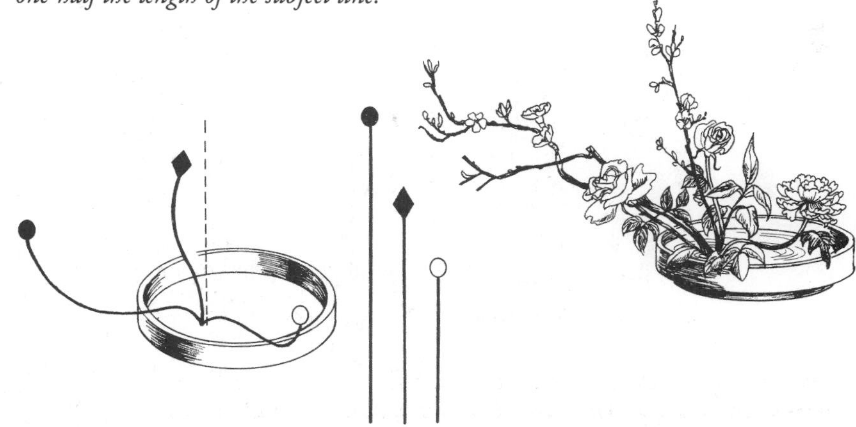

Figure 16-21. A moribana *design made in the slanting style is dominated by the outward bending of the subject line. Often, the secondary and object lines will vary in length.*

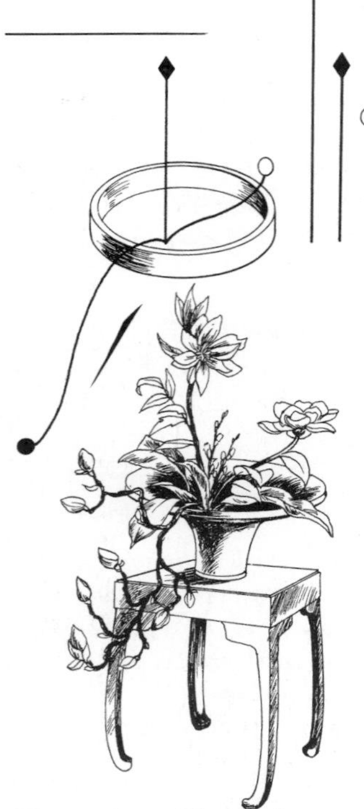

Figure 16-22. This rhythmic moribana *arrangement of magnolias is made in the cascading style. As shown, the tallest, subject line cascades down below the rim of the container. The secondary and object stems differ in length and angling.*

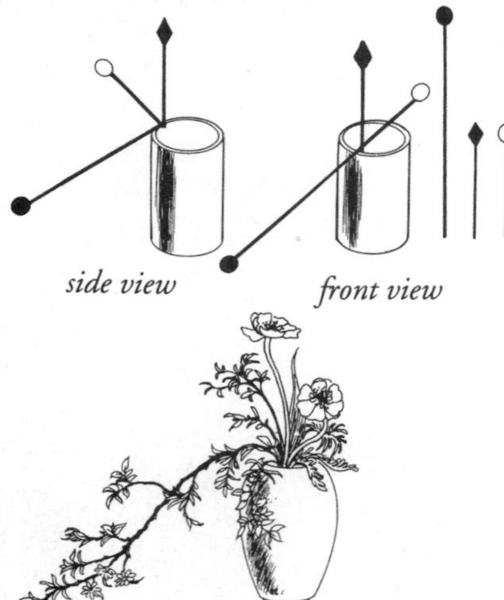

Figure 16-23. This tall vase, nageire design is made in the cascading style. The subject line cascades downward below the container rim, dominating the entire design. The other elements are generally each one-half the length of the subject line.

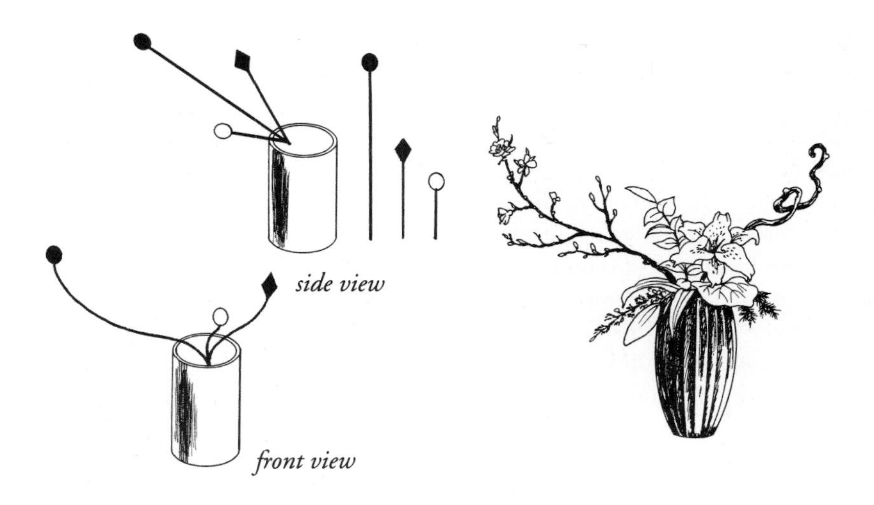

Figure 16-24. The contrasting style nageire *design provides great interest because the subject and secondary stems extend outward in opposite directions. They are unequal in length and are angled at slightly different levels. The object line serves as a focal point.*

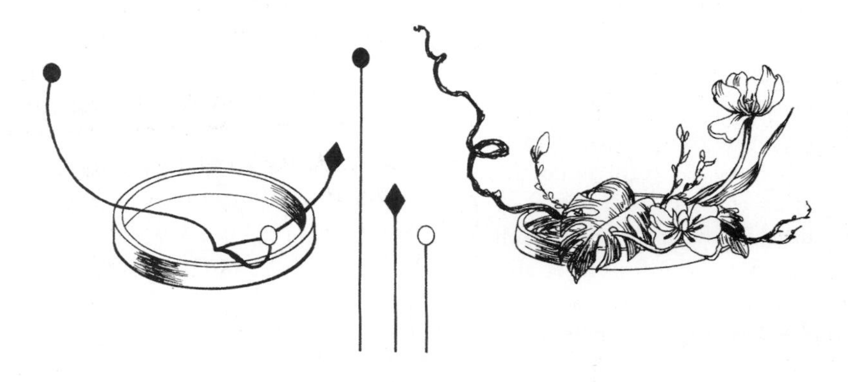

Figure 16-25. Moribana *designs may also be made in the contrasting style. The subject and secondary lines extend outward in opposite directions. The object element provides visual rest in the center of the design.*

In the vertical or heavenly style, the three elements form a tall, narrow triangle. All the stems, extending upward, are exaggerated in length to create a dramatic effect. The secondary line is generally one-half of the subject stem, while the object or tai stem is one-third the length of the subject stem and is located near the container's rim (see figures 16-26 and 16-27).

Abstract or Freestyle

Japanese arrangements classified as abstract or freestyle, often called ***jiyu-bana,*** emphasize the form and texture of the plant material within the design. This design style was introduced following the end of World War II when the Japanese adopted the culture from the Western world. Rather than adhere to strict rules of construction, these arrangements express unique individualism rather than depicting traditional nature scenes. Often these designs combine flowers with glass, raffia, metal, plastic, feathers, and other nonfloral materials. Often vines, leaves, and branches are manipulated into unnatural forms and placed in vases in unusual patterns (see figure 16-28).

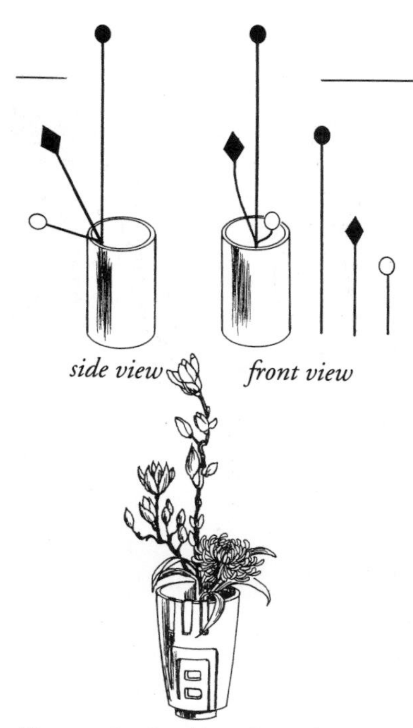

Figure 16-26. Although similar to the upright variation, the vertical or heavenly style places the three elements in a closer, vertical format. Often the lengths of the stems are all exaggerated to create dynamic lines. Because nageire *designs use tall, upright vases, the vertical emphasis is strengthened.*

Figure 16-28. The freestyle or abstract style of design, called jiyu-bana, *rather than following traditional rules, emphasizes plant forms and textures. Often plant stems are arranged in unusual patterns. Nonfloral items, such as glass, raffia, and metal are often incorporated into designs.*

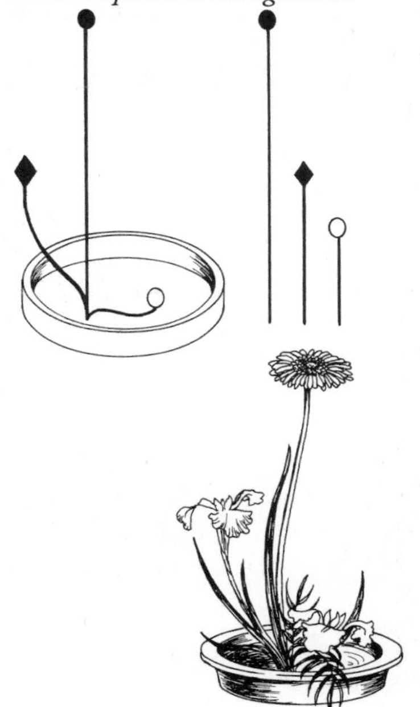

Figure 16-27. A moribana *vertical or heavenly style incorporates a tall, vertically placed subject line. The secondary and object lines are placed to form the classic asymmetrical triangle.*

Figure 16-29. With Ikebana designs, remember "less is more!"

REVIEW

The oriental influence has had a considerable impact on floral design. The use of uncluttered line, asymmetric form, negative space, and simple elegance similar to the ancient designs of China and Japan can be incorporated in floral arrangements of today. In order for you to become proficient in oriental design, it is important to become familiar with the various styles and design techniques and gain skills through practice. Exploring the principles of oriental arrangement will allow you to create distinctive and inspirational designs. Knowledge and experience in oriental floral arrangement will give you greater freedom and confidence in expanding your own design style.

TERMS TO INCREASE YOUR UNDERSTANDING

abstract style
Chinese style
classic style
heika
formal style
freestyle
ikebana
Ikenobo
informal style
Japanese style
jiyu-bana
kenzan
kubari
moribana
nageire
naturalistic style
object line
Ohara
oriental style
rikka
secondary line
seika
shin
shoka
soe
Sogetsu
subject line
tai
tokonoma

TEST YOUR KNOWLEDGE

1. Describe the differences in the Chinese and Japanese styles of design.
2. What are some common rules of all Japanese designs?
3. Name the basic variations of the naturalistic, informal styles of design.
4. What are some characteristics of the jiyu-bana or freestyle design?
5. What are the advantages in knowing oriental design styles and techniques?

RELATED ACTIVITIES

1. From floral magazines, collect pictures of floral arrangements with an oriental flair. Tell what elements are characteristic of the Chinese or Japanese styles.
2. Sketch the five variations of the nageire and moribana styles of design.
3. Select appropriate plant materials, accessories, and a container to make a naturalistic moribana design.

Chapter 17

Contemporary Design Styles and Techniques

Often misinterpreted as weird, bizarre, and undesirable, ***contemporary designs*** are those that are currently in fashion, popular, and representative of leading trends in creativity. Contemporary design encompasses ***traditional design*** and ***classic design, naturalistic design, linear design, modernistic design,*** and ***experimental design.*** Often, certain styles of arrangement, such as experimental designs, are considered contemporary and advanced because they are new, nontraditional, and unfamiliar. This chapter focuses on contemporary designs that are complex or advanced in nature, requiring specialized skills, mechanics, and techniques. Experienced, creative floral designers are interested in advanced, contemporary arrangement styles because these designs offer distinctive, artistic alternatives. ***Advanced design*** provides a creative challenge and requires floral designers to keep current in their knowledge of design styles and techniques.

Figure 17-1. This striking, contemporary design featuring miniature heliconia, gerbera and banksia demonstrates dynamic relationships between textures, colors, and forms. A qualified contemporary designer has confidence to break away from the expected. From Design with Flowers *by Herbert E. Mitchell..*

CLASSIC DESIGN STYLES

Classic floral arrangements are generally mass bouquet designs and continue to be popular and in fashion due to their simplicity and elegance. Often called traditional, classic floral designs are versatile and may be displayed in a variety of ways. Traditional mass floral arrangements include circular, oval, triangular, and fan-shaped bouquets. Advanced classic styles include ***mille fleurs design, Biedermeier design, phoenix design,*** and ***waterfall design.***

Mille Fleurs Design

The arrangement style mille fleurs, sometimes referred to as mille de fleurs, means "a thousand flowers" and refers to having an allover, multicolored pattern of many flowers. Having emerged in the mid-19th century in Europe, this traditional style of arrangement incorporates many different flowers and colors. Generally fan-shaped or rounded, these designs express opulence and abundance with the many varieties of brightly colored flowers together in one arrangement in juxtaposition (see figure 17-2).

Step 1. For these traditional designs, choose a container to support the physical as well as visual weight of the vast numbers of flowers that will be used.

Step 2. If floral foam is used to hold all of the flowers in place, secure the foam in the container. (Do not use floral foam if you select a clear vase.) Without the use of floral foam, a natural grid will soon develop as a tall vase is filled with flowers, by ***lacing*** stems. The multiple stems secure each other and the placement and angling of the flowers becomes more fixed and distinct.

Figure 17-2. A mille fleurs *"a thousand flowers" design contains many different flowers and colors in a multicolored pattern. This design is a contemporary and creative interpretation of the classic, tight and structured* mille fleurs *arrangement. From* Design with Flowers *by Herbert E. Mitchell.*

Step 3. Choose a great variety of flower types and colors in order to achieve the "thousand flowers" appearance. Begin by adding taller, line flowers in a radiating pattern.

Step 4. Next, add rounded, mass flowers within the boundaries of the line flowers to achieve fullness and the desired colorful pattern.

Step 5. Add the final placements, whether they are line, mass, form, or filler flowers, to add forms, colors, and textures to certain parts of the arrangement in order to give an allover floral pattern. Remember the flowers should be styled in a random, loose, and airy fashion to provide a feeling of natural opulence.

Figure 17-3. The Biedermeier *floral arrangement is compact, rounded, and consists of concentric rings. Carnations, sweet William, and roses exemplify this classic style of design. From* Design with Flowers *by Herbert E. Mitchell.*

Biedermeier Design

The Biedermeier style originated in Austria and Germany during the postwar years 1815 to 1848. It is associated with a heavy style of furniture, similar to French Empire and English Regency designs.

A Biedermeier floral arrangement is generally compact, rounded, or slightly conical in shape, and consists of concentric rings of flowers as shown in figure 17-3. Each circular row is composed of the same flower. Each floral ring contrasts to adjacent rows of materials. The contrast of colors, forms, and textures with each row promotes visual interest.

Many adaptations of the Biedermeier style are possible with spiral patterns or looser, mixed flower placements, still in planned alternating patterns. Often, berries, leaves, nuts, small vegetables, and other materials are placed in concentric rings alternating with rows of flowers, creating further contrast and interest.

Step 1. Select a compote container. Fill with floral foam. Next, with a knife, contour the foam's edges into a rounded, pyramidal, or slightly conical shape. Secure the foam in the container with waterproof tape.

Step 2. Begin by placing a ring of leaves (such as salal, galax, or small pieces of leatherleaf) into the floral foam outward, over the rim of the container (see figure 17-4).

Step 3. Next, add a ring of flowers above the row of leaves. (Cut flower stems to about two inches and insert the stems into the foam until the flower heads gently rest against the surface of the foam.)

Step 4. Continue adding concentric rings of contrasting flower types, colors, and textures. The rows of flowers must be close together to hide the floral foam and tape. Filler flowers, such as tiny sprigs of babies breath or statice, may be added in between concentric rings already in place to help conceal any mechanics and to create additional contrasts between rows.

Step 5. Complete the design by placing a single rose at the top of the design.

Phoenix Design

The inspiration and name for this style of design comes from the ancient Egyptian mythological bird, the phoenix. Legend tells of the lone phoenix bird that lived in the Arabian

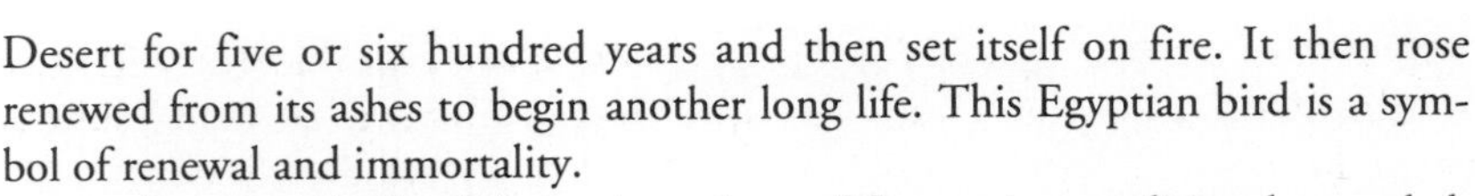

Desert for five or six hundred years and then set itself on fire. It then rose renewed from its ashes to begin another long life. This Egyptian bird is a symbol of renewal and immortality.

A phoenix-style floral design has a base of flowers in a traditional, rounded, compact shape. Bursting from its center rise tall, flowering branches or line flowers, which represent renewal and strength (see figure 17-6). This distinctive design is often used as a party centerpiece or home decoration.

Step 1. Select a compote or other container and fill with floral foam that extends slightly above the container rim. Contour the foam's edges with a knife. Secure the foam in place with waterproof tape.

Figure 17-4. Generally filling a compote container, the Biedermeier floral design consists of concentric circles of foliage and flowers. Starting at the base of the design, insert the same type of leaves or flowers in tight rows. A rose or other fragrant flower is generally placed at the top, completing the design.

Figure 17-5. Legend tells of the ancient Egyptian, mythological phoenix bird that rises from fire and ashes to begin a new life.

Figure 17-6. Like the ancient mythological phoenix bird rising from a pile of ashes, a burst of tall branches emerging from a rounded area of flowers suggests renewal and strength. Eminent spring flowering branches radiate out from a rounded, clustered bouquet at the design's base.

Step 2. As shown in figure 17-7, insert tall, flowering branches or line flowers or other linear material into the center of the foam. These stems should radiate out at the top.

Step 3. Next, insert flowers and foliage at the base of the design to form a traditional round, compact arrangement. Place the flowers at the top of this round design close to the base of the flowering branches in order to hide floral foam and blend with the rising branches.

Figure 17-8. A cascading waterfall gives inspiration for the waterfall floral design.

Figure 17-7. After securing foam into the container, insert tall material deep into the center of the foam. Stems should radiate in a tall, linear pattern.

Waterfall Design

The waterfall design, representative of a waterfall can be traced back to the early 1900s. Initially created for bridal bouquets in Europe, this pendulous style of floral arrangement, often depicted in paintings of the ***art nouveau*** period, has been rediscovered and become a popular container design. Romantic and naturalistic, these flowing designs are a contemporary version of the traditional floral cascade.

This style is characterized by a downward flow of materials, often heavy with foliage. Nonbotanical elements, such as feathers and yarn, are frequently added to give an untidy, undisciplined appearance. Representative of splashing, glistening water, reflective materials such as small fragments of mirror and metallic thread are often incorporated into the bouquet to give the appearance of splashing sunlight. Flowers and foliage that are long, trailing, and pliable are essential to form the downward cascade. Bear grass, sprengeri fern, plumosa fern, conifers, vines, ivies, twigs, and string smilax are examples of useful materials for creating the long, curving form of the waterfall design. The materials flow from the center of the design out and over the container edges. Depth is created through the layering of materials. Layers of foliage alternate with layers of flowers. Colors and textures also alternate within the design, displaying diversity in materials. The waterfall must have adequate room to cascade downward. A tall container is generally needed. However, this design style is also effective flowing over the edge of a pedestal, table, mantle, or shelf. Designs that are elevated provide dramatic results.

Step 1. Select a pedestal vase, urn, or other tall container. Secure floral foam in the container. Foam must extend several inches above the container rim to allow horizontal and downward stem positioning.

Step 2. Select flowers and foliage that have long and flowing curves. Begin the design by placing long, curving foliage to form a cascade on one side of the design. (Waterfall designs may be constructed in different ways. In asymmetrical designs, the materials cascade predominantly on one side of the container. Whereas in symmetrical designs, the materials cascade all the way around the container, displaying equal balance).

Step 3. Place other foliage around the rim of the container, extending it downward (see figure 17-9). The side opposite the long cascade may have a much shorter cascade. This will help balance the design visually.

Step 4. Add layers of flowers—some should face outward and others should face downward.

Step 5. Cover the central base of the design with short-stemmed flowers and foliage. Continue adding layers of flowers and foliage that allow the central portion of the design to flow into the cascading areas. Nonfloral, reflective materials may be added to give the design the image of shimmering light. The waterfall should seem to flow from the center of the container; however, some lines may cross to add interest. Arrange materials so that the design will be pleasing from all sides and rich in depth and texture.

Figure 17-9. Waterfall design steps of construction from (a) to (d). a. Select a compote or other tall container. Secure floral foam above the container rim to allow for downward positioning of stems. b. Place long, curving foliage deep into the foam. These should form a flowing cascade down one side of the container. Next, place shorter foliage around the rim of the container. c. Add layers of flowers flowing down the foliage cascade. Next, add other, shorter flowers around the container rim. Flowers should face outward in all directions, with a gradual transition downward in the cascaded area. d. Cover floral foam with more layers of foliage and flowers. Nonreflective materials may be added to the cascade to give a shimmering appearance.

NATURALISTIC DESIGN STYLES

Often appearing wild and uncultivated, floral designs termed naturalistic or natural are based on nature. Natural designs do not appear contrived or artificial, but represent a slice of the outdoors. These designs emphasize the beauty of flowers without manipulation. Containers must harmonize with the flowers and other materials in the design. Botanical, vegetative, and landscape designs all reflect some aspect of nature.

Botanical Design

Botanical design is considered a new and contemporary American floral arrangement style. This design represents nature through the study of the life of a plant through the close-up look of a bulb flower. Generally focusing on one kind of bulb flower, the parts of the plant, including the buds, blossoms, foliage, stems, bulb, and roots are visible. The plant's natural environment is often depicted in the design's base with stones, mosses, and other bulbed flowers.

Step 1. Select a low container or basket. It should be wide and deep enough to allow for insertion of small live plants (in their soil and containers). Cut a block of floral foam, contour the edges, and place it into the container.

Step 2. In order to show the bulb and roots of the plant, it is necessary to secure the bulbed flower into the foam with a wooden pick as shown in figure 17-10. Insert the sharp end of the pick into the bulb, and insert the opposite end of the pick securely into the floral foam. Add more of the same type of flower into the foam adjacent to the bulb and roots to make a natural clustering of flowers.

Figure 17-10. Botanical design steps of construction from (a) to (c). a. Select a low container or basket. Select plants and basing materials to form a natural-looking environment. b. Contour edges of the floral foam. Insert a sharp wooden pick or other mechanical aid diagonally into the bulb. (This will allow the essential bulb and roots to be visible in the design.) c. Place materials into the container in a natural, clustered pattern. Complete the design by basing the foundation area with mosses, stones, and other materials.

Step 3. Next, add other bulbed flowers and potted plants to the design to form a natural setting. These flowers and plants should be subordinate to the main bulbed flowers that show the bulbs and roots.

Step 4. Conceal the mechanics (such as flower pots and floral foam) with stones, mosses, and twigs. The base should form a natural-looking environment, as shown in figure 17-11.

Figure 17-11. A botanical design represents the life cycle of a plant, in this case a bulb, making visible the bulb, roots, stem, leaves, and blossoms.

Vegetative Design

Vegetative design presents plants as they grow in nature. This natural design simulates a small slice of nature. Flowers and other materials are arranged in a container as they might be found in a natural setting. (Taller growing flowers are placed high in the design, and shorter growing flowers are arranged low in the design.) Flowers and foliage are selected according to seasonal compatibility. Plants that grow together in nature are juxtaposed with one another in a parallel or radial style. These designs should have visual interest on all sides and may easily be used as a centerpiece.

Step 1. Select a low container. Secure floral foam in the container. Contour the foam edges to provide a less rigid block shape.

Step 2. Work from the top of the design downward. Do not place the tallest flowers in the center of the design. Arrange them off center for a more natural appearance. Arrange materials on all sides.

Step 3. Because vegetative designs are a glimpse of the outdoors, do not alter flowers, buds, leaves, or stems. Leave flowers as they would be found outside. Blemishes, mature blossoms, weeds, and thorns remain in these designs.

Figure 17-12. Vegetative designs present plants juxtaposed together as they grow in nature.

Step 4. Layer the heights of blossoms and alternate textures and colors. Bunch similar materials together.

Step 5. Complete the design by adding mosses, rocks, twigs, clumps of grass, and other materials compatible with the flowers, season, and slice of nature chosen.

Landscape Design

Although appearing similar to the vegetative style, a ***landscape design*** arrangement depicts a larger area of nature.

Figure 17-13. Landscape designs depict a larger panoramic view of a natural setting or landscape.

Flowers, branches, and foliage may represent parts of a natural landscape or a groomed garden. Trees, bushes, flowers, and the ground are all represented and organized in color groupings. Patches of foliage, rocks, bark, sand, and moss are placed as they would be found in nature in the design's base to give visual relief from the groups of color. Like the vegetative design, all materials selected must grow in the same environment and during the same season. In contrast to all-sided vegetative designs, landscape arrangements are generally one-sided.

Step 1. Because landscape designs represent a larger, panoramic view of nature, often it is necessary to select a large and low rectangular, oval, or rounded container. Secure floral foam in the container.

Step 2. Place taller materials in the back of the design. These branches represent tall, far-off trees. Materials are generally grouped with contrasts in colors and textures, the same way plants are often found in nature. Do not arrange materials symmetrically. Asymmetrical positioning is more natural.

Step 3. Complete the design by placing moss, rocks, twigs, and other materials at the base.

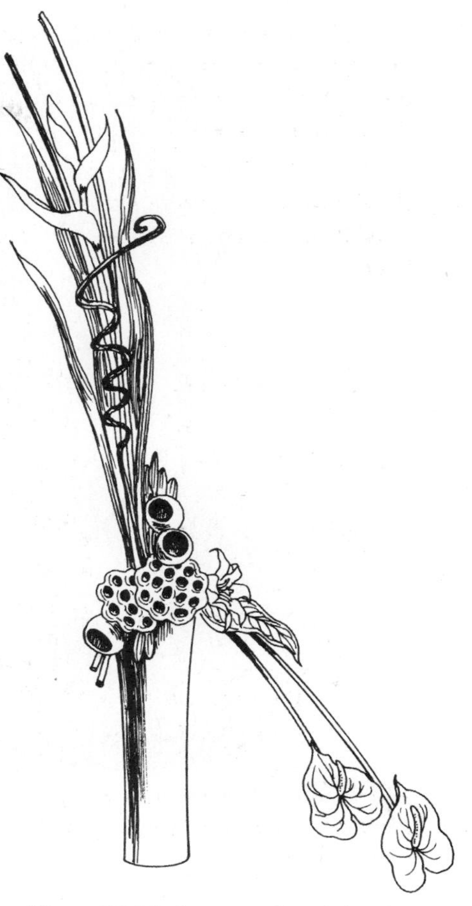

Figure 17-14. A western line design is a dramatic and distinctive arrangement, based on an open, scalene triangle.

LINEAR DESIGN STYLES

Contemporary designs that are termed linear emphasize line and visual movement. Clean, taut lines combined with essential negative space make these designs distinctive and impressive. Form, proportion, and rhythm are necessary principles of design in linear arrangements. Distinctive, advanced linear styles include ***western line design, parallel systems, new convention,*** and ***formal linear.***

Western Line Design

A western line is a general term for symmetrical and asymmetrical line arrangements. A western line design, as shown in figure 17-14, is an open and striking arrangement. The framework, height and width, are all patterned similar to a scalene triangle; yet, the arrangement's negative space is essential to allow the design distinction. The body of the arrangement is not filled in. Often the height of the design is emphasized while the stems creating the width plummet downward.

Parallel Systems Design

A parallel systems arrangement consists of clusters or groups of flowers and foliage. Developed by European floral designers in the late 1980s, this linear style has become a popular alternative to mass style designs. Each group in this design style consists of one type of flower or greenery. Generally, the plant material is set in a vertical pattern with negative space between each section. The negative or empty spaces allow the eye to travel through the arrangement.

This style of design uses the technique of ***parallelism*** in which all stem placements in each group are parallel to each other. The materials within each group may be repeated several times to increase the height and width of each cluster. Using flowers in a parallel manner or parallelism is a technique that may be used in a number of different styles such as vegetative, landscape, new convention, and ***abstract design.***

The composition's base is generally covered with groups of leaves, mosses, stones, and other materials to form a decorative or vegetative look. Or, the base may be left clean, with perhaps just a pool of water or a bed of stones, appearing formal and "architectural." All materials in the design should stay within the container edges. The container should be

Figure 17-15. A parallel systems arrangement consists of clusters or groups of flowers and foliage. Each group consists on one type of flower. Generally the plant material is inserted in a vertical pattern with negative space between groups.
Florists' Review.

simple in design. Most often the container is low and rectangular or oval, but other shapes may be used. A parallel systems design is generally asymmetrical but may also be symmetrical.

Although unique and distinctive, parallel systems arrangements are versatile in their use. Because of the open areas within these designs, they work well as centerpieces. Also, they may be made on a grander scale for use in large areas.

Step 1. Select a simple, low container. Secure a foundation area in the container with needlepoint holders or floral foam (see figure 17-16).

Step 2. Choose flowers and foliage that have a clean, linear appearance to form the various groups. Visually divide the design in sections and establish groups within the composition. Each grouping should be made from one type of flower or foliage. Remember to allow adequate negative space between each cluster. It is best to stagger the heights and spacing of the various groups.

Step 3. Complete the design by concealing the mechanics. This may

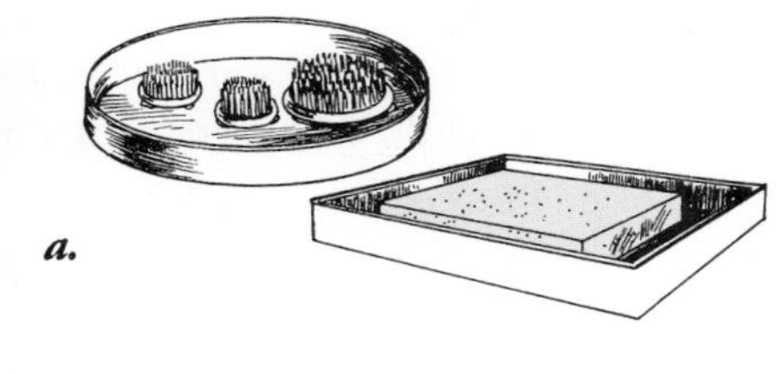

a.

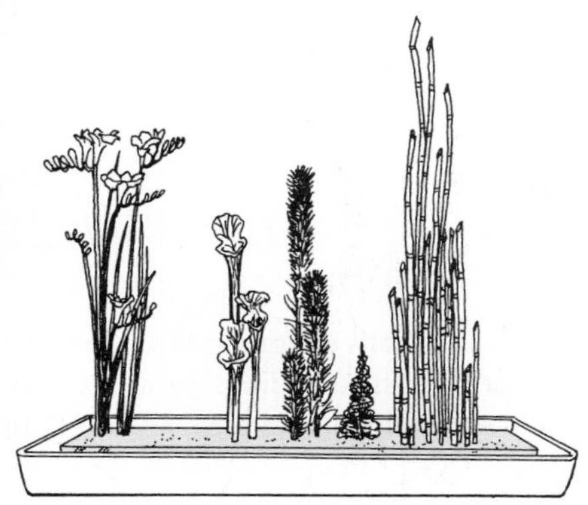

b.

c.

Figure 17-16. Parallel systems design steps of construction from (a) to (c). a. Select a simple, low container. Secure a foundation area of needlepoint holders or foam. b. Choose flowers and foliage with clean, linear stems. Group similar flowers and foliage and allow plenty of negative space between differing materials. c. Complete the design using a variety of basing techniques.

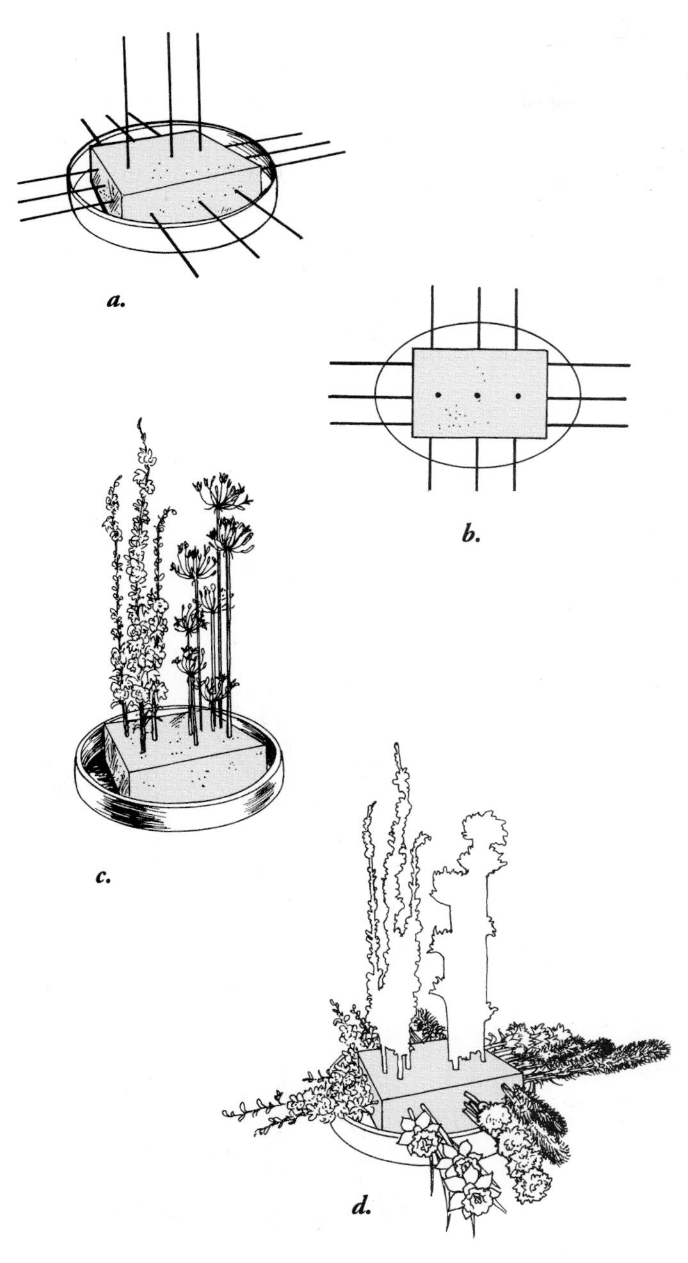

Figure 17-17. New convention design steps of construction from (a) to (d). a. Select a low container. Plan the vertical lines and horizontal lines before construction. b. When viewed from above, the horizontal lines extend outward at 90-degree angles to the sides, front, and back. c. Begin the design, similar to a parallel systems arrangement, by placing groups of material in a vertical, parallel pattern. d. Next, insert stems into the sides of the floral foam. These horizontal groups should be placed at right angles to the vertical groupings. Horizontal extentions are generally reflections of the vertical flowers and colors. Complete the design by using a variety of basing techniques.

be accomplished by using a variety of basing techniques, such as ***clustering, pillowing, pavéing, layering,*** and ***terracing.*** The design should appear neat and organized.

New Convention

Although similar to the parallel systems arrangement, new convention designs not only incorporate vertical groupings of plant material, but horizontal groupings as well. The vertical flower and foliage clusters are repeated low in the design. These horizontal lines are placed at sharp right angles, at the base of the vertical group they reflect. The horizontal groups form linear extentions to the front, back, and sides and are often made in the same materials or similar colors to the adjacent vertical clusters. The horizontal lines are shorter than the vertical lines, using less material. Not all the vertical groups need to be reflected in the horizontal positioning at the base. Negative space must exist between the vertical sections and the horizontal extentions. The combination of vertical groups juxtaposed with horizontal groups results in distinct visual strength and drama.

Containers for these unique designs are generally low and rectangular, but other shapes may be effective. Several basing techniques such as layering, terracing, and pavéing with mosses, leaves, and flower heads are often patterned in rows, parallel to the horizontal stems.

Step 1. Select a low, rectangular container. Secure floral foam in the container. The foam must extend above the container rim to allow for horizontal stems.

Step 2. Begin by visually sectioning and planning the design (see figure 17-17). Place linear materials to form the vertical groupings. Stagger stem heights within each group. The various clusters should also be at different heights.

Step 3. Next, insert stems into the sides of the floral foam, extending at right angles to the vertical groups. The horizontal lines should be shorter than the vertical groups and extend out in the front, back, and sides over the container rim.

Step 4. Complete the base of the design with short flower heads, mosses, and leaves. Use a variety of basing techniques such as layering, terracing, and clustering, to provide textural variety and visual interest (see figure 17-18).

Formal Linear

As the name suggests, forms and lines are dominant in the formal linear style. Often referred to as ***high style*** design or

Figure 17-18. A new convention design incorporates vertical and horizontal lines.

Step 1. Select a container that will present the materials and style of design in an appropriate presentation. Secure floral foam in the container. Floral foam should extend above the container rim to allow for horizontal and downward positioning of stems.

Step 2. Select linear materials to establish the height of the design. Group similar flower types together.

Step 3. Next, add other groups of flowers and foliage. Allow plenty of space between the various groups in the design.

Step 4. Complete the design by filling in the central area of the design and concealing any floral foam. A variety of basing techniques may be used, such as clustering, layering, and terracing.

art deco, these asymmetrical arrangements emphasize shapes, angles, and clean lines. Through the use of a minimum of shapes and quantities, and the use of negative space, the beauty of the flowers, foliage, and stems are accentuated. Generally, similar materials are grouped together to emphasize shape, line, color, and texture. Lines may be vertical, horizontal, diagonal, or curved. Textures are contrasted to allow for the importance of each. Often created with exotic, tropical flowers and foliage, this style of design allows for importance to their unique shapes, colors, patterns, and textures. The entire design must be kept neat and organized with clean lines. The concept of "less is more" is essential to this distinctive design.

Containers are often tall but may be low. A variety of vase shapes are effective for the formal linear style. The container should reflect the high style feeling, allowing it to become an integral part of the entire design.

Figure 17-19. Formal linear design steps of construction. Based on distinctive forms and clean lines, formal linear designs are often termed "high style" and "art deco." Through the use of a minimum of shapes and quantities and the use of negative space, the beauty of the flowers, foliage, and stems are accentuated.

MODERNISTIC DESIGN STYLES

Modern, contemporary trends in floral arrangement are often termed modernistic designs. These designs may reflect contemporary fashion, colors, and even attitudes. Often termed experimental, modernistic designs are often trendy and faddish; yet, for a time they may fill a certain need. Unique modernistic floral styles include ***sheltered design,*** pavé, ***new wave,*** and abstract design.

Sheltered Design

A sheltered design is "protected" within the container. Often, all materials are arranged below the container rim and may only be viewed by peering down inside. These

Figure 17-20. A sheltered design is "protected" inside the container.

Figure 17-21. Sheltered design steps of construction from (a) to (c). a. Select a container that will allow materials to be arranged within the rim. Secure a small piece of floral foam. b. Begin by grouping similar flowers and leaves together. Cut stems short, allowing the heads to rest on the top of the foam. Use a variety of basing techniques. c. Complete the design by concealing any exposed floral foam. Add stones, mosses, sand, twigs, and other materials.

designs, although seeming less dramatic and showy than most styles, require a closer, sometimes longer look. Often slower-paced, sheltered designs offer an artistic alternative to privacy and protection from the outside world.

Step 1. Select a container that will allow you to arrange materials within the container edges. Secure floral foam with an anchor pin or use a needlepoint holder (see figure 17-21).

Figure 17-22 (a). Sheltering steps of construction from (a) to (c). a. A simple technique, sheltering, may be incorporated into a design using bear grass. Insert a grouping of bear grass into one side of the floral foam. b. The tips of bear grass are gathered together using a wired wooden pick. c. After completing the floral bouquet, insert wooden pick with the gathered bear grass into the opposite side of the design.

Step 2. Group similar flowers and leaves together. Cut stems short, allowing the flower heads to rest on top of the foam. Use a variety of basing techniques such as clustering, ***grouping***, and pavéing.

Step 3. Complete the design by concealing any floral foam. Stones, mosses, sand, twigs, and other materials may be added to the base of the container around the flowers and buds. Often, a shallow pool of water with a few stones will give the entire arrangement a restful, protected appearance.

Sheltering is a technique that may be incorporated into many design styles. As shown in figure 17-22, this contemporary technique can be used to create a protected, covered feeling. Bear grass, raffia, curly willow, and other linear materials lend themselves to simple sheltering techniques. Although covering or hiding part of a floral design is a strange concept, it is the covering that actually invites the viewer to discover what is underneath the shelter. The covering often suggests intimacy, mystery, or discernment and creates visual drama.

Figure 17-22 (b). Sheltering provides a covered, protected appearance, inviting the viewer to take a closer look.

Pavé Design

The word pavé (pah-VAY) refers to a setting of jewelry in which gems are placed closely together so that no or little metal is visible. Borrowed from jewelry making, a pavé floral design and the pavé technique refer to flowers, leaves, and other materials arranged closely together in a flat, jewel-like pattern so that no floral foam is visible. The tight clustering emphasizes contrasts in colors and textures.

Step 1. It will be helpful for you to plan out your pavé design pattern first on paper. Select a flat or low container. Next, cut a floral foam block in pieces to fit inside the container, as shown in figure 17-24. Secure the foam into the container with anchor pins, waterproof tape, or pan glue. (Note: Often the pieces of floral foam will fit snuggly into the container and remain secure enough without additional aids.)

Step 2. Select flower heads, leaves, stones, mosses, and other materials that will create the desired pattern. Cut flower stems to a length of about an inch. Insert stems down into the foam. Flower heads should appear to be resting on the top of the foam. Similar materials should fit snuggly against each other with no floral foam showing through.

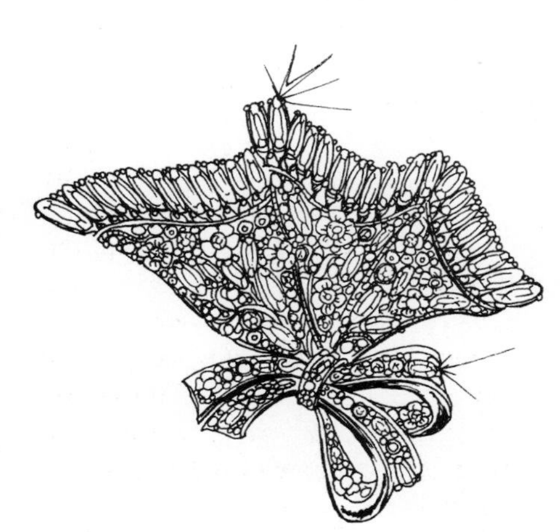

Figure 17-23. Borrowed from jewelry-making, pavé refers to gems placed close together so that no or little metal is visible.

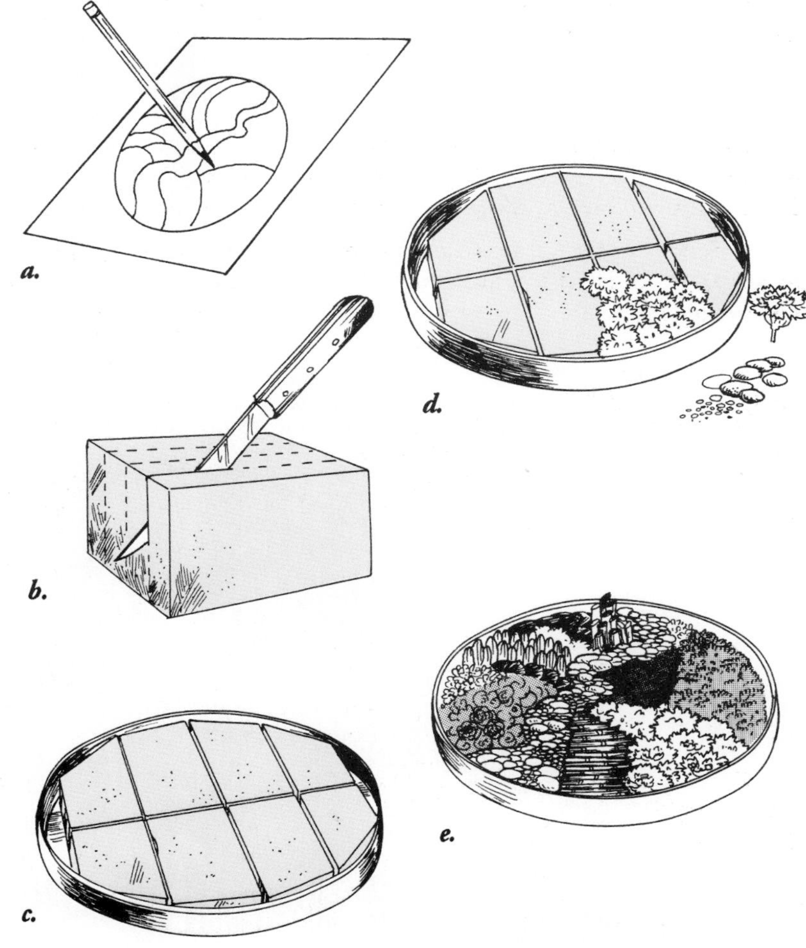

Figure 17-24. Pavé design steps of construction from (a) to (e). a. Plan design on paper first. b & c. Cut a floral foam block in pieces to fit the inside of the container. d. Select flower heads, leaves, stones, mosses, and other materials that will create the desired pattern. e. Continue adding materials in groups. Contrast flower types, colors, and textures for added visual interest. No foam should be visible.

Step 3. Continue adding flowers, leaves, stones, and other materials in the desired pattern. Contrast flower types, colors, and textures for added visual interest. The various sections should fit against each other, with no floral foam visible.

New Wave

New wave refers to any various new or experimental trends or movements. In floral design, new wave is a style of design featuring materials that have been changed with paint and glue, and altered and manipulated in other ways (often folded, bent, braided, curled, etc.). Flowers, foliage, and various materials are presented in unusual and bizarre configurations. Discordant and conflicting lines, colors, and geometric shapes are blended together. Accessories and containers are generally peculiar in themselves, which adds to the eccentricity of the entire design. No rules exist for balance and proportion. Textures and patterns are often overemphasized. Ordinary materials are used in unexpected ways adding to the visual drama.

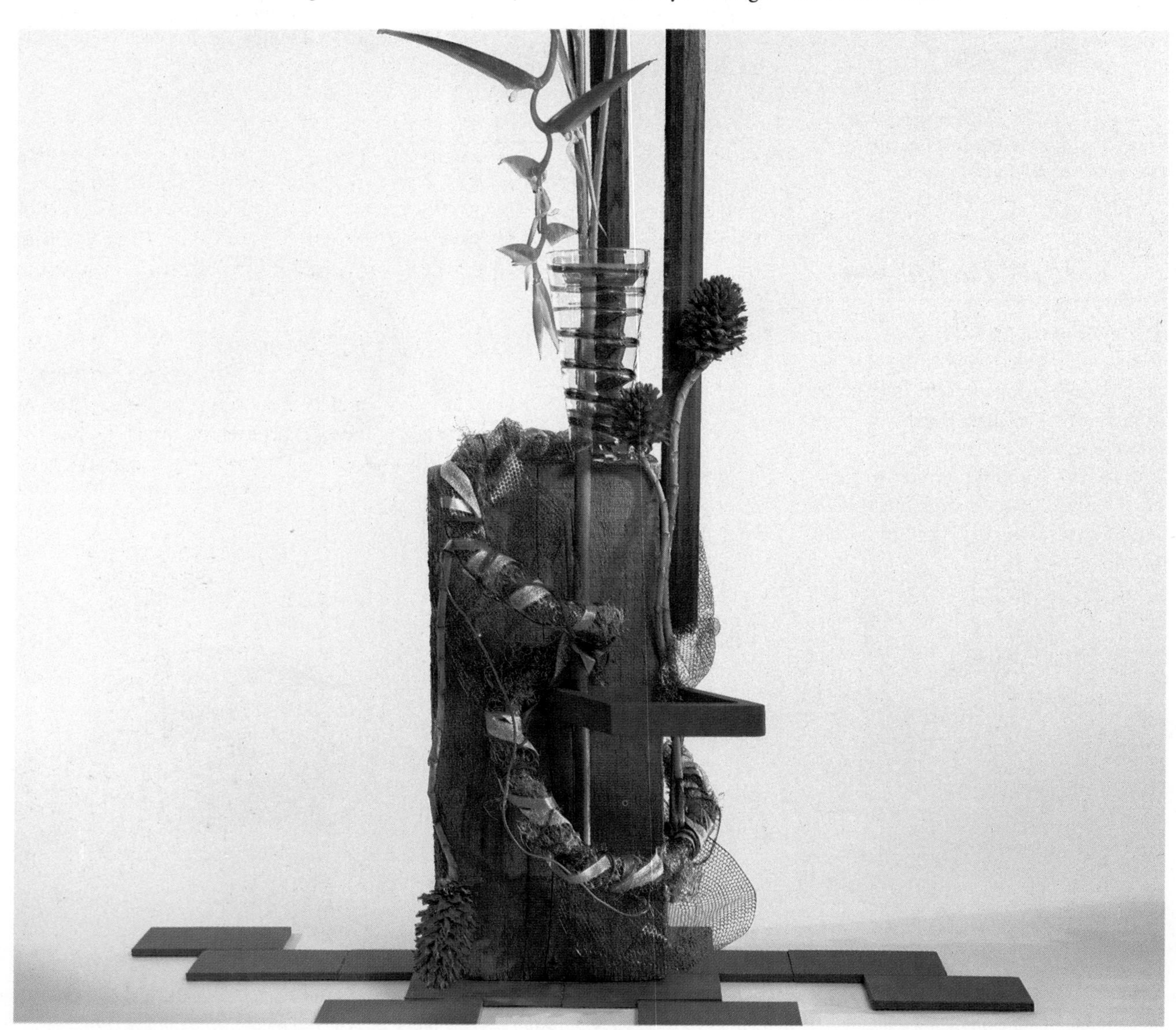

Figure 17-25. Often termed freeform, an abstract design is a nonrealistic floral presentation that emphasizes shape, color, and texture. Often nonfloral materials are used to emphasize geometric forms.
AFS.

Abstract Design

Often termed ***freeform design,*** an abstract design is a nonrealistic floral presentation that emphasizes shape, color, and texture. Nonfloral materials, such as metals, wires, plastics, glass, and mirrors are often used to emphasize geometric forms. Often stems are crossed, petals or leaves may be stripped from stems, or materials are presented upside down. Often, abstract designs become ***interpretive designs,*** reflecting the feelings and ideas of the designer.

Figure 17-26. Basing is the process of placing materials at the foundation area of a design. The base of the design is generally contrasted with tall groups of materials.

ADVANCED DESIGN TECHNIQUES

There are a number of ***techniques*** associated with contemporary, advanced floral design. It is essential to know the names of these techniques or "methods of construction" and how they are carried out in order to make successful, beautiful, and distinctive designs.

Basing

Basing is the process of placing materials at the foundation area of a design. Many different techniques, such as clustering, pillowing, layering, and terracing, may be used to conceal the floral foam at the base of a design (see figure 17-26). Basing provides visual stability and balance at the site in which taller stems emerge. The contrasting colors, textures, and shapes provide maximum visual appeal.

Juxtaposing Materials

For greatest visual impact in color, form, and texture, similar materials may be juxtaposed together. Several techniques are used in contemporary and advanced designs. Many of these methods, such as pillowing and layering create an attrac-

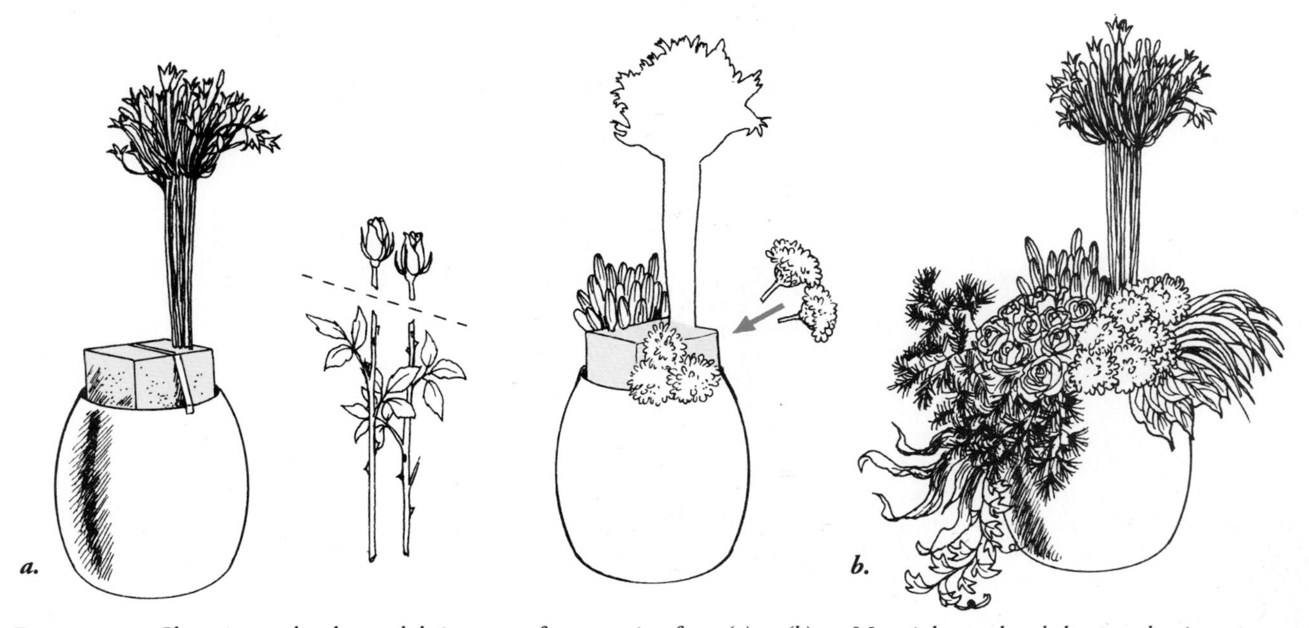

Figure 17-27. Clustering and a clustered design steps of construction from (a) to (b). a. Materials are placed close together in groups. Some stems may be tall, while other stems are cut short. b. Arrange flowers in groups to increase visual contrast of color, texture, and form as materials may lose their identity and function as a mass grouping of texture and color.

tion area low at the base of an arrangement, while other techniques, such as grouping and ***zoning*** stimulate visual interest at higher levels in a design.

Clustering, pillowing, ***tufting,*** and the pavé method share similar application methods; however, there are slight differences. Clustering is the process of placing materials closely together, usually at the design's base. However, materials may be clustered tightly together higher in the design, as shown in figure 17-27. Clustering causes materials to lose their individual identity and function as a mass grouping of texture and color.

Pillowing is a specialized form of clustering and a popular basing technique. The clusters of flowers at the base of a design are arranged to form rounded hills or pillows. These rounded bunches of flowers form an unusual surface (see figure 17-28). Again, colors and textures are emphasized.

Tufting is also a type of clustering, using bunches of short flower stems in a design to create a tufted, airy look. A tufted design is made more interesting with the addition of a few taller flowers, foliage stems, or curly branches rising from the arrangement (see figure 17-29).

The pavé basing technique is a tight clustering method in which the surface of the flower bunches remains flat, rather than rounded, as with pillowing and tufting. Flowers, leaves, and other materials are clustered snuggly against each other, emphasizing color and texture groups.

Terracing, layering, and ***stacking*** are all basing methods in which similar materials are arranged on top of each other, rather than side by side, as in clustering. As shown in figure 17-30, similar materials are placed snuggly on top of each other, with little space between individual leaves and flowers. These basing techniques create an unusual pattern that extends a visual invitation to "follow the path" and look further at the arrangement (see figure 17-31).

Grouping and zoning are methods of bunching similar type materials together to emphasize forms and colors. In contrast to clustering techniques in

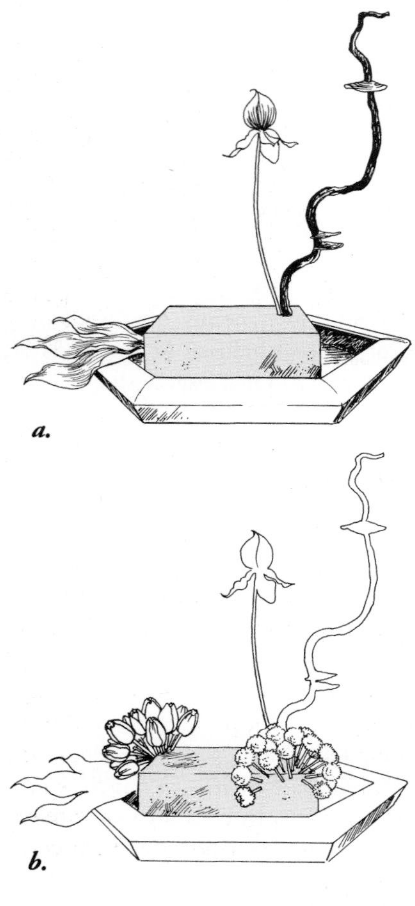

Figure 17-28. Pillowing and a pillowed design steps of construction from (a) to (c). a. Pillowing is a specialized form of clustering. To make a pillow design, the groups of flowers and foliage should form rounded pillows. b. The rounded forms should display an unusual surface. c. Complete the design by snuggling various clusters together in a tight pattern.

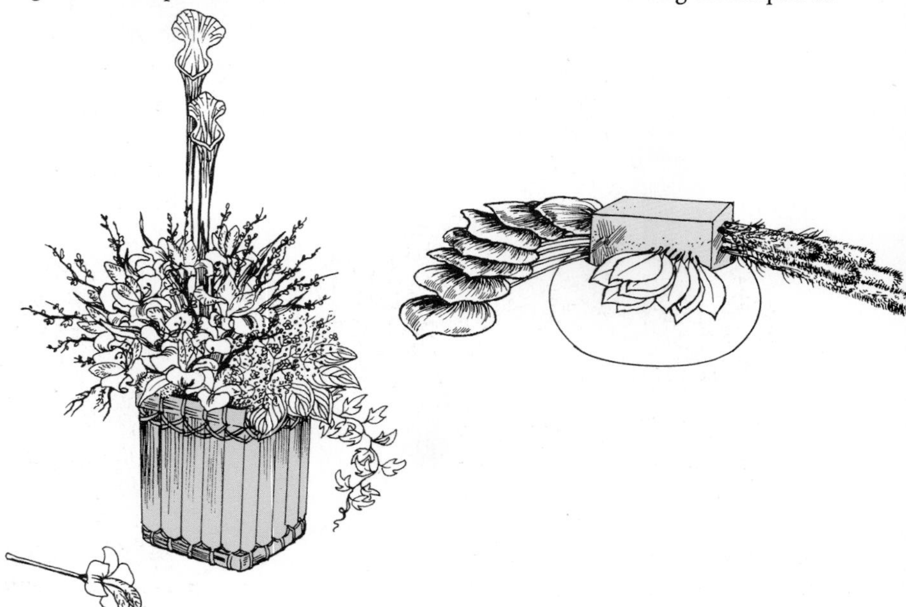

Figure 17-29. A tufted design is made more interesting with the addition of a few taller flowers.

Figure 17-30. Other basing techniques use materials clustered on top of each other, rather than side-by-side.

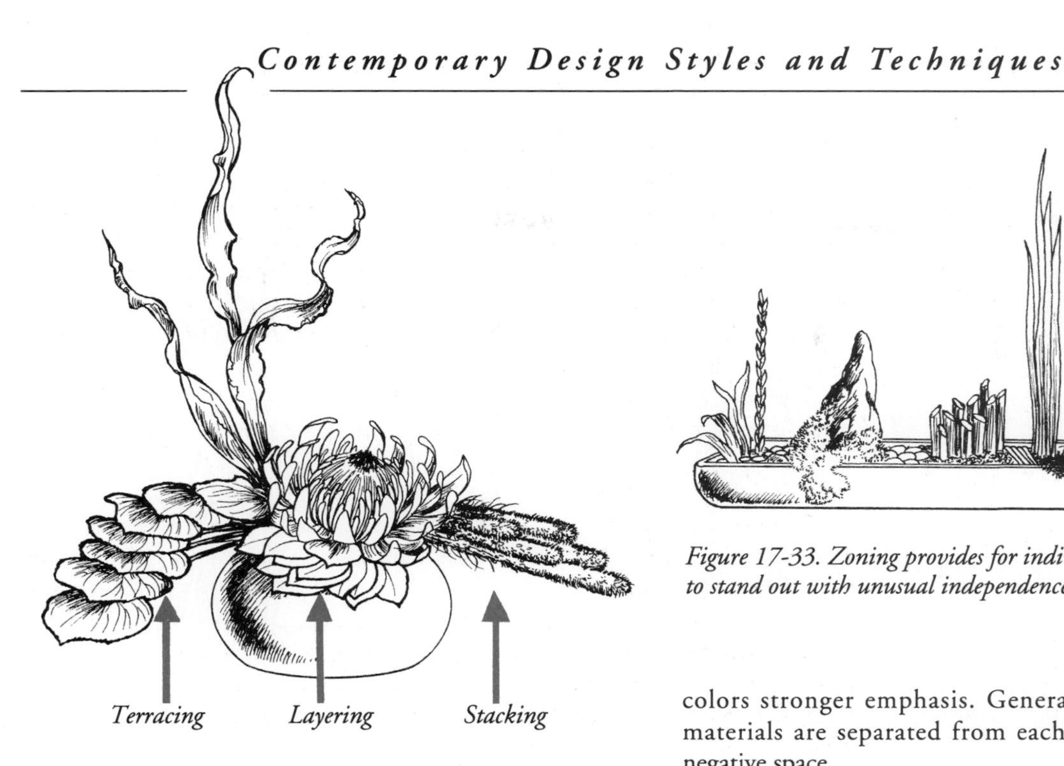

Figure 17-31. Terracing, layering, and stacking techniques create an unusual pattern to increase visual movement in a design.

Figure 17-32. Grouping, a technique used in many design styles, emphasizes shapes, colors, and textures.

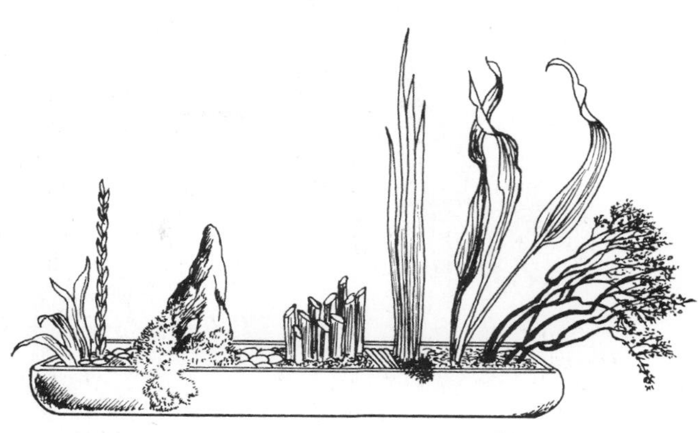

Figure 17-33. Zoning provides for individual shapes and colors to stand out with unusual independence.

which flowers loose their identity, grouping and zoning allow flowers continued visibility because they are arranged in a looser, less compact bunch. As shown in figure 17-32, grouping or gathering similar flower and foliage types closely together gives individual forms and colors stronger emphasis. Generally, taller groups of materials are separated from each other with open or negative space.

Although similar to grouping, the technique of zoning places loose groups of materials in areas or zones within a composition. The quantity of material forming each group is generally restricted to allow a less compact bunch. Each group is slightly isolated from another through the use of negative space. Individual shapes and colors stand out with unusual independence (see figure 17-33).

Uniting Materials

Several uniting or tying techniques, such as ***banding, binding, bundling,*** and ***wrapping*** may be used to unite or join materials together (see figure 17-34). Although the terms become confusing, these techniques all serve to combine materials. Banding is a method of tying materials together to draw attention to a certain area or element and is often merely decorative. Banding provides increased ornamentation and generally does not serve the purpose to physically unite materials together.

Binding, in slight contrast to other tying techniques, is physically joining or fastening stems together. Although this method is functional and serves an actual purpose of holding stems together and in place, the binding of stems with ***raffia,*** ribbon, bear grass, and other materials is an attractive addition to floral arrangements. A ***hand-tied bouquet*** is a bouquet made in one hand while the other hand arranges flowers and foliage within the bouquet (see chapter 13). The stems are held between the thumb and index finger and form what is called the ***binding point.*** The bind-

Figure 17-34. Tying techniques serve a variety of ornamental and functional purposes in advanced, contemporary designs.

ing point is the area where the bouquet is physically tied together with string, ribbon, or raffia.

Bundling is tying or wrapping similar materials together into one unit (such as wheat into sheaves or single cinnamon sticks into a larger group of several) and placing the bundled materials into a floral design.

Wrapping is a technique in which fabric, ribbon, raffia, metallic cord, and other materials are used to cover, coil, or twine a single stem or group of materials to achieve a decorative effect.

Strengthening Visual Movement

Several advanced techniques, such as ***framing, shadowing,*** and ***sequencing*** may be used to increase rhythm and stimulate visual movement. Framing an arrangement is a technique in which material is placed in the perimeter of a design (see figure 17-35). Although the flowers or branches

Figure 17-35. Framing an arrangement is a technique in which material is placed in the perimeter of a design. These framed designs provide visual emphasis to the enclosed, central material.

Figure 17-36. Shadowing is a method of repetition that promotes visual interest and movement. Identical materials are placed closely behind and below taller, front materials.

that outline and frame a design initially lead the eye away from the focal area, the viewer is drawn back to the enclosed space. Framing enhances and calls attention to the materials in the central portion of the arrangement.

Shadowing is a method of repetition. Also sometimes referred to as ***mirroring,*** this technique increases visual depth. Identical materials are placed closely behind and below the taller, front flower or leaf, forming a shadow (see figure 17-36). Shadowing strengthens and draws attention to individual forms, colors, and textures.

When the materials in a floral design move in a progressing pattern of change, often referred to as sequencing, the eye is gradually lead from one area to another (see chapter 5). Sequencing may be easily accomplished through the gradation or transition of materials, especially color (from light to dark, or from one hue to another), size (from large to small), and height (from low to high). A gradual change in elements provides visual flow, increasing visual interest and drama.

REVIEW

Because contemporary floral arrangement styles are closely connected with the styles of interior design, fashion design, and other types of design, it is vital for you as a floral artist to be aware of changing trends. Whether the trend in floral design is boldly tropical and exotic, quietly vegetative and natural, linear and high style, or peacefully sheltered, floral bouquets are indeed an art form that can express the times and feelings of a generation of people. Contemporary designs are in essence, modern period-style floral arrangements.

Because contemporary designs are often unfamiliar and complex, specialized skills, mechanics, and techniques are often required. Advanced, contemporary floral design provides a creative challenge to floral artists and requires designers to keep current in their knowledge of design styles and techniques.

Figure 17-37. Skill, knowledge, and experience are essential when constructing a floral design with many expensive, tropical flowers and foliage.

TERMS TO INCREASE YOUR UNDERSTANDING

abstract design
advanced design
art deco
art nouveau
banding
basing
Biedermeier design
binding
binding point
botanical design
bundling
classic design
clustering
contemporary design
experimental design
formal linear
framing
freeform design
grouping
hand-tied bouquet
high style
interpretive designs
juxtaposition
lacing
landscape design
layering
linear design
mille fleurs design
mirroring
modernistic design
naturalistic design
new convention
new wave
parallelism
parallel systems
pavé
phoenix design
pillowing
raffia
sequencing
shadowing
sheltered design
sheltering
stacking
techniques
terracing
traditional design
tufting
vegetative design
waterfall design
western line design
wrapping
zoning

TEST YOUR KNOWLEDGE

1. Describe contemporary floral design. Name some current, advanced design styles.
2. Compare and contrast the naturalistic styles: botanical, vegetative, and landscape.
3. Why do modernistic designs exist when they sometimes are only popular for a short period of time?
4. Name and describe various basing techniques.
5. What are some advanced techniques that may be used to increase visual movement? How can these same techniques be used in traditional mass and line designs?
6. How do sheltering and framing differ?

RELATED ACTIVITES

1. Arrange a parallel systems design. Incorporate at least four different basing techniques.
2. Construct a Biedermeier design. Use a combination of flowers, leaves, berries, and small fruits.
3. Practice a variety of tying techniques, including banding, binding, bundling, and wrapping.
4. Visit a high style florist shop. View fresh and everlasting designs and notice what advanced designs and techniques are used.
5. Choose a style of design to construct and incorporate the sequencing technique.
6. Collect photographs from floral magazines of contemporary, advanced designs. Organize into a resource reference notebook.

Chapter 18

Wedding Flowers

Because weddings are a traditional part of the florist business, floral designers must be knowledgeable of various wedding designs and be trained in many specialized arranging techniques. A professional wedding flowers consultant must have the ability to consult with brides-to-be to plan all of the floral decoration details of the ceremony, reception, and other events connected with the wedding. Although weddings occur year-round, promotion and advertising are key elements in increasing sales and building business for the retail florist.

Figure 18-1. The receiving line is a traditional part of the wedding festivities.
Fresh from the Altar, *Jessica Hayllar (1858–1940), Christie's, London.*

FLORAL ROMANCE

The history of romance identified with flowers has a rich heritage in bridal flowers. History records the use of flowers to celebrate romantic occasions as early as the 16th century. Flowers have been a messenger of love and romance since the beginning of time. The romantic implications of flowers became popular during the 1800s in the Victorian Era. A courtship started by sending a flower nosegay. Lovers selected blossoms discreetly to convey a proper romantic message. Courtships flourished through sending nosegays that expressed progressive messages of love.

Figure 18-2. An aggressive florist, capturing the attention of the wedding market, follows engagement announcements and sends brides-to-be a mailing, introducing its bridal services and products.
Photo by Patricia Hennessey.

The romantic and sentimental meaning of certain flowers became the essential consideration in selecting blossoms to be used in a Victorian nosegay. Specific colors conveyed explicit romantic messages. Romantics selected flowers only for the meaning associated with them. Luxurious reds, blues, and purples, favored colors in the Victorian era, conveyed love and passion. Red roses, for example, expressed ardent love. White symbolized innocence and purity. Lighter color values expressed sincerity and fidelity. Each variety of flower had an expressed meaning and people selected flowers to articulate a message of romance.

After World War II, wedding flowers became more of a fashion statement than an expression of romance. With the emergence of floral design schools and the "professional art" approach to design, instructors, students, and floral designers emphasized "artistic styling" of wedding party flowers. The various styles of bouquets became the focal point of bridal work. Designers placed less importance on the romance and symbolism of flowers and paid more attention to the mechanical construction of the bouquets.

The focus of this chapter is to present an overview of types and uses of bridal bouquets and wedding decorations. Emphasis is placed on how a retailer attracts wedding customers through promotion and advertising and how professional wedding consultants meet the needs of prospective brides. As wedding styles change, so must the professional florist. The wedding consultant must have practical experience along with vision to translate the bride's wishes into beautiful floral designs.

PROMOTION AND ADVERTISING BY THE RETAIL FLORIST

An essential part of becoming a profitable wedding florist is selling the wedding. Prospective brides must first be attracted to the flower shop and then convinced to purchase flowers there. Selling wedding flowers to a targeted clientel requires specialized selling techniques, including unique promotion and advertising strategies.

Attracting Wedding Customers

While word-of-mouth advertising is an essential ingredient in successfully promoting the sale of wedding flowers, appropriate, aggressive advertising must be targeted at the wedding market and promotional efforts must reach out to individual customers.

Promotion is a show-and-tell activity. In-store promotions help to attract customers to the floral shop. These promotions may include window and interior displays simply serving as reminders to customers of wedding flowers. Or, promotion may be more elaborate with the staging of a bridal show in the floral shop. All of these promotional efforts help a flower shop promote its wedding products and services. A shop can also cooperate with a bridal shop, a photographer, a caterer, a limousine company, and others who serve the wedding market to present a bridal extravaganza in a large area, such as a shopping mall or hotel. This helps the flower shop become more visible to a large population of prospective brides in a short period of time.

Effective promotion includes a presentation of the actual product with personal interaction or printed communication with brides-to-be. Giving door prizes, donating flowers for a bridal program, or doing permanent flower bouquets for mannequin displays are forms of public forms of promotion, outside the floral shop, more specifically ***publicity.***

Advertising is communication through the use of media (newspapers, direct mail, brochures, radio, and television). The market for wedding flowers is a clearly identified customer audience. Therefore, the more direct the advertising, the better a florist's chance for producing positive results. An aggressive wedding florist follows engagement announcements and sends the bride-to-be a mailing introducing its bridal service (see figure 18-2).

Another form of advertising, targeted at the wedding market, involves ***networking*** with other professionals who offer wedding services. Networking is interacting with related businesses to share information and services. A joint effort in attracting prospective brides through promotional and advertising methods is a less expensive and effective form of advertising. Other professionals serving the bride and groom include photography studios, bakeries, bridal salons, tuxedo shops, catering companies, department store bridal registries, and jewelry stores.

THE WEDDING CONSULTATION

The wedding consultation is the process of helping a bride select appropriate flowers for her wedding. This extensive planning session is conducted by a ***wedding consultant*** who is experienced in sales and is knowledgeable about flowers, bouquet styles, ceremony and reception decorations, and profitable pricing. Further, the professionally trained wedding consultant is competent in color theory and knows how to incorporate this color knowledge in planning distinctive weddings. The qualified floral consultant stays up-to-date with bridal fashions.

A competent wedding consultant is the foremost factor in developing bridal business. The consultant communi-

Figure 18-3. Details of the floral designs for a wedding are discussed at the bridal consultation, which is conducted by an experienced designer. Generally, the bride-to-be and her mother are present.

cates the image of the flower shop as he or she presents bridal services to prospective customers. The way a wedding consultant introduces a shop's wedding flower services influences buying confidence and a bride's willingness to spend appropriately for the proper flowers. A consultant should be a good listener. A qualified consultant understands the bride's personal desires for her wedding and helps her make decisions about her wedding flowers. The consultant should encourage the bride to share her thoughts and ideas about the flowers for her wedding.

The consultation is generally conducted by appointment only, to ensure that the florist will be able to devote full attention to the customer. When setting up the consultation appointment, general information about the wedding, such as dates, times, and locations of the ceremony, reception, and other events are recorded on the order form.

At least one hour should be allowed for this lengthy planning session. In order for the meeting to go undisturbed, the consultation should be conducted in an area of the shop away from telephones and in-store customers. Set up a table and enough chairs to accommodate the consultant, the bride-to-be, and one or two other people (usually the mother or a friend of the bride-to-be). Some consultations include the bridegroom and his parents. It is important, however, to remember that too many people present at the consultation may lead to confusion and frustration for the bride and the consultant.

During the consultation, the florist will use a wedding order form, bridal selection books, photograph albums of floral designs, and sometimes silk flower replicas to help the bride select design styles.

The Wedding Order Form

During the consultation, the florist gathers information and records it on a preprinted wedding planner. This order form lists the major design and accessory categories for weddings. A variety of preprinted forms are available or may be printed by the individual florist (see figure 18-4). The purpose of this form is to

Flowers for the Wedding of

The Bride ______ **The Groom** ______
Address ______ Address ______
Phones: ______ or ______ Phones: ______ or ______

Ceremony Location: ______ Date: ______ Time: ______
Reception Location: ______ Date: ______ Time: ______
Rehearsal Dinner Location: ______ Date: ______ Time: ______

	EXPENSES* of the BRIDE	GROOM
THE BRIDE		
Color & style of gown		
Style of bouquet & flowers		$
Floral headpiece		$
THE BRIDAL PARTY		
"Honor Attendant"		
Color & style of gown		
Style of bouquet & flowers	$	
"Bridesmaids" (Number ___)		
Color & style of gowns		
Style of bouquet & flowers	$	
"Flower Girl"		
Color & style of gown		
Style of bouquet & flowers	$	
Floral headpieces	$	
BOUTONNIERES		
Groom	$	
Best Man ___ Fathers ___		$
Groomsmen, Ushers (Number ___) ___ Grandfathers ___		$
Ring Bearer ___ Other ___		$
CORSAGES		
Bride's Mother		$
Groom's Mother		$
Grandmothers (Number ___)		$
Others (Number ___)	$	
DECORATIONS FOR THE CEREMONY		
Main Altar	$	
Aisle & Pew Decorations	$	
Foliage (agmts, rental greens, etc) ❑ Canopy ❑ Candelabra (Altar) ❑ Candelabra (Aisle) ❑	$	
Kneeling Bench ❑ Aisle Runner (50 ft. ❑) (75 ft. ❑) (100 ft. ❑) (125 ft. ❑) (150 ft. ❑)	$	
Other	$	
RECEPTION		
Cake & Cake Table	$	
Centerpieces ___ Other ___	$	
REHEARSAL DINNER		
Centerpieces ___ Other ___		$
Subtotal	$	$
Sales Tax	$	$
TOTAL	$	$
Deposit	$	$
Balance	$	$

Acceptance by: ______
Date: ______

*Traditional expenses are often altered to meet individual circumstances.

Figure 18-4. Obtaining and recording the details of a wedding on a wedding order form is the first step in helping a bride plan the flowers for her wedding. A variety of forms are available from floral suppliers, while many florists choose to have wedding forms custom printed.

organize the wedding plans and provide information to all the designers, who will eventually prepare the floral designs. Most order forms also indicate who is financially responsible for each item.

Items of Discussion

Starting with general information—names, addresses, and telephone numbers—as well as the specifics of the ceremony and reception—dates, times, and locations, the qualified wedding consultant controls the sequence of discussion. The first floral designs discussed are the flowers for the bride and her attendants and others in the bridal party and family. After completing the decisions about personal flowers, the consultant moves to ceremony decorations, then reception decorations, and concludes with home flowers, rehearsal dinner flowers, and other additional decorations.

Figure 18-5. Generally the bride gives a lot of thought to her bouquet. The bouquet should reflect the gown and formality of the ceremony.
© The John Henry Company

Bride

The first floral item discussed is the bridal bouquet. Generally, the bride has definite ideas and has given a lot of thought to the kinds of flowers and the style of bouquet she will carry. A description of the gown will help determine appropriate floral styles. For example, an elaborate full-skirted gown with a lengthy train is accented beautifully when the bride carries a ***cascade*** or ***crescent bouquet.*** The style of bouquet and the colors should be decided before selecting specific flowers. Some flowers, because of shape, size, or longevity, will work better in certain bouquet styles. Seasonal availability of some flowers is also a factor and may affect the bride's selection. The consultant should remind the bride of other floral pieces to coordinate with the bridal bouquet, such as a floral piece for the bride's hair, hat, or veil. The florist may suggest a ***toss bouquet*** in matching colors for the bride to throw to the unmarried girls, following the ceremony or reception.

Bridesmaids

Some brides select the same style bouquet for both the bride's and the attendants' bouquets. When the styles of bouquets are similar, the bridal bouquet usually features different flowers or colors than the bridesmaids' bouquets. Often, brides select contrasting bouquet styles for the bride, the honor attendant (maid or matron of honor) and the bridesmaids. Generally, the bridesmaids' bouquets are modified versions of the bridal bouquet and coordinate with their gowns. Hair pieces may be needed for the bridesmaids and should be made to coordinate with bouquets. The number of attendants and their first names should be recorded on the order form. This will lessen confusion

Figure 18-6. The presence of a flower girl will add special meaning to the wedding. Flowers designed for young flower girls must be small in scale, easy to carry, lightweight, and secure.
© The John Henry Company

for both the bride and the consultant, especially when there are several bridesmaids.

Flower Girl

Because the age and size of the flower girl varies, her age and height should be recorded on the order form. This will help the florist prepare an appropriately sized design for the flower girl. If a ***chaplet*** is needed for the flower girl, it is helpful to have a head circumference measurement as well. A miniature version of the bridesmaids' bouquets is common for the flower girl, as shown in figure 18-7. A small basket with loose petals or a small floral arrangement are popular selections for flower girls. Floral items should be lightweight and securely constructed.

Figure 18-7. The bridesmaids' bouquets are often designed as a modified version of the bride's bouquet. Often the bouquet for the maid of honor is slightly different in color or size. The flower girl may carry a scaled-down bouquet, similar in color and style as the bridesmaids' bouquets.

Figure 18-8. The groom's boutonniere generally is different from the rest of the boutonnieres and is made of a flower or small cluster of flowers to match the flowers in the bridal bouquet. Boutonnieres for the best man and other groomsmen may coordinate with the bouquets of the bridesmaids. Boutonnieres for the fathers, grandfathers, and ushers are similar to each other, coordinating with the wedding colors. In contrast to the other boutonnieres, the one for the ring bearer should be made smaller in scale. Often a floral piece is made to accent the floral pillow he carries.

Boutonnieres

Boutonnieres are traditionally worn by the male members of the wedding party and close relatives. The groom's boutonniere is discussed first and may or may not be more elaborate than the others. Generally, the groom's boutonniere is different in some way from all other boutonnieres. The flowers selected for his design are selected to match those in the bride's bouquet. The boutonnieres for the ***groomsmen*** may coordinate with the bridesmaids' bouquets. Boutonnieres for the fathers, stepfathers, ushers, grandfathers, brothers, and other men, including the organist, vocalist, clergyman, servers, and those helping with registration are often the same type, lending a uniform look and easing distribution prior to the ceremony. The boutonniere for the ring bearer should be smaller, proportionate to his age and height. A ring pillow is often carried by a ring bearer in a ceremony. A floral and ribbon accent is constructed similar to a corsage and pinned to the pillow.

Figure 18-9. The "going-away" corsage may be constructed as part of the bridal bouquet. This 1955 bride elected to have a removeable cattleya orchid corsage, which was constructed in the center of a simple, cascading bouquet. The "going-away" corsage and boutonniere are floral remembrances worn by the bride and groom as they leave for their honeymoon.
True Love by Tenney Originals. All rights reserved. Used with permission.

Corsages

Corsages are ordered for the various female family members and close friends. Often more elaborate, the corsages for the mothers should coordinate with their gowns. Over-the-shoulder designs are popular; however, the style, color, and fabric of the gown will influence the style of corsage and where it is worn. A wristlet or purse accent is sometimes selected.

When the bride-to-be desires to have her ***going-away corsage*** made a part of the bridal bouquet, it is made so that it may be removed from the bouquet. The going-away corsage is a floral remembrance worn by the bride as she leaves with her new husband for the honeymoon (see figure 18-9). Generally, however, a corsage becomes badly worn by the end of the wedding events. It is advisable to make a separate corsage for her going-away outfit.

The consultant should remind the bride that other close female relatives and supporting members of the ceremony and reception traditionally are recognized and remembered with corsages. Designs for stepmothers, grandmothers, sisters, other female family members, are often designed the same to ease distribution of the flowers prior to the ceremony or reception. Corsages may also be ordered for the organist, vocalist, servers, hostess, and those who assist with the registration. Because colors of dresses for all the women are unknown, designs are generally made to match the wedding colors. Often the florist will include an extra corsage and boutonniere with the wedding order.

Ceremony Decorations

The ceremony sets the ambiance for the entire wedding. The floral decorations for a wedding ceremony may be elaborate or simple. Whether the ceremony takes place in a church, chapel, hotel, home, or outdoors, when the guests arrive, they should quickly catch the spirit of the wedding through the floral decorations. During the consultation,

Figure 18-10. The floral decorations for the ceremony help guests to quickly catch the spirit of the wedding.
© The John Henry Company

the florist must ask several questions in order to envision appropriate floral designs and determine the types, sizes, and number necessary. Questions regarding the size of the ceremonial site, the height of the ceiling, the style of architecture, and regulations of flower use will provide important answers, relative to the types of designs necessary. Churches often have rules that must be followed in decorating. Some churches require that their flower guild decorate the sanctuary or chapel. An experienced consultant is familiar with any of the ceremony decoration limitations and works within these guidelines. The consultant will know the specific features of the various local churches and offer suggestions to the bride.

The focal point of any ceremony decoration is the area where the bridal party processional culminates and where the couple exchanges vows. All decorations should carry the eye to and support this point of visual attention.

Figure 18-11. Ceremonial decorations may be elaborate and formal or simple and informal. All decorations lead the eye to the center altar table. Altar decorations serve as a background focal area for the bride and groom.

In discussing ceremony decorations, the florist should encourage the bride to think of where the guests enter the church or ceremonial site and where they will be seated. Planning decorations for the entryway, registration area, and the aisle is important. These other decorations will help support the theme of the flowers surrounding the altar area and provide visual impact and cohesiveness. A typical large, formal church ceremony may include many elaborate decorations for the altar area and the aisle. As shown in figure 18-11, all decorations lead the eye to the center altar table, the point where the wedding ceremony takes place.

The ***altar bouquets*** must have strong impact to focus the attention of the audience to the center of the altar and the bridal party. These decorations should be large enough to be seen by all the wedding guests. This is often a great distance. One large floral design may be placed on the altar table or on the rail directly behind it. Generally, a pair of large altar bouquets, one placed on each side of the altar table, is needed. These provide dignity and help to visually frame the bride and groom during the ceremony. ***Candelabras*** are popular altar decorations, especially for formal evening ceremonies where the soft glow of candlelight provides a romantic feeling. Often candelabras are decorated with floral and ribbon accents. Large potted plants or large foliage arrangements are often selected to help decorate large altar areas.

Decorating the center bridal ***processional*** aisle, which leads to the point of the ceremony, completes a ceremony decoration effectively. Depending on the type of marriage ceremony and the place where it will be held, decorating the processional aisle might include garnishing the ends of pews, chairs, and benches. Often decorated with bows and floral accents, these decorations provide a feeling of unity between the couple and their guests during the ceremony. As shown in figure 18-12, ***pew decorations*** may be made simple or elaborate. Rather than having decorations on all pew ends, they are generally placed only

Figure 18-12. Pews are often decorated with simple bows. Other more elaborate pew designs have become popular. These are made to complement the altar decorations. A variety of foam holders to attach to the pews are available.

Figure 18-13. A canopy or arch may be needed for certain weddings. A Jewish ceremony requires a canopy or Chuppah under which the bride and groom stand. The canopy is sometimes elaborately decorated with flowers, foliage, and vines.

on every second or third pew, with ribbon draping between the bows. Simple bow decorations are generally attached with chenille stems. A variety of holders are available that provide either a water tube or a floral foam cage in which to design floral arrangements. ***Aisleabras*** or aisle candelabras are effective decorations that add formality to an evening ceremony. The aisleabras are posts that hold a single candle enclosed in a glass hurricane cover. (Because of fire safety, these candles are often battery operated.) Aisleabras are often decorated with flowers and ribbon.

For outdoor ceremonies where guests stand, a lovely bridal processional aisle can be created with freestanding single candelabras decorated with flowers. If candles are not appropriate, aisleabras can serve as aisle posts, decorated with flowers and ribbons draped in-between the posts to create a bridal processional aisle.

A ***canopy*** or ***arch*** may be desired for certain wedding ceremonies. In the traditional Jewish wedding, the bride and groom are married under a canopy or ***chuppah*** (see figure 18-13). In most cases, the canopy is a freestanding unit. Many flower shops specializing in wedding flowers have canopies as part of their rental equipment. Some canopies are metal frames that need to be totally covered with flowers and foliage. Other canopies are covered with fabric or vines and require floral embellishment. Wherever the canopy is used, in a synagogue, hotel, country club, or outdoors, it is the focal point of the decoration. A floral arch may be used in a variety of ways for the wedding ceremony. Often an arch is placed at the rear of the center aisle, and visually frames bridal party members during the processional and ***recessional.***

The floral consultant may wish to make further suggestions to the bride-to-be for floral decorations at the site of the ceremony. Depending on the church or other setting, the registration area, windows, and other areas may be decorated with floral designs to add touches of beauty and accent throughout the ceremony area.

For the home wedding ceremony or one held at a country club, often the fireplace is the focal point. Depending on the formality of the wedding, the fireplace can be decorated with a central arrangement and accented with flower and foliage garlands. Garlands placed throughout the ceremony area help to unify the wedding color theme.

Reception Decorations

The wedding reception follows the wedding ceremony. It is a time of celebration with friends and family. A reception is a joyous occasion and may be simple with cake and punch or more elaborate with a formal dinner and danc-

Figure 18-14. Balloons add a festive and unique atmosphere to wedding festivities. Smaller balloons are complimentary accessories in floral decorations.
Balloon Supply of America (BSA). Designed by Judy Gorman and Dottie Blanchard.

Figure 18-15. Floral centerpieces for buffet tables may be made one-sided or all-sided, depending if guests serve themselves from one or both sides of the buffet. Often the punch bowl is given floral accents and the table food trays are decorated with loose flowers and petals. Table skirt decorations and garlands add beautiful and distinctive accents to tables.
© The John Henry Company

ing. Simple receptions sometimes require only flowers for the cake. The formal dinner reception calls for an array of floral decorations to help make the celebration beautiful and festive.

Regardless of the size of the reception, the professional consultant pulls a theme of flowers and colors through the ceremony, the bridal party, and the reception. This coordination of a flower theme and color scheme gives the entire occasion visual unity and a flair of memorable style.

Typical decorations for the reception include flower accents for the cake and cake table, gift table, serving and buffet tables, punch and champagne table, guest tables, and the ***head table.*** A floral arrangement will also be beautiful at the guest book or registration area of the reception. Larger floral designs, arches, and other decorations are often needed to visually frame the bride and groom. These larger designs will create a beautiful backdrop and setting for photographs.

The wedding cake is a focal point and one of the most imporant components in the reception decorations. As shown in figure 18-17, wedding cakes may be decorated in a variety of ways. The style of cake will determine the style and amount of floral accents. If the bride desires a porcelain or blown-glass figurine on top of the cake, it may be accented with flowers. Floral decorations may also be added between the cake layers and around the base to unify and enhance the entire cake. Flowers may be loosely arranged between cake layers, on top of a paper doily, mirror, or plastic separator, to keep the flowers off the icing of the cake. The base of the cake may be decorated with a garland of flowers and foliage, a layering of loose flowers and foliage, or preconstructed flower-accent clusters. The cake knife and server may also be accented with flowers and ribbon. Cake tops are a popular choice and may be easily made in lace-trimmed holders (see figure 18-18).

Figure 18-16. The wedding cake is a focal point at the reception. Flower cake tops, garlands, and floral clusters help to accent and beautify the cake and table.
© The John Henry Company

Figure 18-17. Wedding cakes may be decorated with fresh flowers in a variety of symmetrical and asymmetrical designs. Often the cake knife and server are decorated with matching floral accents.

Figure 18-18. Cake tops are often constructed in specially made lace-trimmed cups. First, a small row of foliage is inserted to conceal most of the foam. Flowers, foliage, and fillers are then added to form the desired shape and size of floral arrangement that will accent, not overpower, the cake.

At a sit-down lunch or dinner, generally there is a separate table for the bridal party, called the head table. This table is a central element in the decoration. The placement of the table in the room and the decorations should make the bridal party table stand out with prominence. Often the floral decorations designed for this table are more elaborate than those for the guest tables, giving emphasis to the bride and groom. Flower arrangements used on the table should be kept low so guests do not have difficulty seeing the bridal party. Though the style of decorations on the guest tables will be different than those on the head table, they should be created with the same flowers and colors. The guest tables continue the theme of the reception throughout the reception room. The size and shape of the guest tables will often determine the type of centerpieces needed.

The punch and champagne table is an area of activity at the reception. Often the punch bowl is decorated with flowers around its base, similar to the cake. The toasting glasses for the bride and groom may be decorated with bows and floral decorations tied to the stems of the glasses.

The buffet table is also an area of activity. Floral arrangements and loose flowers and petals are commonly used to decorate and accent the buffet table and food trays. The consultant needs to know if guests will be serving themselves from one side or from both sides of the buffet. This will help the designer know whether to make centerpieces all-sided or one-sided. Generally, the buffet floral designs are elaborate, giving emphasis to the table.

Table skirt decorations are beautiful accents to the buffet and other tables. They provide accenting spots of color throughout the reception. Garlands of ribbon are commonly used to decorate the tables. Table skirt decorations are usually pinned to the skirting of the table and must be secure so they will remain in place as guests pass by the tables.

Because a ***receiving line*** is generally part of a formal or semiformal reception, decorations will help to emphasize the bridal party as they greet guests. Special floral arches will accent and frame the bride and groom. Receiving line decorations are generally large in scale. Personal flowers are also worn and held by the bridal party in the receiving line. These corsages, boutonnieres, and bouquets are usually the same flowers used at the ceremony; however, sometimes new flowers may be needed.

Additional decorations may be desired, depending upon the spaciousness of the room or area of the outdoor site. Most wedding order forms list additional areas that may be decorated with flowers, such as the registration table and lobby area (if there is one), the powder room, the dance floor and bandstand area, the gift tables, the pool, tent, and the get-away car or carriage. Other floral items needed at the reception may include a toss bouquet for the bride, personal flowers for the band members, emcee, vocalist, and service personnel—waiters, waitresses, hostess, and bartender.

Rehearsal Dinner

Many couples choose to have a rehearsal of the wedding ceremony. Following the rehearsal, a dinner is generally hosted and paid for by the groom's parents. Floral centerpieces may be needed for the rehearsal dinner. The consultant may also suggest small ***hand-tied bouquets*** for the bride and her attendants to carry during and to keep after the rehearsal.

Conclusion of the Consultation

When the consultant and bride have discussed all of the floral items needed, rental options, prices, and terms of payment are discussed. It is best for the florist to offer three different prices for each item and an explanation of the differences in floral designs for the various prices.

According to traditional wedding customs, the groom pays for the bride's bouquet, the men's boutonnieres, and the mothers' and grandmothers' corsages. The bride is traditionally responsible for paying for the bridesmaids' and flower girl bouquets, the groom's boutonniere, corsages for all females (except the mothers and grandmothers), and decorations for the ceremony and reception.

The professional florist seldom gives a prospective bridal client any of the specific floral details and prices in writing until the bride engages the shop to create the flowers for her wedding. This procedure discourages the practice of going to a professionally qualified bridal florist to plan the wedding, then taking these plans to a less qualified florist for fulfilling the plan. Many professional flower shops will not give a final estimate on a wedding until they know they have the job.

Because weddings are labor-intensive, the cost of servicing the wedding and reception—delivering wedding flowers, installing and removing large floral decorations, and decorating the cake, cake table, and other on-site designing and set up—must be added into the cost of the wedding.

Most professional flower shops charge a fee for the wedding consultation. When a bride completes her flower plans and makes an appropriate deposit (usually 20 percent) to reserve the wedding date, the florist generally applies this consultation fee to the total.

STYLES OF BOUQUETS

In wedding work, the term "style" connotes the form or the physical appearance of the bouquet. Form in floral design is the combination of the three-dimensional structural characteristics of the design that creates its individual, physical identity. For example, a colonial bouquet is a round, circular form. This form gives the bouquet its structural characteristics. All names of bouquets that describe the style—colonial, cascade, crescent, arm, clutch, and others—identify the form or the physical appearance of the bouquet.

The terms ***classical, traditional,*** and ***contemporary*** further explain the style of a bouquet. The classical interpretation of any style imposes a formal, structured approach to the design. In classical styles, certain flowers are more appropriate than others. Calla lilies, for example, are an excellent choice for a classical ***arm bouquet.*** Orchids, stephanotis, and trailing ivy strands are ideal for a classical, structured cascade bridal bouquet.

A traditional style is a conventional or "typical" bouquet. It is not as formal as the classical style. A traditional colonial bouquet for the bride could be created with white roses, white button pompons, white miniature carnations, and baby's breath. A traditional cascade bouquet for a bridesmaid might feature miniature carnations and roses.

In floral design, contemporary style identifies relaxed form instead of structured form. Contemporary styling allows freedom in both flower selection, colors, and presentation. Usually, the contemporary execution of any bouquet style incorporates generous areas of negative space giving the design a relaxed, airy appearance. A contemporary arm bouquet for a bride might include a combination

Figure 18-19. The bridal bouquet draws attention to the bride and provides accent to her gown, as shown by this casual gathering of spray roses.
Courtesy of Jackson & Perkins

of white delphinium, pink larkspur and roses, two or three gerbera, Queen Anne's lace, and a stem of white lilies. Contemporary style is more expressive and less structured than either the classical or traditional. A contemporary hand or ***clutch bouquet*** might be an informal gathering of a single stem or a wide assortment of seasonal flowers in a pleasing color harmony.

Bouquet Construction

Construction techniques for many bouquet styles are similar. Bouquets are generally designed in one of three ways: 1. flowers may be tied together with ribbon or string; 2. flowers may be individually wired and taped and then assembled together; or 3. flowers may be inserted into floral foam holders. Some bouquets require several construction methods in combination. Most floral designers use a bouquet holder with saturated floral foam for creating bridal and attendant bouquets (see figure 18-20). Flowers can be placed on natural stems directly into the wet foam. The moisture in the foam keeps flowers fresh. Sometimes, the designer wires and tapes some flowers and then places them in the foam. When using wired and taped stems, and flowers with fragile stems that do not hold fast in the floral foam, place these materials on wooden picks and then insert the pick into the foam. The moisture in the foam expands the wood pick and anchors it in the foam. It is essential to anchor all stems securely in the foam. Otherwise, flowers might fall out of the bouquets as the bridal party proceeds down the aisle. The steps of construction for assembling a cascade bouquet in a bouquet holder can be applied to many other styles of design (see figures 18-21 and 18-22).

Perennial Favorites

Although wedding bouquet styles have changed with fashion, many bouquet styles are classic and have remained popular. These perennial favorites include the colonial, cascade, crescent, clutch or hand-tied, and arm bouquet.

The ***colonial*** or round bouquet is perhaps the most widely selected bou-

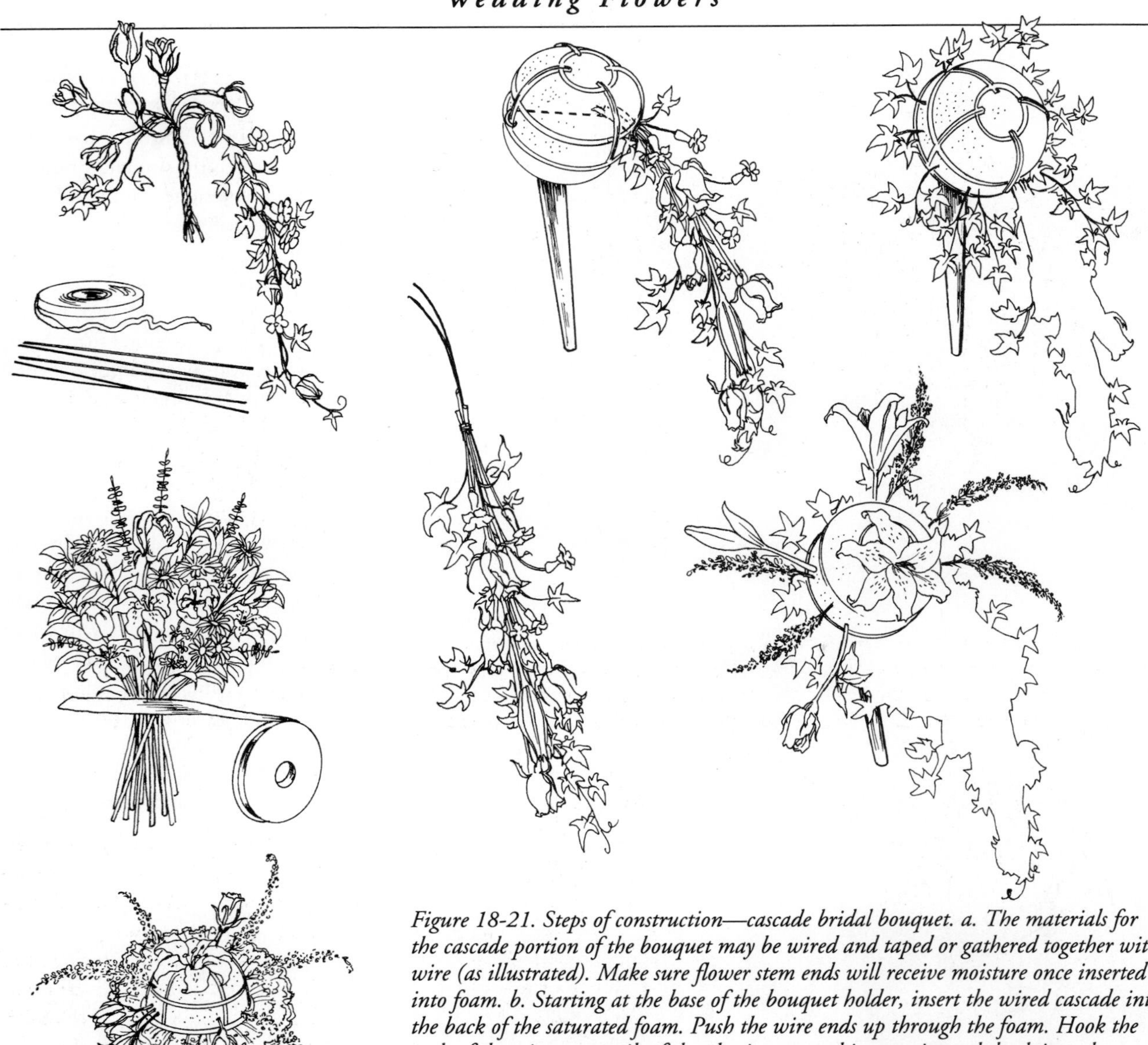

Figure 18-21. Steps of construction—cascade bridal bouquet. a. The materials for the cascade portion of the bouquet may be wired and taped or gathered together with wire (as illustrated). Make sure flower stem ends will receive moisture once inserted into foam. b. Starting at the base of the bouquet holder, insert the wired cascade into the back of the saturated foam. Push the wire ends up through the foam. Hook the ends of the wire over a rib of the plastic cage and insert wire ends back into the foam. c. Add greenery into the back of the foam, all the way around the holder in a circular pattern. Blend the background framework of the bouquet into the cascade. d. Add larger, focal flowers to create emphasis. Add other flowers to help create the perimeters of the bouquet.(Continues. See figure 18-22.)

Figure 18-20. Bouquets are generally constructed in one of three ways: wiring and taping, tying, or inserting flowers into foam bouquet holders. Bouquet holders keep flowers fresh and design time is dramatically reduced. Lace collars are available—often used for round and colonial bouquets.

quet for both brides and bridesmaids. This style is fashioned after the ***nosegay*** designs of the English-Georgian and Victorian periods (see figure 18-23). Many other bouquet styles are some form or variation of the colonial. Often, colonial bouquets are circled with a collar of lace.

Featuring dominant descending line movement, the cascade is a popular bouquet style for brides and bridesmaids (see figure 18-24). The "traditional" cascade is the most preferred form of this style. An orchid or other favorite flower might center the bouquet with cascading lines of stephanotis and other

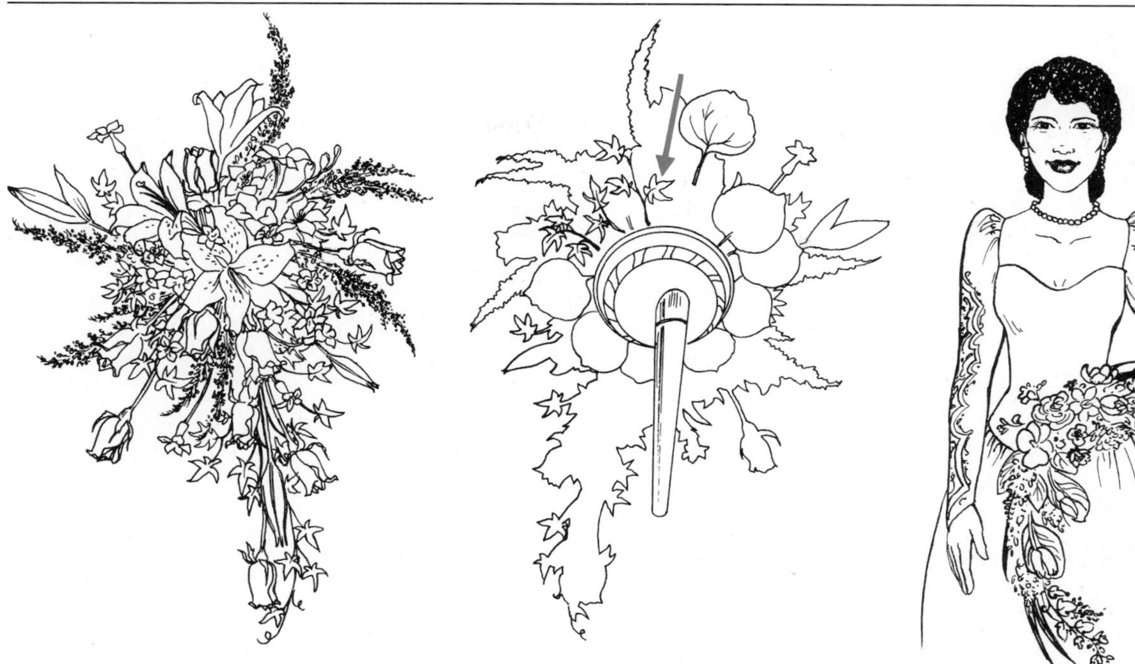

Figure 18-22. Steps of construction—cascade bridal bouquet (continued) from left to right : e. Complete the bouquet by adding flowers, fillers, and foliage. Blend the flowers and colors of the cascade in with the rest of the design to achieve unity. f. If a lace collar is not used, place foliage in the back of the bouquet to conceal mechanics.

Figure 18-25. Distinctive and glamorous, a crescent bouquet has an asymmetrical, curving cascade.

Figure 18-23. Colonial bouquets are round and generally simple, with a minimum of flowers.

Figure 18-24. Classic and formal, the cascade bouquet remains popular with brides today.

flowers flowing gracefully from the orchid center. The bride favoring white roses often selects the traditional cascade style for her bouquet.

The crescent style is a variation of the cascade and shaped in a crescent-moon form, as shown in figure 18-25. Although not as popular as the colonial, cascade, or clutch bouquets, the crescent form is an excellent choice when working with a limited number of flowers with distinctive shapes. Whenever a unique or expensive flower is the point of interest, the crescent style provides maximum visual value with an unobstrusive flair of design.

Considered the most natural bouquet for the bride and her attendants, the clutch or hand-tied bouquet is a casual gathering of blossoms tied into position with ribbon or string. Providing a natural, garden-picked

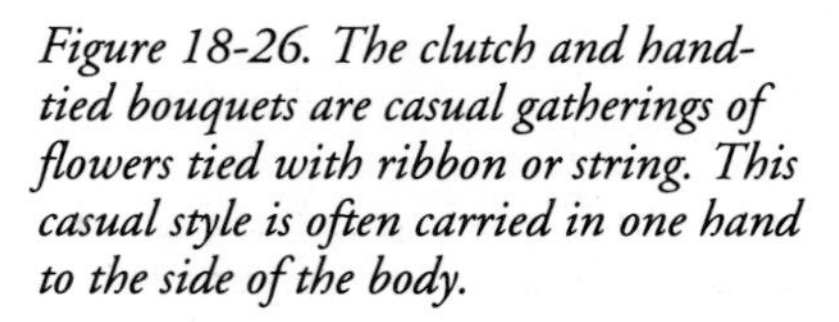

Figure 18-26. The clutch and hand-tied bouquets are casual gatherings of flowers tied with ribbon or string. This casual style is often carried in one hand to the side of the body.

Figure 18-27. To make a clutch or hand-tied bouquet, hold the gathering of flowers and tie with a ribbon at the binding point (where the stems cross each other).

Figure 18-28. A presentation, arm, or queen's bouquet is a vertical grouping of flowers, tied at the stems with a bow. It is generally held along the forearm and across the waist.

appearance, this style of bouquet has gained new popularity in recent years (see figure 18-27).

Another perennial style, the arm bouquet, is favored by brides who object to structured styling and prefer a more natural, but luxurious presentation of flowers (see figure 18-28). Also called a ***presentation bouquet,*** in its purest form, the arm bouquet is a gathering of flowers on natural stems that gives the appearance of being picked from the garden and placed in the arm. Perfectly appropriate for both the bride and her attendants, the arm bouquet can be created with almost any flower and presented in classical, traditional, or contemporary style.

Specialty Bouquets

Bridal and attendant's bouquets with unconventional themes appeal to some brides. These novelty or ***specialty bouquets*** are not of major importance in most flower shops. Still, a professional floral designer must be familiar with novel designs and be qualified to suggest and create them when appropriate. Some unique specialty designs include the ***basket bouquet, wreath bouquet, muff, prayer book*** or ***Bible bouquet, fan bouquet,*** and ***parasol bouquet.***

A basket of flowers is a preferred bouquet for brides who want an informal flower look for an afternoon wedding (see figure 18-29). The basket bouquet has a long-standing tradition for being the best selection for attendants to carry

Figure 18-29. Natural and casual, basket bouquets filled with garden flowers are charming for outdoor summer weddings.

in garden and outdoor weddings. Because it can be used after the ceremony on the reception table, the basket bouquet has gained popularity for indoor ceremonies.

Symbolic of the wedding ring and never ending love, the wreath bouquet is generally made on a wire ring in an appropriate size for the bride and her attendants. All flowers and foliage used in a wreath bouquet are generally wired and taped and then attached to the ring. Foliage strands are often intertwined, providing a fuller, more natural appearance.

Placing a cluster of flowers on a muff is a high-fashion approach to a bridal bouquet. Most common in the winter months, it is usually the choice for an informal second marriage, a small wedding, or any ceremony where the bride wears a suit (see figure 18-31). The cluster of flowers for the muff bouquet is created like a corsage and pinned to the muff.

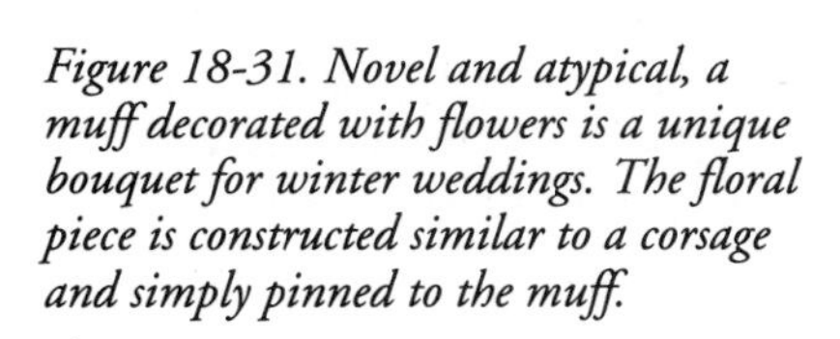

Figure 18-31. Novel and atypical, a muff decorated with flowers is a unique bouquet for winter weddings. The floral piece is constructed similar to a corsage and simply pinned to the muff.

Figure 18-32. Occasionally a bride elects to carry a Bible or prayer book during the wedding ceremony. Flowers may be made in small, cascading designs to beautifully accent the book. The flower piece is wired and taped together and then carefully tied to the Bible with ribbon.

Figure 18-30. Classic and versatile, the wreath bouquet symbolizes the wedding rings and never-ending love.

The prayer book and Bible bouquet require a miniature bouquet that can be completed and then easily anchored or tied to the book (see figure 18-32). For this style of design, flowers must be wired and taped and then assembled into a bouquet.

Fans may be decorated simply or elaborately with flowers and used as a bouquet (see figure 18-33). A variety of fans are available at floral supply and wedding stores. Flowers are generally wired and taped, assembled into a bouquet similar to an exaggerated corsage, and then attached to the fan. Foam holders are also available. These unique designs may have a variety of cascading floral treatments or smaller floral accents.

Brides who seek a novel bouquet for a garden or other outdoor wedding sometimes select a parasol bouquet (see figure 18-34). Floral supply and wedding stores stock several styles of umbrellas that can be embellished with flowers. Some umbrellas are designed to be carried closed while others are decorated with flowers and carried over the shoulder in a full-open position.

Figure 18-33. Fans may be accented with flowers, providing a novel, keepsake bouquet for the bride.

Figure 18-34. Appropriate for outdoor, garden weddings, decorated parasols are sometimes carried by the bride, bridesmaids, and flower girl. Floral designs are most often constructed similar to corsages and boutonnieres and attached to open or closed parasols.

SERVICING THE WEDDING

The purpose of servicing a wedding is to make certain that all flower details are presented effectively. Depending on where the ceremony is held, the responsibilities of servicing a wedding differ. Some churches have wedding guilds that service all weddings. In other settings, the florist plays an important role and attends the rehearsal and may actually direct the wedding.

In all decorating, whether it is a church, club, hotel, or home, the florist must respect and protect the property. Carelessness in spilling water, scratching furniture, and installing decorations that are not appropriate for the surroundings create serious problems and a negative image for florists. When furniture and furnishings need to be moved to install decorations, it is important to receive approval to make these alterations before moving ahead with the decorations.

In a typical situation, the florist who services a wedding sees that the ceremony decorations are in order, makes sure the candles are lit at the proper time, and pins on the corsages and boutonnieres. (Personal flowers should be name tagged, to ease distribution.) The florist may even see the bridal party down the aisle. This final touch of personal service cements customer satisfaction. The florist who services weddings is seen at the event and this exposure generally brings many referrals.

The florist who services the wedding ceremony also generally services the reception. Placing arrangements on various tables, decorating the cake and cake table, placing designs and decorations throughout the reception room, and distributing personal flowers are services carried out by the florist.

Another aspect of wedding service is the labor required to install and remove ceremony and reception decorations. Depending on the requirements, these decorations require the services of a driver and often designers. The labor involved must be added in the cost of the wedding at the time of the consultation.

REVIEW

The florist who wants to build wedding flowers business must develop a market niche through advertising and promotional efforts. Like all other fashion-oriented markets, the style of wedding flowers changes. The professional florist keeps up-to-date with all trends. To be successful in the segment of the wedding flowers market that can be profitable for a full-service flower shop, a florist must offer bridal customers the newest ideas. A professional floral designer is familiar with all the basic styles of wedding bouquets. Each style may be expressed in classical, traditional, and contemporary variations, depending on the requirements of the wedding. A qualified wedding consultant applies knowledge, creativity, skill, and productivity in creating flowers for the wedding party, the ceremony, the reception, and other events connected with the wedding. The desire and ability to serve a bride professionally and competently are the key characteristics that enable a florist to be a wedding flowers specialist.

TERMS TO INCREASE YOUR UNDERSTANDING

advertising
aisleabra
altar bouquet
arch
arm bouquet
Bible bouquet
basket bouquet
candelabra
canopy
cascade bouquet
chaplet
chuppah
classical style
clutch bouquet
colonial bouquet
contemporary style
crescent bouquet
fan bouquet
going-away corsage
groomsmen
hand-tied bouquet
head table
muff bouquet
networking
nosegay
parasol bouquet
pew decorations
prayer book
presentation bouquet
processional
promotion
publicity
receiving line
recessional
specialty bouquet
table skirt decorations
toss bouquet
traditional style
wedding consultant
wreath bouquet

TEST YOUR KNOWLEDGE

1. Define promotion, publicity, and advertising as they apply to developing wedding flowers business. Give three examples of each.
2. Name qualifications necessary for a competent bridal consultant.
3. How do the classical, traditional, and contemporary styles in wedding bouquets differ?
4. List the basic flower arrangements and bouquets that would be used for a typical church wedding and reception.
5. What are key considerations in servicing weddings?
6. Name other professional businesses who may participate in networking to achieve greater wedding business.

RELATED ACTIVITIES

1. Select a picture of a bridal gown and an attendant's gown from a current bridal magazine. Plan the flowers for the bride, attendants, the ceremony, and the reception. Choose a flower and color theme and carry this theme throughout the wedding.
2. Using bridal and floral industry magazines as your source, make a reference manual illustrating different examples of various bouquet styles.
3. Hold a mock wedding consultation, discussing an entire wedding.
4. Construct a colonial or cascade bouquet using a foam holder. Make a matching boutonniere.
5. Make a clutch or hand-tied bouquet.

Chapter 19

Sympathy Flowers

The custom of using flowers in the burial ritual dates from the beginning of civilization. Though the manner in which flowers were used in the funeral ritual changed as society developed, flowers remained an important element in the process of honoring and burying the dead. Almost all cultures value flowers during the ritual of the funeral.

IMPORTANCE OF SYMPATHY FLOWERS

Research studies funded by the American Floral Endowment and the Society of American Florists

Figure 19-1. Flower decorations have long played an important role during the time of grief and mourning.
Elaine *by Toby Edward Rosenthal, American 1848–1917, oil on canvas, 1874, 38 9/16 x 62 1/2 in., Gift of Mrs. Maurice Rosenfeld, 1917.3.*

Figure 19-2. Flowers have long been used to ease the grief and honor the deceased. Transported by train, large funeral sprays are carried by children to the burial site.
Scofield, Utah Mine Disaster, May 1900. Wadsworth Photo Archives. All rights reserved. Used with permission.

Information Committee provide important information about the value and role of flowers and plants in the funeral service and bereavement process. To develop this data, the research team interviewed funeral directors, grief therapists, consumers, and individuals who had recently lost a loved one. Even though the floral industry has experienced a decline in sympathy flower sales, the results of the survey confirm that flowers continue to be an important part of the funeral ritual.

The results showed that 93 percent of bereaved families have purchased flowers for funeral services for a loved one. Eighty-five percent of consumers have sent flowers as an expression of their sympathy. More than half of the funeral directors surveyed said that flowers are an important expression of sympathy to the family. Most of the consumers surveyed send sympathy flowers to comfort the living. Most of the bereaved said they felt that sending flowers shows they care. Eighty percent of grief therapists who were surveyed thought receiving flowers from family and friends is an aid in the grieving process. Over half of the funeral directors agreed that the bereaved identify flowers as a significant comforting aspect of the funeral. Every funeral director surveyed said that friends who come during visitation look at the flowers and talk about them. Flowers are a comforting topic of discussion with those who come to the funeral home. Families admire the flowers and carefully read the attached cards.

In summarizing this data, the research analysts concluded that flowers at a funeral home and funeral service play an important role in softening the sorrow, comforting the living, and honoring the life of the deceased.

Sympathy flowers are a traditional way to express love and understanding. The beautiful flowers at the funeral services live on in the memories of the bereaved. These comforting memories help friends and family deal with grief. Flowers add warmth and comfort to an otherwise cold and painful atmosphere. To many, flowers have spiritual significance and symbolize eternal life, peace, and a god. Floral tributes symbolize love and caring and remind the bereaved that their sadness and sorrow are shared by others. Because love and sympathy are difficult emotions to express to the bereaved, flowers are a form of communication that transcends the heart.

Many families find comfort in sharing the flowers after the funeral services with hospitals, nursing homes, and churches. This kind gesture and service to others of distributing the flowers helps many with their grieving. Many authorities in both religion and psychiatry recognize that grief is a real process that must be worked through personally. Flowers play a significant role in helping to comfort the bereaved.

Figure 19-3. Extravagant massing of flowers in the late 1800s and early 1900s was a common custom at funerals to honor the life of a deceased loved one (circa 1890).
Photo courtesy of the Daughters of the Utah Pioneers Museum, Salt Lake City.

TRENDS AND REGIONAL DIFFERENCES

Significant trends in the style of sympathy flowers have transformed. Designs tend to be smaller than in the past, with emphasis on style. Ornate sympathy designs, while still preferred, are not nearly as popular as they were twenty-five years ago. There is an increasing trend for unusual and distinctive design in funeral sprays. Creative and artistic sympathy designs are on the rise due to the increasing availability of unusual, imported flower varieties and newer, more efficient mechanics.

Regional customs influence the type of designs prepared for a funeral. It is important to realize that not all people and regions of population like the same type of funeral ***sprays.*** Strong floral traditions and certain floral designs are popular in some regions, while unheard of in others. What might be appropriate and customary in one region may be inappropriate for another. For example, in the southern states, favored designs are ***easel*** designs, ***set pieces,*** and ***flat sprays,*** while florists in other parts of the country may rarely receive orders for flat sprays. Favored styles and construction techniques will vary not only from region to region, but from shop to shop.

Funeral Flower Changes

From the floral industry perspective, most information about the use of flowers at funerals begins in the early 1900s. By today's standards, sympathy flowers in the early days of the floral industry were elaborate and massive. Florists hand-tied ***casket sprays*** or made them in bases of moistened *sphagnum moss* using quantities of flowers, trailing garlands of *plumosus* and *sprengeri,* and intricate ribbon treatments.

Emotionally expressive floral designs and set pieces, such as broken and bleeding hearts, broken wheels with a missing spoke, empty chairs (see figure 19-4), gates ajar (representing the gates of heaven), and clocks to show the time of death, were once popular and sent by family and friends to funeral services. Traditionally, the viewing of the deceased took place at the family home, not the funeral home. Hand-tied sprays expressing the sympathy and concern of friends lined the walls of rooms in the home.

Prior to World War II, sending flowers at the time of death was instinctive. It was not unusual for a flower shop to have over one hundred orders for sympathy flowers for individual funerals. However, after World War II, several factors led to significant changes in sympathy flowers and the funeral flower business.

With the arrival of formal design education, the nature of floral design changed. Before training in structured floral design, florists created bounteous fan-shaped bou-

Figure 19-4. Historically, emotionally expressive floral sprays were popular, as shown by this "empty chair" embellished with abundant masses of flowers (circa 1890).

Photo courtesy of the Daughters of the Utah Pioneers Museum, Salt Lake City.

quets. Whether a designer hand tied a spray or created a design in a container stuffed with branches of foliage, the finished product was typically fan-shaped and often over-laden with flowers and ribbon. As floral design became more sophisticated, designs were more mechanical, less spontaneous, and not as showy.

The introduction of wettable floral foam revolutionized floral design. Basket and vase arrangements began to replace hand-tied sprays. This change of design style required funeral directors in many areas of the country to display arrangements, rather than massing walls with sprays.

"In Lieu of Flowers"

To capitalize on what some considered to be the "extravagance" of the traditional funeral ritual, some organizations began using the phrases "in lieu of flowers" and "please omit flowers." The purpose of these promotional efforts was to influence people to make a memorial contribution to an organization instead of sending flowers. When a printed death notice includes the in-lieu-of-flowers phrase, most people who would send flowers do not. Though people acknowledge this request to omit flowers, many feel deprived of communicating sympathy with their choice of expression.

Organizations within the floral industry, particularly the ***Florist Information Committee (FIC)*** activated in the early 1950s, have addressed the problem of please-omit flowers. They have worked with the charitable organizations that encourage memorial contributions in lieu of flowers, and with newspapers that request omitting flowers. Although the practice of printing these phrases is decreasing, florists must confront the situation and deal with it in a positive manner. Alternative phrases, such as "All expressions of sympathy, including flowers and charitable contributions, will be gratefully accepted," will honor the family's request for a charity and will allow friends to select the manner in which they wish to express their sympathy.

The emerging preferences for no visitation, creamation, private burial, and a memorial service tend to elimi-

Figure 19-5. The importance of having flowers present in times of bereavement is shown in this early photo of family and friends honoring their deceased loved one with large floral sprays.
Scofield, Utah Mine Disaster, May 1900. Wadsworth Photo Archives. All rights reserved. Used with permission.

nate sending sympathy flowers. However, sending flowers to the homes of the family of the deceased is a positive, growing area of sympathy flower sales.

Terminology

Many terms are used to describe various sympathy designs displayed at funeral services. Identical floral pieces may be called a spray in one region, an easel in another, and a set piece in another region. Overlapping terminology can be confusing for florists. It is important to become familiar with the various names of funeral pieces and other funeral terms to lessen confusion between the consultant and clients and among florists.

SELLING SYMPATHY FLOWERS

Sympathy designs are an important part of year-round business for many floral shops. Competition among floral shops for sympathy flower business is strong. To increase sales, many florists have begun subtle promotional efforts. With the help of national organizations, business-minded

Figure 19-6. Birds of paradise are the primary flowers in these family pieces. The casket spray, lid inset, floor arrangement, and pedestal arrangement all coordinate with each other in colors, flowers, and style. Floral arrangements harmonize with the casket and often are personalized to honor the deceased.
© The John Henry Company.

florists have discovered methods of tastefully advertising sympathy tributes.

Promotion

While it is never appropriate to advertise for funeral business, it is acceptable to promote sympathy flowers in general. Small and distinctive advertisements in newspapers that express honest statements are standard methods of promoting sympathy business. The Society of American Florists has composed a variety of simple sympathy advertisements, that distinctively promote expressions of sympathy through flowers (see figure 19-7).

Figure 19-7. Small, distinctive newspaper advertisements expressing honest statements, are a tactful way of promoting sympathy flowers.
The Society of American Florists.

Consultation

While most orders for sympathy tributes from friends and family are received over the telephone, the main floral pieces, such as the casket spray, a standing easel, and other "family pieces" are generally ordered by the immediate family of the deceased during a consultation at the floral shop.

Soon after the death of a loved one, the closest family members meet with the floral consultant to discuss flowers for the funeral services. The ideal consultation area should provide a comfortable, private setting for the bereaved family. The consultant must remember this is an extremely difficult time for grieving families. The consultant should remain sympathetic to the situation and needs of the family. Often family members do not have specific designs in mind. Books with colorful photographs will help the selection process. The florist must display professional, tactful selling skills to help the family make floral decisions. The professional floral consultant is knowledgeable in funeral traditions and religious customs, flower varieties and availability, funeral design styles, and pricing.

Because funeral services are generally held within three or four days of the death, the florist must make certain the flowers the family desires are on hand or will be readily available. This may present a problem for flower shops that are not located near a wholesaler or are not able to get certain flowers within a day or two. The consultant must be flexible in suggestions and attempt to meet all the wishes of the family.

Order Form

A special order form for funeral orders is a valuable tool in identifying the family needs during the consultation (see figure 19-9). Order forms that are concise will guide the consultant in asking appropriate and respectful questions. These forms help ease the selection and decision-making process for the family. Finding out that Great Grandma loved purple cattleya orchids and white roses, and that the family has already selected a white

Figure 19-8. Soon after the death of a loved one, the closest family members will meet with the florist to plan the major floral arrangements to be displayed at the funeral. The consultant should be sympathetic with the family. The professional consultant is knowledgeable about the types of designs, funeral customs, flower types and availability, and pricing. The ideal consultation area should be a comfortable, private setting where customers may make floral design selections.

casket can help determine the color, types of flowers, and style of arrangement needed. Certain questions will help personalize the floral sprays and the family will generally appreciate respectful questions that make the flowers more meaningful.

OVERVIEW OF SYMPATHY FLOWER DESIGNS

The purpose of the consultation is to assist family members in their selections concerning the styles, types, and number of sprays and designs to be displayed at the funeral service. The consultant must be aware of the choices available. Knowing who traditionally orders and pays for certain floral pieces will help avoid overlapping. As shown in figure 19-11, a great variety of sympathy tributes are available.

The casket cover is generally ordered and paid for by immediate family members. Large matching ***easel sprays*** and floor designs are ordered by immediate and extended family members (siblings, children, aunts, uncles, and

SYMPATHY ORDER FORM *Bloomworks*

Ordered by
Address
Phone
Name of Deceased
Relationship to Sender ____ Sex ____ Age ____ Religion ____
Favorite Colors ____ Favorite Flowers ____
Hobbies or Special Interests
Type of Service: ❑ Church ❑ Synagogue ❑ Mortuary ❑ Grave Site ❑ Memorial ❑ Crematory
Style of Casket or Urn: ❑ Open Casket ❑ Half Couch ❑ Closed Casket ❑ Urn
Color or Deceased Attire
Name of Mortuary
Address ____ Phone ____
Viewing Time

Funeral Service Location
Address ____ Phone ____
Time

Flowers for the Service:	EXPENSES
Casket (Color & Style):	
Set Pieces (Heart, Pillow, Cross — Color & Style)	
Arrangements (Color & Style)	
Sprays (Color & Style)	
Wreaths (Color & Style)	
Lid Pieces or Body Flowers (Color & Style)	
Corsages (Color & Style)	
Boutonnieres (Color & Style)	
Flowers for the Home	
Subtotal	
Sales Tax	
*Use reverse side of form for additional instructions. Total	

Figure 19-9. An order form designed specifically for funerals will ease the selection process for the florist and the bereaved.

Figure 19-10. Generally, the pall bearers are recognized with a boutonniere. The number of boutonnieres required should be listed on the order form. Often florists provide these boutonnieres complimentary to the family ordering a casket spray.

Figure 19-11a. Traditional sympathy tributes: a. casket spray; b. floor arrangement; c. double-end easel spray; d. pedestal arrangement; e. wall spray; f. set piece emblem design; g. satin-wrapped wreath easel design with floral accent; h. solid heart easel design with floral accent; i. vase design; j. planter dish garden.

Figure 19-11b. Casket spray.
Photo by Brian Yacur.

Figure 19-11c. Double-end easel spray.
Photo by Brian Yacur.

Figure 19-12. A variety of casket floral designs: a. full-couch casket spray (designed to be placed on a casket that is fully closed); b. half-couch casket spray (designed to be placed on the foot of the casket lid, while the head of the casket lid is open); c. casket garland (a classic design that may adorn the opened or closed casket); d. casket scarf (a distinctive tribute, designed on a fabric pall accented with an elongated cluster of flowers). A casket scarf is designed to drape over the casket, either side to side or head to foot.

other close relatives). ***Casket inset*** designs are often from grandchildren. Easel set pieces are given by family members. Specialty set pieces are often given by groups of friends and business associates. Smaller sprays, plants, and plant gardens are given by friends, family, and associates. Generally, easel and other large designs are placed at the head and foot of the casket, and signify close relationship to the deceased.

FLOWERS FOR THE CASKET

The immediate family members of a deceased loved one usually select the floral design to adorn the top of the casket. This design, called the casket spray is generally the most beautiful and elaborate of all the floral designs displayed at the funeral service. Casket sprays may be designed in a variety of shapes and sizes. Designers take a great deal of pride in planning and arranging these elaborate, showy pieces. Smaller coordinating designs, called casket insets or lid insets, are sometimes placed inside the casket.

Casket Spray

A variety of casket spray designs are available and are designed according to the casket the family has selected. Caskets are available in two basic forms: ***full couch*** and ***half couch***. Both can be displayed opened or closed. In contrast to a full couch, a half couch is a casket with a two-piece lid. The head of the half couch is opened during visitation services.

Casket ***saddles*** provide the foundation for the casket spray. Available in various sizes and styles, they hold one or more blocks of saturated floral foam (see figure 9-13).

A full casket spray is a large, elongated design that extends the full length of the casket (see figure 19-14). When the casket is fully open for a viewing, the funeral director displays the spray above the opened lid or in front of the casket. Funeral homes have specialized stands for holding these casket sprays. For a full-open casket, the designer must give careful attention to creating a spray that can be displayed appropriately without creating problems for the funeral director.

Figure 19-13. Molded plastic casket saddles are available in different sizes and styles.

When the casket is closed, the full spray rests on the casket lid. Since the full-couch spray is placed only on the closed lid, the designer has more creative freedom in designing the spray. Some full-couch sprays cover the entire top of the casket with flowers and foliage.

The half-casket spray is used in more areas of the country than any other style of casket spray. Designed for half couches, when the head of the casket is open and the foot of the casket closed, this style of spray can be a simple cluster of flowers that rests on the closed lid, or a fuller design with flowers and greens flowing over the edge and foot of the casket (see figures 19-17). Generally, the half spray is designed to be later shifted to the center of the casket lid when it is closed for the service. A professional designer must know how a casket spray will be displayed for both the visitation and the funeral service. This information influences the styling of the spray.

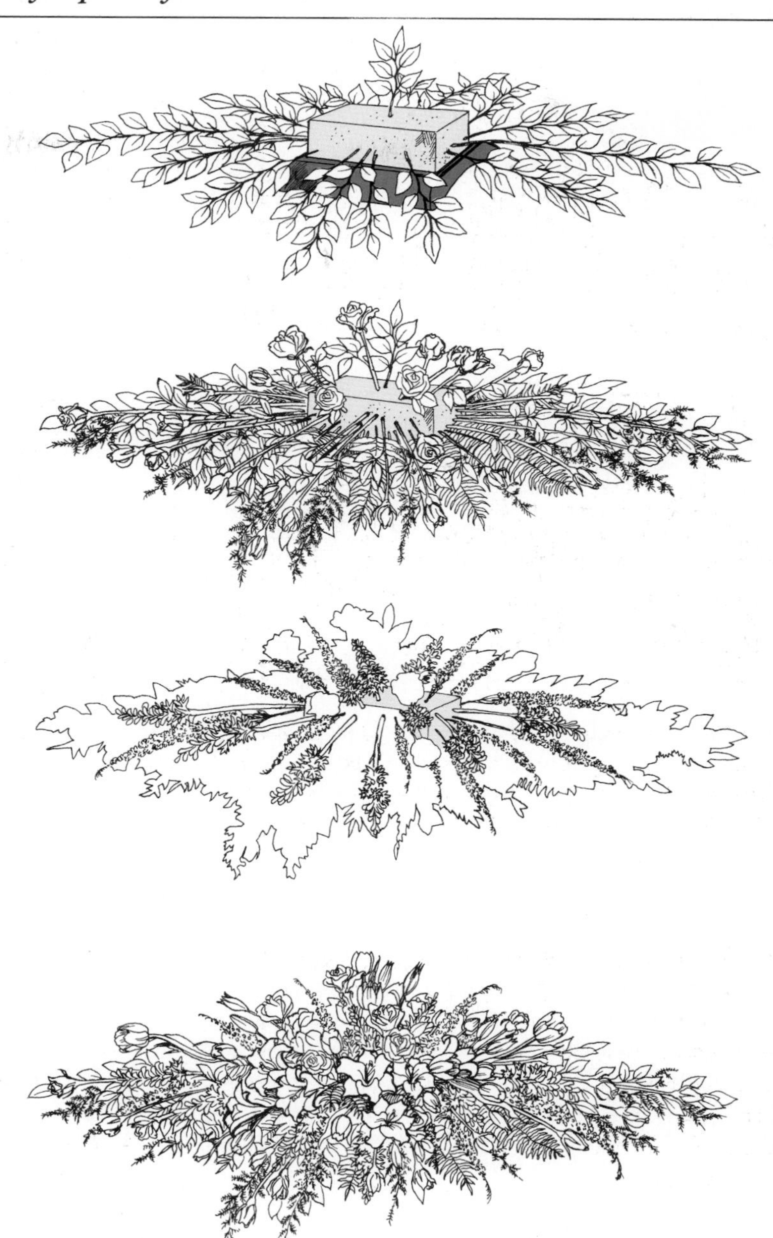

Figure 19-14. Steps of construction of a symmetrical casket spray (top to bottom): Step 1: Prepare casket saddle. Insert foliage into the saturated floral foam to create a base or bed of greens. This foliage base establishes the framework—the basic shape, length, and width of the design. Step 2: Place larger and more open flowers to set the center, focal area of the spray. Then add the feature flowers and fillers. Keep all materials within the basic form of the spray established with the foliage background. Step 3: Position all flowers in the spray to radiate out from the central focal area. Adhering to this radiating pattern is an essential point in construction to achieve visual balance and to create pleasing, flowing lines. Step 4: Complete the spray by adding flowers, gently lifting some blossoms to a high level as they merge into the focal area. This creates a third dimension giving the spray a full, elegant appearance. For visual weight, concentrate larger flower forms in the focal area. Ribbon loops and streamers may be added in the focal area.

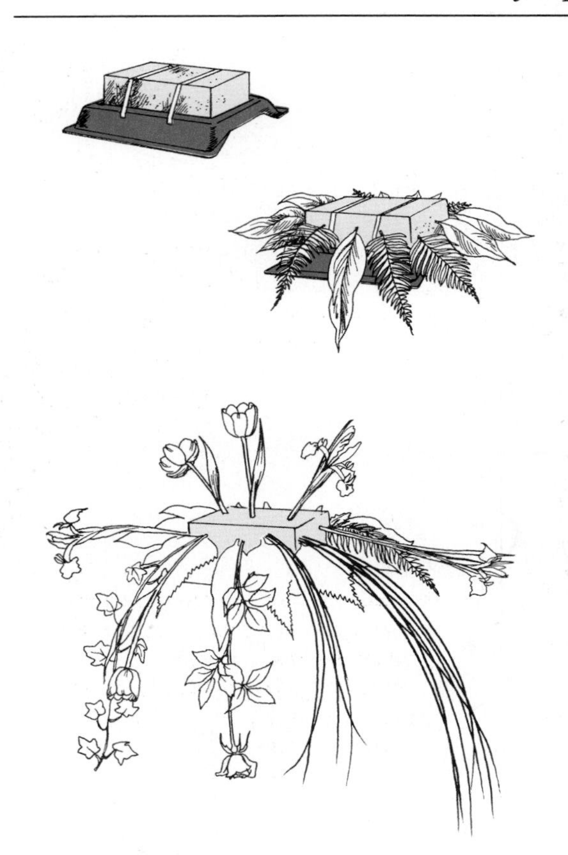

Figure 19-15. Steps of construction—contemporary casket spray (from top to bottom): Many designers create half-casket sprays in an asymmetrical format. Materials move from a high back left area to a cascading lower right area. Flowers, foliage, and ribbons flow over the front and foot of the casket. Step1: Start construction of this spray with an appropriate casket saddle with saturated floral foam. Step 2: Add foliage to cover the base of the saddle. (Do not establish the spray framework with foliage.) Step 3: Add flowers and foliage to establish the height and outer cascade dimensions of the casket spray. This basic framework controls the form and size of the spray.

Figure 19-16. Steps of construction—contemporary casket spray continued (from top to bottom): Step 3: Continue adding flowers and foliage concentrating the larger blossoms and important form flowers in the central focal area. Lift some flowers slightly. This technique of placing some blossoms higher than others creates depth, a third dimension that enhances visual interest. (When all flowers are placed on the same plane and height, the resulting spray is mechanical and visually boring.) Step 4: Finish the spray by placing a concentration of blossoms in the central focal area. In an asymmetrical spray composition, ribbon can be used in the focal area, or incorporated as off-center as illustrated. Script may be added to ribbon streamers.

Ribbons and ***script*** messages are often added to casket sprays and other funeral designs. Ribbon loops and streamers help to decorate and accent designs. When incorporated into designs, ribbon should be used as a design element (to increase the beauty of the style) not as a last thought or to fill in a big, empty space. Although there are many types, generally script is made of gold letters with a sticky adhesive back. It is often added to the ribbon streamers on casket sprays or other large designs. The words of the script indicate the relationship of the deceased to the person giving the floral tribute. For example, "Loving Sister," "Dearest Mother," and "Beloved Father," would denote how the deceased loved one was related to the giver of the tribute.

Special considerations are necessary when there are religious, military, or cultural customs and regulations regarding the use of flowers on a casket. It is important for the consultant to be aware of common restrictions regarding the use of flowers. For example, at military funerals flowers are not placed on the casket because the flag adorns

Figure 19-17. This contemporary casket cover of white anthuriums, lavender orchids, pink roses, and lavender and white dendrobium orchids cascades downward. It is designed to drape over the foot of a half-open casket.

Figure 19-18. With the flag serving as a military tribute, the casket spray is displayed on a high pedestal behind the casket. Red and white flowers, symbolizing bravery and service, are often used with accents of blue to harmonize with the flag.

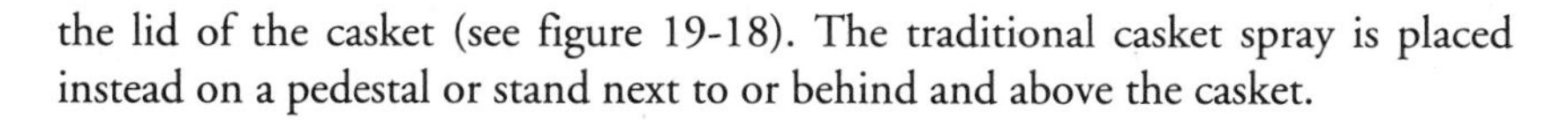

the lid of the casket (see figure 19-18). The traditional casket spray is placed instead on a pedestal or stand next to or behind and above the casket.

Floral Garland

Displayed over the hinge area of an open casket, the ***floral garland*** can be designed to fit the lid area of a half-open casket or extended for a full-open casket. The usual construction for a floral garland consists of tying foliage and blossoms together in a loose, airy, rope-like design. A garland must have a very delicate appearance and be pliable so it can be draped with soft lines. Another method for designing a garland is to first create a garland with foliage, then add flowers with glue or wire.

Casket Scarf

Designed to drape over a very limited area of the casket, the scarf can be used over the front-to-back area of a half-open casket. Or, a scarf can drape from head-to-foot on a closed casket. Typically, a designer makes a ***casket scarf*** with an exquisite fabric like a brocade and adorns it with a garland of flowers, or a cluster of exquisite blossoms like orchids or gardenias.

Casket Blanket

Constructed on a piece of fabric like green burlap, the ***casket blanket*** is an elegant ***pall*** draped over a closed coffin. This blanket may cover the entire top of the casket. Some casket blankets fall to the floor. The usual process for creating a casket blanket is to sew or glue greens to the fabric, and then glue blossoms into the greens.

Casket Inset Designs

Often called insets, lid insets, or lid pieces, small floral pieces are often designed to be displayed inside the casket lid. Traditionally, children or grandchildren send designs for inside the casket. Careful attention must be given to construction techniques for all designs that will be placed inside the casket. Loose and uncovered

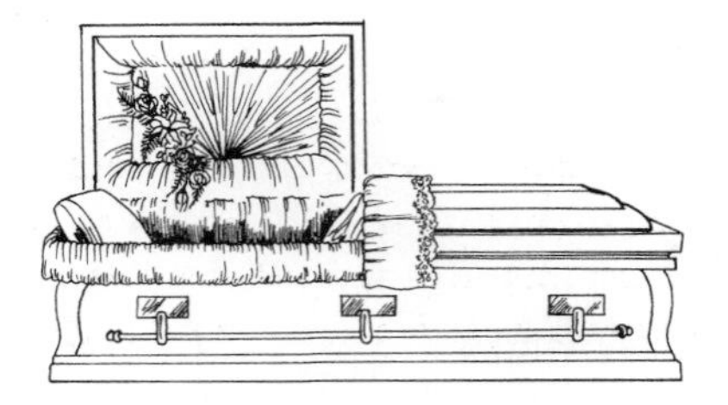

Figure 19-19. Small floral pieces, called inset designs or lid pieces, are designed for placement inside the lid of the open casket. These usually coordinate in color and style with the casket spray. From top to bottom: floral cluster, small heart, small satin pillow with floral accent, small cross, and nosegay.

wires, broken petals, and moisture are unsightly and can damage the lining of the casket.

Pinned in the casket lid or placed in the corner of the casket, the floral cluster inset piece is a very natural and elegant grouping of small flowers, often completed with a simple ribbon tie. Providing accent, these designs generally coordinate in color and style to the casket spray.

Designed to be pinned in the lid of the casket, the floral garland inset is an elongated line of flowers and greens. The construction of the garland inset is similar to a floral garland for the top of a casket, yet miniature in size.

Compared to a floral cluster, which is a natural grouping of flowers, the nosegay bouquet is a designed piece that is placed inside the casket. Usually, the designer creates a nosegay in a bouquet holder with a lace collar, or adds a ruffle of tulle or lace.

Made on a special rosary frame available from most floral supply houses, the ***floral rosary*** is a floral replica of the Catholic symbol of faith. The designer inserts fresh flowers, usually rose buds, into the rosary frame in place of the traditional beads. Customarily, the rosary is draped inside the casket lid.

A small cross inset to be placed in the inside of a casket can be designed by covering a small one-inch-thick hard foam cross with flowers, and edging it with ribbon or lace, or pinning a corsage-like cluster onto a satin cross available from most floral supply houses.

Like the cross, a heart inset for inside the casket can be constructed on a small hard foam form covered with flowers, edged with ribbon or lace and accented with a cluster of flowers. Or, a satin heart-shaped pillow can be used with a small corsage of flowers attached.

The small pillow inset is a satin pillow similar to a ring-bearer's pillow with a cluster of flowers and ribbon streamers attached.

EASEL DESIGNS AND SET PIECES

Floral designs placed on easels are used as floral tributes from relatives and close friends of the deceased person. These standing sprays are placed near the head or foot of the casket during the viewing and funeral services. These elaborate pieces often coordinate in color and style to match the casket spray. Designed to be displayed on a wood or metal tripod, easel designs include sprays, crosses, wreaths, hearts, and set pieces.

Easel Sprays

The term spray describes a variety of one-sided designs created without the usual container. Most designers construct easel spray designs in floral foam. Saturated foam blocks may be placed in a variety of plastic ***cage holders*** and ***spray bars,*** providing convenient bases for sprays. Floral supply houses offer many prepared foam bars and cages for sprays (see figure 19-20). The shape of the spray and style of the design vary greatly. The traditional, radiating, ***double-end spray*** is a popular style (see figure 19-21). Symmetrical shapes, including oval, elongated, and vertical, are traditional shapes. For the double-end sprays, many customers prefer traditional flowers, such as gladiolus, roses, and carnations. In many regions, trends are turning to more contemporary designs with asymmetrical shapes. The use of exotic flowers and foliage harmonize with the style of these designs (see figure 19-22).

Set Pieces

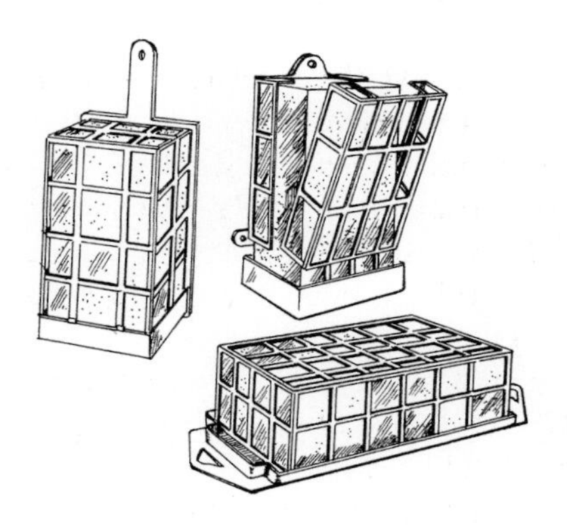

Figure 19-20. A variety of plastic cages and spray bars are available to hold saturated floral foam for easel designs.

Floral tributes designed in special shapes, such as crosses, wreaths, hearts, and pillows are called set pieces. Other floral tribute set pieces may be patterned after emblems of an organization to which the deceased belonged. Easel and set piece designs can be constructed on hard foam forms, wettable foam forms, and vine and twig forms.

The Christian cross, symbolic of the death and resurrection of Jesus Christ, is one of the more traditional funeral tributes for many Christian families. Often created on a large, hard, foam cross form, the easel cross can be a ribbon-wraped cross with a central cluster of flowers, constructed with solid flowers and accented with a center cluster, or a foliage cross embellished with blossoms. Crosses may be made from twigs, wood slabs with bark, mosses, and other natural materials. Wettable foam cross bases in various sizes are also available, making construction for solid flower crosses more convenient and time efficient.

Like the cross, wreaths may be created on hard foam forms or wettable foam wreath bases. When used as a sympathy tribute, the classic wreath form is symbolic of eternal life and never-ending love. The conventional creative design approaches are to cover the wreath form with solid flowers, to cover the wreath

Figure 19-21. Steps of construction—double-end symmetrical easel spray: Attaching a cage holder with saturated floral foam to an easel is a convenient base for an easel spray. For larger, longer sprays, spray bars with solid bases are available.
Step 1: After attaching the caged floral foam to an easel, insert line flowers or foliage to establish the framework. Place the stems in the back of the foam block. Heavy flowers that hang downward may need the extra security of wire or tape, to prevent them from falling out. Step 2: Next, add larger, showy flowers in the center of the design to create the focal point. Placing them in a diagonal pattern will lessen the symmetry of the double-end spray. Step 3: Add other flowers to supplement the mass and fill out the pattern set by the line flowers. Add some mass and line foliage to conceal the mechanics and foundation of the easel. Step 4: Complete the easel design by adding filler flowers and foliage. Ribbon loops and streamers may be added. Remember to cover any floral foam with foliage. This includes the back of the design as well.

Figure 19-22. Steps of construction—asymmetrical, contemporary easel design (clockwise from top left): Step 1: After attaching the wet caged floral foam to an easel, insert the key flowers and foliage, establishing the size dimension and primary lines of the spray. Do not fill in negative spaces in this contemporary design style. Step 2: Add background leaves to conceal the easel mechanics. These leaves establish the form of the composition. Step 3: Complete the design by establishing the focal point with a large, showy flower. Add flowers in groups, for greater visual impact of texture, color, and form. Unusual ribbon treatment will continue the contemporary style.

Figure 19-23. The pillow of comfort or pillow design is a popular easel set piece in many regions of the country. Symbolic of home, family, love, and comfort, this design is generally ordered by immediate family members. This chrysanthemum-covered pillow is accented with a floral spray of gerberas, snapdragons, carnations, asparagus ferns, and ribbon loops and streamers.
© The John Henry Company.

form with ribbon and add a dominant cluster of flowers, or to add floral accents to a vine or twig wreath (see figure 19-26).

Easel tributes may also be designed in the shape of a heart. A variety of heart set pieces are possible, including solid, open, and broken hearts (see figure 19-27). A large heart form covered with flowers or greens, with an elegant cluster of feature flowers, represents the traditional easel heart design. Both solid and open heart frames are available in hard foam and wettable foam forms. As shown in figure 19-28, construction techniques will vary according to the foam type and style of design desired.

More stylized heart designs can be created by covering a hard foam form with leaves, such as galax or salal, and

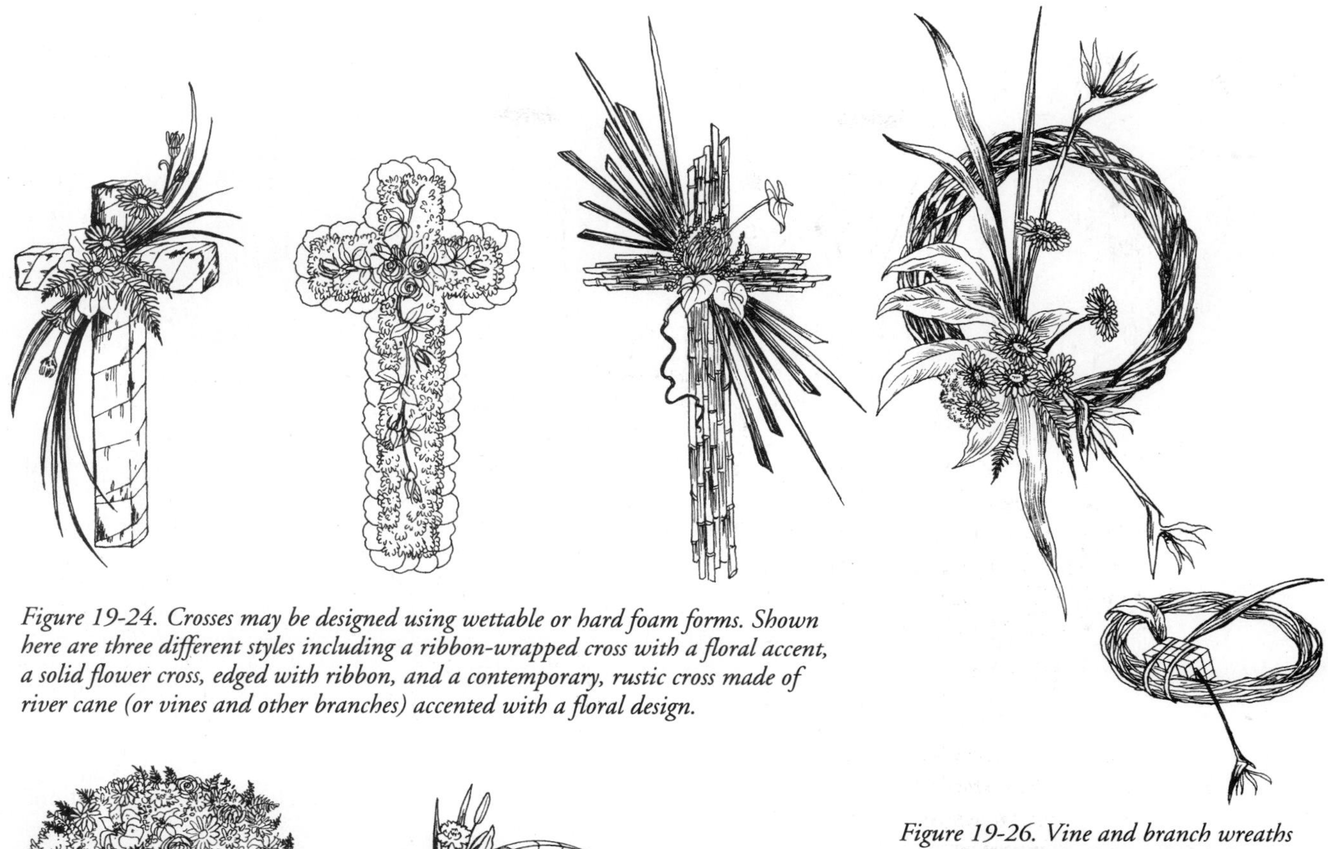

Figure 19-24. Crosses may be designed using wettable or hard foam forms. Shown here are three different styles including a ribbon-wrapped cross with a floral accent, a solid flower cross, edged with ribbon, and a contemporary, rustic cross made of river cane (or vines and other branches) accented with a floral design.

Figure 19-26. Vine and branch wreaths provide a rustic look for a floral cluster. The cage holder must first be secured with tape, wire, and/or glue in the area of emphasis.

Figure 19-25. Wreaths may be designed using wettable or hard foam forms. The top left wreath of solid flowers is made in a wettable foam form. Foliage is first added to the wreath edges to conceal the foundation, then flowers may be added in a mille fleur design. Or, one type of flower, such as carnations or chrysanthemums may be used to cover the wreath foundation. The top center wreath is a satin-wrapped hard foam form. The cage holder of saturated floral foam is secured at the focal point area on the wreath. A floral cluster accents the wreath.

adding an accent cluster of distinctive flowers. Open hearts can be covered with foliage or ribbon before creating a featured flower cluster. With less flowers, these styles of heart designs help to reduce cost and make the designs more affordable.

Pillows are a long-time favorite floral tribute in many regions. To many, they represent family, home, and comfort. To some they signify peaceful rest. These designs may be constructed using hard or wettable foam forms. They are often trimmed in pleated ribbon and lace.

Organizational tributes are special easel set pieces ordered by members of a club, business, religion, or school to which the deceased

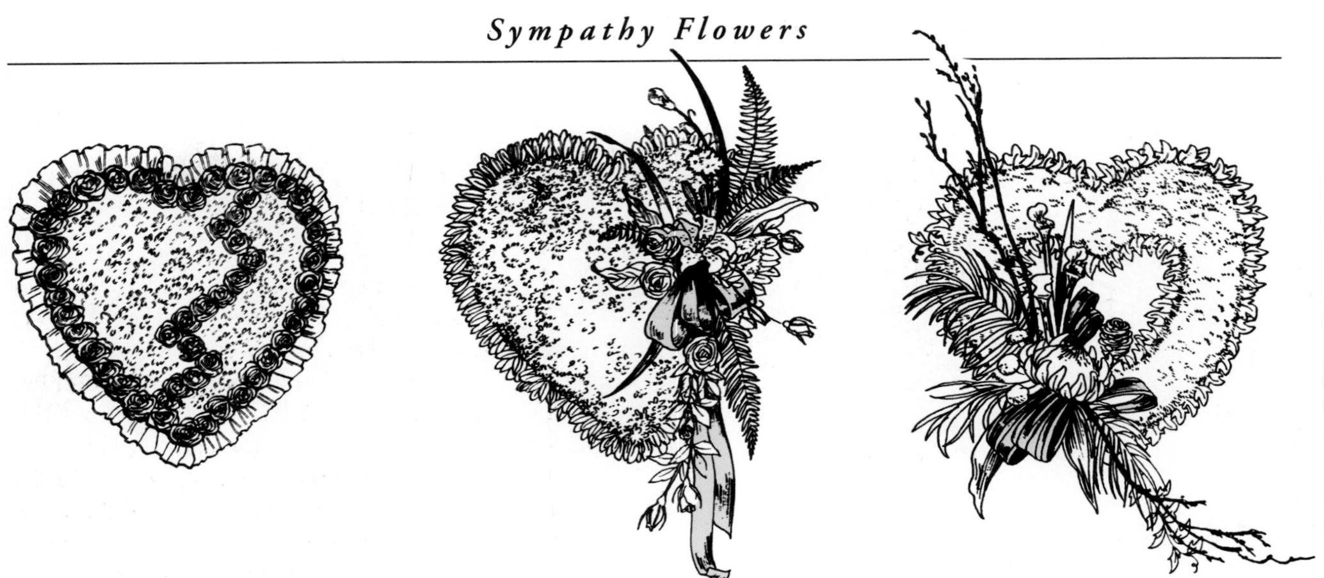

Figure 19-27. A variety of hearts may be designed in wettable or hard foam forms and placed on easel frames. (Left) The broken heart is made on a solid heart form. It is a solid work design of white carnations or chrysanthemums without a floral cluster. Instead, a zigzag line of red roses or carnations is placed in the heart to create a symbolic broken or cracked heart, in the event of a deceased loved one. (Center) A solid heart with a floral accent. When red roses are used in the cascading design, on top of a white heart, the easel spray is often called a bleeding heart. (Right) An open heart design has the inside of the form removed. This open heart has solid floral base with foliage edges, accented with a contemporary flower spray.

Figure 19-28. Construction techniques vary when designing solid flower set pieces. (Left) Traditional construction techniques include the use of hard foam forms. First pleated ribbon or leaves may be added to the form's edges with pins. Then the stems are removed from carnations or chrysanthemums and pinned or picked into the hardfoam to form a flat floral surface. (Center) When using a cluster of fresh flowers that need to be in moisture, anchor a caged piece of saturated floral foam to the heart. With a glue gun, attach two wooden picks on the back of the heart in the area where the floral foam will be wired onto the heart. These wooden picks keep the wires from cutting through the hard foam. (Right) Wettable foam forms expedite design time. Short-stemmed flowers and leaves are inserted directly into the foam.

belonged. These set pieces are floral tributes patterned after the organization's insignia or emblem (see figure 19-29). When working with a recognized emblem symbolic of a specific organization, it is vital to follow the design and colors of the symbol as closely as possible.

Many organizational hard foam emblem forms are available from floral supply houses. Using preformed insignias is the most economical procedure for making set pieces. However, often unusual set piece bases must be cut from hard foam. This process can be difficult and is labor intensive. If it is necessary to cut an emblem from hard foam, first make a pattern and tape it to the hard foam with tape, or draw the pattern directly onto the foam (see figure 19-30). After cutting the pattern out with a small hand saw, it is important to smooth the edges with a small piece of hard foam to finish the emblem base.

Figure 19-29. Organizational floral tributes are special easel set pieces patterned after the organization's insignia or emblem. Shown here are: (Top) Eastern Star Emblem, (Center) Kiwanis Emblem, (Bottom) Masonic Blue Lodge Emblem.

OTHER EXPRESSIONS OF SYMPATHY

Floral designs as sympathy tributes are commonly made in containers, vases, papier–mâché, and baskets and are ordered by family members, friends, and others who associated with the deceased and his or her family. Container and vase arrangements include everything from the papier–mâché container to plastic, glass, and ceramic vases. Typically, ***pedestal arrangements*** are sent by close family members and are placed at the head and foot of the casket on a pedestal. Sometimes, people join together to send a pedestal arrangement as a combined group expression of sympathy. Often, these larger designs will coordinate in color and style to the other family pieces (the casket spray, matching easel, and floor piece).

There are many different types of ***funeral baskets*** used for sympathy arrangements. Some people prefer to send flowers in a wicker basket. The fireside basket is a popular choice for sympathy flowers (see figure 19-31). Tray bas-

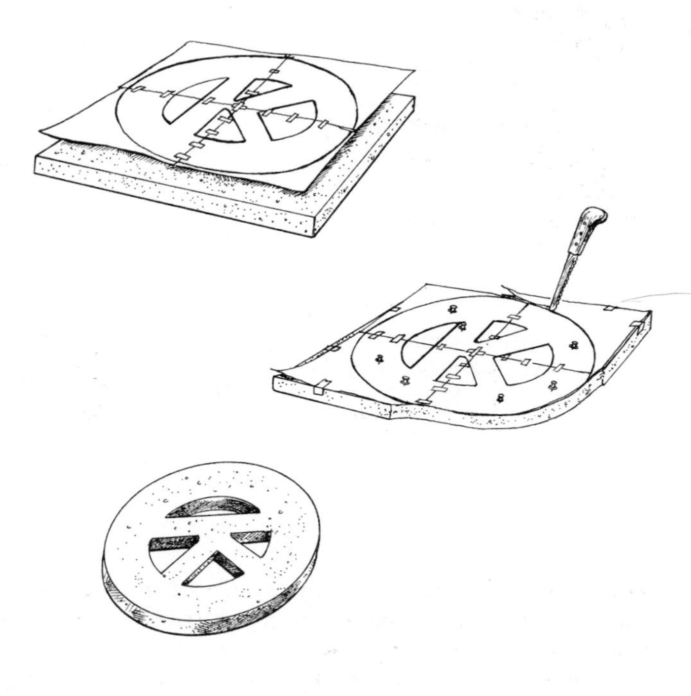

Figure 19-30. If necessary to cut an emblem from hard foam, first make a pattern and tape it to the foam. Or, draw the pattern directly onto the hard foam. Use a small hand saw to cut the hard foam. Then, smooth the edges with a small piece of hard foam to finish the emblem.

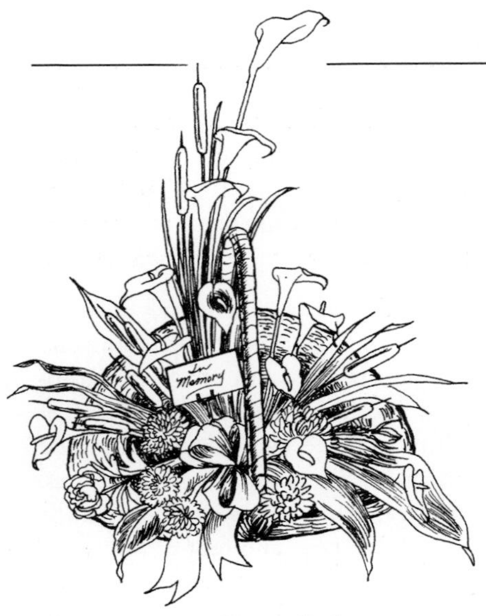

Figure 19-31. A fireside basket is a traditional floral tribute. It is important to use a basket that is stable and mechanics that are secure. These large designs are often placed on the floor beneath the casket at the funeral service.

kets and other types of baskets are especially favored by those who prefer to express their sympathy with a garden-style design.

The traditional sympathy flowers basket is the most preferred selection in most states. Different styles of plastic and metal baskets and wire basket frames that hold a papier–mâché container are the common types of sympathy baskets.

papier–mâché containers and baskets are often used by florists. papier–mâché containers are popular because they are inexpensive, inconspicuous, readily available, and biodegradable. These containers may be easily covered with fabric, moss, and floral paints. In areas where the papier–mâché container is popular, arrangements are typically large, showy, and fan-shaped with a centered ribbon bow. In other regions, vase designs include the western line approach with symmetrical and asymmetrical balance or a creative high-style design.

Flat sprays (see figure 19-32) are

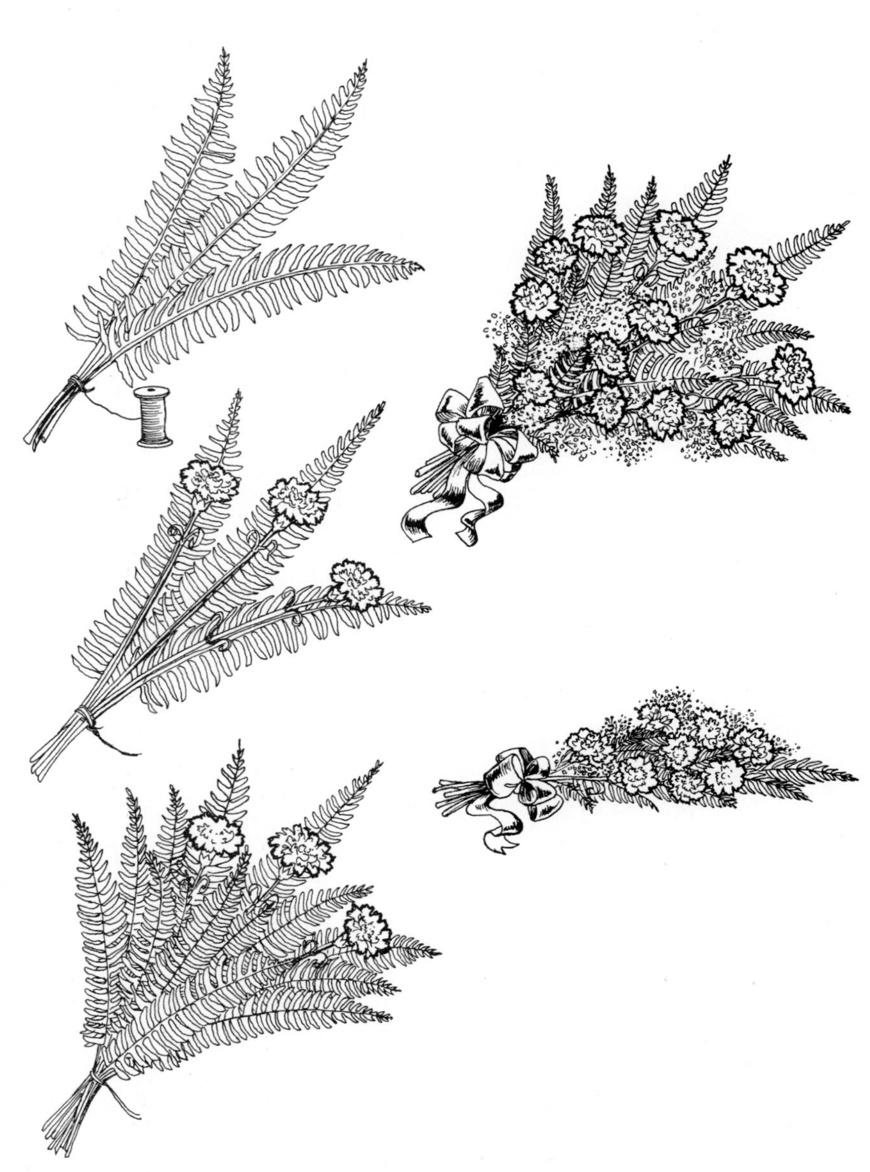

Figure 19-32. Steps of construction—flat spray (top left to bottom right): Flat sprays are a traditional tribute in certain regions. These designs can be either designed in hard foam or tied together with spool wire, as illustrated. Often these elaborate hand-tied sprays are used to decorate graves, not only during funeral services, but year-round.
Step 1: Begin the flat spray by tying three fern fronds together with spool wire. Step 2: Place three flowers on top of the fern fronds and tie together. Step 3: Place more fern to the sides and on top of the flowers, for a fuller appearance. Step 4: Add more flowers in a radiating pattern. Add fillers and more fern fronds. Step 5: Complete the tied spray by adding a bow at the base.

popular expressions of sympathy for funerals in some areas. They also are made year-round to decorate graves in many regions. These designs, while fairly inexpensive, are showy gatherings of flowers and foliage. In regions where the flat spray is a popular choice to express sympathy, many flat sprays may be grouped together to form a larger ***wall spray.***

There is a growing trend of sending sympathy flower expressions to the home of the family of the deceased. Even though flowers sent to the funeral chapel are important and meaningful, some customers believe that flowers sent directly to a family member is a more personal expression of sympathy.

Figure 19-33. A floral gift for the home, this peaceful design, which includes candles, adds special warmth. *©The John Henry Company.*

Any style of basket or vase arrangement appropriate for a home setting is suitable. Home designs should be kept small and elegant and easy to display. Plant gardens with green and flowering plants are an excellent choice for the home. Or, plant gardens with foliage plants accented with fresh flowers are sometimes preferred. Flowering and green plants are suitable expressions of sympathy that may be enjoyed by the family for a long time.

MAINTAINING IDEAL WORKING RELATIONS WITH FUNERAL DIRECTORS

The floral industry and the funeral industry are interrelated businesses. As a professional florist, it is important to maintain close contact with funeral directors in order to build the foundation for an ideal working relationship. Florists should work closely with funeral directors to develop a positive rapport and a healthy sympathy flower business.

Referrals

Funeral directors are in a position to make recommendations to their customers about flower shops and floral tributes. Since the funeral director generally meets with the family of the deceased first, he will often suggest to his clients the flower shop he prefers. When a funeral director has a good working relationship with a particular shop, he will refer his clients to that professional shop.

Survey Results

Surveys of funeral directors identify common complaints that impede positive relationships between florists and funeral directors. Complaints included poor flower quality, instability of floral tributes, late delivery of flowers to the funeral home, unmarked tributes and not enough information on enclosure cards, wobbley containers, and water leakage.

Flowers and Mechanics

A professional flower shop has a responsibility to provide fresh, high-quality sympathy tributes. Wilted blossoms or loose flowers that fall out of designs disappoint the sender, the family members, and those attending the service. The

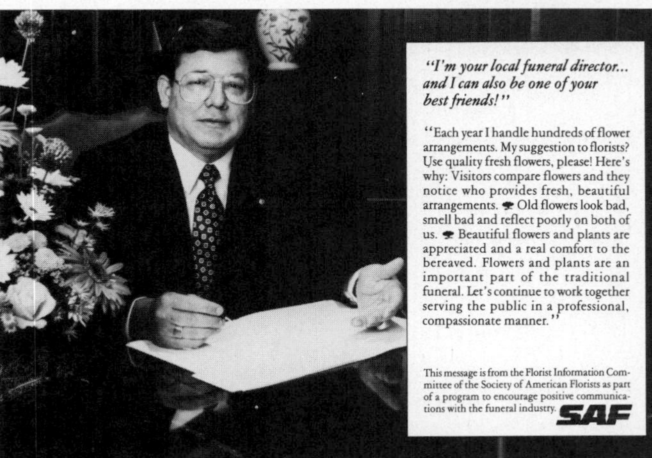

Figure 19-34. Because the floral industry and the funeral industry are two very interrelated businesses, friendly working relations must be established and maintained. Problems can be overcome by keeping lines of communication open and understanding each other's role in serving the bereaved. The Society of American Florists.

anthers on lilies must be removed before arrangement or delivery because pollen stains are a serious problem for funeral homes.

A sympathy design must be mechanically stable. When admiring floral tributes, people like to touch the flowers. Also, floral arrangements are moved several times before, during, and after services. In order to avoid problems, it is vital that the florist makes sure the construction of each design will hold up when people touch the flowers and when the tributes are moved around. Sympathy flowers that are difficult to display because of their instability and designs that fall apart create very serious problems for funeral directors.

Delivery of Floral Tributes

Late deliveries were the major complaint of funeral directors in surveys conducted regarding florists. Funeral directors need time to display the flowers before the family arrives. It is important to deliver flowers on time and with the correct information securely attached. Most funeral homes have certain hours for delivery of flowers. If a floral tribute must be delivered late, the florist should call the funeral director and explain the situation. When delivering flowers to the funeral home, truck radios must be turned off and delivery personnel must be neat and courteous.

Identification of Sympathy Tributes and Enclosure Cards

The florist is responsible for clearly identifying each floral tribute with the name of the deceased. Enclosure cards should not be tied or stapled to ribbons. Florists should write a description of the floral piece and include the sender's full name and address on the enclosure card. After the service, the funeral director removes all the cards from the floral arrangements and gives them to the family. It is up to the florist, not the funeral director, to make sure the complete information identifying the design is printed on the back of the enclosure cards (see figure 19-35). Full names and addresses are essential. The florist who provides this information to the family is providing an important service to the family and the funeral director.

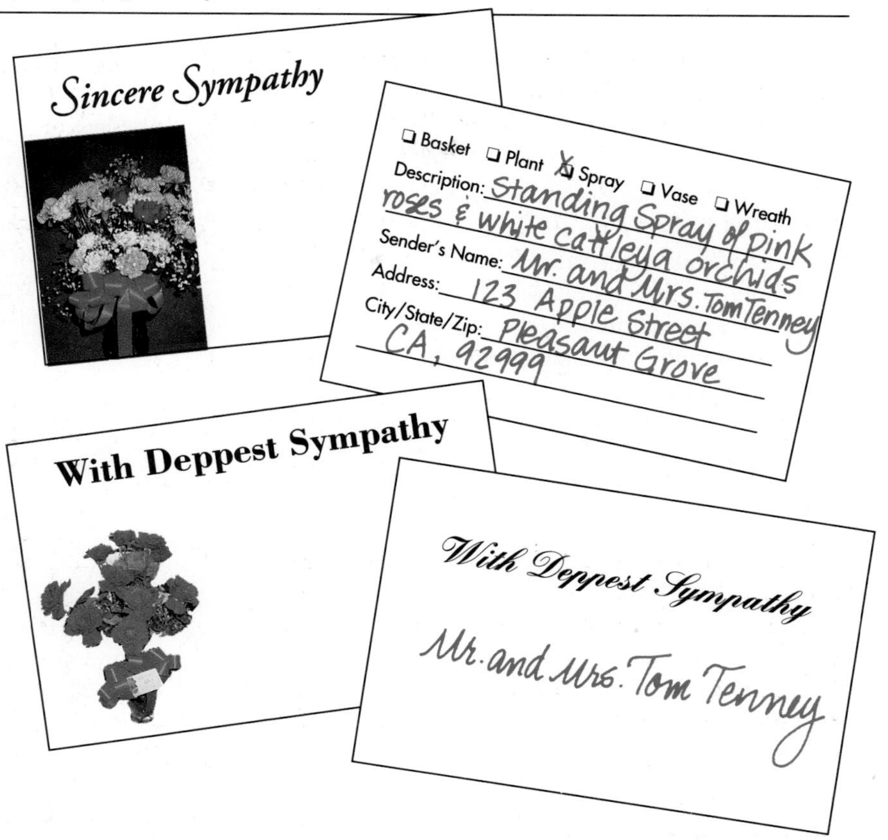

Figure 19-35. The florist should make every effort to complete the information printed on the back of enclosure cards. This will aid the family in sending acknowledgments.
Photos by Denise Dupras

Containers and Water

Excessive water and leakage cause costly damage to funeral home carpets and tables. Top-heavy baskets and arrangements that fall over may cause serious problems. Funeral directors identified fireside baskets as hard to handle and display due to over-arranging and instability. Some spray bars and floral cages are difficult to hang and may also leak water.

Since these are the major complaints funeral directors express, the professional flower shop avoids creating these irritating and costly problems. A concerned florist interested in sympathy flower business calls on his local funeral directors to receive a first-hand evaluation of what he can do to work closely with each funeral home. When a problem develops, the professional flower shop should do whatever is required to correct it immediately.

A funeral director is an important public relations arm for a flower shop. When a florist shop provides professional service, a funeral director often refers clients to the shop. By building a reputation for professionalism in the sympathy flowers market and working closely with funeral directors, a florist develops an influential bond with the funeral homes the shop services.

SERVICING THE FUNERAL

The essential consideration in servicing a funeral is making sure that all flowers meet the needs of the customer and the requirements of the funeral home. It is important to understand any size restrictions or limitations for floral sprays. For example, funeral directors with newer hearses cannot accommodate casket sprays that exceed certain height limits (see figure 19-36).

Most funeral directors are self sufficient. Nevertheless, it is important for a professional florist to inform funeral directors that the services of a flower shop are available when needed. Some funeral directors need assistance in displaying and moving flowers when there are a large number of floral tributes. For significant customers, freshening the casket spray before a service is an essential detail in servicing the funeral.

Figure 19-36. Understanding size and height restrictions set by the funeral director is important to ensure the casket adorned with a floral spray will fit into the hearse properly.

When family flowers are moved to a church for the funeral service, the professional florist should make sure the designs are in good condition in order to encourage a bond of loyalty with the family and the funeral director. Flower shops that work closely with funeral directors often loan pedestals and other display equipment. By keeping in personal contact with a funeral director, a flower shop knows the individual service needs of each funeral home and can respond accordingly.

REVIEW

Sympathy flowers have a rich heritage in the burial ritual. Even with in-lieu-of-flowers and please-omit-flowers promotional efforts, flowers still play an important role in softening the sorrow, comforting the living, and honoring the life of the deceased.

Flowers are a significant source of sales for many professional flower shops. Simple and distinctive advertising promotes the sale of sympathy floral tributes. The florist plays an important role in making the final tribute of a loved one beautiful and meaningful.

Staying in close contact with the funeral directors in a florist's delivery area is an essential responsibility. This commitment to work closely with funeral homes is a critical factor in both building and maintaining sympathy flowers business. Personal service rendered by the florist builds a positive working relationship with the funeral director and promotes a positive reputation for the flower shop, resulting in referrals and added sales.

TERMS TO INCREASE YOUR UNDERSTANDING

accent cluster
cage holder
casket blanket
casket inset
casket scarf
casket spray
corner piece
double-end spray
easel
easel spray
flat spray
floral garland
floral rosary
Florist Information Committee (FIC)
foam forms
full couch
funeral basket
half couch
"in lieu of flowers"
organizational tributes
pall
pedestal arrangement
saddle
satin forms
script
set piece
spray bars
sprays
wall spray

TEST YOUR KNOWLEDGE

1. Summarize the current trends in sympathy designs.
2. What are the classifications of sympathy flowers from a design perspective?
3. Explain the effect of the phrase "in lieu of flowers" on the use of sympathy flowers. What are some appropriate alternative phrases?
4. What are the significant points discovered by floral industry research about the positive impact of sympathy flowers?
5. What are the significant regional differences in the styles of sympathy flowers commonly displayed at funerals?
6. What are common complaints of funeral directors about the florist and floral tributes?
7. What are the all-important considerations a florist must follow in maintaining a positive working relationship with a funeral director?
8. Name ways in which sympathy flowers can be tactfully promoted.
9. What are some appropriate questions the florist may ask the family in order to personalize a casket design or other floral tribute?

RELATED ACTIVITIES

1. Invite a funeral director to speak on the use and importance of flowers at funeral services. Discuss ways of improving floral tributes and flower shop service from the funeral directors point of view.
2. Follow the obituaries in a local newspaper for one week. Record the number of obituaries and the number requesting "in lieu of flowers" and "please omit flowers" or any suggestions for memorial donations to charitable organizations. Write an editorial individually or as a group, regarding these requests and the positive impact of flowers. Provide the editor with appropriate alternative phrases.
3. Invite a professional floral designer to speak on sympathy tributes and demonstrate efficient and stable mechanics.
4. Practice constructing tied flat sprays.
5. Sketch a variety of set piece designs, including traditional tributes and organizational tributes.
6. Conduct a mock consultation and plan the family floral tributes.
7. Plan and make a casket spray and matching easel design. Work individually or as a team.

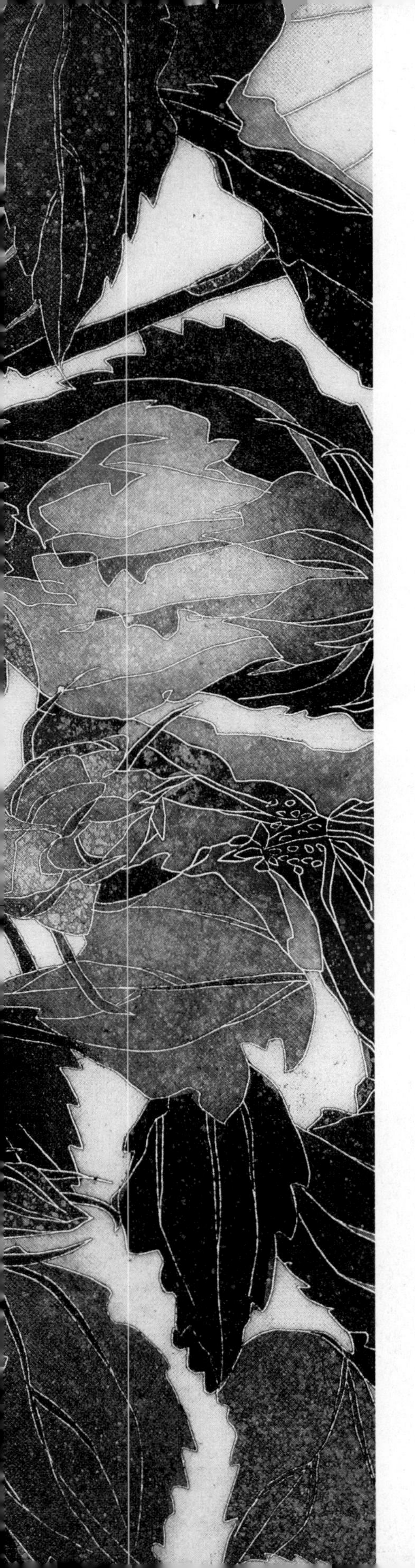

Section 5

THE FLORAL INDUSTRY

Chapter 20

Harvest and Distribution

Figure 20-1. Flowers for personal enjoyment offer an expanding opportunity for merchandising fresh flowers. To develop a market for non-occasion, everyday use, a retailer must make flowers readily available, provide a wide selection of quality flowers, and present them so they are attractive and easy to purchase.
Flower Council of Holland.

The floral industry has become a worldwide network of ***production*** and ***distribution*** of floral crops. Areas of production are increasing throughout the world, making more flowers and a greater selection available all year long. Much research on the ***postharvest technology*** of cut flowers and foliage has influenced proper care and handling techniques. The ***harvesting, packing,*** and ***shipping*** of floral products all contribute to the longevity of cut flowers and foliage. The speed of shipment and the channels of distribution, that of moving flowers from the ***grower*** to the ***final consumer,*** are varied, depending on many factors, such as where the flowers are grown and new production sites, the distance of transit, and the method of shipping. The quality and longevity of cut flowers are largely predetermined before getting to the final consumer. Proper care and handling at each point of distribution is vital. It should lead to longer-lasting flowers and repeat consumers.

THE WORLD FLOWER MARKET

In the past twenty years, there have been dramatic shifts in the production and distribution of fresh flowers. Before the global explosion of fresh flower production, the United States was nearly self sufficient. Most flowers consumed in the United States were grown domestically. Historically, fresh flowers were grown locally and consumed in a limited geographical area around the grower.

Many factors have influenced changes in the production of floral crops. The improvement of transportation and the increase in air transportation for the shipment of floral products to ***domestic*** and ***foreign markets,*** have resulted in a ***worldwide market.*** Many new production areas in Third World countries have increased production of floral crops worldwide. And, much research in the postharvest care and handling combined with advances in crop selection and cultivation methods have all contributed to increased cut flowers supplies year-round.

Production

Floral crops are grown in practically every country of the world. Though over twenty countries worldwide produce and export flowers, Holland, Colombia, Israel, and Italy are the major producers, accounting for more than 90 percent of the worldwide cut flower exports. Internationally, Holland has been the dominant player in the production and distribution of fresh flowers. The Aalsmeer Flower Auction is the focal point for distribution around the world. Holland's geographical location, close to the major fresh-flower-consumption areas of the world, gives this country distribution leadership.

Other countries that are prominent producers of flowers include Kenya, Spain, France, the United States, South Africa, Central America, and Jamaica. In the United States, the primary areas of flower production are California, Colorado, Ohio, Florida, and Hawaii. Jamaica, Mexico, Guatemala, Costa Rica, and Honduras are the flower hubs in Central America. It is important to note that Mexico is rapidly developing a strong presence in the fresh flower

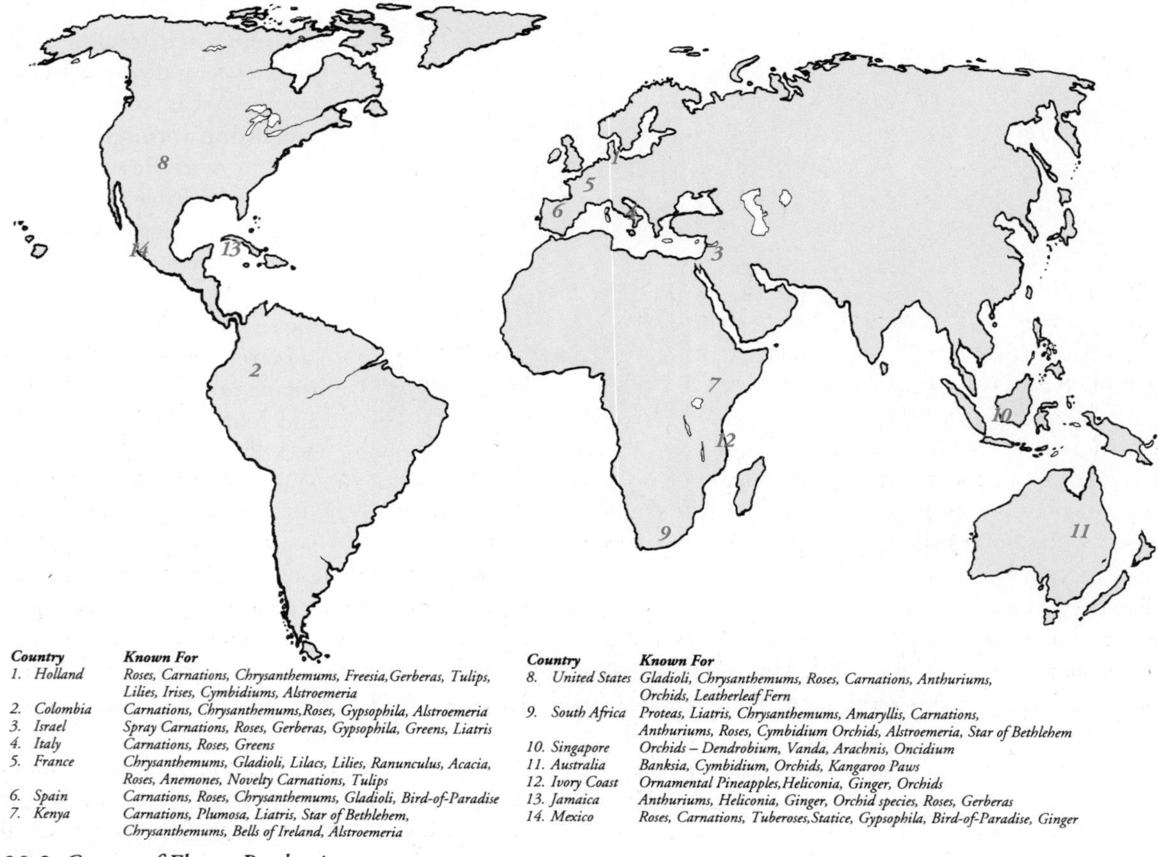

Figure 20-2. Centers of Flower Production

Figure 20-3. Flowers are harvested from growing fields for worldwide distribution.
Photo courtesy of Marcia M. Bales. Used with permission.

market in the United States. Although many South American countries are increasing cut flower production, Colombia is the country's flower production capital, shipping throughout the world. Holland, France, Spain, and Italy are the major fresh flower production areas in Europe. In the Middle East and Africa, Israel, Kenya, and South Africa grow and export flower products. In the Asian region, Thailand, Singapore, and Australia are the key countries with a meaningful presence in the world flower market.

Exports of floral products do not necessarily coincide with production areas. Some countries are ***importers*** of floral crops and then resell them for export. For example, Holland, the world's largest ***exporter*** of floral products, is not the world's largest producer. Holland is second, behind the United States, which is the world's largest ***producer*** of floral crops. Although the United States is the primary producer of floral crops, they rank low as an exporter of cut flowers.

Consumption

The countries importing or consuming the greatest number of floral products include West Germany, the United States, France, Switzerland, Holland, Great Britain, Austria, Belgium/Luxembourg, and Sweden. Holland is the leader in supplying floral products to these and other importing countries, with the exception of the United States and itself. The United States has a large domestic ***consumption*** for the floral products it produces, making exports less important. Holland, in contrast, produces to export. Since Holland is a hub for worldwide exporting of fresh flowers, many countries ship their floral crops to Holland for exporting to other countries, not for consumption in Holland.

From a marketing viewpoint, the fresh flower market in the United States has emerged from a domestic market to a worldwide market. Countries throughout the world consider the United States to be a major opportunity for consumption of the products they grow. The consistent, but slow, growth of the market for fresh flowers makes the United States an opportunistic market for many foreign producers.

Seasonal Availability

The abundance of domestically grown and imported cut floral products offers year-round availability for florists in the United States. Staple crops, such as roses, carnations, and chrysanthemums, are constantly available because they are produced by so many countries. A shortage of these flowers in one area due to ***seasonal availability,*** weather, or other factors may temporarily affect prices, but will not generally effect the supply. A producer in another area will make up for any flower shortage.

Many seasonal crops, such as tulips, daffodils, and gladioli, have been bred to be produced year-round helping make most flowers available to the worldwide market all year long. However, some crops, primarily those grown outdoors, are still seasonal and limited to the plant's natural growth and flowering cycles. Spring-flowering branches, holly with red berries, and acacia are still only available at certain times of the year.

Tropical flowers and cut greens are available year-round due to the opposite growing seasons in the northern and southern hemispheres. The summers of the southern hemisphere provide flowers for the northern hemisphere during their winters. Production from local, domestic, and foreign sources in the global market, provides ***wholesalers, retailers,*** and consumers with a wide variety of cut flowers and foliage every day, all year long.

Figure 20-4.

The Far Side cartoon by Gary Larson is reprinted by permission of Chronicle Features, San Francisco, CA.

HARVEST

Systems for harvesting flowers vary according to individual crops, growers, production areas, and marketing systems. The immediate postharvest of flowers for commercial use involves ***grading, bunching, sleeving, conditioning,*** and packing.

Throughout the world, flower products are generally harvested by hand in greenhouses, growing structures, and open fields. Using shears or sharp knives, flowers are usually harvested in the early morning and then processed for distribution the same day. Simple mechanical aids are used for certain crops. For example, special rose shears that grip the flower stem after it has been cut, allow the rose to be withdrawn with one hand from the bush (see figure 20-5). Other tools are designed to harvest long-stemmed flowers without workers having to stoop.

Flowers have inherent genetic factors that effect lasting qualities. Some cut flowers just last longer than others, due to their genetic makeup. However, whatever the typical time frame is for the life of a cut flower, it can be dramatically increased with correct ***harvesting*** conditions.

All flower crops need to be harvested at the proper stage of development to ensure longevity as they move through the various distribution channels. The time of harvest and the maturity of the flower bud at harvest are both factors that effect the postharvest quality and longevity of a floral crop.

Time of Harvest

Research indicates that in the late afternoon flowers have higher carbohydrate levels. When cut during this time, they last longer because they contain stored energy.

Figure 20-5. Specialized shears hold a rose stem after it has been cut, enabling the worker to remove the flower single handedly from the plant, simplifying the task of harvesting. Photo by Michael Reid, University of California.

Commercially, it is not practical for most growers to cut flowers in the late afternoon. Typically, a grower harvests in the early morning and then processes the flowers for distribution the same day. Early-morning harvest allows time for flowers to be properly processed and to arrive fresh and newly cut to the wholesaler or other place of distribution.

Stage of Bud Development

How flowers are handled after harvest and the changes that occur within them during the postharvest period are strongly influenced by their basic structure. The ideal stage for flowers to be harvested is different for each crop. Once cut, the flower must continue to grow and develop, if necessary, to reach maximum quality. If flowers are harvested prematurely, they will not open properly. Some flowers that are cut too soon may open, but will be smaller. Often petals will not be as colorful. In contrast, if flowers are harvested when too mature, they will not ship well. Storage and vase life are drastically reduced.

Although there are proven proper stages of cutting for most flowers, bud-opening solutions have made it possible to cut some flowers, such as carnations, at an even tighter bud stage. Flowers in the tight bud stage ship better, suffer less damage, and usually last longer.

Grading

The grading of fresh flower crops is an important factor in controlling quality. Although not all flower types are graded, all flowers are checked for quality before packing for shipment. Because there are no United States federal or state standards for cut flowers, producers, wholesalers, and retailers may have their own internal grade standards; they are highly variable. Roses, carnations, and gladioli are flower crops most often available by grades. Grading is

Figure 20-6. Roses are packaged in bunches of twenty-five and are graded and priced according to stem length, color, and quality.

generally based on stem length. But stem length may have little relationship to flower quality, vase life, or usefulness. Other factors, such as stem straightness and stem strength, flower size, vase life, uniformity, freedom from defects, and foliage quality, are useful in order to determine grade for many flower types.

Roses are graded and sold according to the stem lengths, in four-inch increments: those under 10 inches in length, 10 to 14 inches, 14 to 18 inches, 18 to 22 inches, and so on, up to 30 inches and over. Roses are identified as shorts, mediums, longs, fancy, and extra fancy. Roses are also graded according to stem strength and uniformity.

The price of flowers fluctuates based on grade. Although there is some inconsistency in grading, grade standards for various crops are fairly established within the floral industry. Grading has set standards that allow industry members—growers, wholesalers, and retailers—to communicate the quality and characteristics of products. Grading also allows a retailer to purchase the quality and stem length of roses or other flowers needed. For example, when a retailer needs roses for corsage work, the extra-fancy, long-stemmed roses are not necessary. Conversely, when the longest stems and highest quality are needed, grades make it possible for retailers to get what they desire.

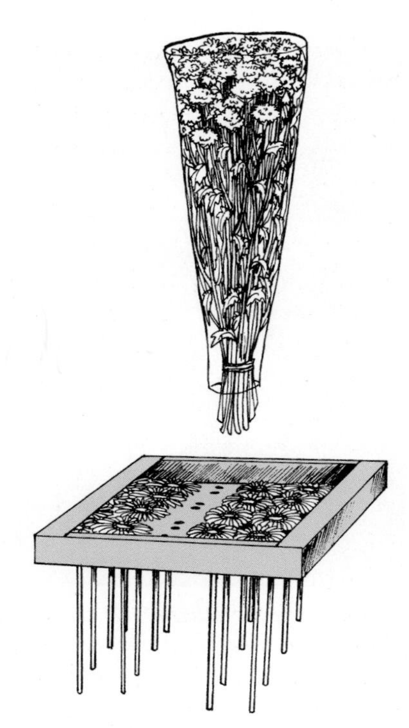

Figure 20-8. Special packaging protects flowers and stems from damage throughout the channels of distribution. Many flowers are tied together in bunches and covered with protective thin plastic sleeves. Gerberas are usually packaged and shipped in boxes that contain specially designed inserts that support and protect the heads.

Bunching

Flowers are generally bunched, except for orchids, anthuriums, and a few other specialty flowers. The growing area, market, and species are all factors that determine the number of flowers in a bunch. Single-stem flowers, such as roses and carnations, are generally grouped in bunches of twenty-five. Bunches of ten and twelve stems are common for many flower types. Spray-type flowers, such as spray chrysanthemums and spray carnations, are bunched by the number of open flowers, by weight, or by total bunch size.

Figure 20-7. In this Southern California production of statice, harvesting, grading, bunching, and packing are all carried out in the field, lessening damage by decreasing handling.
Photo by Michael Reid, University of California.

Bunches are tied together with string, paper-covered wire, or elastic bands. Many are sleeved to protect and separate the flower heads, prevent tangling, and identify the grower or ***shipper*** (see figure 20-8). A variety of materials are used for sleeving, including waxed and unwaxed paper and perforated, unperforated, and blister polyethylene (thin plastic).

When flowers are graded and bunched while in the field or greenhouse, damage through multiple handlings is reduced. Once flowers are graded and bunched, they may be treated with ***chemical solutions*** or placed in storage.

Chemical Solution Treatments

There are a variety of solutions or chemical treatments that flowers may be placed in after harvest. Each has a specific role in increasing the longevity and quality of cut flowers.

Generally, flowers are placed in the proper flower-preservative solution immediately after harvesting. Some ***growers*** do not follow this recommended procedure because a picker cannot harvest a flower and immediately place it into a preservative solution in a cost-effective manner.

Rehydration

Rehydration treatment is done to quickly hydrate flowers. This chemical solution consists of deionized water that is acidified with citric acid to a pH near 3.5. Wetting agents and a germicide are also added to the water; however, no sugar is added to the solution.

Pulsing

Pulsing solutions are formulated to extend the storage and vase life of flowers. Freshly harvested flowers are pulsed for a short time or for longer periods in different types of solutions. The main ingredient in most pulsing solutions is sucrose, ranging from 2 to 20 percent, depending on the flower being pulsed.

Some flowers are also pulsed with silver thiosulfate (STS) to reduce the detrimental effects of ethylene. Flowers are pulsed for various time lengths, depending upon the crop and the solution. Ranging from ten-second dips to several hours or a day, pulsing increases the longevity and enhances the quality of cut life.

Bud-Opening Solution

Bud-opening solution speeds the opening of bud-cut flowers before they are sold. These solutions contain sugar and a germicide. If the sugar concentration is too high, damage to foliage may result. This treatment is used in conjunction with relatively warm temperatures and high light intensities.

Tinting

Tinting or artificial coloring of flowers, may be done in two ways: through the stem, as with carnations, or by dipping the flower heads, as with daisies. The carnations to be tinted are allowed to dry, in order to get "thirsty" so they will drink up the tinting solution quickly. The tinting process is stopped and the flowers are put into a preservative solution before the flowers reach the desired color. This is because the dye still in the stem is flushed into the flower by the preservative solution.

PACKING

Packing flowers for shipping is a specialized science. Depending on the crop, the grower, the distance of transport, cost, packing method, and other factors, the materials used for packaging will vary. It is essential that boxes be packed in ways that minimize transport damage.

Specialty and delicate flowers are often packed in layers of shredded newspaper, paper wool, or wood wool to minimize friction and prevent damage. Other flowers are packed with chopped ice or polar packs to keep them cool.

For wilt-sensitive tropical flowers, such as anthurium and heliconia, the shredded newspaper may be moistened to reduce water loss. Many flowers, such as bird of paradise, may be individually protected by paper or sleeves. The stems of orchids are generally placed in small vials containing preservative solution.

Some delicate, specialty flowers are packaged in small boxes for protection. Gardenias are generally packaged three to a box. The gardenias inside the box are protected with a plastic lining to prevent moisture loss. Twenty-five stephanotis blossoms are packaged in small boxes and generally cushioned with moistened, shredded paper. Gerberas are usually packaged and shipped in boxes that contain specially designed inserts that support and protect the heads while keeping the stems straight and outstretched (see figure 20-9). Flowers, such as gloriosa lilies, are packaged in sealed plastic sleeves that have been filled with air. These "pillows" prevent the unique flower heads and floral parts from being crushed and damaged.

Boxes

Boxes are available in several standard sizes for the shipment of cut flowers. Most boxes for cut flowers are long and flat, restricting the depth of flowers. This helps keep flowers from crushing each other. Corrugated fiberboard, polyurethane-sprayed (insulated) fiberboard, and polystyrene (Styrofoam™) boxes are commonly used box materials.

Flower heads are usually placed at both ends to utilize space efficiently. To prevent sliding and transport damage, most packers anchor the layers of flowers with one or more

Figure 20-9. Colorful gerberas, packaged in protective boxes, make a showy display.

cleats, (crosswise supports spanning the width of the box). Foam or newspaper-covered wood cleats are placed over the flowers, pushed down, and stapled into each side of the box. Flower heads are placed several inches from ends of the box to eliminate mechanical damage to the flower heads during loading and unloading.

Some flowers, such as gladioli and snapdragons are packed in upright boxes, called ***hampers,*** to prevent unsightly geotropic bending. Other flowers packed upright in water are also shipped in hampers (see figure 20-10).

Figure 20-10. Some flowers are shipped standing upright inside specially made crates or hampers to prevent unsightly geotropic bending.

Precooling

Boxes are usually precooled before shipping. ***Precooling*** is a process that replaces the warm air in a box of flowers with refrigerated air. Lowering the air temperature in a box of flowers keeps them in a dormant stage of development. This curbs deterioration during transit. Care must be given to how flowers are packed within the box so that airflow is not impeded.

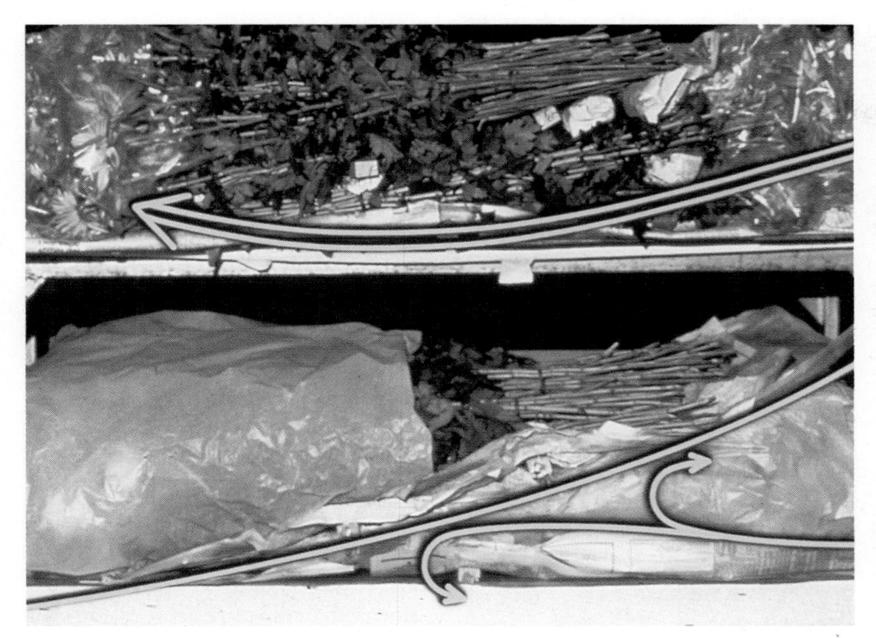

Figure 20-11. Proper precooling of cut flowers requires that the flowers be packed so that airflow (shown by the arrows) through the box is not impeded. The upper box is correctly packaged and airflow is not blocked. In the lower box, excess paper packaging prevents proper airflow. Precooling forces cool air into the box and replaces the warm air, generated by respiring flowers.
Photo by Michael Reid, University of California.

Packed boxes that have not been precooled, do not cool well if they are just placed in a refrigerated room or on a refrigerated truck. The rapid respiration of the flowers combined with the insulating properties of a packed box, result in heat buildup inside the boxes. Flowers generate enough heat from respiration to literally "cook" themselves.

Forced-air cooling of boxes with closeable flaps or vent holes on both ends is the most effective method for cooling cut flowers prior to transit. Forced-air precooling is accomplished by drawing cooled air, with the proper relative humidity for the particular flower, through the box to remove the excess heat. Boxes are positioned in front of a precooling fan unit. Most packed boxes can be cooled in less than an hour. However, the temperature of the products inside the boxes should be closely monitored during precooling. Precooling may be repeated at any time during the time of distribution, especially if the packed flower box has been left out of refrigerated conditions.

Ethylene

Because all plants produce ethylene gas (see chapter 10), there is a natural buildup of ethylene while flowers are packed tightly together in boxes over lengthy periods of time. Specialized trucks used to ship flowers have ethylene scrubbers or filters that remove ethylene from the circulating air inside the trailer. Fruits and vegetables emit high amounts of ethylene. For this reason, flowers should not be shipped with fruits and vegetables.

SHIPPING

Shipping is the process of transporting floral products from one point to another point in the channels of distribution. The several links between distribution points are major factors that influence the longevity of cut flowers and foliage. It is estimated that two-thirds of the transport of cut floral products is completed by truck and about one-third by air. The remainder of shipping is completed by a combination of courier, bus, rail, and in-house means. Most flowers are shipped ***drypacked,*** without water. The speed and the conditions in which flowers are transported effect their quality and longevity.

Shipping by Air

Shipping by air usually provides the quickest and most frequent service of transportation. Shipping by air is the primary and sometimes only means of transportation of floral products from most foreign countries. Air shipping is also used to transport flowers within the United States from growers and ***brokers*** to wholesalers. While shipping by air usually costs more than other modes of transportation, the time from delivery of floral products to the airport near the broker or wholesaler is generally twenty-four hours or less. The proper packing of flowers for air travel is essential. Often, the shipping of flower boxes is delayed, and sometimes transit takes forty-eight hours or more. Other problems include the loss of space, transfer problems, and lack of climate

Figure 20-12. Holland is the largest exporter of cut flowers. Flowers are shipped by air from Holland to countries all over the world.
Flower Council of Holland.

Figure 20-13. Refrigerated trucks are the most common mode of transporting floral products along the channels of distribution.
American Floral Marketing Council.

control. Upon arrival at the destination airport, flowers should be unloaded and shipped to the next point of distribution as quickly as possible.

Shipping by Truck

Trucks are the most common mode of transporting floral products to and from points throughout the channels of distribution. A wide variety of trucks, equipped with proper shipping conditions, transport floral products from grower to shipper, grower to wholesaler, and wholesaler to retailer. Transport by truck is certainly slower than air carriage to major markets, but the benefits of long- and short-distance refrigerated shipping cause it to account for the majority of floral-industry transportation.

Most shippers who transport floral products use environmentally controlled trucks. These refrigerated trucks provide the constant, proper temperature and humidity control throughout transit. Many trucks provide ethylene-scrubbing units to prevent and reduce the buildup of ethylene, which may prove to be a problem for flowers that are inside boxes and in transit for several days.

Many factors effect the outcome of flowers that are shipped by truck, including whether or not flowers are precooled or recooled, the cooler facilities on the truck, delays of inspection, and loading and unloading procedures. Upon arrival at the wholesaler or retailer, flowers should be unloaded and cared for as quickly as possible.

Time Frame

The typical time frame from grower to retailer varies depending on the distance floral products must travel and the mode of transportation (see table 20-1). It is important that flowers move as quickly as possible through the distribution channel to allow the final consumer to have more time to enjoy the flowers. The time from grower to retailer and the quality of care and handling during that time will increase the longevity of flowers for the final consumer—making happy, repeat customers.

TABLE 20-1: TYPICAL TIME FRAME OF FLORAL PRODUCTS FROM GROWER TO RETAILER

BY TRUCK

Day 1	Harvesting and Conditioning
Days 1–2	Packing and Shipping
Days 2–4	In Transit
Days 3–5	Delivery to Wholesaler
Days 3–8	Delivery to Retailer

BY AIR

Day 1	Harvesting and Conditioning
Days 1–2	Packing and Shipping
Days 2–4	Arrival at Airport
Days 2–4	Arrival at Wholesaler
Days 2–6	Arrival at Retailer

DISTRIBUTION

Traditionally, flowers have moved from a grower to a shipper or broker, to a wholesaler, to a retailer, and from the retailer to the final consumer. Today, however, the distribution channel is changing. As shown in figure 20-14, elimination of some interim points of distribution has resulted in flowers moving more directly from the grower to the final consumer.

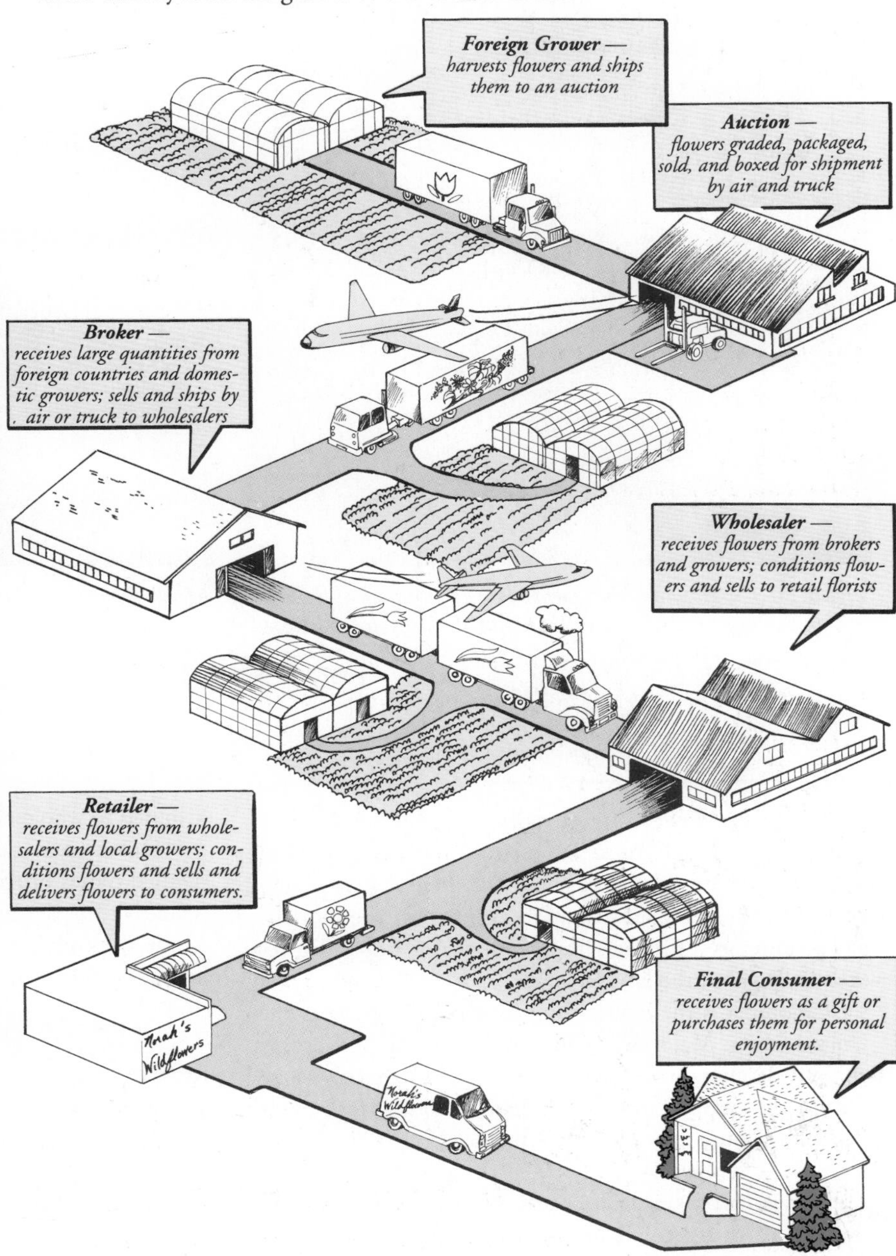

Figure 20-14. The shipping routes of floral products into the United States form this channel of distribution.

Figure 20-15. Holland, the world's largest producer of floral crops, has the biggest total area covered by glass in the world. Over 7 billion cut flowers and 350 million potted plants are cultivated per year in glasshouses with a total surface area of 10,000 acres. Flower Council of Holland.

Foreign Growers

Imported cut flowers arrive daily in the United States mostly by air. Some trucks are used to import flowers and foliage from Canada and Mexico, but these shipments are generally targeted for distribution points near the borders. Truck-shipped cut foliage from Central America is distributed throughout the United States.

Colombia is the largest shipper of cut flowers into the United States. Major crops include roses, carnations, spray chrysanthemums, baby's breath, and alstroemeria. These flowers are shipped mostly to Miami, Florida, to brokers who redistribute the flowers to wholesalers or to other brokers.

The second-largest source of imported cut flowers into the United States comes from Holland, which specializes in tulips, gerbera, freesia, roses, carnations, chrysanthemums, lilies, irises, orchids, and alstroemeria. Shipments sent by air to major airports in the United States are distributed to wholesalers.

Other countries generally ship floral products to a United States broker, who acts as a sales agent and representative for the producing country or for a cooperative group of producers. These floral products are then distributed to wholesalers.

Domestic Growers

Depending on the size of the grower, the volume of crops produced, and the preference of the grower, crops generally move directly from the field or greenhouse to the wholesaler. Sometimes, however, especially for small, seasonal growers, flowers may be directly sold to the retailer or final consumer.

Figure 20-16. Flower auctions in Holland are unique. The flowers and plants to be sold are delivered by the growers early each morning to the auctions. After inspection, floral products are automatically transported by trolleys to the auction room. Here the buyers listen through earphones to information on the respective flowers or plants that are for sale. The price to be paid for the products is determined by a large auction clock on the wall, opposite the seats. The buyer stops the hand of the auction clock with the push of a button, which he must press as quickly as possible to secure the purchase of flowers before another buyer bids. The sooner the buyer pushes the button, the higher the price. Buyers must have experience, sound judgment, and a quick response. *Flower Council of Holland.*

Auction

An ***auction*** is in reality a centralized wholesale facility where many growers bring their floral products, early each weekday, to sell through open-market bidding by brokers, wholesalers, and retailers. At the auction, the products are inspected and graded for quality. Inferior products are not allowed to go through the auction for sale. Products retained for resale in the auction are given a lot number.

These numbered products are carried by a trolley system through the auction room. Buyers sit at desks and chairs in a theater-like room. Each buyer has a purchasing number. As flowers are paraded across the stage, buyers are able to purchase flowers through a bidding process.

At the world-famous Aalsmeer Auction in Holland, a large "clock" on the wall marked with prices has a rotating hand. The hand swings around to a price (for a particular flower that is being shown for sale) and then begins to drop to lower prices. When the price suits a buyer, and he decides he wants to purchase the lot of flowers on the auction, he presses a button on his desk (before the other buyers) and the sale is recorded electronically.

At the end of the auction, each buyer is presented with an invoice for all he has purchased and his account is settled in cash before he leaves. As shown in figure 20-17, employees at the auction warehouse organize and package floral products that have been purchased. Once flower boxes are packed, they are stored in the building until they

Figure 20-17. At the auction warehouse, millions of flowers are organized in buckets, placed on trolleys, and sold at the auction. After being sold, floral products are packaged for distribution. Eighty percent of the flowers and plants auctioned in Holland are exported.
Flower Council of Holland.

are loaded onto trucks for shipment. Flowers that are sold through the auction may have been cut that very morning. Once sold at the auction, they are delivered to an airport for worldwide shipping, often the very same day. Flowers are generally shipped to brokers and wholesalers in the United States.

Broker

A broker functions as the marketing hub by handling the functions between buyers and sellers. Sources of delivery are coordinated all over the world. A broker represents one grower or a group of growers. The grower packs the flower products at the point of production and then sends them to a central point of redistribution, for example, brokers in Miami, Florida. Brokers purchase flowers in large quantities for their own inventory. These are sold by salespeople who call on wholesalers and other brokers. When brokers presell floral products, the grower packs for these specific orders. Brokers move the products in box lots exactly as packed by the grower.

Brokers often have refrigerated facilities for storing boxes of flowers shipped to them by their growers. Often, the broker never even sees or physically handles the floral products, and never processes flowers or opens boxes to repack in smaller quantities. Brokers sell and ship from boxed, stored inventory.

Wholesaler

A wholesaler is the traditional link between the grower, the broker or shipper, and the retail flower shop. Many growers, especially regional growers, sell directly to wholesalers. Products usually arrive in the early morning by truck, coming from an airport or directly from a grower. Early-morning arrival allows time for processing and resale the very same day the floral products are received. When shipments arrive at the wholesale house, they are generally unloaded into a cold-storage area, around 40 to 50 degrees Fahrenheit. Boxes are opened and inspected for damage or insects. Once inspected, they are laid out on tables, to be processed or repacked for immediate delivery.

Wholesalers have multiple coolers, set at various temperatures. The general holding coolers for most flowers and foliage are set at 36 to 38 degrees Fahrenheit. The cooler for tropical flowers is set at 50 degrees Fahrenheit. Sometimes a separate cooler for roses is set at 34 degrees Fahrenheit. All of the coolers maintain high humidity levels and are equipped with ethylene scrubbers.

Figure 20-18. Wholesalers receive floral shipments daily from all over the world. When shipments arrive at the wholesale house, the boxes are opened, and the flowers are conditioned. Retailers are able to select from an abundant array of flowers. Evergreen Wholesale, Inc. Used with permission.

Figure 20-19. Wholesalers generally have multiple walk-in storage coolers set at various temperatures. Flowers are stored in buckets of preservative solution.

Many large wholesalers who process thousands of flowers each day have sophisticated water systems that remove impurities from the water. Preservative water is often dispensed through special tanks.

A wholesaler assembles products from many different sources and makes them available to retailers. The whole-

Figure 20-20. This wholesale operation brings many different distributors together under one roof, provides a great variety of floral products, competitive pricing, and convenience to the retail customer.
Boston Flower Exchange. Used with permission.

saler buys in bulk quantities directly from growers and brokers, breaks the bulk, sorts shipments into smaller lots, and then distributes products to retailers. Large retail florists often buy in box lots of one product—a box of five to six hundred carnations or roses or an entire box of spray chrysanthemums. Typically, however, retailers buy their flowers in a bunch, not a full box of one type of flower.

Most retail outlets in the United States purchase their floral products from a wholesaler. However, some large shops, chain retailers, and mass merchandisers are able to buy directly from growers. Wholesale purchasing offers many advantages to retail florists. Small quantities may be purchased through wholesalers, rather than having to purchase an entire box of the same kind and color of flower. A great selection of flowers is available at one location. Wholesalers extend credit to retailers, welcome special orders, offer delivery, and back the quality of the flowers.

Figure 20-21. Because flowers are grown in both hemispheres, tropical foliage and flowers, like these hanging and upright heliconia and ginger, are available from many countries year-round.

Retailer

Wholesalers are the immediate source for most flower retailers. These florists buy from their wholesaler several times each week, each morning, or several times each day. While retailers may visit the wholesaler to select their own flowers, as shown in figure 20-23, orders are usually placed by phone and then delivered by the wholesaler to retail flower shops.

By the time fresh flowers and foliage reach the retail store, they may have been cut for several days or a week. If the floral products have received proper care and handling during harvest and distribution, they will be in good condition. However, to continue the longevity and quality of the flowers, it is vital that boxes are opened at the retail level immediately upon receipt. Deliveries should be checked for accuracy and inspected for insects or diseases.

Cut flowers should be properly processed (see chapter 10) and immediately placed in warm hydrating and preservative solutions and placed in the cooler. Flowers have the ability to revive and become vital again with proper care treatments.

Figure 20-22. The wholesale florist spends much time on the phone, ordering (buying) flowers from growers and brokers located around the globe and calling (selling to) local retailers, on a daily basis.
Couresty of Evergreen Wholesale, Seattle, WA. Photo by Raymond Gendreau.

Storage

Flowers are stored in the cooler at the retail floral shop. They may also have been placed in storage facilities along the route of distribution. Proper storage conditions are vital for increasing the length of vase life for the final consumer. Low temperatures and high humidity reduce the rate of respiration, eliminate heat buildup, delay the development of flowers, and reduce the effects of ethylene gas (see chapter 10).

Storage coolers are those that are used to store flowers, rather than dis-

Figure 20-23. While most retailers order flowers over the phone to have them delivered, some choose to visit the wholesaler to select their own flowers.

play them. A typical storage cooler is a walk-in unit with solid walls (no windows). Storage coolers are used for dry-pack storage of flowers and foliage. Proper temperatures and humidity levels are essential to keep dry-pack flowers and greens without a water supply fresh. Generally, flowers are stored in buckets of preservative solution while in the cooler. Floral arrangements awaiting delivery are also stored in the cooler at retail shops and at some wholesale outlets (see figure 20-24).

Final Consumer

The final consumer is the last person to receive flowers through the channels of distribution. The consumer must be made aware of continued proper care and handling methods to help flowers last as long as possible (see chapter 10). Consumers generally underestimate the potential vase life. Without proper care, flowers wilt and die prematurely. The retailer should include a packet of floral preservative with each flower order, whether the flowers leave the retail shop with a customer or the flowers are delivered. Care tags should also be included with every sale of cut flowers. Care tags provide general care information to educate the consumer and prolong the vase life of flowers. Satisfied customers will return for more flowers, which ultimately keeps the floral industry strong, healthy, and alive.

"Being discharged today! I just spent $8 on these flowers."

Figure 20-25.
Herman © 1985 Jim Unger. Reprinted with permission of Universal Press Syndicate.

Figure 20-24. Floral arrangements and bunched flower groupings are ready for delivery to retail mass merchandisers and floral shops.

MARKETING FLOWERS

Marketing is everything a business does to find, develop, and keep customers. Many businesspeople confuse marketing with the movement and sale of products and services. Marketing addresses fundamental questions about the customer. Whether a grower, broker, wholesaler, or retailer, the first questions this businessperson must ask are, "Who are the specific customers I want to find, develop, and keep?" and "What products and services must I offer to develop this potential?" Until a businessperson identifies a market niche to serve, there may be difficulty building a solid customer base.

Many people in flower businesses approach marketing from a nonproductive point of view. This error in marketing judgment creates a serious problem in developing opportunities for the business. The typical question these people ask is, "How do I find customers to buy what I

Figure 20-26. National advertising and promotional efforts make the general public more aware of flowers for nonoccasion uses.
American Floral Marketing Council.

want to sell?" They focus on selling the product instead of developing customers.

Flower business in the United States is occasion oriented. Most ***advertising,*** often incorrectly identified as marketing, focuses on selling specific ideas for specific occasions. This advertising does not develop customers. It attracts buyers for specific ideas or specific occasions. Occasion buyers are seldom developed into everyday buyers. The future-minded florist understands that the long-range opportunity for a flower shop depends on developing loyal customers, not just selling arrangements.

Advertising

The major advertising in the floral industry is wire-service oriented. Advertising funds collected by wire services, generate orders for specific flower arrangements for certain occasions. Most of these dollars are used to advertise arrangement specials including a container and accessories the florist must purchase from the wire service. Some advertising is more generic, promoting flowers for everyday use instead of particular occasions.

The American Floral Marketing Council (AFMC) collects funds from retailers, wholesalers, growers, and floral manufacturers and suppliers. Compared to the wire services that mandate collection of advertising funds from its members and subscribers, AFMC contributions are voluntary. These advertising funds are used to create a more general awareness of flowers, especially the everyday enjoyment of flowers (see figure 20-27).

There is a serious void in advertising and ***promotion*** by retail florists.

Figure 20-27.
American Floral Marketing Council.

Large florists often use newspapers, radio, television, and direct mail to develop sales. Small florists often struggle with the availability of funds, the needed advertising expertise, and the time to devote to furthering their business opportunity. This lack of focus on developing their business makes many flower shops vulnerable to competition. It also inhibits the growth of not only the retail industry but the floral industry as well.

Public Relations

Public relations includes everything that creates an image awareness of the flower shop or of flowers in general. Unless an advertisement presents a specific idea, for a specific reason, and at a specific price range, it is public relations, not advertising. Using the "flowers-for-all-occasions" theme in paid media, for example, is public relations, not necessarily advertising. This creates an image awareness but does not ask the customer to buy. Advertising always asks a customer or prospect to buy.

Promotion

Promotion is the process of advancing the awareness of a flower shop, its products, and its services through special events and product demonstrations. A Christmas or spring open house is a form of promotion. The florist who gives a demonstration to a civic group is using promotion to advance the awareness of product, creative talent, and the flower shop. Promotion includes giving flowers for door prizes and other give-away events.

Marketing, advertising, public relations, and promotion do specific tasks in building a flower business, and thus, the floral industry. Unless consumers are made aware of flowers and reasons why they should make purchases, the floral industry will not thrive.

REVIEW

Once a nation that produced nearly all the flowers sold in the United States, the United States is now served by worldwide producers. Flowers are imported into the United States from nearly every part of the world. This plentiful supply of fresh flowers creates a competitive market. Techniques used for harvesting, grading, bunching, conditioning, and packaging determine the quality and longevity of fresh flowers. The stage at which a flower is harvested depends on the market serviced. Today's rapid shipping with climate-controlled conditions provides for an abundant array of flowers daily from all over the world.

The traditional channel of distribution moves flowers from grower to broker, to wholesaler, to retailer, to the final consumer. Using appropriate care and handling techniques all along the route of distribution, ensures both a quality product and longer vase life. Consumers often judge the quality and the price/value relationship of fresh flowers by how long they last. Care and handling involves using proper and consistent care.

TERMS TO INCREASE YOUR UNDERSTANDING

advertising
auction
broker
bud-opening solution
bunching
chemical solutions
cleats
conditioning
consumption
distribution
domestic market
drypacked
exporter
final consumer
foreign market
grading
grower
hampers
harvesting
importer
packing
postharvest technology
precooling
producer
production
promotion
public relations
pulsing
rehydration
retailer
seasonal availability
shipper
shipping
sleeving
tinting
wholesaler
worldwide market

TEST YOUR KNOWLEDGE

1. Name the major foreign countries that produce flowers for exporting.
2. What are the key considerations in harvesting fresh flowers?
3. What are the two types of grading?
4. List the essential procedures that must be followed in conditioning flowers to ensure maximum quality and longevity.
5. Why is precooling important?
6. Explain the traditional distribution channel.
7. Explain the process and function of the Holland Auction.

RELATED ACTIVITIES

1. Visit a wholesale house in your area. Notice what care and handling procedures they use.
2. Visit a retail store. Volunteer to unpack and properly condition boxes of flowers delivered from the wholesaler. As you unpack the flowers, take note of where the flowers were grown or where they came from.
3. Visit a local grower and learn of harvesting techniques.

Chapter 21

The Retail Flower Shop

A flower shop is a link in the distribution system that moves products from the source of production or manufacturer to the ultimate or final consumer. As a method of distribution, a flower shop provides customers with quality products and value-added services, desirable and beneficial to the consumer, which differentiate a flower shop in its marketplace.

Changes in the traditional levels of distribution create a very competitive marketplace. A flower shop must be an aggressive marketer of its products and services. Though the art of professional floral design is an essential element in the success of a flower shop, it is not the sole key to success. The florist who prospers and grows understands the vital importance of combining quality products, desirable floral design, and professional, value-added customer service, with a consistent ***marketing*** program.

The major function of a flower shop is to add value to the products sold. This added value encompasses a wide spectrum of professional services including, but not limited to design, delivery, creativity, and personal attention to customer needs. The more exclusive, unique, and desirable the services provided are, the more value the consumer places on them.

Figure 21-1. Flowers have long been purchased for personal enjoyment.
Crossing the Street *by Giovanni Boldini, 1875. Sterling and Francine Clark Art Institute, Williamstown, Massachusetts.*

Figure 21-2. A full-service, professional flower shop provides everything a customer needs in flowers, services, and flower-related products. The majority of flower shops in the United States are full-service, professional shops. The layout of any flower shop and the manner in which the areas of a shop are used for sales, display, flower receiving, designing, delivery, and accounting impacts the efficiency and the profitability of an operation.

The type of operation a flower shop owner wants to develop and the target market to be reached dictates the style and location of the shop. There are many unique responsibilities and functions within a floral shop. The process of operating a profitable full-service shop involves serveral distinct functions, including ***visual merchandising,*** efficient shop layout, marketing, sales, buying, pricing, wire service, and delivery.

TYPES OF FLOWER SHOPS

Flower shops are set apart by the products and services they provide consumers. There are four broad classifications of flower shops. This categorizing of flower shops does not suggest that one type of flower shop is superior. In the marketplace, there are successful flower operations in all four categories. General classifications include: full-service, professional flower shops, specialty flower shops, limited-service flower shops, and flower ***merchandisers.***

Full-Service, Professional Flower Shops

Full-service implies that the florist provides everything a customer needs in flowers, services, and flower-related products. ***Professional*** indicates that a flower shop owner and all employees are thoroughly trained in all aspects of the flower business and competently qualified to serve customers. The majority of flower shops found in the United States are full-service, professional shops.

A full-service flower shop provides a complete range of flower products. It services all the flower needs of its cus-

"Have you got any others with more spikes?"
Figure 21-3.
Herman © 1978. Jim Unger. Reprinted with permission of Universal Press Syndicate. All rights reserved.

tomers for all occasions, special events, everyday use, and personal enjoyment. The full-service, professional flower shop provides delivery service for their customers as well as out-of-town delivery through a ***wire service.*** With a broad product line that often includes gifts, cards, potted plants, silk and dried designs, gourmet foods, fruit baskets, balloons, and other floral accessories, the full-service florist attracts customer attention, because of the great range of products and services provided.

Specialty Flower Shops

Targeting a particular need in the marketplace, a ***specialty flower shop*** specializes in a profitable aspect of floral design. A specialty shop may focus on a particular floral need, such as everlasting, silk and dried designs, wedding flowers, or high-style and party work. Because of the limited products offered, specialty shops are most successful in affluent locations and in close proximity to other businesses targeting the same market in other ways. For example, floral shops specializing in weddings are often located near a catering company, wedding reception hall, bridal salon,

Figure 21-4. Limited-service flower shops, found in many grocery stores and hospitals, are designed for customers who purchase flowers or gifts on impulse. These floral shops provide little service since they are limited in space.

Figure 21-5. Generally located in heavy-traffic areas, flower merchandisers specialize in selling loose-cut flowers by the stem or bunch.

or wedding photographer. Or, a florist specializing in high-style and party work may be located in a downtown area adjacent to large, luxurious hotels and convention centers.

Limited-Service Flower Shops

The ***limited-service flower shop*** is a business that narrowly defines the range of products and services it provides, compared to a full-service florist servicing all the floral needs of their customers. A limited-service flower shop is designed for customers who purchase flowers or gifts on impulse. These floral shops provide little service, since they are most often limited in space. Often these shops, such as those found in hospitals, hotels, and supermarkets, are operated by large full-service shops. Floral designs are often arranged at a full-service shop and delivered to the small limited-service shop.

When a florist does not accommodate customers with a complete selection of seasonal flowers plus specialty and exotic flowers, it limits flower choices. The flower department in a supermarket that does not provide delivery restricts its range of services. Some flower shops choose not to do weddings, parties, decorations, and other labor-intensive floral work. This approach to operating a flower shop limits the clientele that a florist attracts and serves. However, limiting the services provided does not restrict the opportunity for success. There is a niche in every marketplace for the specialty or limited-service florist.

Flower Merchandisers

Often called stem shops or cash-and-carry shops, these floral operations are generally located in heavy-traffic areas. The flower merchandiser is characterized by the retailer

Figure 21-6. A rapidly growing trend is the flower merchandiser who offers a full selection of quality flowers at affordable by-the-stem prices.
Photo by Brian Yacur.

Figure 21-7. Some flower merchandisers focus on featuring low price by purchasing flowers in oversupply and promoting them in mass displays.
Photo by Brian Yacur.

specializing in loose cut flowers sold by the stem or the bunch (see figure 21-6). This flower retailer generally does not provide delivery or design services and focuses specifically on selling loose flowers with little, if any, value-added service. Most of these operations do not have refrigeration, requiring flowers to be brought in daily.

LOCATION

Some florists believe that the location of the shop is the most important influence in the success of the business. Others show conclusive evidence that a unique flower shop serving a defined clientele profile with desirable products and services not readily available elsewhere, attracts loyal buyers regardless of location. The most important consideration is always the cost of location in relationship to sales that can be developed.

With the growing trend of shoppers buying from their homes instead of going to a retail store, the florist who develops a distinctive image and reputation in its marketplace builds a large, dependable, and loyal ***customer base.*** Often the value of location is overrated. If a florist's focus is on design work and developing telephone sales, location has little relative value.

There are a variety of location sites typical for flower shops. Whether the florist is located in a free-standing shop, downtown in the central business district, or on a street corner, the location must fit the particular style or type of shop and service the targeted market.

Figure 21-8. Many flower shops are located in shopping malls where heavy foot traffic brings in many customers.
Bentley Square, *Salt Lake City, Utah.*

Free-Standing Flower Shop

A shop in a single unit building is a free-standing business. An old home converted to a flower shop, a flower retailer sharing a single building with another business, and a location designed and built exclusively for the flower shop are all examples of free-standing locations. Free-standing locations usually have a more distinctive physical image, provide convenient parking, and offer the florist many advantages in merchandising products.

Strip-Center Flower Shop

A ***strip-center flower shop*** combines several businesses that adjoin one another and comprise a small shopping complex. Most strip-centers adjoin residential areas. Because of their accessibility, parking facilities, and rent structure, the strip-center location is desirable for many flower shops. The typical strip-center does not provide the same foot traffic as the well-established shopping mall. However, the difference in facility cost makes the strip-center location cost-effective for many flower shops.

Shopping Mall

On the surface, the shopping mall appears to be the ideal location for a flower shop because of the foot traffic it should provide. Yet, the failure rate of flower operations in shopping malls is high. Unless a flower shop has unlimited capital, it has difficulty surviving in a new shopping center. Often it takes several years for a florist to develop the required foot-traffic in a new mall. For the florist interested in a mall location, it is often better to negotiate with an established center where foot traffic is established and defined.

The rent and additional facility charges are usually highest in a shopping mall. This cost-of-operation disadvantage can create critical financial difficulties for a flower shop. Most florists operating successfully in a mall location carry gifts, cards, plush animals, novelties, and other flower and gift-related products (see figure 21-9). It is not unusual for their non-flower sales to be higher than flower sales. This type of shopping mall operation requires a higher inventory investment as well as astute inventory turn-over ***management.***

Business Complex

Some very successful flower shops are located in office buildings and other types of business complexes. Usually, these shops have a clearly defined client base determined by the immediate tenants. Shops in business complex locations often combine the full-service features needed by business accounts with merchandising loose fresh flowers for personal, office, and home use. The business-complex florist frequently has a valuable advantage of capturing loyal customers who have a very high repeat-purchase pattern.

Downtown Location

Although the general trend in some areas of the country, except in smaller towns, is for flower shops to locate in suburban areas instead of downtown, the downtown flower shop is still a viable business. Generally, florists in downtown locations tend to be long-established businesses with a following of loyal clientele. These floral shops often focus on high-style designs for local hotels and businesses.

Figure 21-9. Most shopping mall flower shops carry gifts, cards, plush animals, novelties, and other flower and gift-related products. It is not unusual for their non-flower sales to be higher than their flower sales.
Bentley Square, *Salt Lake City, Utah.*

Figure 21-10. Downtown flower shops tend to be long-established businesses with a following of loyal clientele. These floral shops often focus on high-style designs for local hotels and businesses.
Especially for You, *Salt Lake City, Utah.*

Figure 21-11. Supermarkets and mass merchandisers often feature floral departments in their stores. Large floral departments often have on-site designers.

Floral Department

Supermarkets and mass merchandisers with aggressive flower merchandising programs feature floral departments in their operations. Newer supermarkets often have large floral departments prominently positioned to command visibility. Floral departments usually concentrate on merchandising flowers by the stem and in prepackaged assortments to attract impulse buyers. Sometimes, they feature complete floral services with full-time, on-premise designers who create custom designs.

PRODUCT PRESENTATION AND SHOP LAYOUT

The layout of a flower shop must be convenient and serviceable to customers and enable employees to work efficiently. A typical flower shop has an area for display and sales, a work area for receiving and designing flowers and other merchandise, storage areas for supplies and flowers, an office area, and a convenient area for processing orders and loading them into vehicles for delivery.

A planned foot-traffic pattern is necessary, enabling customers to move easily throughout a floral shop without getting confused or feeling trapped. Ideally, the window display attracts customers into the store, the in-store displays promote sales and add-on sales, and the service areas (the sales counter and consultation area) service customers efficiently.

Visual Merchandising

The ultimate goal of the retail florist is to sell flowers and other merchandise quickly, efficiently, and profitably. Efforts to attract customers to the shop and create interest in the flowers and merchandise is known as visual merchandising. The lighting, setting, color, mood, and other elements of display all contribute to the success of selling products.

The way a florist displays its products to attract customers, establishes and communicates the visual image of the flower shop. This image is identified as "the merchandising personality" of the business. Through visual communication, this merchandising personality customers and potential buyers that the retail store is a desirable place to shop.

Figure 21-12. The primary goals of displays are to attract attention, create interest, turn interest into desire, and generate sales. Visual merchandising through attractive and creative displays establishes "the merchandising personality" of the flower shop.
Knud Nielsen Company, Inc., Evergreen, Alabama. Products available through local wholesale.

The primary purpose of any display is to capture attention and motivate people to buy. To understand visual merchandising, it is first important to realize the purpose and primary goals of display. An effective visual display will achieve four primary goals:

1. attract attention
2. create interest
3. turn interest into desire
4. generate sales

First, consumers must be attracted to the floral shop and then to the flowers and other merchandise, before sales can take place. Window displays attract initial attention. However, visual merchandising does not stop at the front window. The consumer, once enticed into the shop, must be attracted to the merchandise by the in-store displays.

Interest in the flower shop and its products are created after captivating attention. When flowers and other merchandise are displayed in creative and artistic ways, customers become more interested in them. Displays must create a desire to buy.

When flowers and products are displayed in ***vignettes,*** they are emphasized and the impact of the display is enhanced. The visual presentation of products sways the buying attitude of customers and prospects. Through visual merchandising, product presentation educates the viewer with new ideas and new uses for the products displayed. Helpful information, such as the name of the flower, the price, and other useful or interesting information will assist the customer in making a buying decision.

Figure 21-13. A walk-by window display must command attention, connect with the viewer, and create the desire for the individual to come into the shop.
Photo by Brian Yacur.

Window Display

The purpose of ***window displays*** is to communicate visually with people who pass by the shop. In locations where many prospective clients go by in automobiles (versus foot traffic), the window display must be bright, bold, and communicate a message in a split second. Immediate visual impact is the only appropriate approach in creating a window display in a high automobile-traffic location. The name of the flower shop must be very prominent and easily read. The objective of this type of window display is to catch the fleeting eyes of those passing by in an automobile and to declare loudly and effectively the name of the business so it is remembered.

A window display for the flower shop in a walk-by location should capture and hold attention. The function of the display is to influence the viewer to stop, and entice this prospective buyer to come into the flower shop. Designing a beautiful window just to be beautiful does not meet the objective. A window display people look at but do not respond to has little marketing value. Key questions that must be addressed when planning a window display are: What is the purpose of this window display? and How will it entice customers to come into the shop?

Whether a window display is geared for automobile traffic or for foot traffic, it should be planned and designed with the principles and elements of display (similar to the principles and elements of design) in mind: balance, focal point, scale and proportion, viewpoint, rhythm and motion, harmony, and unity in theme. Elements of color, lighting, space and depth, merchandise, and signs all contribute to the success of a window display.

In-Store Display

Once in the store, product presentation should move the customer throughout the store in a preplanned sequence. For example, some shops position the display refrigerator with fresh flowers at the back of the store. This attraction pulls the customer through the store.

Figure 21-14. In-store displays should move the customer throughout the store.

By planning key displays that attract attention at various points through the store, customers move from one key display to another. Ideally, the traffic flow initiated by interior displays should expose as much merchandise and as many ideas as possible.

Large flower shops and those presenting a wide range of gifts and accessory merchandise often plan displays in vignettes (see figure 21-15). A vignette presentation makes a distinctive statement for a limited line of merchandise. It is a specific area defined by panels or other methods for containing a display. A vignette presents a line of merchandise, a theme, compatible items, or new ideas. Rather than having a central look or theme that goes throughout the store, using vignette displays enables a florist to present many different looks and themes.

All merchandise on display should be clearly priced. Unpriced merchandise frustrates customers and prospective buyers. Ideally, displays should include merchandise and designs in a wide range of prices. Then, customers and prospects who shop the store recognize that the business offers products in all price ranges comfortable to all buyers.

Presenting the full-range of services a flower shop provides is another important consideration in product presentation. The flower display refrigerator should offer an attractive display of fresh flowers and designs. Attractive displays promoting bridal and wedding and party work intrigue many customers. If a shop creates fruit baskets and gourmet food gifts, the ideas should be visually presented in compelling displays. Engaging presentations with silk flower designs and compositions with dried materials often create impulse sales. The purpose of interior display is to show customers and shoppers the full range of products and creative services a professional flower shop offers.

Service Area

When visual merchandising has accomplished its primary goal, the next natural step is servicing the sale. The service area includes the sales counter and the consultation area. Many shops do not have the space needed for separate

Figure 21-15. A vignette display makes a distinctive statement by presenting a single line of merchandise, a distinct look, or a seasonal visual emphasis.
Photo by Brian Yacur.

areas of service. Ideally, the sales counter includes the cash register and wrapping supplies, while the consultation area includes selection guides with a place to sit down. These areas in the shop help customers make selections and purchases.

For servicing telephone customers, some flower shops have phones at all work stations. Large shops usually have an area or room for taking and receiving telephone orders. Materials needed for taking and receiving orders over the phone, such as order forms, pens, selection guides, and telephone books, must be readily at hand.

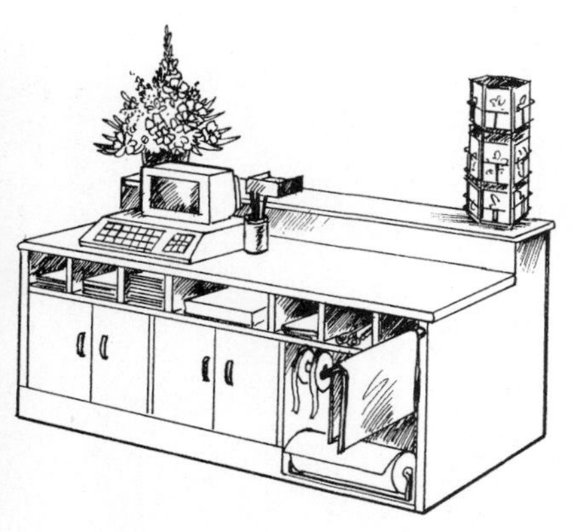

Figure 21-16. The sales counter of a flower shop often includes a wrapping station where cut flowers and potted plants are packaged. This area must remain clean and uncluttered with tools and supplies organized for efficiency.

Work Area

A well-planned flower shop makes working efficient and cost-effective. In designing the plan for a flower shop, cautious attention must be given to the flow of products into and out of the operation.

The work area for designers must be organized. Conditions must be comfortable for employees. Tools and supplies must be organized and at hand. The cooler, sink, and ***hardgoods*** must be located near the design tables to lessen the time needed to gather supplies and prepare arrangements. Organized and uncluttered work stations increase productivity.

Ideally, flowers and other merchandise should arrive in the rear work area or the ***receiving area.*** The office is sometimes in this area to organize and log shipments and deliveries. Everything required for proper care and handling should be convenient, including containers, the appropriate underwater cutting equipment, sinks, and water.

Figure 21-17. The work area for designers must be organized and comfortable for employees. Tools, supplies, and equipment needed to accomplish certain tasks and jobs must be easily accessible.

The ***delivery area,*** also at the rear of the store, is where outgoing orders are prepared for delivery. This is where extra packaging for cold weather may be added and care tags, cards, and water levels are checked. In large shops, the delivery area is filled with shelves where arrangements are placed according to where they will be delivered—the hospital, hotel, funeral home, and various parts of the city.

Storage

Adequate storage space for hardgoods and seasonal supplies is essential. Items will last longer and keep cleaner. When extra supplies, seasonal items, and window-display props are organized and out of the way, workers are able to free themselves of boxes and clutter and focus on the jobs at hand.

Refrigeration

The type of refrigeration unit a shop has often is determined by the volume of customers. Whether a shop has mostly phone customers or has a significant percentage of in-store customers will influence the type of cooler necessary. Whether flowers are sold mostly by the stem or in arrangements also will influence the type of cooler a shop needs. Coolers fall into two categories: storage and display. Large walk-in storage coolers hold flowers in buckets of preservative solution for use in arrangements, store boxes of fresh greenery, and keep arrangements cool and fresh while awaiting delivery, .

Display coolers are a form of visual merchandising. These coolers exhibit flowers by the stem, floral arrangements ready for sale, and floral arrangements ready for

Figure 21-18. Display coolers are a form of visual merchandising. They exhibit flowers by the stem and arrangements for sale.

Figure 21-19. Regardless of size, every full-service flower shop has six distinct roles and responsibilities that must be addressed: owner, manager, sales, design, delivery, and accounting.

delivery. A variety of sizes and types of display coolers are available, meeting the needs of every kind of floral shop and mass merchandiser.

EMPLOYEES AND RESPONSIBILITIES

In every flower shop, regardless of size, sales volume, and number of employees, there are six distinct roles (see figure 21-19):

1. Every shop has an owner.
2. Someone must manage the business.
3. Products must be sold.
4. Most items require a designer to complete.
5. In most shops, the majority of floral orders are delivered.
6. In all shops, someone must be responsible for the accounting.

In smaller shops, the same person fills several of these roles. While in larger operations, several people may complete each area of responsibility.

MARKETING

Marketing is everything a flower shop does to find customers, serve them professionally, assure unconditional satisfaction, develop customers with more frequent buying patterns, and maintain customer loyalty. The primary purpose of marketing is to create and keep customers.

Marketing includes all advertising and promotion directed to finding, developing, and keeping customers. Marketing and advertising should be focused on known buyers and targeted prospective buyers to influence them to buy, whether the media is direct mail, newspaper, radio, television, door-to-door flyers, statement stuffers, or product samples (see figure 21-20). Thanking customers for their business, reminder services, and everything else a florist does to assure satisfaction and secure buying loyalty are included in marketing as well.

More flower shops are design oriented, rather than market oriented. Yet, without marketing, a florist fails to develop a loyal customer base to buy the design talent. In many flower shops, the lack of attention to aggressive marketing stymies the growth and future of the business. Florists often say that they depend on word-of-mouth advertising to build their operations. While positive word-of-mouth marketing is important for success, it fails to provide enough consistent impact to influence sufficient potential customers to make a flower shop successful.

Credit cards are a distinct advantage in operating a profitable flower shop. In many areas, florists complete the majority of their sales with customers who use credit cards. This operational advantage becomes a serious marketing disadvantage unless a flower shop maintains an up-to-date customer list. Because of competition, a flower shop must keep in consistent contact with its customers through direct mail to develop and keep customer loyalty.

Recent consumer research conclusively identifies that the primary reason a customer leaves one flower shop to buy from another shop is because of lack of attention. When their purchases and buying loyalty are not acknowledged, customers feel that a flower shop does not appreciate or need their business. These important buyers do not become bonded to a specific retailer if the florist does not keep in contact with them. Unless ***customer bonding*** exists between a customer and a specific flower shop, the customer looks for another florist who will acknowledge their purchase loyalty.

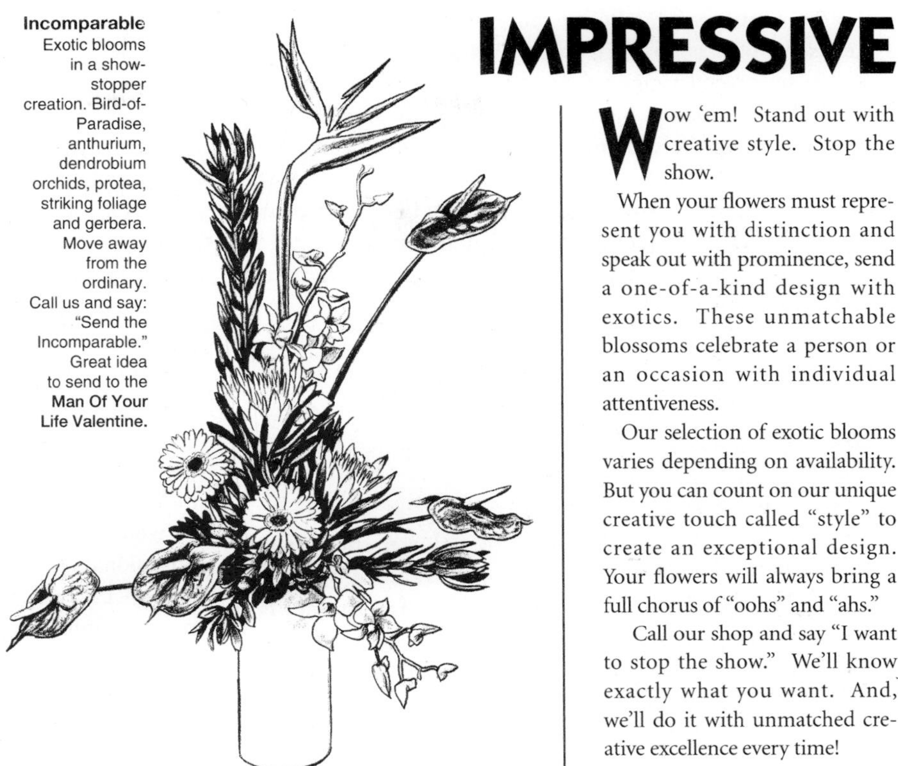

Figure 21-20. In the competitive marketplace of flower retailing, keeping consistent contact with customers and key prospects by promoting the products and services a shop offers is essential for growth and success.

Figure 21-21. Offering complete credit-card services is a necessary requirement in servicing customers professionally and in improving the flower shop's cash flow. Photo by Brian Yacur.

To be successful, a flower shop must have a significant and solid base of loyal customers who consistently make purchases several times a year. The vital necessity of this loyal customer base is often overlooked by the retailer who does not recognize the importance of consistent customer communication. Marketing is the complete cycle of attracting, serving, satisfying, and keeping loyal customers who buy flowers frequently. The keeping-customers aspect of marketing is as important, and perhaps more important, than seeking new customers.

SALESMANSHIP AND CUSTOMER RELATIONS

The moment a customer steps into a flower shop or is greeted on the telephone, ***customer service*** begins. Customer service does not conclude until the customer is unconditionally satisfied.

Customer service goes far beyond the tangible services given. The attitude of the employee performing the service makes a greater lasting impression in the customer's mind than the actual service. Ideally, every flower shop employee provides flawless service with a positive, friendly attitude.

It is the customer, and only the customer, who makes a flower shop successful. As trite and overworked as this statement might be, it is still a fact of business success. Everyone who works in a flower shop must understand that customers make all jobs possible. Customer service is a means of ensuring business success and work opportunity for every employee.

Personal Requirements for Selling

Salespeople must be properly trained to effectively and professionally present the products and services a flower shop offers. Training salespeople is an essential management responsibility. Salespeople must be friendly and enthusiastic and be able to communicate well. To serve customers competently, salespeople must be knowledgeable about the products available, design suggestions, the unique services a flower shop provides, and appropriate price ranges. A flower shop salesperson plays a consequential role in the success of the business.

Selling is the process of influencing a buying decision. Selling is giving the customer the confidence to exchange his or her money for the products and services a flower shop provides. This is the voice-to-voice interaction of the salesperson and the customer on the telephone, or the face-to-face meeting of customer and salesperson in a store situation.

Two types of salespeople work in flower shops. The ***passive salesperson*** looks upon the selling responsibility as recording what the customer requests—simply taking the order. The passive salesperson gives little effort helping the customer make an appropriate buying decision and influencing the purchase. In contrast, the ***assertive salesperson*** recognizes that professional selling is a process of asking questions, listening to the customer's responses, and then recommending the appropriate suggestion to fulfill the buyer's specific needs.

It might seem that the passive salesperson is more customer-oriented, less aggressive, and, therefore, desirable to the buyer. Actually, the opposite is true. Few customers are well-informed about flowers, floral design, and the professional service a florist provides. Many buyers are hesitant to call a flower shop because they don't want to appear uninformed and feel embarrassed by their lack of flower knowledge.

The assertive salesperson creates a more comfortable buying climate for the customer. By asking questions, listening to the responses, and then offering the customer appropriate options, the assertive salesperson focuses on the customer's needs. Most customers welcome personalized interest and suitable suggestions. They want options from which to make a buying decision. This caliber of professional selling is an important value-added service the professional flower shop provides.

Telephone Selling

In most flower shops, the majority of orders, approximately 80 percent of the sales, come to the shop over the telephone. The business image these buyers develop for a flower shop is the result of what they hear and perceive from talking to the telephone salesperson. Additionally, some 65 percent of the orders placed by a customer are delivered by the flower shop and never seen by the buyer. Customer surveys conducted by some florists revealed that over 50 percent of the buyers had never been in the flower shop. The primary image these buyers perceived for the shop was communicated by the telephone salespeople. This is the major reason professional, caring selling skills are essential in every successful flower shop.

Telephone sales include taking traditional orders from customers and receiving and sending wire service orders. It is essential for salespeople to be familiar with the steps of

Figure 21-22. In most flower shops, the majority of sales come over the telephone. A telephone salesperson must be professionally trained to present effectively the flower shop and the products and services it provides. Photo by Brian Yacur.

FLOWERS BY CINDY
123 Lassonde Way
Albany, New York 12205
Phone 555-4567

DELIVER TO			PHONE NO.
ADDRESS			DELIVERY DATE
			S | M | T | W | T | F | S _____ A.M. _____ P.M.
WIRE ❑ IN ❑ OUT	ASSOCIATION	CODE NO.	CALL TAKEN BY
FLORIST			PHONE NO.
ADDRESS			

❑ ARRANGEMENT ❑ SPRAY ❑ CORSAGE ❑ CUT FLOWERS ❑ PLANT

TAX	
OCCASION TOTAL	
CARD	
CHARGE TO	ORDERED BY
ADDRESS	DATE OF ORDER
	PHONE NO.
CREDIT CARD NO.	EXP. DATE

❑ CASH ❑ C.O.D.
❑ CHARGE ❑ NEW ACCOUNT

Figure 21-23. Order forms must be used for traditional orders and wire-service orders. A variety of forms are available that keep information organized. These forms remind salespeople of all the information that must be gathered before closing the sale.

taking telephone orders to work efficiently while on the phone.

The salesperson should answer every phone call promptly. A friendly greeting that includes the name of the shop is essential. The salesperson answering the phone should personalize the call by telling the customer his or her name. The initial greeting might be, "Good afternoon. Aloha Flowers. Julia speaking. How may I help you?" This approach to selling helps direct the conversation. Questions, such as, "Do you have a color preference?" or "Would you like those arranged in a basket or vase?" will help to quickly determine the desired item and will aid the customer in describing what he or she wants.

Next, the salesperson discusses prices by offering a range of choices. The customer should be made aware of any delivery, wire, or service charges that will be added to the price of the order.

The sales staff then discusses the type of card and the message the customer would like included. It is important to have names spelled correctly.

Then, the salesperson obtains complete delivery information: the date for delivery and the recipient's name, address, and phone number. Often there are

Figure 21-24. Customers who come to a flower shop for personal service deserve the attention of a professional, well-trained salesperson.
Photo by Brian Yacur.

special delivery situations that must be discussed. Other information, such as hospital room number or business office hours, will assist in efficient delivery.

When closing the sale, the salesperson must discuss the method of payment and complete all the necessary information on the order form. This includes the customer's complete name, address, and home and work phone numbers. When the customer uses a credit card, vital information must be obtained: the type of card, the account number, and the expiration date. The salesperson should repeat numbers back to ensure accuracy. When the method of payment is established, the salesperson should ask the customer if anything else is needed or if there are any questions. The salesperson should thank the customer for the order, invite future purchases, and end the call. It is important to let the customer hang up first, in case the customer has any last requests or questions.

The procedure for receiving a wire service order is similar to taking traditional orders from customers; however, there are a few exceptions. The caller will identify the order as a wire service order, then give the name of the shop, city and state, and wire service account number. The caller proceeds to give a description of the desired floral item, and a second choice or an alternative style of arrangement. The caller gives the purchase price, the delivery fee, and then gives the total price (purchase price plus delivery). The information regarding delivery, including the date, and the recipient's name, address, and phone number are given next. The type of card and the desired message is then given to the salesperson. The call is ended by exchanging names. Other information needed on the order form include the date the call was received and the name of the salesperson who received the call.

In-Store Selling

Although telephone sales account for the majority of most flower shops' sales, in-store customers contribute to the success of a business. Competent and friendly sales personnel help ensure happy and buying customers. Personal interaction with customers allows the salesperson the opportunity to adjust the floral item or service to the needs of the customer. This may be done by observ-

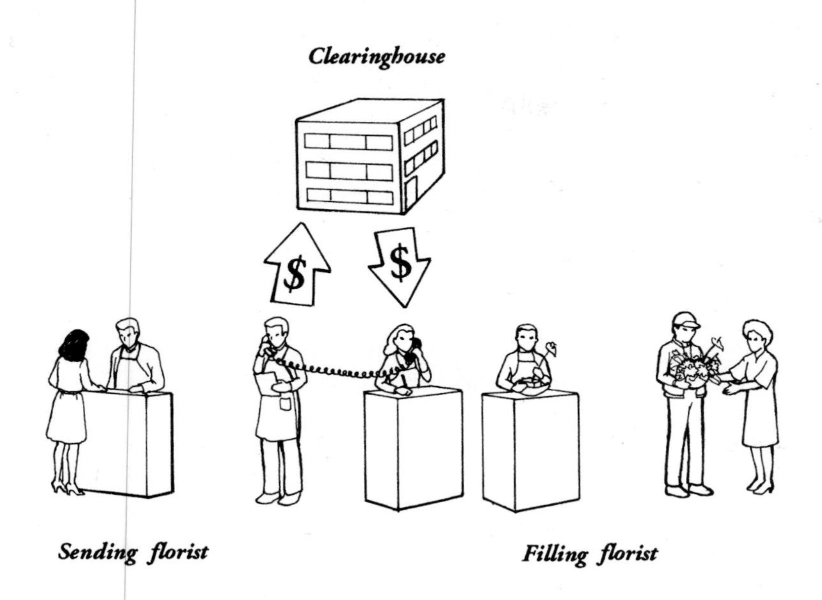

Figure 21-25. Many wire-service companies make it possible to send flowers anywhere. A customer in one city selects and pays for a design to be sent to someone who lives in another city, state, or country. The florist who takes the order (sending florist) chooses a florist from the wire-service directory. He then calls that flower shop (the filling florist) who records the information, designs and fills the order, and is responsible for delivery to the recipient as instructed. Each wire service has a head office or clearinghouse, which processes the transactions that take place between florists and ensures that each flower shop receives proper payment. The sending florist charges for the arrangement, a delivery fee, service fee, and applicable tax. Depending on the wire service, the sending florist retains the service charge and about 20 percent of the order. The clearinghouse retains a small percentage, and the filling florist receives about 80 percent. The filling florist is required to give 100 percent value.

ing and listening to the customer. Customers coming into a floral shop may be classified in three ways: decided customers, undecided customers, and browsers.

The ***decided customer*** has a definite floral need. This customer already may know exactly what he or she wants to buy, including flower types, colors, and style. For instance, a customer may want to send a floral arrangement to her father, who loves bird of paradise and other tropical flowers, to celebrate his birthday. Another customer may want to send an arrangement of two dozen pink roses and baby's breath to his wife, who just had twin girls. The decided customers are the easiest to sell and do not need to be persuaded to buy.

The ***undecided customer*** has a floral need, but has not determined exactly what he or she wants. This customer is the one who needs the most personal assistance from the sales staff. For instance, a customer wishing to send something to her sick friend in the hospital, or the young man who needs to order some kind of corsage for his date for the prom, may both need to look at a selection book and the flowers on display. Both may need helpful suggestions from the sales staff.

Many customers who enter a flower shop are interested in flowers and the merchandise on display, but do not have a specific need to buy at the time. The window display, advertising, or word of mouth may have enticed them to come into the store. When approached by the sales staff, ***browsers*** often respond that they are "just looking." These customers are hardest to sell. They may be looking for an idea or a bargain, or shopping prices. They are not interested in being pressured to purchase anything. This person is a potential customer and may be persuaded to make a purchase while browsing or return to make a purchase. For instance, a bride-to-be may be anonymously shopping around, looking at several floral shops' merchandise, style of design, prices, and service. Just by the way she is treated as an "unknown" customer may be the key factor that brings her back to a certain shop to order her wedding flowers.

WIRE SERVICE

The mechanics of ***flowers-by-wire*** are a mystery to the person who is not familiar with the process. A flower shop in one city sells an order for delivery in another city, state, or country. The customer pays for the specific flowers ordered, a delivery charge, a service charge, and appropriate sales tax. The florist who sold the order, the ***sending florist,*** sends the order to a florist in the city of delivery. This flower shop, the ***filling***

florist, fills and delivers the order as specified by the sending florist.

The commonly used term flowers-by-wire is misleading. In the early days of the floral industry, flowers were transmitted by wire or telegraph; today, out-of-town orders are transmitted by telephone, computers, and FAX machines. The transfer of orders between the shop that originates the order and the one that fills and delivers it, represents significant sales volume for many flower shops.

There are many wire-service companies competing for a flower shop's business. Florists' Transworld Delivery Association (FTD) is a member-owned organization. Therefore, florists who use the FTD service are members of FTD. All other wire services are independent businesses. American Floral Services (AFS), Teleflora, Carik Services Inc., Redbook Florist Services, Florafax International, and others also provide the services that enable florists to transfer orders. Florists subscribe to these wire services. A flower shop can be a member of FTD and subscribe to one or more other wire services.

Each wire service is comprised of florists who, as members or subscribers, send and receive floral orders to other shops on a daily basis. Each wire service has a head office or ***clearinghouse,*** which processes the transactions that take place between florists. Each month, the wire service prepares a statement for each flower shop. The clearinghouse acts as a bookkeeping service, ensuring florists that each shop will receive proper payment for transactions.

In recent years, another dimension has developed in flowers-by-wire. A service called 1-800-FLOWERS and similar services market with national television advertising and other media to influence buyers to call them directly to place their flower orders. Though these services are only sales agents that transfer orders to flower shops for filling, they function as sending flower shops.

BUYING AND PRICING

Most florists buy products from their local wholesalers and from out-of-town suppliers. Both perishable and hardgoods products are available from many sources. The buyer for a flower shop must locate the very best supply sources for the specific operation. With perishable products it is essential for the buyer to be connected with reliable sources that understand the shop's requirements for selection, quality, freshness, and price. Buying involves researching markets and suppliers, gaining product knowledge, developing a purchase plan, and controlling inventory.

After making a desirable purchase, ***profit control*** begins with pricing (determining the retail price of products). There are no general guidelines or infallible formulas for pricing. Every flower shop has specific pricing requirements based on the type of operation, expenses, merchandise, and profit requirements. Therefore, profitable pricing demands that the owner or manager understands the complete costs of operating the business. Without this knowledge, pricing is only an estimate that seldom produces a profit.

Retail florists typically price their products and services on the basis of a predetermined ***ratio markup,*** which is figured on the cost of merchandise sold. For example, a 2-to-1, 3-to-1, and 4-to-1 markup on an item that costs $1 wholesale would give a $2, $3, and $4 retail price, respectively. The problem with this markup system is its failure to control and maximize net profits. This system does not consider the impact of operating expenses, overhead costs, and the shop's net profit goals, relative to the ultimate retail price. Attaining desired net profit goals necessitates retail pricing policies that utilize a ***percentage markup*** to fully cover overhead and operating expenses, the cost of merchandise sold, and the individual shop's net profit goals.

Because floral design is a manufacturing process, the products and labor that go into any design must be controlled. This means charging a profitable price for every item included in a design, plus an appropriate charge for design labor.

Profitable flower shop managers use several methods of ***design control.*** For standard, ***production designs*** in which the same design with the same contents is repeated, a content list and labor charge for each design is prepared. From this design proforma, the designer knows the exact amount of materials to select.

Every shop uses production design throughout the year. Many holiday designs featured by a flower shop are production designs. Some of these designs are dictated by wire services. Other shops prefer to create their own featured holiday specials rather than promote designs available in every flower shop. In any production design, the selling price, the contents, and the design labor charge are predetermined before production begins. For example, producing ten centerpieces for a party requires a content plan for each design. The designer then repeats this plan ten times. Many areas of professional design require a prototype and a master design plan that is repeated several times.

Preplanning designs enables the manager to predetermine the exact amount of materials to order. By making a design plan for each item required in a wedding or other special event, the manager can intelligently plan product needs and control the use of materials. Unless a flower shop manager carefully plans and controls the materials required for a certain occasion, the job could be a profitless experience. Likewise, the amount of labor that goes into weddings, parties, and other events must be anticipated and charged for when pricing the work. Equally important, the labor expended must be controlled to make the job profitable.

Most everyday floral designs sold and created by the typical flower shop are produced to presold price points, for example, $20, $35, and $50. The designer who understands profitable design knows that the contents of the design must be predetermined based on the selling price. The contents are then selected and placed in the design (see figure 21-26). The sequence is essential. The first step is to figure out the exact contents of the design based on the selling price. Then, select the perishable materials and place them in a container with water on the design bench while gathering the hardgoods. Next, create the design with the preselected materials. This disciplined process for creating each design controls the exact contents that go into a design. Without content control, it is impossible to operate a profitable flower shop.

Some orders are "open," meaning the designer decides both the contents and the price. Even for the ***open order,*** materials must be preselected, the selling price must be determined based on these materials, and an appropriate charge for design labor must be accessed before the designer creates the design.

DESIGN CONTROL WORKSHEET

Invoice Reference G17567 Designer Joel Date 3/27/94

Type of Arrangement/Notes: Carnations and Lilies in Basket-SmartStyle

Retail Selling Price $ 50.00

Quantity	Description	Price Each	Retail Value
	Labor @ 20 %		$ 10.00
	Container: Whitewashed Basket		5.00
	Filler: 1 Block Floral Foam		1.50
4 pcs	Leatherleaf	.25	1.00
6 stems	Eucalyptus	.50	3.00
2 stems	Star Gazer Lilies	5.00	10.00
11	Deep Red Carnations	1.50	16.50
1/2 bu	Heather	6.00/bu	3.00
		TOTAL RETAIL VALUE	$ 50.00

TOTAL LABOR AND MATERIALS AT RETAIL VALUE

Total Labor	$ 10.00
Total Fresh and Greens	33.50
Total Container	5.00
Total Hardgoods	1.50
Total Plants	
Total Silk	
Total Dried	
Total Holiday Specialties	
TOTAL RETAIL VALUE	$ 50.00

Figure 21-26. A designer must be responsible for planning the contents of every flower arrangement so the design makes a responsible profit for the flower shop. A design worksheet enables the designer to plan the contents of a design. It becomes a control for the profitable use of both products and design labor.

The most risky and unprofitable way to produce a design is to create it without content preplanning. This procedure requires the designer to count and price everything used after the design is completed. Not only is this method time-consuming, it is an inaccurate control process that creates profit problems for many flower shops.

Even more disastrous is the design procedure of creating a design to look like a certain price range without counting the materials used. Floral design is a manufacturing process. Therefore, a designer must be responsible and accountable to charge for every item used in a design plus charge an appropriate fee for labor.

DESIGNING

A serious problem in many flower shops is the undisciplined work habits of designers. The profitable designer works with an explicit control system requiring that the contents of each design are recorded and priced before creating it. Those designers who create and then try to justify the design to the price without an itemized list of contents seldom make a responsible profit for the flower shop.

Equally important as design control and pricing for profit, a designer has a responsibility to the customer. The design created must be appropriate from the customer's perspective and satisfy without compromise both the sender and the recipient.

Customer research identifies the two key factors that influence customer satisfaction in floral design. First, customers react positively to a design that is pleasing and "comfortable to the eye." This means that unless the creative expectations of the customer are well known, the more conservative approach to professional design is the best decision. Second, flower shop customers expect a design to provide appropriate visual value. A floral design must look the price.

Both customer-satisfaction factors are subjective. A design favorable to one customer might be unattractive to another. The expectations for appropriate visual value are sometimes difficult to define. One customer may want a design to be as large and showy as possible, while another customer, spending the same amount of money, may expect the design to be quietly elegant. Understanding these differences in customer expectations is an essential responsibility of a floral designer.

DELIVERY

Although it is an essential value-added service for a professional flower shop, delivery is a difficult function to manage. Most customers have immediate, same-day delivery needs. Since a flower shop is a manufacturing operation, orders must be sold, designed, and delivered within a short period. Managing the flow of orders so they can be delivered as required is a challenging task demanding time-related efficiency. Seldom is there a systematic flow of orders in the typical flower shop. The nature of a florist's on-demand service means that employees must be responsive to the pressures to produce within the delivery requirements of the day.

Like a salesperson, the individual who delivers floral arrangements is an important spokesperson for a flower shop. The quality of the shop and the implied quality of the products and services provided are communicated by the appearance of the delivery person and the delivery vehicle, as well as the attitude with which the delivery is completed.

Florists use three primary methods of delivery: self, ***delivery cooperatives,*** and commercial. Some florists deliver all their orders with their own trucks and drivers. These retailers do not want to release the control and responsibility for delivery to another florist or an outside service. They believe that it is important for their truck bearing their shop name to make all deliveries (see figure 21-27).

Figure 21-27. Most floral shops use their own delivery van to make all deliveries. It usually bears their shop's name. Especially for You, *Salt Lake City, Utah.*

These florists prefer to have complete control over the delivery of their orders and the personnel who deliver them.

Other flower shops belong to pool or cooperative (co-op) delivery associations. In this method of delivery, a florist brings its deliveries to a central location by a specific time. Deliveries are sorted according to delivery areas. Each florist in the pool completes deliveries to a defined area. Florists who belong to delivery pools and co-ops find that this method lowers the cost of delivery.

Some florists use commercial delivery services. These specialized delivery companies pick up deliveries at a flower shop at certain times each day. Usually, a commercial delivery service covers a far greater area than the delivery area of the florist or delivery cooperative. This enables a flower shop to deliver to an expanded area. A commercial delivery service usually provides delivery for less cost than a florist making its own deliveries. However, florists using a commercial delivery service maintain at least one vehicle for special delivery requirements.

The cost of delivery is a significant expense in operating a flower shop. Although most shops charge for delivery, seldom does this charge cover the actual cost.

MANAGEMENT

There are three critical responsibilities of managing a flower shop. The successful manager must:

1. supervise people
2. control the buying, use, and movement of products
3. administer the flow of money into and out of the business

People, products, and money are the three key areas of flower shop management.

People

Supervising people begins with hiring—having the right people—and then properly training each employee to carry out his or her responsibilities efficiently and productively. Since a professional, full-service flower shop sells, manufactures, and delivers, people must be qualified to fulfill their responsibilities with cost effectiveness. Employees must be accountable for profitable performance.

Operating a profitable flower shop requires focus on the two major areas of cost: labor cost and product cost. In every flower shop operation, failing to control these two costs causes the business to be unprofitable.

To control the cost of labor, employees must be competent to produce commensurate with their wages. A profitable operation measures the productivity of employees by monitoring what they do (see figure 21-28). The manager must know how much a salesperson needs to sell each week to be a profitable employee. Records must be maintained to verify performance. Also, appropriate production requirements must be determined for designers and records must be kept to evaluate performance. In small flower shops, there is a significant crossover between jobs. In these situations, adequate performance standards must be established to evaluate performance. Every profitable flower shop knows the exact relationship that must be maintained between payroll costs and production. The management responsibility is to control labor so that this cost-to-production relationship is kept in profitable balance.

An effective manager communicates clearly. Regardless of the size of the flower shop and the number of employees, people want honest and dependable answers about expectations, evaluations, and rewards. Finally, managing people requires the ability to inspire, motivate, and develop people.

Product

A flower shop owner or manager controls the purchase and movement of two types of products: perishable and hardgoods. Products must be purchased, priced for profit, and then managed astutely through the product turnover cycle.

Buying the right product, whether perishable products or hardgoods, is only the beginning of product management. Every purchase of product represents an investment. The primary purpose of this investment is to provide a profitable return. Many shops call this management of the inventory responsibility profit control. Because of the perishable nature of many products sold in a flower shop, controlling inventory is a major management responsibility in operating the business profitably.

Money

A flower shop must have ample money to support the operation. In accounting terms, this is known as ***working capital.*** To operate efficiently, a flower shop must have adequate cash flow to have money available to buy products, pay suppliers, and cover the operating expenses, including payroll.

PRODUCTION REPORT

Employee____________________ Date __________

ORDERS SOLD	ORDERS DESIGNED	ORDERS FILLED FROM DESIGN INVENTORY	PLANT ORDERS FILLED	SPECIALIZED DESIGN	
				TYPE	DOLLARS

OTHER WORK ACCOMPLISHED NOT RELATED TO ORDERS:

PROJECT	TIME SPENT

HOURS WORKED ________

SUMMARY

Orders Sold $ ________

Orders Designed $ ________

Orders Filled From Design Inventory $ ________

Plant Orders Filled $ ________

Specialized Design $ ________

TOTAL FOR DAY $ ________

Figure 21-28. All flower shop employees must be accountable for productivity. Profitable flower shops require each employee to account for the work accomplished by keeping a daily production report.

Responsible money management requires a written financial operating plan. Often, a florist is not skilled in money management. Administering the finances of any flower business demands timely attention to all financial details. Unless a shop is large enough to maintain an employee trained and qualified in financial management, it is imperative the business work with an outside accountant or accounting service. Also, the accounting work must be completed on a timely basis each month. The cost of this financial service is a wise investment, pays for itself, and returns valuable dividends.

REVIEW

The retail flower shop is a method of distribution that connects the final consumer with flower products and creative services. As a link in the distribution channel, a flower shop must enhance the product with value-added services that customers find desirable.

The type of operation a flower shop owner wants to develop and the target market to be reached dictates the style of the business. In the marketplace, there are specific niches of opportunity for full-service flower

Figure 21-29.
Herman © 1981, Jim Unger. Reprinted with permission of Universal Press Syndicate.

shops, specialty shops, limited-service flower shops, and flower merchandisers. The type of business often prescribes the ideal location for the shop.

Although knowledge and skill in design work are essential, they are not the key to running a successful flower shop. Other vital factors include knowledge, experience, implementation of visual merchandising and marketing, efficient shop layout, employee responsibilities, salesmanship, customer relations, buying and pricing, and wire and delivery services.

TERMS TO INCREASE YOUR UNDERSTANDING

assertive salesperson
browsers
clearinghouse
customer bonding
customer base
customer service
decided customer
delivery area
delivery cooperatives
design control
filling florist
flowers-by-wire
full-service
hardgoods
limited-service flower shop
management
marketing
merchandiser
open order
passive salesperson
percentage markup
professional
profit control
production design
ratio markup
receiving area
selling
sending florist
specialty flower shop
strip-center flower shop
undecided customer
vignette
visual merchandising
window display
wire service
working capital

TEST YOUR KNOWLEDGE

1. Explain the function of a retail flower shop.
2. What is value-added service?
3. Outline the differences among a full-service shop, a specialty shop, a limited-service shop, and a flower merchandiser.
4. What are the four primary goals of displays?
5. What is the general sequence of taking a traditional order or receiving a wire service order?
6. How do passive and assertive salespeople respond to selling and customers?
7. How are in-store customers classified?
8. Describe how a wire service functions.
9. What are the key responsibilities of a designer in creating a profitable design?

RELATED ACTIVITIES

1. Visit several different types of flower shops in various locations. Describe the pros and cons of your observations.
2. Plan a window display around a month, season, or occasion. Establish how it might be different for automobile and foot traffic.
3. As a group, present merchandise in vignette. Discuss how the primary goals of displays are met.
4. Visit a floral shop. Notice how key displays influence traffic flow in the store.
5. Practice salesmanship through roleplay with several types of customers and with passive and assertive salespeople.
6. Fill out a design control worksheet. Make the planned arrangement.

Chapter 22

Careers and Continuing Education

F*loriculture* or the ***floral industry*** offers a variety of rewarding and profitable careers. As with other industries today, there is an increasing need for trained professionals. Those who have business skills, horticultural knowledge, and communication or "people" skills are in demand by employers throughout the floral industry.

Segments of the floral industry include the retail florist, the ***wholesale florist,*** and the grower. Some of the more known traditional positions include the floral designer, the wholesale salesperson, and the crop production supervisor. Various newer occupations have developed in recent years, within each area of the industry, such as ***commentators, brokers,*** and genetic engineers. Floriculture offers many specialized occupations that require various levels of education and experience.

Because this book is geared to design, the emphasis on careers in this chapter will be on-the-job opportunities for the ***floral designer.*** A brief overview of other floral industry occupations, those at the retail, wholesale, and grower levels will be covered as well.

The success of any career may be strengthened through ***continuing edu-***

Figure 22-1. The floral industry depends on research and education for advancement. There are career oportunities in tissue culture and genetic engineering in the development of new varieties. Post harvest care research helps lead to solutions that will help increase the longevity and quality of flowers.
Flower Council of Holland.

cation. In the floral industry, as with other industries, there are professional and ***trade associations, trade publications,*** educational workshops and seminars, certification and training programs, and college education programs. All of these are designed to increase the knowledge and skills of an individual, as well as improve the profession. To become a talented professional in any occupation, it is important to make a lifelong commitment to learning.

CAREER OPTIONS FOR THE QUALIFIED, PROFESSIONAL FLORAL DESIGNER

In the retail flower industry, there are three areas of employment opportunity for a floral designer—shop

Figure 22-2. There are many career opportunities for the qualified, professional floral designer.

owner, shop manager, and shop employee. The intellectual understanding of floral design principles and techniques and developing practical ***working skills*** as a designer are essential masteries for the individual who chooses professional floral design as a career goal. Once an individual moves beyond the design responsibility into shop management or ownership, then marketing, business management, and people skills are mandatory.

Each area of opportunity for a professional floral designer requires different credentials. Starting with a solid base of knowledge in the principles and techniques of floral design gives the foundation for skill development. Then, experience is the pivotal factor that opens opportunity.

In gathering working experience, it is a good idea to choose an employer carefully. Two types of flower shops make up the business population—the professional and the hobbyist. The career-oriented floral designer needs to connect with a professional flower shop or floral operation. In this work environment, the exposure to the world of floral design, plus the requirements for profitable performance provide the caliber of experience that molds talent.

Opportunities in a Full-Service, Professional Flower Shop

Depending on the size of the business, a flower shop requires design skills in three areas—general design work, production design work, and head or managing design.

The major opportunity for employment as a floral designer is with a full-service flower shop, serving customers with a complete range of creative design work. Many designers prefer this ***work environment.*** It allows them to express their creativity through all types of designing from the everyday hospital design to large wedding and party designs.

A small flower shop needs at least one qualified floral designer, while a larger flower shop might employ two or three designers. A very large operation requires a head or managing designer who supervises a team of five to twenty floral designers. Most full-service flower shops employing more than three designers require a design manager to manage the flow of design work and the productivity of the designers.

Some larger flower shops often employ ***specialized designers*** who focus on one type of work. A shop that experiences a large number of weddings might have one designer who has expertise creating flowers for the wedding party, and another designer who is skilled in wedding reception decorations. In some shops, designated employees design casket sprays and other types of funeral work. The party florist often has a designer talented in planning parties and decorations. Once the party is planned and design prototypes are prepared, a team of designers prepare the decorations. Flower shops that do important volume in silk flower and dried designs, often have one designer who specializes in this work.

Some floral organizations with multiple shops have a ***central design*** location. All designs for delivery are created in the central design location rather than at the individual stores. A central design facility uses a team of designers supervised by a design manager. This method for filling orders enables a florist to maximize production and control the flow and use of merchandise.

Larger flower operations tend to employ ***design specialists*** who concentrate on one or perhaps two areas of design work. For example, one designer makes up vases of roses all day. Another designer arranges creative plant gardens. Still another arranges spring flower gardens with an array of colorful seasonal blossoms. This specialization enables the shop to produce designs more efficiently and cost-effectively.

Smaller flower shops need a ***design generalist,*** or one who can make basic, everyday designs, and is efficient in all areas of design creativity. In this shop environment, most designs are one-of-a-kind. This approach to floral design is creatively rewarding for many designers.

Opportunities in a Limited-Service Flower Shop

The philosophy of the ***limited-service shop*** is to focus on areas of design where the business can make a responsible profit. In general, this type of operation usually avoids labor-intensive work like weddings, funerals, parties, and decorations. Designers who work in this limited-service retailing environment must be happy and creatively fulfilled with a restricted opportunity to use their skills. For some floral designers, this limited approach to design is desirable. It often eliminates the creative and production pressure associated with catering to special events.

Opportunities in a Floral Department

The floral department in a modern supermarket or mass merchandiser usually combines on-site design work with merchandising of flowers and plants. The on-site designer creates ***custom work,*** makes up fresh flower designs for the refrigerator presentation, maintains the displays of flowers available by the stem and in assorted bouquets, and keeps the displays of flowering and foliage plants attractive.

Some floral departments offer a complete selection of designs for hospitals, sympathy, weddings, and parties. Other floral departments limit design work to general designs appropriate for hospitals, birthdays, anniversaries, and everyday celebrations. Floral departments that service the complete flower needs of customers often employ several designers.

Most supermarkets and mass merchandisers work from a ***distribution center*** where bouquets of fresh flowers are packaged and standard designs created in ***production-line design.*** These products, loose fresh flowers, and plants are delivered to stores from this distribution center.

Working in central design at a distribution center is another job opportunity with supermarkets and mass merchandisers. In a well-managed distribution facility, designers do production-line design based on a work plan to achieve maximum efficiency for controlling both materials and labor. Often, production-line designers in a distribution facility create hundreds of the same design.

Opportunities with a Wholesaler

The more aggressive floral wholesale house uses a qualified floral designer to create designs to sell to florists, gift stores, mass merchandisers, and other retailers. Some wholesalers concentrate on designs with silk flowers and dried plant materials. Other wholesalers also create fresh flower designs for their customers. Most wholesale houses that carry Christmas items use a designer to show prospective customers new ideas for creating designs with the seasonal merchandise.

Many retailers find they cannot afford to carry a complete line of silk flowers for creating silk designs. These florists find buying silk designs from their wholesaler is a more profitable way to inventory this type of merchandise. This developing trend for selling pre-made designs means that many wholesalers will be employing professional floral designers.

Figure 22-3. Standard floral designs are mass produced in central-design at a wholesale distribution center.
Evergreen Wholesale Florists Inc., Seattle, WA. Photo by Raymond Gendreau.

Most of the design opportunities available at wholesale houses require outstanding skills in product presentation and display. Besides creating designs for resale, floral designers in this work environment must plan, install, and maintain displays for the entire merchandise line.

Some wholesale houses present design schools and design classes for their customers. Normally, students pay to attend these classes. This means that the instructor must be a qualified teacher-designer. The designer with creative skills, teaching skills, and the ability to communicate effectively is a desirable candidate for working in a wholesale house.

Figure 22-4. Designing display booths for conventions, product presentation, and product demonstration are responsibilities of designers who are employed by manufacturers and suppliers to the floral industry.
Display table designed by Lyndon Holt for Excelsior Plastics Industries.

Opportunities with a Manufacturer or Floral Supplier

Those manufacturers and suppliers who cater to the floral industry often employ full-time, in-house floral designers, or contract for these creative services. The designer's responsibilities in this work environment often include product development as well as floral design. To function in product development, a designer must have extensive experience in floral design plus well-developed creative skills. The ability to express new ideas in products is an essential skill. In addition, in-depth knowledge of the floral industry and trends in products, and insight about opportunities for new product development are prerequisites.

Floral designers who work with manufacturers and floral suppliers must have keen product-presentation skills. Most of the opportunities in this area of employment require experience in presenting products, designing display booths for conventions, and demonstrating products.

Opportunities for the Free-lance Floral Designer

There is an expanding need for ***free-lance floral designers.*** Many flower shops do not maintain a full staff of employees. Instead, they depend on free-lance designers for peak periods of business and special events. Wholesale houses that stage open houses often use free-lance designers. In every area of employment opportunity discussed earlier, free-lance designers are used either exclusively or to supplement the in-house staff.

Being a free-lance designer requires more skill and creative focus than working full-time in one business. A free-lancer is expected to be knowledgeable and experienced in all areas of professional floral design. Since businesses hire free-lance designers for peak periods and specific assignments, they are expected to be extremely productive. Frequently, the work environment is demanding.

OTHER CAREER OPPORTUNITIES WITHIN THE FLORAL INDUSTRY

Although opportunities for the the floral designer have been emphasized in this chapter, there are numerous other occupations at the retail, wholesale, and grower levels. Still other opportunities are available within the floral industry, generally requiring advanced education and experience. Examples of related floriculture professions include a commentator, a writer or an editor for trade publications, a teacher, and a researcher.

The wholesale florist and greenhouse producer also offer career opportunities. Successful individuals in floriculture are those who have experience and knowledge in all aspects of the floral industry. A combination of education and on-the-job training and experience at all levels of the industry will provide you invaluable information to make an informed decision about pursuing a career in the floral industry.

Retail Florist

Positions available at the retail florist include owner/manager, floral designer, assistant designer, salesperson, delivery personnel, interior landscaping and maintenance personnel, and cut flower processor. (For an in-depth look at the retail shop and employee descriptions and responsibilities, see chapter 21.)

Owner/Manager

Most retail floral shops are owner/managed, while larger stores and franchise operations have positions available in many management areas, including purchasing, design and display, financial management, market research, sales management, and advertising.

Before becoming an owner or manager of a retail florist, receiving a formal education with emphasis in ***horticulture*** or floriculture is highly recommended. Courses in business and communication, as well as art and design, will also prove to be beneficial. In addition, attending a floral design school will instill invaluable knowledge and skills. Along with education and design experience, you must obtain training and experience by working several years in a retail florist shop. Experience in all segments—greenhouse, retail, and wholesale—is recommended for a working knowledge of the industry. Education, experience, motivation, and hard work are all prerequisites to become a successful owner or manager at the retail level.

Floral Designer

As outlined earlier, many opportunities are available to qualified, professional floral designers in numerous capacities. Designers must be able to arrange a wide variety of designs, such as corsages, centerpieces, arrangements of all sizes and styles, wedding designs, funeral and sympathy designs, and various special occasion and party designs. In a larger shop, the head floral designer may supervise or manage other designers and assistant designers.

Working in a successful retail floral shop for several years is necessary to become competent and confident in your design abilities and in satisfying customers. Attending a reputable design school is highly beneficial. In addition, knowledge and skills obtained in college courses in floriculture, art, design, communication, and business will prove invaluable in the work environment of a designer.

Salesperson

The sales personnel at the retail level are responsible for taking telephone orders from customers and helping in-store customers with their purchases. Knowledge of flowers and plants is necessary, as well as skills in communications. The sales staff at a retail shop plays a vital role in the success of the business, for they are the people who are in contact with the public and must help to satisfy customers and maintain friendly relations with those who purchase their products.

A successful salesperson will benefit from some vocational or college education, as well as experience gained while working in a retail shop.

Delivery Personnel

The delivery staff at a retail shop is responsible for delivering flowers and plants to their customers. A delivery person is a goodwill ambassador, meaning he or she plays a vital role at the floral shop, acting as its representative. Often the delivery person is the only contact a customer may have with a flower shop. The impression the delivery person makes must be friendly and professional. Often delivery personnel may be required to help set up floral arrangements and background displays at weddings, parties, and other special events.

Delivery personnel are also responsible for picking up flowers, plants, and other merchandise from wholesalers, growers, airports, or other shipping stations. Drivers must have a valid driver's license, know how to drive courteously, and have a good knowledge of the area.

Interior Landscaping and Maintenance Personnel

Some flower shops, especially larger retail operations, may offer services in interior landscaping. An individual who has gained adequate experience and training in the use and care of plants may become an interior landscape designer and consultant.

A college education in horticulture is recommended for those interested in pursuing a career in interior landscaping. Other courses, such as salesmanship, business, art, and design, will prove helpful.

Cut Flower Processor

Large retail operations may employ individuals whose main role is to care for the incoming shipments of cut flowers, foliage, and live plants. While at smaller floral shops, all employees are generally responsible for the care and handling of perishable merchandise. Because flowers and greenery are generally shipped in boxes or wrapped in paper, without water, it is vital that boxes and packages be opened and flowers and greens are processed as quickly as possible. This involves recutting flower stems, placing flowers in buckets of preservative solution, and placing the buckets of flowers in the coolers. Processors help to maintain the highest quality of the flowers for resale to the final consumers. Processors also take care of potted plants and help the sales staff to prepare orders for delivery. Some experience and training at a vocational or technical school in horticulture is beneficial to a flower processor.

Wholesale Florist/Broker

The employment opportunities at the wholesale florist, like the retail florist, are varied. They include owner/manager, stock buyer and control personnel, salesperson, delivery personnel, and cut flower processor. As mentioned earlier, many wholesale houses employ floral designers to display dried and silk merchandise and to arrange fresh and everlasting designs to distribute to retail florists.

Owner/Manager

Small wholesale houses are often owner/managed, while larger companies generally have positions available in many management areas, including financial management, market research, sales management, and advertising. Generally, many years of experience in all segments of the floral industry combined with higher education, motivation, and

Figure 22-5 a & b. Many rewarding positions are available at the wholesale level. A professional wholesale employee is commited to service and gives personal attention to detail.
Evergreen Wholesale Florists Inc., Seattle, WA.
Photos by Raymond Gendreau.

hard work are prerequisites to become a successful owner or manager at the wholesale level.

Stock Buyer and Control Personnel

The stock buyer at the wholesale house spends a great deal of time each day telephoning many commercial greenhouses and floral manufacturers from whom they usually purchase flowers and supplies. Stock buyers are responsible for maintaining a constant supply of cut flowers, cut foliage, plants, containers, and design materials and accessories for their customers. Buyers must locate and purchase high-quality merchandise for the lowest possible prices from growers and manufacturers throughout the world.

As shipments of merchandise are received at the wholesale house, buyers and control personnel inspect each shipment for quality, quantity, and type of flowers and merchandise, making sure merchandise is the same as the products that were ordered and purchased. Most wholesale operations, whether large or small, rely on computer systems to record purchases and sales transactions of the business.

To become a stock buyer at the wholesale level requires several years of experience working at a wholesale operation. It is also helpful to have worked at the grower and retail levels. Work experience at a greenhouse and a retail operation will help a buyer to have the knowledge to order plant materials confidently from the growers, as well as the knowledge of how flowers and supplies are used at the retail level, in order to assist the retail customers. In addition to years of job experience, a successful buyer will benefit from seeking additional training in business, marketing, horticulture, and computers and obtaining a degree at a technical school or college.

Salesperson

The sales personnel at the wholesale level are responsible for taking telephone orders from retailers, as well as helping walk-in retail customers and preparing these orders for delivery. The sales staff at the wholesale operation plays a vital role in the success of the business, for they are the people who must maintain contact and friendly relations with the retail customers who purchase their products.

A successful salesperson will benefit from experience at the retail level combined with some vocational or college education.

Delivery Personnel

The delivery staff at a wholesale operation is responsible for delivering flowers and merchandise to various retail customers and sometimes picking up merchandise directly from growers, manufacturers, airports, or other shipping stations. Drivers must have a valid driver's license for the types of vehicles they will be operating, which often include large and small trucks, large refrigerated vans and trucks, and tractors with semitrailers. At larger operations, the delivery staff is also responsible for working forklifts and organizing incoming shipments.

Cut Flower Processor

At wholesale operations, the staff responsible for caring for and handling the cut flowers, cut foliage, and live plants, as shipments are received, is a vital link in the success of the wholesale florist. Because flowers and greenery are generally shipped in boxes without water, it is imperative that boxes are opened and flowers and greens are processed as quickly as possible. This involves recutting flower stems, placing flowers in buckets of preservative solution, carrying out any specialty care techniques (such as placing flowers in bud-opening or coloring solutions), and then placing buckets of flowers in the refrigeration units. Processors help to maintain the highest quality of the live product for resale to retail customers. Processors also take care of potted plants and help the salespeople prepare orders for delivery to the retail customers. Some experience and training at a vocational or technical school in horticulture is beneficial to a flower processor.

Greenhouse Producer

At the grower levels of field or ***greenhouse producer,*** there are numerous job opportunities. These include grower/crop manager, crop production foreman, greenhouse maintenance foreman, sales manager, and greenhouse worker.

Grower/Crop Manager

The grower/crop manager is responsible for the production practices necessary to grow a specific crop or manage a number of greenhouses. Growers and crop managers are specialists who have spent many years producing cut flowers, greenery, and potted plants. Managers are often responsible for training and overseeing other employees and managing crop production in general.

Many successful growers and crop managers have received training and education in horticulture along with years of experience working at the grower level.

Crop Production and Maintenance Foreman

A crop production foreman at a large greenhouse operation is responsible for coordinating the efforts of the growers and crop managers, overseeing crop scheduling, and main-

Figure 22-6 a, b, & c. There are numerous job opportunities at the grower level, including the harvesting and grading of flowers.

taining the highest quality of flowers and plants. The maintenance foreman is in charge of the greenhouse facility. The crop production and the maintenance foreman work together to keep the greenhouses and crops in the best condition possible. Both individuals together generally maintain a staff of greenhouse workers.

Generally, foremen receive training and experience for many years working in a greenhouse. Also, successful foremen have a degree in horticulture from a university or technical school.

Marketing/Sales Manager

The sales manager works with the crop production foreman in selecting crops and timing their harvest to coordinate with various floral holidays. The sales manager is responsible for selling plants and flowers to various wholesalers and retailers. He oversees the storage and shipping of floral products to buyers. Often the sales manager coordinates a large sales staff.

A solid background in business administration, sales, and communication, as well as knowledge in horticulture, in addition to training and work experience at the grower level for several years is necessary to becoming a successful sales manager.

Greenhouse Worker

A greenhouse worker is generally involved with all aspects of growing crops, such as propagating plants from cutting and seed, preparing soil mixtures, watering, fertilizing, and caring for the plants until they are harvested at maturity. Maintenance of the greenhouse facilities may also be part of the job. Several years of training and experience will allow a greenhouse worker to advance to a more advanced position within the greenhouse operation. A background in horticulture is extremely helpful for a position in working with plants.

CONTINUING EDUCATION

Within the floral industry, as with any industry, it is vital to keep current and continue in education—to strive for excellence in a profession—in order to be a successful professional in a chosen career. It is important to be open to new ideas. Improve yourself by learning new skills and gaining new knowledge. A successful professional invites learning and self-improvement.

Figure 22-7. Attending industry meetings will help you stay current and up-to-date by learning new skills, gaining new knowledge, seeing new design styles and learning from other industry professionals.

Purpose

The sole purpose of ***design knowledge*** is to challenge a designer to think and visualize creatively. Education enables a floral designer to unlock and express the full potential of the individual's creativity. But creativity must have boundaries requiring that knowledge be transferred into know-how and practical application. Unless a designer applies what he or she has learned and puts this knowledge to work every day in the designs produced, design education is meaningless.

Trade Associations

A trade association is an organization or group that is set up by individuals, merchants, or business firms for the unified promotion of their common interests. Generally a nonprofit corporation, a trade association assists the industry in educating its members and promoting its products and services in the marketplace.

In the floral industry, associations often meet specific

needs, such as production, business, or design. Trade associations may also be geared to a specific segment of the industry, for example, growers and producers, manufacturers, wholesalers, retailers, designers, commentators, and educators. Although each trade association may have a different emphasis, all are interested in the sale and promotion of flowers and floral products.

Becoming a member of a trade organization involves some research. Each trade organization offers information packets and membership applications upon request. As you learn more about each organization, it will become clear which associations will serve your particular needs and help the most in promoting the skills in your desired career.

Following is a list of some of the well-known trade associations and organizations in which information and membership applications may be obtained.

American Institute of Floral Designers (AIFD)

The ***American Institute of Floral Designers (AIFD)*** is the premier organization that accredits floral design as a profession. The purpose of AIFD is to establish high standards in professional floral design.

Membership in AIFD is stringent, often requiring several years to achieve. Applicants must demonstrate high professional ability. Any floral designer who meets the application qualifications can apply for AIFD accreditation. The application process requires a portfolio and demonstration of design skills before a membership panel. When a designer meets the uncompromising qualifications and is accepted into membership, the individual makes a commitment to uphold excellence in professional floral design. Those who earn the credentials of AIFD are rightfully respected and share a professional bond with the organization and its members.

AIFD offers extravagant regional and national symposiums. The organization has student chapters (SAIFD) at many college campuses, promoting professional floral design education to floriculture students. AIFD sponsors an Artist in Residence (AIR) program, providing educational workshops and programs to the student chapters.

American Institute of Floral Designers
720 Light Street
Baltimore, Maryland 21230
(301) 752-3320

Society of American Florists (SAF)

Representing the needs and interests of the entire floral industry, the ***Society of American Florists (SAF)*** is a national trade association that provides members with updated information on all aspects of the industry. It provides many services, including programs and workshops, publications, newsletters, and videotapes. It is involved in government relations, promoting the floral industry by lobbying for and against legislation affecting any segment of the industry. Sponsored by SAF, the Florist Information Committee (FIC) works to alleviate any negative press about the industry and its products and instead provide positive information to the public.

Society of American Florists
1601 Duke Street
Alexandria, Virginia 22341
(800) 336-4743 or (703) 836-8700

American Academy of Floriculture (AAF)

The ***American Academy of Floriculture (AAF)*** is a division of the SAF. This organization is dedicated to service in the floral industry. Applicants must meet several high standards. Top qualifications include ten years of floral experience, industry service, and community service. Acceptance into AAF and earning its credentials is acknowledgment of professional leadership and excellence.

American Academy of Floriculture
c/o Society of American Florists
1601 Duke Street
Alexandria, Virginia 22314
(800) 336-4743 or (703) 836-8700

American Floral Marketing Council (AFMC)

The ***American Floral Marketing Council (AFMC)*** is a branch of the SAF that promotes the sale of flowers, plants, and floral products throughout the year for "non-occasion" days and reasons. The entire floral industry benefits from AFMC promotions and advertising that encourage awareness of the importance of flowers and plants and supports flower sales throughout the year.

American Floral Marketing Council
c/o Society of American Florists
1601 Duke Street
Alexandria, Virginia 22314
(800) 336-4743 or (703) 836-8700

Figure 22-8. Floral conventions provide florists with updated information on all aspects of the industry. The Society of American Florists displays new varieties of flowers at its annual, national convention.

Professional Floral Commentators International (PFCI)

Professional Floral Commentators International (PFCI), a division of the SAF, was established to promote communication and understanding within all segments of the floral industry. PFCI is an organization of industry people who want to continually gain industry knowledge, commentate at industry functions, and improve their oral skills. This organization helps educate industry members and consumers about the care and handling of flowers and plants, and promotes education about flowers and floral products. They promote professionalism in commentating at all floral events and introduce new and talented commentators to the floral industry. The credentials of PFCI represent high accomplishment in professional floral commentating. Applicants must complete several rigorous requirements.

Professional Floral Commentators
c/o Society of American Florists
1601 Duke Street
Alexandria, Virginia 22314
(800) 336-4743 or (703) 836-8700

Redbook Master Consultants (RMC)

Redbook Master Consultants (RMC) is comprised of florists and industry people interested in advancing and improving the floral industry through education. They offer educational seminars and workshops for florists. Applicants must have floral education and industry experience and must hold a position in the floral industry.

Redbook Master Consultants
P.O. Box 1706
3307 E. Kingshighway
Paragould, Arkansas 72451 (800) 643-0100

Allied and State Florists' Associations

Allied florists' associations and ***state florists' associations*** are groups of people that serve as a support system for the floral industry in a region or state. Including growers, wholesalers, retailers, and manufacturers of floral products, members of florists' associations are devoted to the support and improvement of floral business. Associations provide educational programs, newsletters, and conventions, helping to keep their members educated and informed of current trends and floral industry events.

Trade Publications

Trade journals, magazines, and supplements offer information of interest to all segments of the floral industry on a weekly, monthly, bimonthly, or quarterly basis. These publications provide information on business, design, care and handling, industry information, new products and sources, classified ads, and much more. Some of these publications are available at no cost to members of certain organizations. Most of the publications are available to floral students at a discounted subscription rate.

Following is a listing of the well-known trade journals, magazines, and supplements that are beneficial to florists.

Dateline: Washington

Published twenty-one times a year by the Society of American Florists (SAF), the American Floral Marketing Council (AFMC), and the Florist Information Committee (FIC).

Society of American Florists
1601 Duke Street
Alexandria, VA 22314
(800) 336-4743
(703) 836-8700

Floral Finance

Published monthly by American Floral Services Inc.

Floral Finance Inc.
8801 S. Yale, Suite 400
Tulsa, OK 74137
(800) 722-9934
(918) 491-9933

Floral Management

Published monthly by the Society of American Florists (SAF).

Society of American Florists
1601 Duke Street
Alexandria, VA 22314-3406
(800) 336-4743
(703) 836-8700

Floral Mass Marketing

Published bimonthly by Cenflo Inc.

Cenflo Inc.
549 W. Randolph Street
Chicago, IL 60661
(312) 236-8648

Florist

Published monthly by Florists' Transworld Delivery Association.

Florists' Transworld Delivery Association
29200 Northwestern Highway
P.O. Box 2227
Southfield, MI 48037
(313) 355-9300

Florists' Review

Published thirteen times a year by Florists' Review Enterprises Inc.

Florists' Review Enterprises Inc.
P.O. Box 4368
Topeka, KS 66604
(913) 266-0888

Flower News
Published weekly by Cenflo Inc.

Cenflo Inc.
549 W. Randolph St.
Chicago, IL 60661
(312) 236-8648

Flowers &
Published monthly by Teleflora.

Teleflora
Teleflora Plaza, Suite 260
12233 W. Olympic Blvd.
Los Angeles, CA 90064
(213) 826-5253

Holland Flower
Published quarterly by the Flower Council of Holland.

Flower Council of Holland
250 West 57th Street
New York, NY 10019
(212) 307-1818

PFD (Professional Floral Designer)
Published six times a year by American Floral Services Inc.

The Professional Floral Designer
Attn: Promotional Services
American Floral Services Inc.
P.O. Box 12309
Oklahoma City, OK 73157-2309
(800) 456-7890

The Retail Florist
Published monthly by American Floral Services Inc.

The Retail Florist
Attn: Promotional Services
American Floral Services Inc.
P.O. Box 12309
Oklahoma City, OK 73157-2309
(800) 456-7890

Figure 22-9. A professional designer expands creative skills by attending floral workshops and design programs that are presented year round.

Supermarket Floral
Published monthly by Vance Publishing.

Vance Publishing
7950 College Blvd.
Overland Park, KS 66210
(913) 451-2200

Educational Programs and Floral Design Schools

Design programs and workshops are presented every month in various parts of the country. These programs are organized and promoted by the wire services, industry associations, wholesale houses, and private groups. A professional floral designer expands creative skills by being exposed to other floral designers. Attending design programs is an excellent way to see outstanding design talent and accumulate a reservoir of design ideas.

Floral design schools across the country offer a variety of educational courses. Even if you have some experience working in a floral shop, the knowledge and skills gained at one of these schools is beneficial. Courses last from one day to several weeks or months. Basic floral design, as well as specialized subjects, such as wedding, sympathy, contemporary, or Japanese design, is offered year-round at various schools. Because tuition and amenities vary, it is important to investigate a floral design school before attending and paying fees. Recommendations from former students or word-of-mouth is the best advertisement for a school.

College Education/Certification Programs

The floral industry depends upon continuing research and higher education for advancement. There are many career options that contribute directly and indirectly to the growth and success of the entire industry, although they are not directly related to the production, distribution, or sale of flowers and plants. Research in plant propagation and postharvest physiology of cut flowers are examples of these career options. Many colleges and universities provide undergraduate and graduate programs in horticulture and floriculture. These programs lead to opportunities in teaching, research, government, and extension work.

A list of colleges, universities, and technical schools offering degrees and certification programs may be obtained by writing to the Society of American Florists.

"How are you getting on with the diet?"

Figure 22-10.
Herman © 1978 Jim Unger. Reprinted with permission of Universal Press Syndicate. All rights reserved.

REVIEW

The floral industry offers numerous, exciting, and rewarding career options. In today's market, there is an increasing need for trained, educated individuals.

Obtaining knowledge and skills in business, floriculture, and communication will provide a solid base in becoming a competent professional.

Formal training and education can be augmented and strengthened through continuing-education programs, professional associations, and trade publications. The floral designer needs to keep up-to-date with trends and design styles and stay creatively sharp. By reading industry publications, attending design programs, and interacting with other designers, a professional floral designer can keep expanding his or her knowledge about professional floral design.

Figure 22-11. Committed to excellence, owners, partners, and employees at the wholesale level work together to build a business based on the principles of service and quality. Dedicated professionals make the difference in helping increase cut flower longevity. Evergreen Wholesale Florists Inc., Seattle, WA. Photo by Raymond Gendreau.

TERMS TO INCREASE YOUR UNDERSTANDING

allied florists' associations
American Academy of Floriculture (AAF)
American Floral Marketing Council (AFMC)
American Institute of Floral Designers (AIFD)
broker
central design
continuing education
commentator
custom work
design generalist
design specialist
distribution center
floral designer
floral industry
floriculture
free-lance floral designer
greenhouse producer
horticulture
intellectual design knowledge
limited-service shop
production-line design
Professional Floral Commentators International (PFCI)
Redbook Master Consultants (RMC)
Society of American Florists (SAF)
specialized designers
state florists' associations
trade association
trade publication
wholesale florist
work environment
working skills

TEST YOUR KNOWLEDGE

1. What are the opportunities for a professional floral designer?
2. What are the three areas where design skills are required in the full-service, professional flower shop?
3. Why does a large flower shop prefer specialized designers?
4. What are the advantages for a designer in a limited-service flower shop?
5. How does a wholesaler use a professional floral designer?
6. What career opportunities are available within the floral industry?
7. Why is continuing education important in floral design?
8. Name some of the industry associations and trade publications.

RELATED ACTIVITIES

1. Visit a full-service flower shop, a limited-service flower shop, and a floral department in a supermarket. List the advantages and disadvantages you observe for a career opportunity in each.
2. Look through a recent industry publication that includes design work. Pick out six different designs and identify how studying these compositions improves your floral design knowledge.
3. By appointment, visit a grower or wholesale operation and learn of employee responsibilities. Compare the various job positions at these operations.
4. Attend a design program, workshop, or convention. Write a short report about what you have learned.

Appendix A

FLOWERS

A

ACACIA (see Plate 10)

(a-KAY-sha)
Family: Leguminosae (Fabaceae)
Name Origin: Named from the Greek word *akis* (sharp point), referring to the thorns.
Species: *dealbata, longifolia;* others
Common Names: acacia, mimosa, wattle
Availability: October through March
Description: Trees with clusters of fragrant, ball-shaped yellow flowers with gray-green, finely cut leaves. Branches of fluffy clusters work well as fillers. Cut flowers dry out and lose their fuzzy appearance as they age; in storage, cover with plastic to retain moisture.
Vase-life: 4 to 5 days

ACHILLEA

(ah-kil-LEE-ah)
Family: Compositae (Asteraceae)
Name Origin: Named after Achilles of Greek mythology, who is said to have used it medicinally.
Species: *filipendulina, millefolium;* others and hybrids
Common Names: yarrow, milfoil
Availability: July through September
Description: Flat-headed, yellow corymb flowers over a feathery foliage. Also available in pink, red, and white.
Vase-life: Long lasting, 7 to 10 days

ACHILLEA (Yarrow)

ACONITUM (Monkshood)

ACONITUM

(ak-ah-NEE-tum)
Family: Ranunculaceae
Species: *napellus;* others
Common Names: monkshood, aconite
Availability: April through October
Description: Tall spikelike racemes of deep blue, hooded (helmet-shaped) florets. Also available in light blue, white, and cream varieties. Monkshood is extremely toxic. Wash hands after using to get rid of toxic substances.
Vase-life: 5 to 10 days

AGAPANTHUS (see Plate 1)

(ag-a-PAN-thus)
Family: Amaryllidaceae
Name Origin: From the Greek words *agape* (love) and *anthos* (flower).
Species: *africanus (A. umbellatus), orientalis;* others and hybrids
Common Names: lily of the Nile, African lily
Availability: March through August
Description: Large umbels of blue, funnel-shaped flowers in various tints and shapes. Also available in white. The stem of the agapanthus adds a strong line element with its bold-shaped yet airy flower head. Mixes well with many flowers in many different design styles.
Vase-life: 4 to 7 days, with individual florets blooming continuously over several days

AFRICAN DAISY see *Gerbera*
AFRICAN LILY see *Agapanthus*
AFRICAN CORN LILY see *Ixia*

ALCHEMILLA

(al-ke-MIL-la)
Family: Rosaceae
Species: *mollis;* others
Common Name: lady's mantle
Availability: Spring to early fall with peak supplies May through July
Description: Feathery sprays (compound cymes) of small, yellowish or greenish flowers. Useful as an airy filler.
Vase-life: Long lasting, 7 to 10 days

ALCHEMILLA (Lady's Mantle)

ALSTROEMERIA
ANEMONE
AGAPANTHUS
ALLIUM

The Farside © 1986.

ALLIUM (see Plate 1)

(AL-ee-um)
Family: Amaryllidaceae
Name Origin: From the Greek name for garlic.
Species: *giganteum, sphaerocephalon;* others
Common Names: allium, onion flower, garlic, flowering onion
Availability: April through September
Description: Bulbs producing globe-shaped umbels of purple and pink flowers on leafless stems. Also available in blue, white, and yellow varieties. Most smell like onion when cut, but odor generally dissipates.
Vase-life: 5 to 7 days

ALPINIA (see Plate 2)

(al-PIN-ee-ah)
Family: Zingiberaceae
Species: *purpurata, zerumbet;* others
Common Names: ginger, ostrich plume, torch ginger, shell ginger
Availability: Year-round
Description: The flower head consists of shiny red or pink bracts at the end of a heavy, thick stem, often up to 36 inches long. Ginger provides a strong vertical line. The colored bracts provide emphasis. If flowers appear wilted, immerse the entire stem and bracts in room-temperature water for 15 to 30 minutes.
Vase-life: These flowers are incredibly long lasting, up to 3 weeks.

ALSTROEMERIA (see Plate 1)

(al-stre-MEAR-ee-ah)
Family: Alstroemeriaceae
Name Origin: Named after Baron Claus Alstroemer (1736-94).
Species: *heamantha, pelegrina;* others and hybrids
Common Names: alstroemeria, alstro, Peruvian lily, Inca lily, lily of the Incas
Availability: Year-round
Description: Clusters of delicate, trumpet-shaped flowers, borne at the end of short flower stalks that spray off a single stem. Available in many colors with many intermediate colors. Most varieties are freckled or streaked with contrasting colors.
Vase-life: Long lasting, up to 2 weeks, with individual flowers lasting 5 to 7 days each. Alstroemeria is ethylene sensitive, with leaves often yellowing prematurely. Remove lower leaves.

AMARANTHUS

(am-a-RAN-thus)
Family: Amaranthaceae
Name Origin: From the Greek word *amarantos* (unfading), referring to the long- lasting flowers.
Species: *hypochondriacus, tricolor, caudatus;* others and hybrids
Common Names: amaranth; prince's feather *(A. hypochondriacus);* Joseph's coat *(A. tricolor);* love lies bleeding, cat's tail, tassel flower *(A. caudatus)*
Availability: Summer and Autumn
Description: Erect and brushlike flower racemes and spikes up to 6 inches long. Available in red, green, and cream. The species *A. caudatus* and hybrids are slender, drooping red flower racemes up to 16 inches long.
Vase-life: Long lasting, 7 to 10 days

AMARYLLIS (see Plate 3)

(am-a-RIL-is)
Family: Amaryllidaceae
Name Origin: Named after a shepherdess in Greek mythology.
Species: *belladonna;* See *Hippeastrum* for the larger trumpet-shaped flower that is commonly called *Amaryllis.*

ANTHURIUM
STRELITZIA
Bird of Paradise
ALPINIA
Ginger

AMARANTHUS (Amaranth)

Common Names: belladonna lily, naked lady lily, cape belladonna
Availability: May through November
Description: Trumpet-shaped flowers, each 2 to 3 inches across on a single thick, leafless stem. Available in pink, red, and white. Belladonna lilies are fragrant and attractive in many design styles, especially parallel and vegetative designs where their forms are apparent.
Vase-life: Long lasting, up to 10 days

AMAZON LILY see *Eucharis*

AMMI (see Plate 21)

(AM-me)
Family: Umbelliferae (Apiaceae)
Species: *majus;* (similar are the species *Daucus carota* and *D. sativus)*
Common Names: Queen Anne's lace, bishop's weed; wild Queen Anne's lace *(Daucus)*
Availability: Year-round
Description: Delicate, white, compound umbels, 3 to 6 inches across. Queen Anne's lace is more airy than *Daucus* varieties. Both *Ammi* and *Daucus* work well as fillers.
Vase-life: 3 to 5 days

ANANAS (see Plate 21)

(a-NA-nas)
Family: Bromeliaceae
Species: *nanus, bracteatus stivus, comosus;* others
Common Names: ornamental pineapple, dwarf pineapple
Availability: Year-round
Description: Large or small pink, white, or variegated head resembles the edible pineapple. Provides an immediate focal point through shape and texture.
Vase-life: Long lasting, 1 to 3 weeks

ANEMONE (see Plate 1)

(a-NEM-oh-nee)
Family: Ranunculaceae
Name Origin: Named after Adonis, also called Naamen, a handsome young man of Greek mythology who was loved by Aphrodite. He was killed by a wild boar and his blood is said to have given rise to the blood-red flowers.
Species: *coronaria;* others and hybrids
Common Names: windflower, lily of the field, poppy anemone
Availability: October through May
Description: Cup-shaped solitary flowers open up flat. Petals surround a dark center. Available in vibrant colors of red, pink, blue, purple, as well as white. Many anemones curve toward the light.
Vase-life: 3 to 7 days. Keep cool to prevent premature wilting and drooping.

ANETHUM

(a-NAY-thum)
Family: Umbelliferae (Apiaceae)
Species: *graveolens*
Common Name: dill
Availability: Year-round
Description: Compound umbel clusters of yellow flowers, similar to Queen Anne's lace, on stems up to 36 inches

AMARYLLIS
Belladona Lily
CALLISTEPHUS
China Aster
ASTER
ASTILBE

tall. A useful, long-lasting filler.
Vase-life: 7 to 10 days

ANIGOZANTHOS (see Plate 14)

(a-nee-go-ZAN-thus)
Family Name: Haemodoraceae
Name Origin: From the Greek words *anoigo* (to open) and *anthos* (flower), referring to the flowers that are widely open.
Species: *flavidus, pulcherrimus, rufus;* others and cultivars
Common Name: kangaroo paw
Availability: Year-round
Description: Unusual red, purple, green, or yellowish, fuzzy flowers borne in one-sided racemes and spikes. Each blossom has a covering of short, colored fur providing a delicate texture contrast. Combines easily with other exotic flowers.
Vase-life: Long lasting, up to 2 weeks

ANNUAL DELPHINIUM see *Consolida*

ANTHURIUM (see Plate 2)

(an-THUR-ee-um)
Family Name: Araceae
Name Origin: From the Greek words *anthos* (flower) and *oura* (tail), referring to the tail-like inflorescence or spadix.
Species: *andraeanum, scherzeranum*
Common Names: anthurium, tail flower, flamingo flower, painted tongue, painter's palette, peace lily
Availability: Year-round
Description: Anthurium flowers consist of a modified, shiny, colorful leaf called a *spathe.* The spathe is usually heart-shaped or arrow-shaped, in reds, pinks, white, and tricolors. The true flowers are on a cylindrical, long *spadix* that is usually yellow, but sometimes appears in other colors. The size of the spathe varies from 2¾ to 6 inches or more. These flowers lend a dramatic focal point and look best when used alone or with other exotic flowers.
Vase-life: Anthuriums are extremely hardy cut flowers, lasting 2 to 3 weeks or more; however, care must be taken in handling anthuriums as they bruise easily. Anthuriums are accustomed to humid and warm conditions. Mist often with water and do not store below 45°F. If bracts look wilted, immerse entire stem in water for 30 to 90 minutes.

ANTIRRHINUM (see Plate 24)

(an-tee-RYE-num)

AQUILEGIA (Columbine)

Family Name: Scrophulariaceae
Name Origin: From the Greek words *anti* (like) and *rhis* (snout) referring to the appearance of the individual flowers.
Species: *majus;* cultivars
Common Name: snapdragon
Availability: Year-round
Description: A spiked, terminal raceme with florets 1½ inches long that are tubular with rounded upper and lower lips (when squeezed together they snap open and shut). Available in many colors. Snapdragons are geotropic (bend away from gravity); store or arrange in a vertical position to prevent curvature. Removing the very top bud can help deter bending.
Vase-life: Capable of a long vase-life of up to 2 weeks; sensitive to ethylene

AQUILEGIA

(ak-wi-LEE-ji-a)
Family: Ranunculaceae
Name Origin: From the Latin word *aquila* (eagle). The flower petals resemble the claws of eagles and other birds of prey.

MOLUCCELLA
Bells of Ireland
BOUVARDIA
RUDBECKIA
Black-eyed Susan
TRITELEIA
Brodiaea

Species: *caerulea, canadensis, chrysantha, flabellata;* others and hybrids
Common Name: columbine
Availability: April through September
Description: Terminal, bonnet-shaped flowers consisting of 5 petals, each with its own protruding spur. The flowers are 1 to 3 inches long. Available in white, pink, yellow, blue, and purple.
Vase-life: 3 to 5 days

ARUM LILY see *Zantedeschia*

ASCLEPIAS

(as-KLEE-pee-us)
Family: Asclepiadacea
Name Origin: From Asklepios, the Greek god of medicine, referring to medicinal properties.
Species: *tuberosa;* others
Common Names: milkweed, blood flower, silkweed
Availability: Year-round, with peak supplies June through September
Description: Tiny, waxy, orange flowers appear in rounded, umbellate cyme clusters. Adds mass and accent to designs; can also be used as a filler flower.
Vase-life: 3 to 7 days; florets continue to open; remove old or dying florets. Stems exude a milky sap (hence the name milkweed); seal stems in hot water.

ASIAN LILY see *Lilium*

ASTER (see Plate 3)

(AS-ter)
Family: Compositae (Asteraceae)
Name Origin: From the Latin word *aster* (star), referring to the flowers.
Species: *cordifolius, ericoides, novi-belgii;* others and hybrids
Common Names: Monte Cassino aster (*A. ericoides* 'Monte Cassino'); Michaelmas daisy *(A. novi-belgii)*
Availability: July through December
Description: Numerous daisylike flowers ¾ to 1 inch across in double or single types. Available in a wide variety of colors, usually with a yellow center. The flowers are clustered racemes, corymbs, or panicles.
Vase-life: Long lasting, up to 10 days

ASTILBE (see Plate 3)

(a-STIL-bee)
Family: Saxifragaceae
Name Origin: From the Greek words *a* (without) and *stilbe* (brilliance), meaning the individual flowers are extremely small.
Species: *chinensis;* others and hybrids
Common Names: false spirea, goat's beard, meadowsweet
Availability: Year-round
Description: Pyramidal feather plumes. Flowers are borne in loose, pyramidal panicles on slender stems. Available in white, pink, and red. Excellent filler flower.
Vase-life: 5 to 7 days

ASTRANTIA

(a-STRAN-tee-ah)
Family: Umbelliferae (Apiaceae)
Name Origin: From the Latin word *aster* (star) referring to the starlike flowers.
Species: *major;* others and hybrids
Common Name: masterwort
Availability: April to September
Description: Delicate, star-shaped flowers in clustered umbels. Available in white, pinks, and reds. Useful as a mass or filler flower. Up close Astrantia has a slight, sometimes unpleasant, odor.
Vase-life: 4 to 7 days

B

BABY'S BREATH see *Gypsophila*

BACHELOR'S BUTTON see *Centaurea*

BANKSIA (see Plate 20)

(BANK-see-ah)
Family: Proteaceae
Name Origin: Named after botanist Sir Joseph Banks (1743-1820).
Species: *ashbyi, baxteri, coccinea, collina, ericifolia, menziesii, speciosa;* many others.
Common Names: protea, bird's nest, giant bottlebrush
Availability: Year-round
Description: Conelike, dense terminal spikes in red, oranges, and yellows. Provides an immediate focal point. Foliage has textural appeal.
Vase-life: Extremely long lasting, 2 to 4 weeks

BARBERTON DAISY see *Gerbera*

BELLADONNA LILY see *Amaryllis*

BELLFLOWER see *Campanula*

CIRSIUM
CAMPANULA
Chimney Bells
CAMELLIA
IBERIS
Candytuft
ZANTEDESCHIA
Calla Lily

BELLS OF IRELAND see *Moluccella*
BIRD OF PARADISE see *Strelitzia*
BIRD'S NEST see *Banksia*
BLACK-EYED SUSAN see *Rudbeckia*
BLANKET FLOWER see *Gaillardia*
BLAZING STAR see *Liatris*
BLOOD FLOWER see *Asclepias*
BLUEBELL see *Scilla*
BLUE LACEFLOWER see *Trachymene*
BLUE THROATWORT see *Trachelium*

BOUVARDIA (see Plate 4)

(boo-VAR-dee-ah)
Family: Rubiaceae
Name Origin: Named after Dr. Charles Bouvard (1572-1658).
Species: *longiflora* hybrids
Common Name: bouvardia
Availability: Year-round
Description: Small, tubular flowers with spreading starlike petals in terminal cyme clusters. Generally fragrant. Available in white, pinks, orange, and reds. Excellent filler flower. Often used in wedding and corsage work.
Vase-life: 7 to 10 days; remove lower leaves; sensitive to water stress

BRAIN FLOWER see *Celosia*
BRODIAEA see *Triteleia*
BROOM see *Cytisus*
BUTTERFLY ORCHID see *Oncidium*
BUTTON MUM see *Chrysanthemum*
BUTTON SNAKEROOT see *Liatris*

CALENDULA

(ka-LEN-dew-la)
Family: Compositae (Asteraceae)
Name Origin: From the Latin word *calandae* (first day of the month), referring to its long flowering period.
Species: *officinalis;* many hybrids
Common Name: pot marigold

CALENDULA (Pot Marigold)

Availability: April through November
Description: Daisylike double-petaled head flowers. Available in bright yellows and oranges.
Vase-life: 5 to 7 days

CALLA LILY see *Zantedeschia*

CALLISTEPHUS (see Plate 3)

(ka-LIS-te-fus)
Family: Compositae (Asteraceae)
Name Origin: From the Greek words *kallos* (beautiful) and *stephanus* (crown), referring to the showy solitary flower heads.
Species: *chinensis* hybrids
Common Names: China aster, aster
Availability: June through September
Description: Large, solitary flower heads. Available in a variety of shapes and colors.
Vase-life: 5 to 10 days; remove any excess foliage for longer vase-life.

CALLUNA see *Erica*
CALYCINA *(Thryptomene)* see *Erica*

CHYRSANTHEMUM—Some of the most common inflorescence forms grown commercially.

CAMPANULA GLOMERATA

CAMELLIA (see Plate 5)

(ka-MEL-ee-a)
Family: Theaceae
Name Origin: Named after pharmacist George Joseph Kamel (1661-1706), who studied the Philippines flora.
Species: *japonica, reticulata, sinensis;* others
Common Name: camellia
Availability: January and February
Description: Solitary waxy flowers. Available mostly in pinks, reds, and white. Single flowers are excellent for wedding flowers and corsages. Larger flowering branches add interest to large designs.
Vase-life: Single flowers that are prepackaged, only 1 to 2 days. However, woody branches with flower buds can last more than a week.

CAMPANULA (see Plate 5)

(kam-PAHN-ew-lah)
Family: Campanulaceae
Name Origin: From the Latin word *campana* (bell), referring to the bell-shaped flowers.
Species: *glomerata, persicifolia, pyramidalis;* others
Common Names: bellflower, chimney bells
Availability: April through August
Description: Whether clustered or spikelike, these work well in mixed summer arrangements. *C. glomerata* has a bold clustered shape that works well in contemporary design styles.
Vase-life: 5 to 7 days

CANDYTUFT see *Iberis*
CAPE JASMINE see *Gardenia*
CARNATION see *Dianthus*

CARTHAMUS (see Plate 24)

(CAR-tha-mus)
Family: Compositae (Asteraceae)
Species: *tinctorius*
Common Names: safflower, false saffron
Availability: June through November
Description: The flowers are about 1 inch across with a green globular center, where thin orange petals emerge. Safflowers add interesting texture to designs. Useful as an accent or filler.
Vase-life: Long lasting, 1 to 2 weeks. Removal of excess foliage increases vase-life.

CAT'S TAIL see *Amaranthus*

CATTLEYA (see Plate 19)

(KAT-lee-ah)
Family: Orchidaceae
Name Origin: Named after horticultural patron, William Cattley (early 1800s).
Species: *aurantiaca, bicolor, intermedia, labiata;* others and hybrids.
Common Name: corsage orchid
Availability: Year-round
Description: The flowers measure 3 to 6 inches across. Available in white, lavender, pink, and yellow. Their shape is exotic and interesting, with showy sepals and petals. A broad ruffled lip is in the center. Generally used for corsage and wedding flowers but can be incorporated into arrangements. Generally packaged singly with individual water tubes.
Vase-life: Long lasting, 5 to 10 days with direct water supply

CROCOSMIA
CENTAUREA
Cornflower
NARCISSUS
Paperwhites
NARCISSUS
Daffodil
DELPHINIUM

CELOSIA

CELOSIA (crested variety)

CELOSIA

(see-LO-si-ah or see-LO-shee-ah)
Family: Amaranthaceae
Name Origin: From the Greek word *keleos* (burning), referring to the brightly colored flowers.
Species: *argentea, cristata, plumosa;* hybrids
Common Names: brain flower, cockscomb, woolflower
Availability: July through November
Description: Available in reds, yellows, and oranges. *C. cristata* with its crested shape adds interesting form and texture to contemporary designs. The feathery plumes of *C. plumosa* work well as fillers and accents.
Vase-life: 5 to 7 days

CENTAUREA (see Plate 7)

(sen-ta-REE-a)
Family: Compositae (Asteraceae)
Name Origin: From the Greek word *kentaur* for Centaur, who is said to have used it medicinally.
Species: *cyanus;* others
Common Names: cornflower, bachelor's button, bluebottle
Availability: February through September
Description: Small, thistlelike head flowers. Available in blue, pink, and white. Useful as a filler.
Vase-life: 5 to 7 days

CHAENOMELES (see Plate 22)

(kee-NAHM-el-eez)
Family: Rosaceae
Name Origin: From the Greek words *chaina* (to gape) and *melon* (apple), referring to the belief the fruit was split.
Species: *cathayensis, japonica, speciosa*
Common Name: flowering quince
Availability: January to April
Description: Clusters of pink, red, or white flowers on woody branches. Works well in oriental designs.
Vase-life: 3 to 10 days

CHAMELAUCIUM (see Plate 28)

(cham-ee-LAW-si-um)
Family: Myrtaceae
Species: *uncinatum*
Common Name: waxflower
Availability: September through May
Description: Heathlike shrubs with star-shaped flowers in white, pinks, lavenders, and bicolors. Densely clustered along woody branching stems. Needlelike foliage. An excellent filler flower.
Vase-life: 7 to 10 days

DIANTHUS—Some of the most common types grown commercially.

CHELONE (Turtlehead)

CHELONE

(kee-LO-nee)
Family: Scrophulariaceae
Name Origin: From the Greek word *chelone* (turtle), referring to the corolla that is shaped like a turtle's head.
Species: *obliqua*
Common Names: turtlehead, snakehead
Availability: July through October
Description: Pink flowers similar in appearance to snapdragons; however, florets are clustered closer together in terminal, spikelike racemes.
Vase-life: 5 to 7 days

CHIMNEY BELLS see *Campanula*
CHINA ASTER see *Callistephus*
CHINCHERINCHEE see *Ornithogalum*
CHRISTMAS ROSE see *Helleborus*

CHRYSANTHEMUM (see Plate 6)

(kris-ANTH-e-mum)
Family: Compositae (Asteraceae)
Name Origin: From the Greek words *chrysos* (golden) and *anthos* (flower).
Species: *morifolium, frutescens, coccineum, parthenium, carinatum;* others and cultivars.
Common Names: mum or florist's chrysanthemum *(C. morifolium)* includes spray varieties (anenome, button, daisy, pompon, spider, and starburst) as well as single-head varieties (Fuji, incurve or football mum, and mefo); daisy, marguerite daisy *(C. frutescens);* Shasta daisy *(C. maximum);* feverfew *(C. parthenium);* pyrethrum *(C. coccineum)*
Availability: Year-round
Description: Available in a wide variety of shapes, colors, and textures. Smaller spray types work well as fillers and accents. Larger solitary flowers provide mass and emphasis.
Vase-life: Long lasting, 1 to 3 weeks; Marguerite daisies are heavy drinkers. Remove wilted foliage from stems. To prevent shattering, spray the back and top of flower heads with an aerosol designed for this purpose.

CIRSIUM (see Plate 5)

(SIR-cee-um)
Family: Compositae (Asteraceae)
Species: *japonicum;* others and hybrids
Common Names: plume thistle, plumed thistle. *Cnicus benedictus* is a similar thistlelike flower with spiny, leafy bracts.
Availability: May through November
Description: Small thistlelike flower heads. Available in pinks and purples. Useful as a filler. Provides interesting accent because of texture and shape.
Vase-life: Long lasting, 1 to 2 weeks

CLARKIA (see Plate 13)

(KLARK-ee-ah)
Family: Onagraceae
Name Origin: Named after Captain William Clark (1770-1838).
Species: *amoena, concinna;* others
Common Names: godetia, satin flower, farewell-to-spring. *Godetia* (goh-DEE-shee-ah) comprises a subgenus of *Clarkia* and is sometimes listed separately.
Availability: May through August
Description: Clustered funnel-shaped flowers about 2 inches across with a papery texture. Available in a wide range of colors and bicolors. Godetia provides emphasis or mass, and can also be used as a filler.
Vase-life: 5 to 7 days

CNICUS see *Cirsium*
COBRA HEAD see *Sarracenia*
COCKSCOMB see *Celosia*
COLEONEMA see *Diosma*

EREMURUS
Foxtail Lily
CYNARA
Globe Artichoke
DIGITALIS
Foxglove

COLUMBINE see *Aquilegia*
CONE FLOWER see *Rudbeckia*

CONSOLIDA (see Plate 16)
(kon-SO-li-da)
Family: Ranunculaceae
Name Origin: From the Latin word *consolida* (to make whole), referring to its medicinal properties.
Species: *ambigua (Delphinium ajacis); orientalis (D. orientale); regalis (D. consolida). Consolida* florets differ slightly from *Delphinium* in having the two upper petals united into one, and in lacking the lower two petals.
Common Name: larkspur
Availability: June through September
Description: Spikelike racemes in blues, lavenders, pinks, and white. Excellent for adding mass. Useful as a line flower.
Vase-life: 7 to 10 days

CONVALLARIA (see Plate 18)
(kahn-val-AIR-ee-ah)
Family: Liliaceae
Name Origin: From the Latin word *convallis* (valley).
Species: *majalis*
Common Name: lily of the valley
Availability: Year-round
Description: Fragrant and delicate white or pink bell-shaped florets along short, terminal, one-sided racemes. Popular for wedding and corsage work. Also an excellent accent in contemporary or vegetative design styles.
Vase-life: 2 to 6 days

CORNFLOWER see *Centaurea*
CORN LILY see *Ixia*
CORSAGE ORCHID see *Cattleya*

COSMOS
(KAHZ-mohs)
Family: Compositae (Asteraceae)
Name Origin: From the Greek word *kosmos* (beautiful).
Species: *bipinnatus, sulphureus;* others
Common Name: cosmos
Availability: June through August
Description: Solitary flowers with a single row of petals in white, yellow, reds, and pinks surrounding a yellow center. Useful for adding mass and work well as filler in summer bouquets.
Vase-life: 4 to 6 days

COSMOS

COSTUS
(KAWS-tus)
Family: Zingiberaceae
Species: *megalobractiatus, pulverulentus, spicatus;* others
Common Names: spiral ginger, kiss of death, spiral flag
Availability: Year-round
Description: Spirally twisted conelike bracts, available in red, orange, yellow, pink, and white. These unusual flowers provide emphasis.
Vase-life: 5 to 10 days or longer

CRANE LILY see *Strelitzia*

CROCOSMIA (see Plate 7)
(kro-KOS-mee-ah)
Family: Iridaceae
Name Origin: From the Greek word *krokos* (saffron) and *osme* (smell). When dried, they smell of saffron.
Species: *masoniorum*
Common Names: montbretia, coppertip
Availability: June through November
Description: Trumpet-shaped flowers in one-sided, spikelike patterns. Available in shades of scarlet, orange, and red. Useful in many design styles as a filler, line, or accent.
Vase-life: 1 to 2 weeks

ACACIA
EUCALYPTUS PODS
FORSYTHIA
Golden Bells
EUPHORBIA
Scarlet Plume

CYMBIDIUM (see Plate 19)

(sim-BID-ee-um)
Family: Orchidaceae
Name Origin: From the Greek word *kymbe* (boat), referring to the hollowed lip.
Species: hybrids
Common Name: cymbidium
Availability: Year-round
Description: Spray of orchid flowers. Usually sold singly with individual water tubes. Wide variety of colors. Useful for corsage and wedding flowers but can be incorporated into arrangements.
Vase-life: Extremely longlasting, 1 to 3 weeks with good water supply

CYNARA (see Plate 9)

(SIN-ah-rah)
Family: Compositae (Asteraceae)
Name Origin: From the Latin name for these perennial herbs.
Species: *scolymus*
Common Name: globe artichoke
Availability: July through October
Description: Tight green, thistlelike spiney heads, 2 to 4 inch in diameter, from which purple and blue flowers emerge. These add interesting texture to designs. Useful as a focal point, accent or filler.
Vase-life: 7 to 10 days

CYTISUS

CYPRIPEDIUM see *Paphiopedilum*

CYTISUS

(SIT-is-us)
Family: Leguminosae (Fabaceae)
Name Origin: From the Greek word *kytisos,* the name for these and similar shrubs.
Species: *canariensis (Genista canariensis);* for *C. scoparius (G. scoparia)* see Appendix B: Foliage under *Cytisus.*
Common Names: genista, broom, sweet broom.
Availability: January through May
Description: Small, fragrant pea-shaped florets on wiry, leafless branches. Available in white and yellow, along with other dyed colors of pinks and purples. Useful as a filler and for soft curving lines in many design styles.
Vase-life: 1 to 2 weeks

D

DAFFODIL see *Narcissus*

DAHLIA (see Plate 15)

(DAHL-yah)
Family: Compositae (Asteraceae)
Name Origin: Named after Swedish botanist Dr. Anders Dahl (1751-89).
Species: many hybrids and cultivars
Common Name: dahlia
Availability: July through November
Description: The wide variety of shapes, sizes, colors, and textures of these head flowers allow many design uses. Can add mass, provide emphasis, and accent. Smaller varieties work well as fillers.
Vase-life: Long lasting, 1 to 10 days

DAISY see *Chrysanthemum*
DANCING DOLL see *Oncidium*
DAUCUS see *Ammi*

DELPHINIUM (see Plate 7)

(del-FIN-ee-um)
Family: Ranunculaceae
Name Origin: From the Greek word *delphis* (dolphin), referring to the shape of the flowers.
Species: *elatum, D. x belladonna, cardinale, grandiflorum;* others. For *D. ajacis,* see *Consolida ambigua* and *C. orientalis.*

GAILLARDIA
Blanket Flower
FREESIA
Blanket Flower
GARDENIA
GERBERA

Common Name: delphinium
Availability: Year-round
Description: Spikelike racemes available in white and various tints and shades of blue, lavender, and pink. Useful as a line flower or in adding mass. Some varieties work well as fillers. Many design uses for a variety of styles.
Vase-life: Varies greatly with species, from 3 to 4 days to 2 weeks; florets can easily shatter, so handle with care.

DENDROBIUM (see Plate 19)

(den-DRO-be-um)
Family: Orchidaceae
Name Origin: From the Greek words *dendron* (tree) and *bios* (life), referring to their epiphytic growth (an epiphyte is a plant that grows on another plant but is not a parasite and produces its own food by photosynthesis).
Species: *bigibbum, nobile, phalaenopsis;* others and hybrids
Common Names: dendrobium, Singapore orchid
Availability: Year-round
Description: Spray orchids in lavenders, pinks, yellow, and white. Spray racemes provide slightly curving lines. Commonly used with exotic or tropical flowers. Individual blossoms are used for corsages, boutonnieres, and wedding designs.
Vase-life: 7 to 10 days

DESERT CANDLE see *Eremurus*

DIANTHUS (see Plate 8)

(die-ANTH-us)
Family: Caryophyllaceae
Name Origin: From the Greek words *Di* (Zeus) and *anthos* (flower).
Species: *barbatus, caryophyllus, chinensis;* hybrids and cultivars
Common Names: sweet William *(D. barbatus);* carnation, clove pink *(D. caryophyllus);* pixie, spray carnation, mini-carnation *(D. caryophyllus* nana); pinks, Chinese pink, annual pink *(D. chinensis)*
Availability: Year-round
Description: Sweet Williams are densely packed clusters of florets. Available in pinks, reds, purples, and white. Useful as a filler or accent in many design styles.
Carnations are one of the floral industry's staples. Available in a wide range of colors and varieties. White and other light colors can be color dyed or tipped with paint. Most are fragrant. Carnations are versatile and offer many design uses. Miniature carnations also come in a wide variety of colors. Generally there are several buds and flowers on each stem. Sprays work well as fillers or in adding mass. Individual flowers work well for corsages, boutonnieres, and wedding flowers.
Vase-life: Carnations and pixies are long lasting, 1 to 2 weeks; sweet William lasts 5 to 10 days; ethylene sensitive.

DIDISCUS see *Trachymene*

DIGITALIS (see Plate 9)

(di-ji-TAL-lis)
Family: Scrophulariaceae
Name Origin: From the Latin word *digitus* (finger), referring to the fingerlike florets.
Species: *purpurea, grandiflora (D. ambigua), lanata*
Common Name: foxglove
Availability: June through September
Description: Tall spikelike, often one-sided, racemes. The downward-facing tubular florets are splashed in the throat with exotic spots. Available in white and various shades of red, purple, pink, and yellow. Useful for adding line to arrangements.
Vase-life: Long lasting, about 2 weeks, with florets continuously opening

DILL see *Anethum*

DIOSMA

(die-OZ-mah)
Family: Rutaceae

DIOSMA (Breath of Heaven)

GLORIOSA
GLADIOLUS
GLADIOLUS
Mini Glad
ECHINOPS
Globe thistle

Species: *ericoides;* sometimes referred to as *Coleonema ericoides*
Common Names: diosma, coleonema, breath of heaven
Availability: February through May
Description: Fragrant, heathlike shrubs with tiny white, pink, lavender, or red flowers in terminal cymose clusters.
Vase-life: 5 to 7 days

DRYANDRA

(dry-AN-dra)
Family: Proteaceae
Species: *floribunda, formosa;* others
Common Name: dryandra
Availability: Year-round
Description: Dense-headed flowers mostly available in yellow, but also available in oranges and reds, with prickly-toothed leaves, similar to their *Banksia* relatives. These flowers provide accent and emphasis.
Vase-life: Long lasting, 2 to 3 weeks

E

EASTER LILY see *Lilium*

ECHINOPS (see Plate 12)

(EK-i-nops)
Family: Compositae (Asteraceae)
Name Origin: From the Greek words *echinos* (hedgehog) and *-opsis* (appearance).
Species: *ritro, humilis*
Common Name: globe thistle
Availability: July through October
Description: Intense blue, white, or metalic blue, round, thistlelike flowers at the end of branching stems. The bold texture of these flowers creates striking effects.
Vase-life: Long lasting, 1 to 3 weeks

EREMURUS (see Plate 9)

(air-re-MOUR-us or e-ray-MEW-rus)
Family: Liliaceae
Name Origin: From the Greek words *eremia* (desert) and *oura* (tail), referring to their desert habitat and the shape of the inflorescence.
Species: *robustus, stenophyllus*
Common Names: foxtail lily, desert candle, king's spear
Availability: May through September
Description: Tall spikelike racemes with small star-shaped florets on leafless stems. Generally yellow, but white, cream, orange, and pink varieties are also available. Excellent for adding line.
Vase-life: Long lasting, 1 to 3 weeks, during which time tiny new florets continuously open

ERYNGIUM (Sea Holly)

ERICA (see Plate 13)

(AIR-i-kah)
Family: Ericaceae
Species: *canaliculata;* others and hybrids. The genus *Calluna* (kah-LOO-nah) is known as common heather. Many other genera, such as *Thryptomene,* are heathlike shrubs and are often called heather or heath.
Common Names: heath, heather
Availability: November through April
Description: Spikelike flower clusters in panicles and racemes, with tiny bell-shaped florets. Available in pinks, purples, white, yellows, and green. Useful as a filler. Taller varieties can create line.
Vase-life: Long lasting, 1 to 3 weeks

ERYNGIUM

(air-IN-jee-um)
Family: Umbelliferae (Apiaceae)
Species: *alpinum, amethystinum, giganteum;* others and hybrids

CLARKIA
Godetia

GYPSOPHILA
Baby's Breath

ERICA
Heather

EUCHARIS (Amazon Lily)

Common Names: sea holly, eryngo
Availability: June through October
Description: Flower head is cone-shaped, surrounded by feathery bracts. Available in silver-purple, blues, some pinks, green, and white. Adds unusual texture and form to arrangements. Well suited for many design styles.
Vase-life: Long lasting, 1 to 2 weeks

EUCHARIS

(YOU-kah-ris)
Family: Amaryllidaceae
Species: *amazonica (E. grandiflora)*
Common Names: Amazon lily, eucharis lily, lily of the Amazon
Availability: Year-round
Description: 3 to 6 individual fragrant, large, white flowers are borne in umbel patterns on leafless stems.
Vase-life: 7 to 14 days; individual flowers continue to open.

EUPHORBIA (see Plate 10)

(you-FOR-bee-ah)
Family: Euphorbiaceae
Name Origin: Named after Euphorbus, the physician to the king of Mauritania, Juba.
Species: *fulgens, pulcherrima, marginata;* others
Common Names: scarlet plume, spurge *(E. fulgans);* poinsettia, Christmas flower *(E. pulcherrima);* snow on the mountain, ghostweed *(E. marginata)*
Availability: September through February
Description: *E. fulgans* has small flowers alongside drooping stems. Adds colorful, graceful lines to designs. Available in orange, red, yellow, pink, and white. *E. marginata* with its green and white color pattern is useful as a foliage. All *Euphorbia* species bleed milky sap when cut or if leaves are removed; it can produce a severe dermatitis in susceptible individuals.
Vase-life: 7 to 10 days; condition in hot water with floral preservative.

EUSTOMA (see Plate 18)

(yew-STOW-mah)
Family: Gentianaceae
Species: *grandiflorum (Lisanthus russellianus)*
Common Names: lisianthus, prairie gentian
Availability: May through December
Description: Anemone-shaped flowers solitary on stems or in branching panicles. Generally 3 to 5 flowers that open on each stem. Available in purples, pinks, reds, bicolors, and white. A variety of single, double, and triple petal forms are available. Adds mass with a soft texture. Single blooms work well for corsages.
Vase-life: Long lasting, 1 to 2 weeks

EVERLASTING FLOWER see *Helichrysum*

F

FALSE DRAGONHEAD see *Physostegia*
FALSE SPIRAEA see *Astilbe*
FEVERFEW see *Chrysanthemum*
FLAME TIP see *Leucodendron*
FLAMINGO FLOWER see *Anthurium*

FORSYTHIA (see Plate 10)

(for-SITH-ee-a)
Family: Oleaceae
Name Origin: Named after Scottish gardener, William Forsyth (1737-1804).
Species: *x intermedia, ovata*
Common Names: forsythia, golden bells

HELICONIA
Lobster Claw

HELICONIA
Torch Heliconia

ANIGOZANTHOS
Kangaroo Paw

Availability: November through March
Description: Small yellow flowers clustered along woody stems, appearing before leaves. Tall stems add colorful line. Short stems work well as fillers. Useful in oriental and contemporary design styles as well as traditional springtime bouquets.
Vase-life: 1 to 2 weeks

FOXGLOVE see *Digitalis*
FOXTAIL LILY see *Eremurus*

FREESIA (see Plate 11)

(FREE-zee-ah or FREE-sha)
Family: Iridaceae
Name Origin: Named after a German physician, Friedrich Freese (1800s).
Species: *x hybrida*
Common Name: freesia
Availability: Year-round
Description: Lovely, fragrant flowers. The florets are funnel shaped and branch asymmetrically off a main stem on upper side of a curved spike. Available in a wide variety of colors. Popular for wedding flowers and corsages. Interesting forms work well as an accent or filler in contemporary design styles.
Vase-life: 7 to 10 days; ethylene sensitive.

G

GAILLARDIA (see Plate 11)

(gay-LARD-ee-ah)
Family: Compositae (Asteraceae)
Name Origin: Named after French magistrate, Gaillard de Charentonneau (18th century).
Species: *aristata* cultivars
Common Name: blanket flower
Availability: July through September
Description: Solitary, large, showy, daisylike bicolored flowers in yellows, oranges, and reds. Bright coloration of these flowers creates emphasis. Gaillardia provide mass in mixed bouquets.
Vase-life: 5 to 7 days

GARDENIA (see Plate 11)

(gar-DEE-nee-ah)
Family: Rubiaceae
Name Origin: Named after a Scottish physician and botanist, Alexander Garden (1730-91), of South Carolina.
Species: *jasminoides* cultivars
Common Names: gardenia, cape jasmine
Availability: Year-round
Description: Fragrant, waxy, multipetaled flowers. Available in white to cream. Commercially they are packaged in special boxes wrapped for high humidity. Generally three to a box; each flower has a support collar made from leaves. Gardenias bruise easily; moisten hands when working with them. Popular corsage and wedding flower.
Vase-life: Short lived, 1 to 2 days

GAYFEATHER see *Liatris*
GENISTA see *Cytisus*

GERBERA (see Plate 11)

(GER-be-rah)
Family: Compositae (Asteraceae)
Name Origin: Named after a German naturalist, Traugott Gerber (1700s).
Species: *jamesonii;* others
Common Names: gerbera, Transvaal daisy, African daisy, barberton daisy, veldt daisy
Availability: Year-round
Description: Daisylike flowers 3 to 5 inches across. Single and double-petaled forms. Available in a wide range of colors and bicolor patterns, usually with contrasting centers. Stems are hollow and leafless. These flowers have many uses. Can easily provide a focal point or add mass. Useful in traditional and contemporary design styles.
Vase-life: 7 to 10 days; weak stems; prone to stem blockage; fluoride sensitive.

GERMAN STATICE see *Limonium*
GIANT BOTTLEBRUSH see *Banksia*
GILLYFLOWER see *Matthiola*
GINGER see *Alpinia*

GLADIOLUS (see Plate 12)

(gla-dee-OH-lus)
Family: Iridaceae
Name Origin: From the Latin word for a small sword, referring to the shape of the leaves.
Species: hybrids
Common Names: gladiolus, glad, sword lily, corn flag
Availability: Year-round; smaller varieties available March through July

IXIA
Corn Lily
HYACINTHUS
Hyacinth
MUSCARI
Grape Hyacinth
DAHLIA
IRIS

Description: Flowers are arranged on a thick stem in a one-sided spike or spikelike raceme. Available in a wide range of colors. Florets, textures, and shapes vary and may be ruffled, fringed, or plain and shaped like orchids, tulips, or roses. The florets of miniature gladiolus are more loosely arranged on the stem. Both are excellent line flowers. Individual florets are useful for corsage work.
Vase-life: 1 to 2 weeks

GLOBE AMARANTH see *Gomphrena*
GLOBE ARTICHOKE see *Cynara*
GLOBEFLOWER see *Trollius*
GLOBE THISTLE see *Echinops*

GLORIOSA (see Plate 12)
(glow-ree-OH-sah)
Family: Liliaceae
Name Origin: From the Latin word *gloriosus* (glorious).
Species: *rothschildiana, superba*
Common Names: gloriosa lily, glory lily, climbing lily
Availability: Year-round
Description: Flowers have reflexed petals that are curled at the margins. Brightly colored red and yellow flowers on leafless stems. Unusual flower form provides emphasis. These flowers work well in oriental and contemporary design styles.
Vase-life: 5 to 10 days, with individual blooms continuously blooming, each lasting 4 or 5 days

GLORIOSA DAISY see *Rudbeckia*
GLORY LILY see *Gloriosa*
GOAT'S BEARD see *Astilbe*
GODETIA see *Clarkia*
GOLDENROD see *Solidago*
GOLDEN SHOWER see *Oncidium*

GOMPHRENA
(gom-FREE-nah)
Family: Amaranthaceae
Species: *globosa, haageana*
Common Name: globe amaranth
Availability: July through September
Description: Round flower heads about 1 inch across on the top of long stems. The flower head is made up of tiny fluffy flowers. Available in white, pinks, purples, and orange.
Vase-life: 5 to 10 days

GONIOLIMON see *Limonium*
GRAPE HYACINTH see *Muscari*
GUELDER ROSE see *Vibernum*
GUERNSEY LILY see *Nerine*

GYPSOPHILA (see Plate 13)
(jip-SOF-i-la or jip-so-IL-la)
Family: Caryophyllaceae
Name Origin: From the Greek words *gypsos* (gypsum) and *philos* (loving), referring to some species favoring gypsum or lime.
Species: *elegans, paniculata;* others and cultivars
Common Names: baby's breath, gyp
Availability: Year-round
Description: Complex panicles or dichasial cymes of white or pinkish white florets. A popular and delicate filler for arrangements. Small clusters work well in corsages, boutonnieres, and wedding flowers.
Vase-life: 5 to 7 days; ethylene sensitive; favors cool temperatures and high humidity.

H

HEATH see *Erica*
HEATHER see *Erica*

HELIANTHUS (see Plate 26)
(hee-li-ANTH-us)
Family: Compositae (Asteraceae)
Name Origin: From the Greek words *helios* (sun) and *anthos* (flower).
Species: *annuus, tuberosus;* others and hybrids
Common Name: sunflower
Availability: Year-round, with peak supplies June through October
Description: Large daisylike head flowers up to 10 inches across with yellow ray petals surrounding contrasting center disk florets. Single and double forms. Available in creams and yellows. Provides mass and emphasis.
Vase-life: 7 to 10 days

HELICHRYSUM (see Plate 27)
(hee-li-KRIS-um)
Family: Compositae (Asteraceae)
Name Origin: From the Greek words *helios* (sun) and *chryson* (golden).
Species: *bracteatum;* others and cultivars

CONSOLIDA
Larkspur
LIATRIS
Gay Feather
SYRINGA
Lilac
LUPINUS
Lupine

Common Names: strawflower, everlasting flower
Availability: July through September
Description: Brightly colored flowers with crisp, papery texture. Available in a wide range of colors including yellows, oranges, reds, pinks, and white. Sizes vary, most common about 2 inches across. Adds interesting texture and mass to arrangements.
Vase-life: Long lasting, 1 to 2 weeks

HELICONIA (see Plate 14)

(hel-i-KO-nee-ah)
Family: Heliconiaceae
Species: *humilis, caribaea, psittacorum, pendula;* others
Common Names: heliconia, lobster claw, rainbow, parrot's flower; many common names with each species
Availability: Year-round
Description: Tropical erect or drooping flowers; brilliant bracts of inflorescence densely packed together on thick stems. Most are bicolored in reds, oranges, yellows, and green. Use alone or with other tropical flowers and foliage. Coloring and unique shape demand attention.
Vase-life: Long lasting, 10 to 14 days

HELIPTERUM

(hee-LIP-tur-um)
Family: Compositae (Asteraceae)
Name Origin: From the Greek words *helios* (sun) and *pteron* (wing), referring to their sun-loving nature and their feathery bristles.
Species: *humboldtianum, manglesii, roseum*
Common Names: Swan River everlasting, immortelles
Availability: May through October
Description: Daisylike, straw-textured head flowers on thin stems. Available in white and pink with yellow centers. Adds mass and also used as a filler.
Vase-life: 5 to 7 days

HELLEBORUS

(he-li-BOHR-us)
Family: Ranunculaceae
Species: *niger, orientalis;* others and hybrids
Common Names: Christmas rose, Lenten rose, hellebore
Availability: Year-round
Description: Large and attractive white, greenish, pink, and purple flowers 2 to 5 inches across. Five petals surround golden anthers. Useful for adding emphasis and mass.
Vase-life: 5 to 7 days

HIPPEASTRUM (Amaryllis)

HIPPEASTRUM

(hip-ee-AS-trum)
Family: Amaryllidaceae
Name Origin: From the Greek word *hippos* (horse), referring to the inflorescence of the species *H. puniceum,* which was likened to the head of a horse.
Species: hybrids
Common Names: amaryllis, Barbados lily
Availability: Year-round
Description: Hippeastrum has 4 to 6 trumpet-shaped flowers clustered together at the top of a thick, hollow, leafless stem in an umbel pattern. Individual flowers are 5 to 9 inches across. Available in white, pinks, reds, oranges, and bicolors. These flowers have distinctive shapes that demand attention and works well in contemporary design styles. Allow space for blossoms to open.
Vase-life: 1 to 2 weeks, while individual florets continue opening

HYACINTH see *Hyacinthus*

HYACINTHUS (see Plate 15)

(hi-ah-SIN-thus)
Family: Liliaceae
Species: *orientalis* cultivars

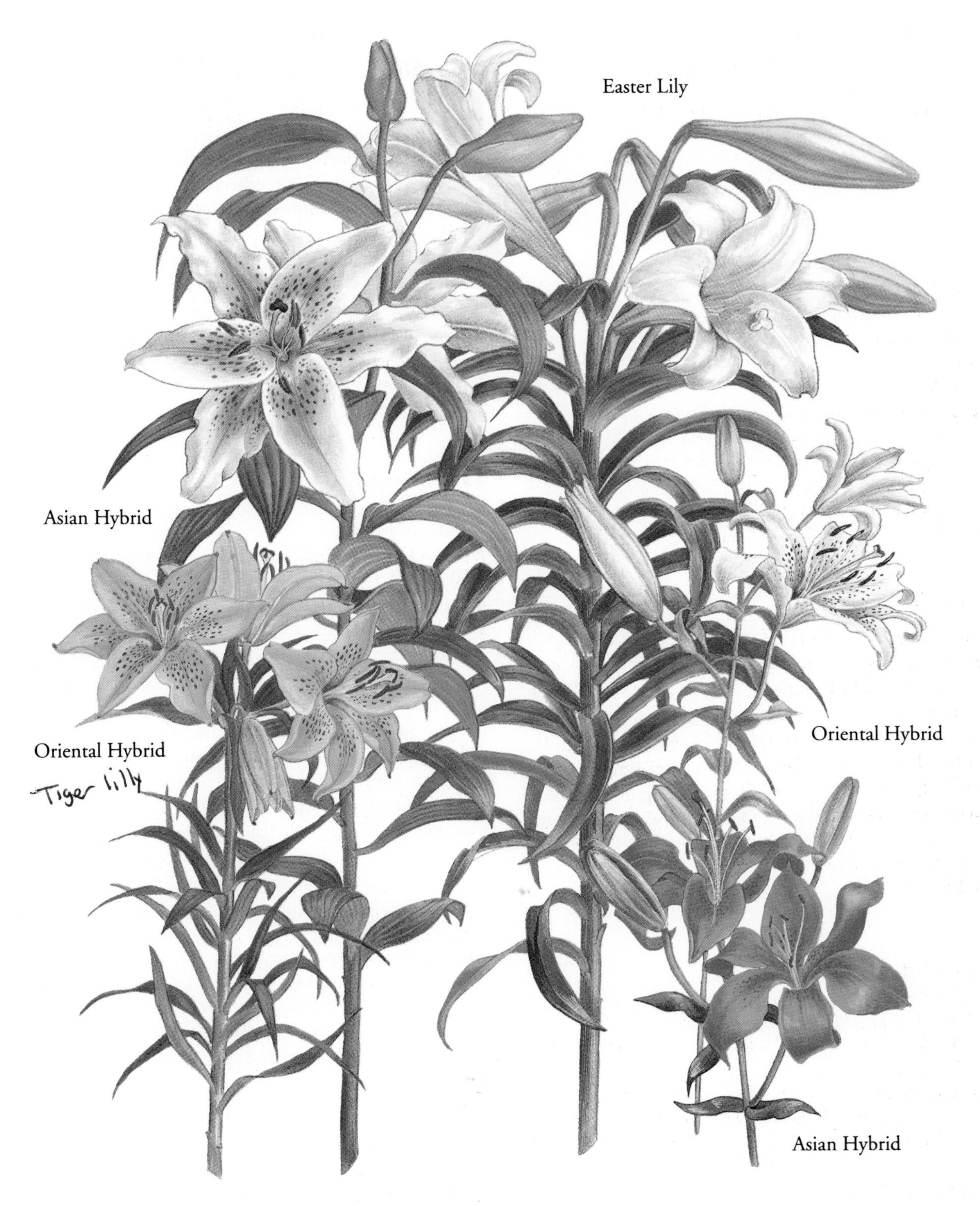

LILIUM—Many hybrids are available, including large and showy Oriental hybrids and small colored Asian hybrids.

HYDRANGEA

Common Names: hyacinth, Dutch hyacinth
Availability: November through April
Description: Compact spikelike racemes with bell-shaped, waxy florets. Available in white, pink, red, blue, purple, and yellow. Extremely fragrant. These flowers are at home with other bulb flowers in traditional spring designs. Also distinctive in contemporary designs, particularly vegetative and parallel styles.
Vase-life: 3 to 7 days

HYDRANGEA

(hi-DRAN-jee-ah)
Family: Saxifragaceae
Name Origin: From the Greek words *hydro* (water) and *aggos* (jar), referring to the fruits that are shaped like cups.
Common Name: hydrangea
Species: *macrophylla, paniculata;* others
Availability: July to October
Description: Large rounded or pyramidal compound clusters of tiny starlike florets in blues, pinks, and white. These flowers provide mass and emphasis in large arrangements.
Vase-life: 5 to 10 days; seal the bleeding latex for longer vase-life

I

IBERIS (see Plate 5)

(i-BEER-is)
Family: Cruciferae (Brassicaceae)
Name Origin: From the Greek word *iberis* (from Iberia).
Species: *amara, umbellata;* others and cultivars
Common Names: candytuft, rocket candytuft
Availability: March through August
Description: Clustered florets that form a convex corymb or elongated raceme inflorescence. Available in white, pink, red, and lavender. Useful as a filler in mixed arrangements. Also works well grouped in contemporary design styles.
Vase-life: 5 to 7 days

INCA LILY see *Alstroemeria*

IRIS (see Plate 15)

(EYE-ris)
Family: Iridaceae
Name Origin: Named after the Greek goddess of the rainbow.
Species: *reticulata, sibirica;* others and hybrids
Common Names: iris, Dutch iris, flag
Availability: Year-round, with peak supplies March through May.
Description: Distinctive flower forms available in blues, purples, yellows, and white. Useful in a variety of design styles. Can be used for emphasis and accent. Allow plenty of space around flowers, as they continue to open. Striking in oriental and contemporary designs.
Vase-life: 2 to 6 days; ethylene sensitive; avoid water loss.

IXIA (see Plate 15)

(IKS-ee-ah)
Family: Iridaceae
Name Origin: From the Greek word *ixia* (bird lime), referring to the sticky sap.
Species: *viridiflora* hybrids
Common Names: corn lily, African corn lily
Availability: March through August
Description: Star-shaped florets in spike or panicle clusters on wiry stems. Available in cream, yellow, pink, orange, and red with mixtures of these colors. Works well as a line flower. Can also be used as a filler in larger bouquets. Especially striking in contemporary design styles.
Vase-life: 5 to 10 days, with individual florets continuously opening

EUSTOMA
Lisianthus
NERINE
Nerine Lily
CONVALLARIA
Lily of the Valley
PHYSOSTEGIA
False Dragonhead

J

JONQUIL see *Narcissus*
JOSEPH'S COAT see *Amaranthus*

K

KANGAROO PAW see *Anigozanthos*
KISS OF DEATH see *Costus*

KNIPHOFIA

(nee-FOF-ee-ah)
Family: Liliaceae
Name Origin: Named after Johann H. Kniphof (1704-63).
Species: *uvaria* hybrids
Common Names: red hot poker, tritoma, torch lily, poker plant
Availability: June through October
Description: Tight, terminal, spikelike racemes packed with overlapping florets of red, orange, and yellow on leafless stems. The scarlet florets become orange and then yellow with age. These tall-stemmed flowers add emphasis and height and look best in contemporary design styles.
Vase-life: 7 to 10 days, with florets continuously opening

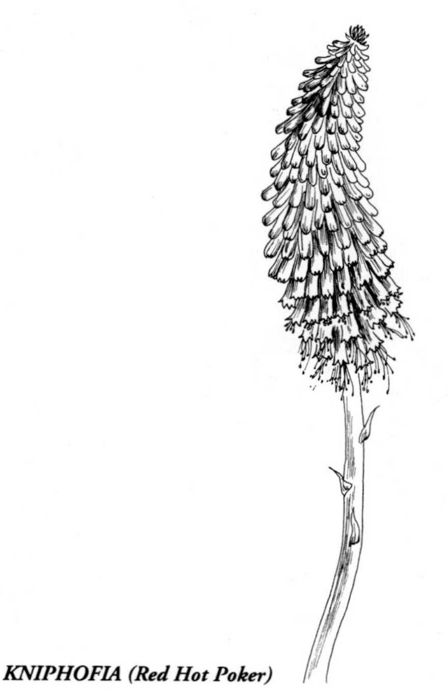

KNIPHOFIA (Red Hot Poker)

L

LADY'S MANTLE see *Alchemilla*
LADY'S SLIPPER see *Paphiopedilum*
LARKSPUR see *Consolida*

LATHYRUS (see Plate 26)

(LATH-i-rus)
Family: Leguminosae (Fabaceae)
Name Origin: The Greek name for the pea.
Species: *odoratus* cultivars
Common Name: sweet pea
Availability: February through September
Description: Sweetly scented flowers with delicate petals in soft colors, including white, pinks, reds, blues, and lavenders. Stems are fairly short with 3 to 7 flowers per stem. Sweet peas can be used to add mass or as a filler. Simple and charming alone or with other spring and summer flowers.
Vase-life: 3 to 7 days

LAVATERA

(la-vah-TER-ah)
Family: Malvaceae
Name Origin: Named after Zurich naturalists, the Lavater brothers of the sixteenth century.
Species: *trimestris*
Common Names: mallow, tree mallow
Availability: June to October
Description: Trumpet-shaped flowers that look like miniature Hawaiian hibiscus flowers. Several flowers appear at the top of tall leafy stems. Available in reds, pinks, and white.
Vase-life: 5 to 10 days

LEI ORCHID see *Vanda*

LEPTOSPERMUM (see Plate 22)

(lep-toe-SPUR-mum)
Family: Myrtaceae
Name Origin: From the Greek words *leptos* (slender) and *sperma* (seed), referring to its narrow seeds.
Species: *scoparium*
Common Names: lepto, tea tree, New Zealand tea tree
Availability: Peak supplies January through April
Description: Clusters of small blossoms on woody stems. Available in white, pinks, and reds. Long branches are excellent for adding line. Smaller stems work well as spiky

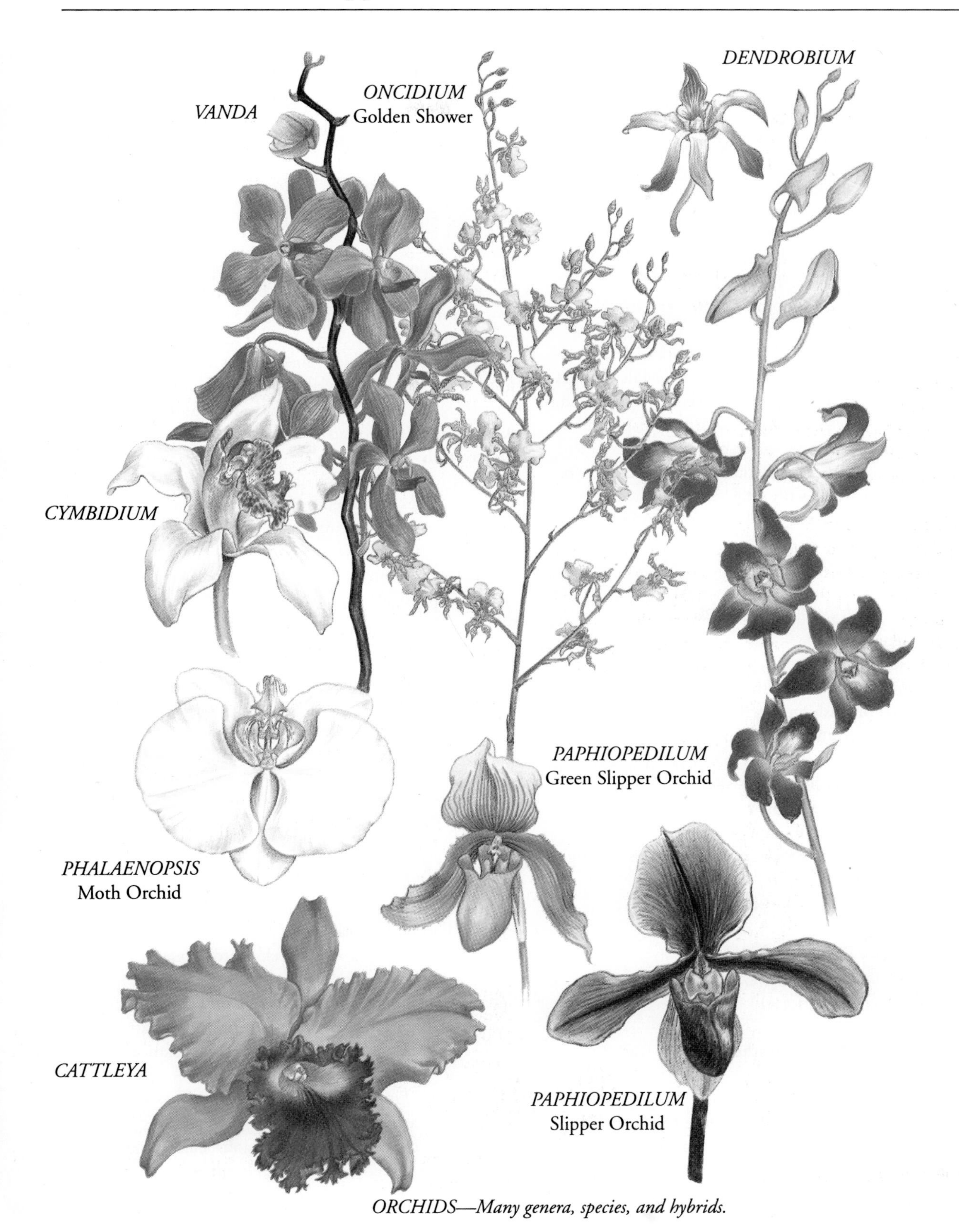

ORCHIDS—*Many genera, species, and hybrids.*

fillers. Provides an elegant line in oriental and contemporary design styles.
Vase-life: Long lasting, 7 to 10 days

LEUCODENDRON (see Plate 20)

(loo-ka-DEN-dron)
Family: Proteaceae
Name Origin: From the Greek words *leukos* (white) and *dendron* (tree), referring to its silvery foliage.
Species: *argenteum;* others and cultivars
Common Names: silver tree, flame tip
Availability: Year-round; some species are limited by season
Description: Flower head consists of stiff bracts surrounding a cone or small inconspicuous flower. Male flowers have terminal, sessile heads; female flowers have terminal, conelike heads. The woody bracts, often mistaken for petals, are generally colorful, most often reds, burgundy, green, and yellows, and combinations of these colors. The stems are densely packed with stiff leaves that radiate out on all sides. These flowers work well with other exotic flowers.
Vase-life: Long lasting, 2 to 3 weeks

LEUCOSPERMUM (see Plate 20)

(lu-co-SPER-mum)
Family: Proteaceae
Species: *cordifolium, reflexum;* others and cultivars
Common Name: pincushion protea
Availability: Year-round; some species are limited by season
Description: The flower heads consist of colorful orange or reddish orange styles that form a domed, globular shape (thus the name pincushion). Solitary flowers top woody stems loaded with stiff leaves. Creates emphasis through texture and form.
Vase-life: Long lasting, 2 to 3 weeks

LIATRIS (see Plate 16)

(lie-AH-tris)
Family: Compositae (Asteraceae)
Species: *spicata;* others
Common Names: liatris, gayfeather, blazing star, button snakeroot, purple poker
Availability: Year-round
Description: Tiny florets appear in dense spikes; opening from the top downward. Available in purple and white. Excellent for creating strong line in arrangements. Works well in mixed bouquets and is striking in oriental and contemporary design styles.
Vase-life: 7 to 10 days

LILAC see *Syringa*

LILIUM (see Plate 17)

(LIL-ee-um)
Family: Liliaceae
Name Origin: The Latin name for these bulbous flowers.
Species: *longiflorum;* others and hybrids and cultivars
Common Names: Lilies bred from *L. auratum,* and *L. speciosum,* (such as *L.S.* 'Rubrum') are called oriental lilies and are generally white, pink, and white with red. Asian lilies are generally the yellows, oranges, reds, and whites; Easter lily *(L. longiflorum)* is white and trumpet-shaped.
Availability: Year-round
Description: The flowers are 4 to 6 inches across. A wide range of forms and colors are available. Flowers appear on short branches at the end of the stem. Some are upright, while others are pendant or outward-facing. The striking form of lilies with their central floral parts provide a focal point. Leave room for lily blossoms to open. Lilies work well in mixed arrangements, wedding, and sympathy designs. Striking in oriental and contemporary designs.
Vase-life: 4 to 5 days per bloom

LILY see Lilium
LILY OF THE FIELD see *Anemone*
LILY OF THE INCAS see *Alstroemeria*
LILY OF THE NILE see *Agapanthus*
LILY OF THE VALLEY see *Convallaria*

LIMONIUM (see Plate 25)

(lee-MO-nee-um)
Family: Plumbaginaceae
Name Origin: From the Greek word *leimon* (meadow), referring to their natural habitat.
Species: *latifolium, perezii, sinuatum;* others and cultivars
Common Names: statice, sea lavender, seafoam statice; *Goniolimon (L. tataricum)* is called German statice.
Availability: Year-round
Description: Tiny white or yellow flowers are surrounded by papery bracts. Flowers are sessile, in panicles or spikes. Bract colors include white and various shades and tints of pink, yellow, blue, and purple. Useful as a colorful filler

Members of the *Proteaceae* family.

adding interesting texture to mixed bouquets.
Vase-life: Long lasting, 1 to 2 weeks; subject to mildew and brown spots; provide air circulation.

LISIANTHUS see *Eustoma*
LOBSTER CLAW see *Heliconia*
LOOSESTRIFE see *Lysimachia*
LOVE IN A MIST see *Nigella*
LOVE LIES BLEEDING see *Amaranthus*
LUPINE see *Lupinus*

LUPINUS (see Plate 16)
(lu-PEEN-us)
Family: Leguminosae (Fabaceae)
Species: *polyphyllus* cultivars
Common Name: lupine
Availability: July through September
Description: Lupine florets resemble pea blossoms and are densely packed on tall, erect spikelike stalks in racemes. Available in a wide variety of colors. These are graceful, old-fashioned flowers that look lovely in mixed summer bouquets.
Vase-life: 5 to 10 days

LYSIMACHIA
(lie-si-MAK-ee-ah)
Family: Primulaceae
Name Origin: Named after King Lysimachos of ancient Thrace. According to legend, the king was able to pacify a bull by using a piece of loosestrife.
Species: *clethroides, punctata*
Common Name: loosestrife
Availability: July to September
Description: Tiny white or yellow, star-shaped florets tightly arranged in a small slender, curving, spikelike racemes at the top of dense foliage.
Vase-life: 5 to 7 days

M

MADAGASCAR JASMINE
see *Stephanotis*

MAGNOLIA
(mag-NOL-ee-ah)
Family: Magnoliaceae

MAGNOLIA

Name Origin: Named after French professor of botany Pierre Magnol (1638-1715).
Species: many species and hybrids
Common Name: magnolia
Availability: February to May
Description: Elegant focal flowers on woody stems. A variety of forms and colors. Beautiful in oriental designs.
Vase-life: 3 to 6 days

MALLOW see *Lavatera*
MARGUERITE DAISY
see *Chrysanthemum frutescens*
MARIGOLD see *Tagetes*
MASTERWORT see *Astrantia*

MATTHIOLA (see Plate 25)
(ma-THEE-oh-lah)
Family: Cruciferae (Brassicaceae)
Name Origin: Named after Italian botanist Pierandrea Mattioli (1500-77).
Species: *incana* and hybrids
Common Names: stock, gillyflower
Availability: January through October
Description: Small, 1-inch florets form rounded, spikelike racemes. Available in a wide selection of colors. These fragrant flowers add mass and line to mixed bouquets.
Vase-life: 3 to 7 days

MEADOWSWEET see *Astilbe*
MICHAELMAS DAISY see *Aster*

PHLOX
ANANAS
Ornamental Pineapple
AMMI
Queen Anne's Lace
RANUNCULUS

MILKWEED see *Asclepias*

MIMOSA see *Acacia*

MOLUCCELLA (see Plate 4)

(mahl-you-SEL-ah)
Family: Labiatae (Lamiaceae)
Species: *laevis*
Common Names: bells of Ireland, shellflower, molucca balm
Availability: Year-round, with peak supplies June through October
Description: Whorls of tiny, white, fragrant flowers are surrounded by curious green shell-like calyces resembling bells, which are often mistaken for petals. The flowers and sepals are clustered along tall stems. Often reserved for St. Patrick's Day, but useful throughout the year for adding line and accent. Striking in all-foliage and contemporary designs.
Vase-life: 7 to 10 days

MONKSHOOD see *Aconitum*

MONTBRETIA see *Crocosmia*

MONTE CASSINO see *Aster*

MOTH ORCHID see *Phalaenopsis*

MUSCARI (see Plate 15)

(mus-KAIR-ree)
Family: Liliaceae
Species: *armeniacum, botryoides;* others
Common Name: grape hyacinth
Availability: January through March
Description: Tiny, tubular white or blue florets densely clustered on short stems. These small and delicate flowers work best in contemporary design styles, such as parallel systems or vegetative, where they can be clustered together for greater impact.
Vase-life: 4 to 7 days

N

NARCISSUS (see Plate 7)

(nar-SIS-us)
Family: Amaryllidaceae
Name Origin: Narcissus was a legendary youth of Greek mythology who was arrogant and fell in love with his own reflection.

NIGELLA

Species: *pseudonarcissus;* others and hybrids
Common Names: narcissus, daffodil, jonquil
Availability: November through April
Description: The single, trumpet-shaped flowers are known as daffodils; other varieties are known as jonquils or narcissus. The tiny white narcissus in clusters are known as paperwhites. Narcissus flowers are available in a great range of forms, sizes, and colors—white, cream, yellows, oranges, and bicolors. Their form adds accent and emphasis to simple and contemporary design styles.
Vase-life: 4 to 6 days; condition narcissus alone, as they secrete sap when cut that is harmful to other flowers; do not recut when mixing with other flowers in designs.

NERINE (see Plate 18)

(near-REEN)
Family: Amaryllidaceae
Name Origin: Named after a Greek sea nymph.
Species: *bowdenii, sarniensis;* others
Common Name: Guernsey lily
Availability: Year-round
Description: Small florets are lilylike and clustered at the top of leafless stems in umbel patterns. There are anywhere from 6 to 12 florets forming each cluster. Available in pinks, reds, and white. These flowers work well in many

SALIX
Pussy Willow
PRUNUS
CHAENOMELES
Flowering Quince
LEPTOSPERMUM
Flowering Branches

design styles including mixed bouquets, oriental, and contemporary. The individual florets are useful in corsages, boutonnieres, and wedding designs.
Vase-life: Long lasting, 1 to 2 weeks

NIGELLA

(nee-JEL-lah)
Family: Ranunculaceae
Species: *damascena;* others
Common Names: love in a mist, fennel flower, wild fennel
Availability: June to September
Description: Nigella flowers are small and flat; several appear at the top of a thin, hairy stem. Egg-shaped seed capsules among the flowers provide accent. Flowers are available in blue, pink, and white. Nigella provides interesting texture.
Vase-life: 5 to 7 days

O

OBEDIENT PLANT see *Physostegia*

ONCIDIUM (see Plate 19)

(ahn-SID-ee-um)
Family: Orchidaceae
Name Origin: From the Greek word *onkos* (tumor), referring to a swelling on the lip.
Species: *pulchellum, splendidum, varicosum;* others and hybrids
Common Names; oncidium, golden shower, butterfly orchid, dancing doll, dancing lady orchid
Availability: Year-round
Description: Masses of tiny orchid flowers in thin, branching stems. The flowers are yellow with speckles of orange, red, and brown. The stems arch with the weight of the florets and provide graceful linear curves in arrangements. These flowers work well with other orchids and exotic flowers. Arching stems are beautiful in oriental designs.
Vase-life: Long lasting, 1 to 2 weeks

ONION FLOWER see *Allium*

ORCHID see *Cattleya, Cymbidium, Dendrobium, Oncidium, Paphiopedium, Phalaenopsis, Vanda*

ORIENTAL LILY see *Lilium*

ORNAMENTAL PINEAPPLE see *Ananas*

ORNITHOGALUM (see Plate 27)

(or-ni-THAHG-ah-lum)
Family: Liliaceae
Name Origin: From the Greek words *ornis* (bird) and *gala* (milk).
Species: *arabicum, thyrsoides;* others
Common Names: chincherinchee, star of Bethlehem
Availability: Year-round
Description: White, star-shaped flowers in tight racemes and corymb clusters at the top of leafless stems. Flower centers are either whitish green or black. Star of Bethlehem is a versatile flower, adding line, mass, or accent. Also useful as a spiky filler. A beautiful, long-lasting addition to mixed bouquets, oriental designs, and contemporary styles.
Vase-life: Long lasting, 2 to 3 weeks

OSTRICH PLUME see *Alpinia*

OUTDOOR GYPSOPHILA see *Saponaria*

P

PAEONIA (see Plate 24)

(pee-OH-ne-ah)
Family: Paeoniaceae (Ranunculaceae)
Name Origin: From the Greek name *paionia,* referring to Paion, who was the physician to the gods.
Species: *lactiflora, suffruticosa*
Common Name: peony
Availability: May through July
Description: Large, fragrant, single flowers from 3 to 8 inches across. Available in a wide variety of forms, including single, double, and anemone types. These giant blossoms are grand all by themselves or in mixed arrangements.
Vase-life: 3 to 7 days

PAINTED TONGUE see *Anthurium*

PAINTER'S PALETTE see *Anthurium*

PAPAVER

(pah-PAY-ver)
Family: Papaveraceae
Species: *orientale, nudicaule;* others and hybrids
Common Names: poppy, Iceland poppy
Availability: May through September
Description: Papery petals surround a contrasting center,

ROSA

***PAPAVER* (Poppy)**

open up almost flat. Flowers are solitary at the top of nodding, wiry stems. Available in a variety of intense colors.
Vase-life: 2 to 5 days; to increase vase-life, sear the stem ends with a flame or dip in boiling water.

PAPERWHITES see *Narcissus*

PAPHIOPEDILUM (see Plate 19)

(paf-ee-oh-PED-il-lum)
Family: Orchidaceae
Name Origin: From the Greek *Paphos,* the site of a temple on Cyprus where Aphrodite was worshipped. Also from the Greek word *pedilon* (slipper).
Species: *bellatulum, fairrieanum;* others and hybrids; greenhouse Cypripediums belong to the genera Paphiopedilum.
Common Names: slipper orchid, lady's slipper
Availability: Year-round
Description: Exotic flowers with dramatic form and coloring. Ideal for contemporary designs.
Vase-life: 5 to 7 days

PEONY see *Paeonia*

PERSIAN BUTTERCUP see *Ranunculus*

PERUVIAN LILY see *Alstroemeria*

PHALAENOPSIS (see Plate 19)

(fal-en-NOP-sis)
Family: Orchidaceae
Name Origin: From the Greek words *phalaina* (moth) and *-opsis* (resembling).
Species: *amabilis, gigantea;* others and hybrids
Common Name: moth orchid
Availability: Year-round
Description: These delicate, elegant orchid flowers are 3 to 4 inches across and flatter-looking than most orchids. Available mostly in white and pink. Generally sold individually. Frequently used in wedding and corsage designs.
Vase-life: 3 to 5 days

PHLOX (see Plate 21)

(floks)
Family: Polemoniaceae
Name Origin: From the Greek word *phlox* (flame).
Species: *paniculata;* others
Common Name: phlox
Availability: June through November
Description: Small florets form dense terminal panicles on leafy stems. Available in white, pinks, reds, lavenders, and bicolors. An excellent filler flower.
Vase-life: 3 to 7 days

PHYSOSTEGIA (see Plate 18)

(fie-soe-STEE-jee-ah)
Family: Labiatae (Lamiaceae)
Species: *virginiana* cultivars
Common Names: false dragonhead, obedient plant, lion's heart, obedience
Availability: July to October
Description: Tubular florets form along a 12- to 24-inch stem, forming a spikelike inflorescence with an unusual shape. Available in white, pinks, and lavender.
Vase-life: 1 week

PINCUSHION FLOWER see *Scabiosa*

PINCUSHION PROTEA see *Leucospermum*

PINEAPPLE, ORNAMENTAL see *Ananas*

PINK see *Dianthus*

PITCHER PLANT see *Sarracenia*

PIXIE see *Dianthus*

SCABIOSA
Pincushion Flower
CARTHAMUS
Safflower
PAEONIA
Peony
ANTIRRHINUM
Snapdragon

PLUME THISTLE see *Cirsium*

POINSETTIA see *Euphorbia*

POLIANTHES (see Plate 28)

(pah-lee-ANTH-eez)
Family: Agavaceae
Name Origin: From the Greek word *polis* (grey) and *anthos* (flower).
Species: *tuberosa*
Common Name: tuberose
Availability: February to October
Description: Clustered spikes of fragrant, waxy, white flowers. Star-shaped florets are about 1 inch across. These flowers can add mass, line, and accent to a number of design styles. Useful in corsage and wedding designs.
Vase-life: Long lasting, 1 to 2 weeks

POPPY see *Papaver*

POT MARIGOLD see *Calendula*

PRINCE'S FEATHER see *Amaranthus*

PROTEA (see Plate 20)

(PRO-tee-a)
Family: Proteaceae
Name Origin: Named after the Greek sea God, *Proteus,* who had the power of prophecy.
Species: *compacta, cynaroides, grandiceps, magnifica, nerifolia, obtusifolia, repens;* others and hybrids
Common Names: protea; king protea *(P. cynaroides);* queen protea *(P. magnifica);* pink mink *(P. nerifolia);* many other common names
Availability: Year-round
Description: Proteas are large focal flowers. They are rounded with colorful bracts in a variety of colors, forms, and textures. They stand singly at the top of woody, leafy stems.
Vase-life: Long lasting, 2 to 3 weeks

PRUNUS (see Plate 22)

(PRU-nus)
Family: Rosaceae
Name Origin: Prunus is the Latin name for the plum tree.
Common Names: Prunus includes cherry, plum, peach, apricot, nectarine, and almond trees and shrubs.
Availability: October through May, depending on species. Can be forced to blossom indoors.
Description: Single or double flowers along woody branches. A wide variety of forms, depending on the species. Available in white and tints and shades of pink. Many types are fragrant. Tall branches add line. Shorter stems used as linear fillers. Blossoming branches are elegant in oriental and vegetative designs.
Vase-life: Long lasting, 1 to 2 weeks

PUSSY WILLOW see *Salix discolor*

Q

QUEEN ANNE'S LACE see *Ammi*

QUINCE see *Chaenomeles*

R

RAINBOW see *Heliconia*

RANUNCULUS (see Plate 21)

(rah-NUN-kew-lus)
Family: Ranunculaceae
Name Origin: From the Latin word *rana* (frog), as many prefer to grow in wet areas.
Species: *asiaticus;* others and hybrids
Common Names: ranunculus, Persian buttercup
Availability: January through May
Description: These single and double flowers resemble small peonies, with centers often a contrasting color or black. Sizes vary from 1 to 4 inches wide. Available in a wide selection of bright colors including white, yellow, orange, red, and pink. Colorful additions to mixed bouquets and contemporary design styles.
Vase-life: 3 to 7 days

RED HOT POKER see *Kniphofia*

ROSA (see Plate 23)

(RO-za)
Family: Rosaceae
Species: hybrids and cultivars
Common Name: rose
Availability: Year-round
Description: Roses are available in a wide range of colors and sizes. The flowers are solitary, corymbose, or panicled on erect stems. The heads are densely crowded with petals. Roses can be grouped simply by head and stem size into large, medium, mini or sweethearts, and spray roses. Roses are popular cut flowers and are useful in many design

LIMONIUM
Statice
STEPHANOTIS
LIMONIUM
Sea Lavender Statice
MATTHIOLA
Stock
LIMONIUM
German Statice

styles. Excellent flowers for corsages and boutonnieres.
Vase-life: Varies greatly, from 3 to 14 days

ROSE see *Rosa*

RUDBECKIA (see Plate 4)
(rude-BEK-ee-ah)
Family: Compositae (Asteraceae)
Name Origin: Named after Olof Rudbeck the elder (1630-1702) and the younger (1660-1740).
Species: *fulgida, hirta;* others and hybrids
Common Names: coneflower, black-eyed Susan, gloriosa daisy
Availability: July through September
Description: Daisylike gold and orange flowers with prominant black, cone-shaped centers. These flowers add mass and accent. They are striking additions to mixed summer bouquets.
Vase-life: 7 to 10 days

S

SAFFLOWER see *Carthamus*

SALIX (see Plate 22)
(SAY-liks)
Family: Salicaceae
Species: *discolor;* others
Common Name: pussy willow; a variety of plant material from the genera *Salix* is commonly used in floral arrangement, including leafy branches with their blossoming catkins and curly or corkscrew willow.
Availability: January through April
Description: Pussy willow flowers are soft and fuzzy catkins. They are generally greyish white and closely spaced along tall woody stems. Useful as line material in a number of arrangement styles, especially traditional springtime bouquets and oriental designs.
Vase-life: Long lasting, 10 to 14 days

SAPONARIA
(sap-oh-NAH-ree-ah)
Family: Caryophyllaceae
Name Origin: From the Latin word *sapo* (soap). A soap can be made from the species *S. officinalis.* Saponaria is also known as *Vaccaria.*
Species: *officinalis;* others
Common Names: saponaria, bouncing bet, soapwort, outdoor gypsophila

SAPONARIA (Outdoor Gypsophila)

Availability: June through September
Description: Small star-shaped flowers at the tips of cymose or paniculate branching stems. Available in white and pink. A delicate and airy filler in mixed arrangements.
Vase-life: 7 to 10 days

SARRACENIA (see Plate 26)
(sa-ra-SEE-nee-ah)
Family: Sarraceniaceae
Name Origin: Named after a French botanist and physician, Michael Sarrasin (1659-1734).
Species: *flava, purpurea;* others and hybrids
Common Names: sarracenia, pitcher plant, swamp lily, cobra head, cobra lily, trumpet
Availability: April through September
Description: Unusual curiosities that have unique form and venation color patterns. Often mistaken for flowers, these leaves with an odd tubular, trumpet shape provide emphasis, line, and accent.
Vase-life: 7 to 10 days

SATIN FLOWER see *Clarkia*

SARRACENIA
Swamp Lily
LATHYRUS
Sweet Pea
HELIANTHUS
Sunflower

SCABIOSA (see Plate 24)

(skab-ee-OH-sah)
Family: Dipsacaceae
Name Origin: From the Latin word *scabies* (itch), for the rough leaves were said to cure the itch.
Species: *altropurpurea, caucasica*
Common Names: pincushion flower, scabiosus
Availability: June through October
Description: Round, single flowers with papery texture. Flowers are 2 to 3 inches across. Available in a wide variety of colors. Useful as a mass flower in mixed bouquets. The unusual texture and intricate circular form also can create a focal point in contemporary design styles.
Vase-life: 5 to 7 days

SCARLET PLUME see *Euphorbia*

SCILLA

(SKIL-lah or SIL-lah)
Family: Liliaceae
Name Origin: From the Greek name for the sea squill.
Species: *sibirica;* others and hybrids
Common Names: squill, blue bells, wood hyacinth
Availability: March through June
Description: Several small blue florets in racemes cluster the top of short stems. Works well in vegetative designs.
Vase-life: 7 to 10 days

SCILLA

SEA HOLLY see *Eryngium*

SEA LAVENDER see *Limonium*

SEDUM (see Plate 27)

(SEE-dum)
Family: Crassulaceae
Name Origin: From the Latin word *sedo* (to sit), a classical name for many succulent plants.
Species: *spectabile, telephium;* others and hybrids
Common Names: sedum, stonecrop
Availability: April through October
Description: Tiny star-shaped flowers that form dense terminal panicles. Available in yellow, pinks, reds, and white. Sedum works well as a filler in mixed bouquets.
Vase-life: 7 to 10 days

SHAMPOO GINGER see *Zingiber*

SHASTA DAISY see *Chrysanthemum*

SHELLFLOWER see *Moluccella*

SILVER TREE see *Leucodendron*

SINGAPORE ORCHID see *Dendrobium*

SLIPPER ORCHID see *Paphiopedilum*

SNAPDRAGON see *Antirrhinum*

SNOWBALL see *Viburnum*

SNOW ON THE MOUNTAIN

see *Euphorbia marginata*

SOLIDAGO (see Plate 27)

(so-li-DAY-go)
Family: Compositae (Asteraceae)
Name Origin: From the Latin word *solido* (to strengthen or make whole), making reference to its medicinal properties.
Species: *canadensis;* others
Common Names: solidago, goldenrod
Availability: May through October
Description: Tiny yellow flowers forming soft panicled or racemed plumes. Solidago works well as a filler.
Vase-life: 7 to 10 days

x SOLIDASTER (see Plate 27)

(so-li-DAS-ter)
Family: Compositae (Asteraceae)
Species: Solidaster is an intergeneric hybrid, from the names of the parents: *Aster* and *Solidago; S. luteus;* hybrids

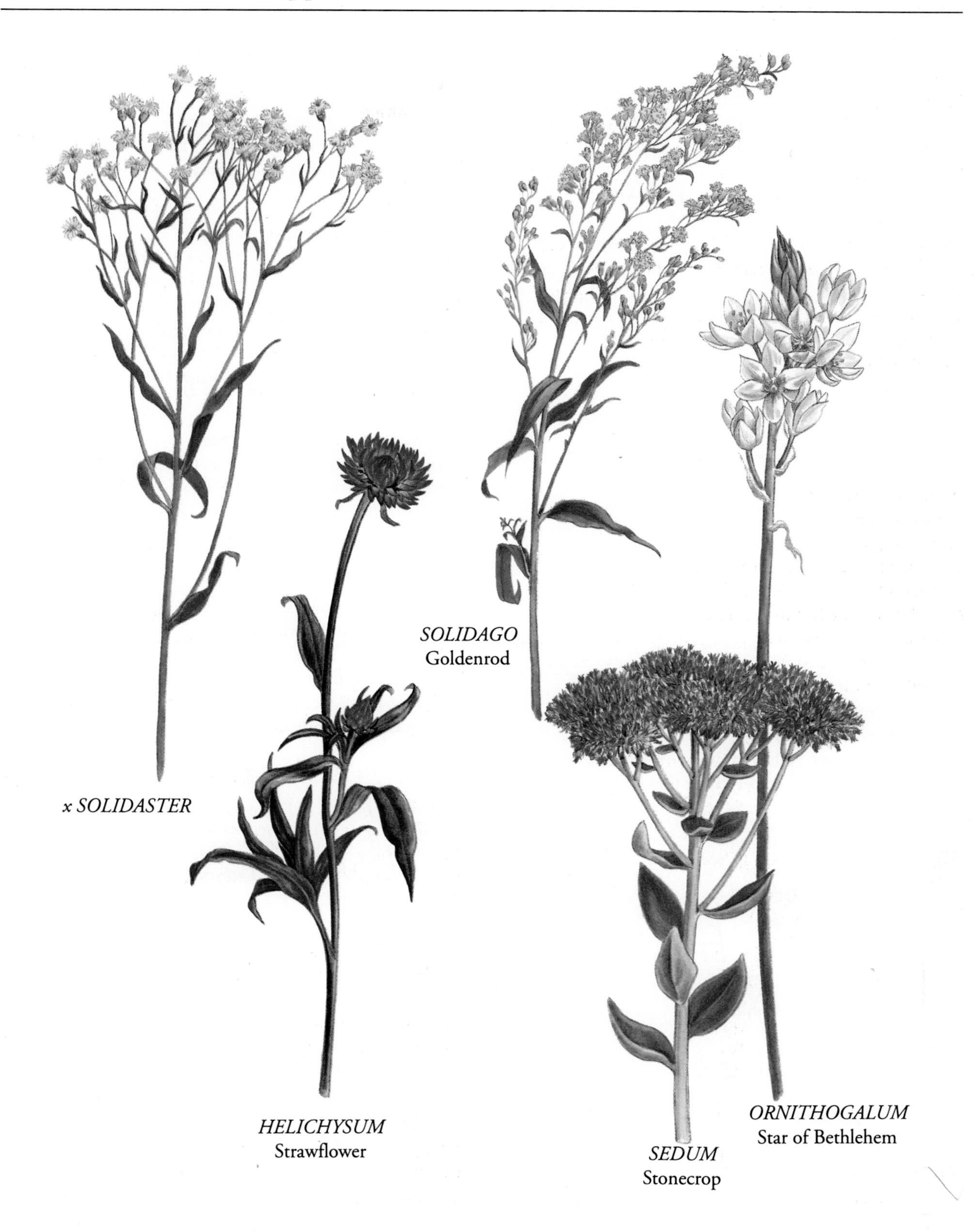
SOLIDAGO
Goldenrod
x SOLIDASTER
HELICHYSUM
Strawflower
ORNITHOGALUM
Star of Bethlehem
SEDUM
Stonecrop

STRELITZIA (Opening up Strelitzia florets and removing old florets)

Common Name: solidaster
Availability: Year-round
Description: Tiny yellow flowers on branching stems. Adds a fluffy texture. Excellent filler flower.
Vase-life: 7 to 10 days

SPIRAL GINGER see *Costus*
SPURGE see *Euphorbia*
SQUILL see *Scilla*
STAR OF BETHLEHEM see *Ornithogalum*
STATICE see *Limonium*

STEPHANOTIS (see Plate 25)

(ste-fa-NO-tis)
Family: Asclepiadaceae
Name Origin: From the Greek words *stephanos* (crown) and *otos* (ear). Greek name for myrtle, which was used to make crowns.
Species: *floribunda*
Common Names: stephanotis, steph, Madagascar jasmine
Availability: Year-round
Description: Small white fragrant flowers. Tubular, star-shaped, waxy blossoms. Flowers are cut off vines and sold stemless. Packaged in air-tight, humid boxes or bags. Stephanotis works well in corsages, boutonnieres, and wedding designs.
Vase-life: 3 to 4 days

STOCK see *Matthiola*
STONECROP see *Sedum*
STRAWFLOWER see *Helichrysum*

STRELITZIA (see Plate 2)

(stre-LITS-ee-a)
Family: Strelitziaceae (Cannaceae)
Name Origin: Named after Queen Charlotte of Mecklenberg-Strelitz (1744-1818).
Species: *reginae*
Common Names: bird of paradise, crane lily; giant bird of paradise *(S. nicolai)*
Availability: Year-round
Description: Showy blossoms are oddly shaped, resembling a bird, and bourn in rigid, boatlike bracts. Stems are tall and thick. These tropical flowers demand attention. Use alone or with other exotic flowers and foliage.
Vase-life: 1 to 2 weeks; beauty and longevity of flowers can be increased by gently lifting out florets.

SUNFLOWER see *Helianthus*
SWAMP LILY see *Sarracenia*
SWEET BROOM see *Cytisus*
SWEET PEA see *Lathyrus*
SWEET WILLIAM see *Dianthus barbatus*
SWORD LILY see *Gladiolus*

SYRINGA (see Plate 16)

(si-RIN-gah)
Family: Oleaceae
Name Origin: From the Greek word *syrinx* (pipe), referring to the hollow stems.
Species: *vulgaris;* others and cultivars
Common Name: lilac
Availability: December through May
Description: Clustered florets in a compound dichasium inflorescence. Individual florets are star-shaped and avail-

POLIANTHES
Tuberose
TULIPA
Tulip
CHAMELAUCIUM
Waxflower

able in white, cream, pinks, and purples. Most varieties are fragrant.
Vase-life: Varies greatly, from 3 to 10 days; remove all foliage from woody stems for increased cut life. Lilacs also perform better when arranged in preservative solution without the use of floral foam. Re-cut stem ends often.

T

TAGETES

(tah-JEE-teez)
Family: Compositae (Asteraceae)
Name Origin: From Tages, an Etruscan deity, the grandson of Jupiter, who sprang from the newly plowed earth.
Species: *erecta, patula;* others and hybrids
Common Names: marigold; African marigold *(T. erecta);* French marigold *(T. patula)*
Availability: July through September
Description: Bright orange and yellow daisylike or carnationlike flowers. Useful for adding mass to mixed summer bouquets. Most have a unique fragrance.
Vase-life: 1 to 2 weeks

TAGETES (Marigold)

TELOPEA (Warwatah)

TAIL FLOWER see *Anthurium*

TELOPEA

(tay-LO-pee-ah)
Family: Proteaceae
Name Origin: From the Greek word *telopos* (seen from afar), referring to the showy flowers.
Species: *speciosissima*
Common Name: Waratah
Availability: May through November
Description: Dense terminal racemes surrounded by colored bracts. The flowers are vibrant red and add dramatic emphasis to designs.
Vase-life: 1 to 2 weeks

THROATWORT see *Trachelium*
THRYPTOMENE see *Erica*
TORCH LILY see *Kniphofia*

TRACHELIUM

(tra-KEEL-lee-um)
Family: Campanulaceae
Name Origin: From the Greek word *trachelos* (neck), referring to supposed medicinal properties.

TRACHELIUM (Throatwort)

Species: *caeruleum*
Common Names: throatwort, blue throatwort
Availability: March through November
Description: Clustered flowers in a dense terminal compound corymbs; available in pink, purple, blue, and white. Excellent for adding mass or can be used as a filler in large bouquets. Entire flower heads can be used for basing contemporary design styles.
Vase-life: 7 to 10 days

TRACHYMENE

(tray-ki-MEE-nee)
Family: Umbelliferae (Apiaceae)
Name Origin: From the Greek words *trachys* (rough) and *meninx* (membrane), referring to the fruit.
Species: *caerulea (Didiscus caeruleus)*
Common Names: didiscus, blue laceflower, laceflower
Availability: June through November
Description: Flat or rounded umbels of delicate blue or white flowers, similar in appearance to Queen Anne's lace. Trachymene works well as a filler or as a graceful mass flower.
Vase-life: 7 to 10 days

TRANSVAAL DAISY see *Gerbera*

TRITELEIA (see Plate 4)

(tri-te-LAY-a)
Family: Amaryllidaceae
Name Origin: From the Greek words *tri* (three) and *teleios* (perfect), referring to the floral parts, which are in threes.
Species: *laxa*
Common Names: triteleia, brodiaea
Availability: May through November, with peak supplies May through August
Description: Flower heads at a glance are similar to Agapanthus. However, funnel-shaped florets form a smaller umbel and are not as compact and globular as those of Agapanthus. A good filler and accent.
Vase-life: 7 to 10 days

TRITOMA see *Kniphofia*

TROLLIUS

(TROH-lee-us)
Family: Ranunculaceae
Species: *Trollius* species
Common Name: globeflower

TRACHYMENE (Didiscus)

Description: Globe-shaped yellow, orange, or white flowers. Useful in adding mass and accent.
Availability: April through July
Vase-life: 5 to 7 days

TRUMPET LILY see *Zantedeschia*
TUBEROSE see *Polianthes*
TULIP see *Tulipa*

TULIPA (see Plate 28)
(TEW-li-pa)
Family: Liliaceae
Name Origin: From the Turkish word *tulband* (turban).
Species: many cultivars
Common Name: tulip
Availability: November through May, with peak supplies January through April.
Description: Single rounded flowers with colorful sepals and petals. Adds mass to mixed spring bouquets. Tulips also work well in oriental and vegetative design styles.
Vase-life: 3 to 7 days

TURTLEHEAD see *Chelone*

VERONICA

V

VACCARIA see *Saponaria*

VANDA (see Plate 19)
(VAN-dah)
Family: Orchidaceae
Species: *coerulea, sanderana, teres, tricolor;* others and hybrids
Common Names: vanda, lei orchid
Availability: Year-round
Description: Sprays with clusters of flat-looking orchids, 1 to 3 inches across. A wide selection of colors with spotted patterns, many of which are fragrant. Many types are used in leis.
Vase-life: 1 to 3 weeks

VERONICA
(ver-RON-ik-ah)
Family: Scrophulariaceae
Name Origin: Named after St. Veronica.
Species: *spicata;* others and cultivars
Common Name: Veronica
Availability: Year-round, with peak supplies June through September
Description: Upright spikelike racemes made of small purple, blue, pink, or white flowers. Useful as a filler. The flower tips curve, adding graceful lines to mixed arrangements.
Vase-life: 5 to 7 days

VIBERNUM
(vy-BUR-num)
Family: Caprifoliaceae
Species: *opulus;* others and cultivars
Common Names: snowball, Guelder rose
Availability: January through May
Description: Showy ball-shaped terminal panicles or umbel-like cyme clusters of white or greenish florets on

VIBERNUM (Snowball)

woody stems. These fragrant flowers add mass and texture to large mixed bouquets. Frequently used in large wedding and church decorations.
Vase-life: 5 to 7 days

W

WARATAH see *Telopea*
WATTLE see *Acacia*
WAXFLOWER see *Chamelaucium*
WILD QUEEN ANNE'S LACE see *Ammi*
WINDFLOWER see *Anemone*

Y

YARROW see *Achillea*
YOUTH AND OLD AGE see *Zinnia*

Z

ZANTEDESCHIA (see Plate 5)

(zan-te-DES-kee-ah)
Family: Araceae
Name Origin: Named after Italian botanist Francesco Zantedischi (born 1797).
Species: *aethiopica, elliottiana, rehmannii;* others and cultivars
Common Names: calla lily, calla, arum lily, trumpet lily
Availability: Year-round, with peak supplies in spring and summer
Description: A striking white, greenish, yellow, or reddish spathe surrounds a yellow cylindrical spadix. Available in a variety of sizes. Stems are thick and long with heart- or spear-shaped and long-stemmed leaves. Distinctive shape and sleek line provide emphasis.
Vase-life: Varies greatly, from 3 to 14 days

ZINGIBER

(ZIN-ji-ber)
Family: Zingiberaceae
Species: *zerumbet;* others

ZINGIBER (Shampoo Ginger)

Common Name: shampoo ginger
Availability: June to September
Description: Florets appear inside waxy bracts; the entire cluster is similar in appearance to a pinecone. Available in pinks, reds, and yellow. Shampoo ginger provide emphasis and combines well with other tropical and exotic flowers.
Vase-life: 7 to 10 days

ZINNIA

(ZIN-ee-ah)
Family: Compositae (Asteraceae)
Name Origin: Named after Johann Gottfried Zinn (1727-59)
Species: *elegans* cultivars
Common Names: zinnia, youth and old age
Availability: May through October
Description: Daisy-shaped flowers available in a variety of colors, forms, and sizes. These mass flowers are colorful additions to mixed summer arrangements.
Vase-life: 5 to 7 days

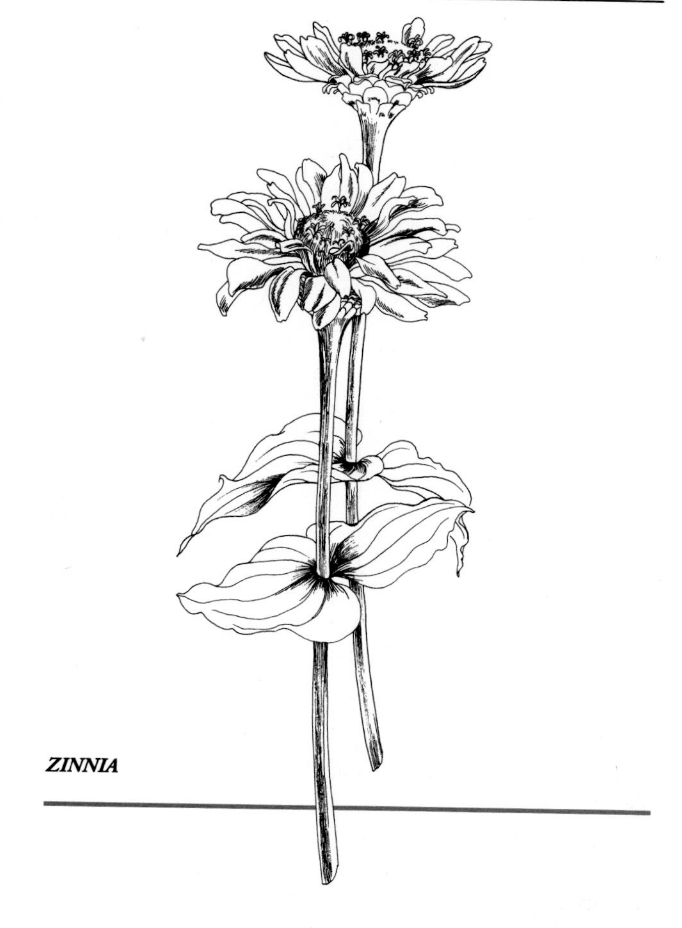

ZINNIA

Appendix B

FOLIAGE

A

ABIES (see *Conifer* illustration)

(AY-beez)
Family: Pinaceae
Species: *alba, procera, balsamea;* many others
Common Names: fir, balsam fir; many others
Availability: Commercially, October through January
Description: A large genus of coniferous evergreen trees. Stiff branches and aromatic foliage. Excellent filler, providing accent with other conifer foliage in winter or Christmas arrangements. Noble fir has bluish, stiff needles with rounded tips that densely cover branched stems. Balsam has dark green, rounded needles on branching stems.
Vase-life: Long lasting, from 3 to 4 weeks

ACACIA

(a-KAY-sha)
Family: Leguminosae (Fabaceae)
Name Origin: Named from the Greek word *akis* (sharp point), referring to the thorns.
Species: *cultriformis*
Common Names: knifeblade acacia, knife acacia
Availability: Limited supplies year-round
Description: Tall branches with bluish knife-shaped leaves and crowded heads of flowers in terminal racemes. Distinctive shape, color, and texture adds line and accent.
Vase-life: 7 to 10 days

ACACIA (Knifeblade acacia)

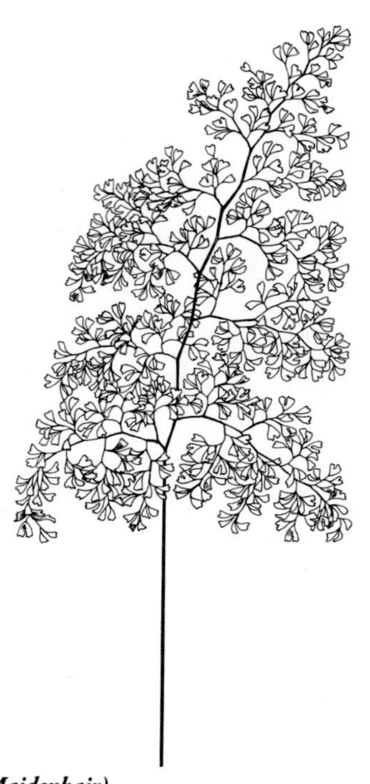

ADIANTUM (Maidenhair)

ADIANTUM

(a-dee-AN-tum)
Family: Polypodiaceae
Name Origin: From the Greek word *adiantos* (unwetted), referring to the fronds' way of repelling water.
Species: *raddianum;* others and cultivars
Common Name: maidenhair fern
Availability: Year-round
Description: Soft green foliage with shiny black leafstalks; useful as a filler or accent in smaller bouquets and wedding designs.
Vase-life: Short vase-life of 2 to 4 days; maidenhair is wilt sensitive; soak in water and mist with water to help regain turgor.

ANGEL'S WINGS see *Caladium*

ANTHURIUM

(an-THEWR-ree-um)
Family: Araceae
Name Origin: From the Greek words *anthos* (flower) and *oura* (tail), referring to the tail-like inflorescence (spadix) of these flowers.
Species: *andraeanum, scherzeranum*
Common Name: flamingo flower foliage
Availability: Year-round, however anthurium foliage is not common, as it is easily damaged while still on the plant.
Description: Heart-shaped like the anthurium flowers. The leaves are dark glossy green and vary in size. These work well with the anthurium flowers and other exotic flowers. Useful as an accent foliage.
Vase-life: 10 to 14 days

ARACHNIODES see *Rumohra*

ARCTOSTAPHYLOS (see Plate 35)

(ark-tuh-STAF-uh-los)
Family: Ericaceae
Name Origin: From the Greek words *arctos* (bear) and *staphyle* (bunch of grapes). Manzanita means "little apple" in Spanish.
Species: *manzanita;* others
Common Name: manzanita
Availablity: Limited supplies year-round
Description: Evergreen shrubs, admired for their mahoganylike, smooth, red to purple bark and crooked branches that twist and gnarl attractively.
Vase-life: 7 to 14 days

ANTHURIUM

ASPARAGUS
Sprengeri Fern

ASPARAGUS
Plumosa Fern

ASPARAGUS
Ming Fern

ASPARAGUS (see Plate 29)

(a-SPA-ra-gus)

Family: Liliaceae

Common Names and Species: string smilax, greenbrier *(A. asparagoides);* ming fern *(A. densiflorus* 'Myriocladus'*);* sprengeri fern, sprenger fern *(A. densiflorus* 'Sprengeri'*);* tree fern *(A. pyramidalis);* plumosa fern, lace fern *(A. plumosus);* others

Availability: Year-round

Descriptions: String smilax is medium green with small elliptical leaves closely spaced along a thin stem that grows around a piece of string, forming delicate garlands. Commonly used to decorate wedding cakes and banquet tables. Useful whenever a soft garland is needed.

Ming fern is light to dark green; filler foliage. The tiny tufts of soft, needlelike leaves appear on branching sprays. Excellent foliage for oriental designs; small tufts work well in corsages and boutonnieres.

Sprengeri fern is generally dark green but new shoots are often light green. Has needlelike leaves densely packed on trailing stems. Often will have small green berries. Useful as a filler and an accent. Can create graceful curving lines in large sympathy pieces. A beautiful accent in bridal bouquets.

Tree fern is medium to dark green and appears in bushy plumes; a soft and airy filler in arrangements. Small pieces of tree fern work well in corsages and boutonnieres.

Plumosa fern ranges from light to dark green, and most types have either a trailing or an upright appearance. This soft, lacy foliage is a delicate filler in arrangements and small pieces work well in corsages and boutonnieres.

Vase-life: All are long lasting, from 1 to 2 weeks

ASPARAGUS ASPARAGOIDES (String Smilax)

ASPIDISTRA (see Plate 30)

(as-pi-DIS-tra)

Family: Liliaceae

Name Origin: From the Greek word *aspideon* (a small round shield), referring to the shape of the stigma.

Species: *elatior*

Common Names: cast iron plant, barroom plant

Availability: Year-round, with limited quantities in the winter months.

Description: Shiny dark green or variegated color stripes and color blotches. These leaves look similar to *Cordyline terminalis* (ti leaves) but have a softer texture and do not have a prominent center midrib. Useful as a background

ASPARAGUS PYRAMIDALIS (Tree Fern)

ASPIDISTRA
Cast Iron Plant
XEROPHYLLUM
Bear Grass
BUXUS
Oregoni Boxwood

foliage; common for large designs.
Vase-life: Long lasting; 3 to 4 weeks

AUSTRALIAN LAUREL see *Pittosporum*

B

BAKER FERN see *Rumohra*
BALSAM see *Abies*
BARROOM PLANT see *Aspidistra*
BEAR GRASS see *Xerophyllum*
BELLA PALM see *Chamaedorea*
BIRD OF PARADISE FOLIAGE see *Strelitzia*
BOSTON FERN see *Nephrolepis*
BOXWOOD see *Buxus*
BRAKE FERN see *Nephrolepis*
BROOM see *Cytisus*
BUTCHER'S BROOM see *Ruscus*

BUXUS (see Plate 30)
(BUCK-sus)
Family: Buxaceae
Species: *sempervirens;* others and cultivars
Common Names: boxwood, box *(B. semperivirens);* Oregonia *(B. species)*
Availability: Year-round
Description: Boxwood has small, oval, dark green, glossy leaves; the leaves densely cover woody stems. Oregonia is similar but with a variegated coloring. Useful as a filler and for providing line.
Vase-life: Long lasting, 1 to 2 weeks

C

CABBAGE PALM see *Sabal*

CALADIUM (see Plate 31)
(ka-LAY-dee-um)
Family: Araceae
Name Origin: From the native name, *kaladi.*
Species: *bicolor* and hybrids
Common Names: angel's wings, elephant's ear, heart of Jesus, mother-in-law plant
Availability: Year-round
Description: Caladium plants are grown for their interesting foliage. They have large, heart-shaped leaves with contrasting, marbled color patterns and venation. Wide range of colors including greens, white, cream, pinks, and reds; provide emphasis and accent.
Vase-life: 3 to 5 days

CALATHEA (see Plate 31)
(ka-lah-THEE-a)
Family: Marantaceae
Name Origin: From the Greek word *kalathos* (basket), referring to the flowers being clustered as if in baskets.
Species: *zebrina;* others
Common Names: zebra plant, peacock leaf
Availability: Year-round
Description: All species have leaves with interesting colors and venation patterns. They work well as accent foliage, and can easily create or enhance a focal point. Calathea harmonizes well with exotic flowers.
Vase-life: 1 to 3 weeks

CAMELLIA (see Plate 31)
(ka-MEL-lee-a)
Family: Theaceae
Name Origin: Named after George Joseph Kamel (1661-1706), who studied the Philippines flora.
Species: *japonica;* others
Common Name: camellia
Availability: Year-round
Description: Handsome, glossy green leaves on long woody branches. An excellent background foliage for large arrangements. Smaller branches are useful as accents or fillers. Individual leaves work well in corsages and boutonnieres.
Vase-life: 5 to 7 days

CAST IRON PLANT see *Aspidistra*
CEDAR see *Cedrus*

CEDRUS (see *Conifer* illustration)
(SEE-drus)
Family: Pinaceae
Species: *atlantica, deodara;* others and cultivars
Common Name: cedar, atlas cedar, deodar cedar; others
Availability: Year-round

CALATHEA
Zebra Plant
CAMELLIA
CODIAEUM
Croton
CALADIUM
Angel's Wings
EUONYMUS

Description: Cedar is flat and lacy with stiff needles in clusters; provides a soft, sweeping line; smaller pieces work well as a filler.
Vase-life: Long lasting, 3 to 4 weeks

CHAMAEDOREA

(ka-mee-DOR-ee-ah)
Family: Palmacae (Arecaceae)
Name Origin: From the Greek words *chamai* (on the ground) and *dorea* (gift), referring to the fruits that are within reach, unlike most palms.
Species: *elegans, oblongata;* others
Common Names: giant palm, parlor palm *(C. elegans);* bella palm, narrow palm *(C. elegans* 'Bella'*)*; others
Availability: Year-round
Description: Palms are medium to dark green. Giant palm has sessile, elongated 1-inch leaves. Narrow palm leaves are 1/2 inch wide. Both are useful in large arrangements. Excellent background foliage. Both types can be trimmed for cleaner geometric or abstract patterns.
Vase-life: 5 to 7 days

CLUB MOSS see *Lycopodium*

CODIAEUM (see Plate 31)

(koh-die-EE-um)
Family: Euphorbiaceae
Species: *variegatum* cultivars
Common Name: croton
Availability: Year-round
Description: Colorful, variegated, bold patterns on leaves that vary in size and shape. Wide range of colors including red, orange, pink, green, yellow, and white. Croton leaves can create a focal point or add accent.
Vase-life: 1 week

CONIFER see *Abies, Cedrus, Juniperus, Pinus*

CORDYLINE (see Plate 37)

(kor-di-LIE-nee)
Family: Agavaceae
Name Origin: From the Greek word *kordyle* (club), referring to the large and fleshy roots of some species.
Species: *terminalis* (formerly *Dracaena terminalis);* cultivars
Common Name: ti (tea) leaf, ti, good luck plant, tree of kings
Availability: Year-round
Description: Long glossy leaves are emerald green or darker green with a reddish margin. Leaf lengths and widths vary. Useful as background foliage, especially with tropical and exotic flowers. Leaves can be curled or cut into many shapes.
Vase-life: 1 to 2 weeks

Conifers grown commercialy—top row, left to right: *PINUS* (pine), *JUNIPERUS* (juniper). Bottom row, left to right: *CEDRUS* (cedar), *ABIES* (fir)

CROTON see *Codiaeum*

CYCAS

(SIE-kas)
Family: Cycadaceae
Name Origin: From the Greek name for a palm.
Species: *revoluta*
Common Names: cycas palm, sago palm
Availability: Year-round

Description: Stiff, short, dark green leaves on straight stems, about 20 inches long; leaflets have margins bent downward with sharp tips. Useful with tropical and large arrangements.
Vase-life: Long lasting, 1 to 3 weeks

CYPERUS

(sie-PEE-rus)
Family: Cyperaceae
Name Origin: From the Greek word for a sedge.
Species: *alternifolius, papyrus*
Common Names: umbrella palm, palm crown *(C. alternifolius);* papyrus, bulrush, paper plant *(C. papyrus)*
Availability: Year-round
Description: Cyperus is medium green and quite showy. Its radiating umbrella appearance adds a striking form to contemporary design styles. Papyrus is also medium green but has a moplike tuft of short, stiff, grasslike leaves at the top of each stem. Unusual foliage that adds interest, texture, and form. Looks best combined with unusual and exotic flowers or in contemporary design styles.
Vase-life: Long lasting, 2 to 4 weeks

CYPERUS PAPYRUS

CYTISUS (see Plate 37)

(SIT-is-us)
Family: Leguminosae (Fabaceae)
Name Origin: From the Greek word *kytisos,* the name for these and similar shrubs.
Species: *scoparius* cultivars
Common Name: Scotch broom, broom
Availability: August through April
Description: Long, needlelike, dark green leaves that appear stiff on woody stems. Ideal foliage for setting the shape of crescent and Hogarth designs. The branches can be gently shaped into curves. Gently use the warmth of your hands to shape; blow hot air on the scotch broom to speed the curving process.
Vase-life: Long lasting, up to 3 weeks

D

DIFFENBACHIA (see Plate 32)

(deef-en-BAHK-ee-ah)
Family: Araceae
Name Origin: Named after J. F. Dieffenbach (1790-1863).
Species: *imperialis;* others and cultivars
Common Name: dumb cane
Availability: Limited; generally leaves are cut from a potted plant.
Description: Handsome leaves have variegated color patterns, spotted and feathered with white, creme, and yellow markings. Diffenbachia adds emphasis and accent to designs.
Vase-life: Varies with species; 1 to 2 weeks

DRACAENA (see Plate 32)

(dra-SEE-nah)
Family: Agavaceae
Name Origin: From the Greek word *drakcina* (dragon).
Species: *cincta, concinna;* others and cultivars
Common Name: dracaena; others. Many properly belong to the closely related genus *Cordyline,* but are sold and grown as dracaena.

Availability: Year-round
Description: Long slender leaves, from dark to light green with variegated striping patterns. These cut foliage add line and accent to floral designs.
Vase-life: Varies with species; 3 to 4 weeks

DRYOPTERIS see *Rumohra*
DUMB CANE see *Diffenbachia*

E

ELEPHANT'S EAR see *Caladium*
ELK GRASS see *Xerophyllum*

EQUISETUM

(ek-wi-SEE-tum)
Family: Equisetaceae
Name Origin: From the Latin words *equus* (horse) and *seta* (bristle).

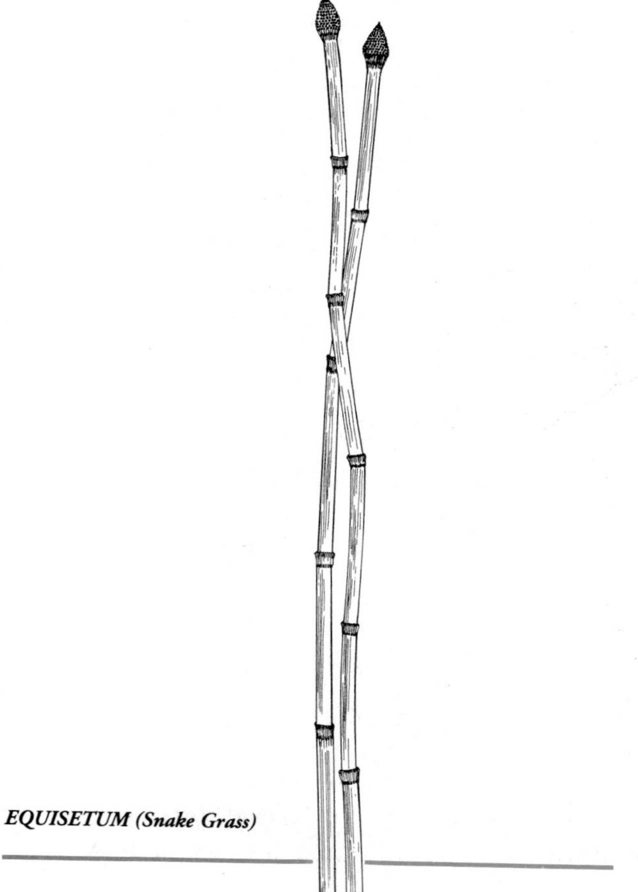

EQUISETUM (Snake Grass)

Species: *hyemale*
Common Names: horsetail, snake grass, scouring rush
Availability: Year-round
Description: Hollow, jointed dark green stems; leaves are mere scales at the joints. Spores are borne in conelike spikes at the top of each stem. Useful for adding line; easily manipulated into abstract linear shapes.
Vase-life: Long lasting, 1 to 2 weeks

EUCALYPTUS (see Plate 32)

(ew-ka-LIP-tus)
Family: Myrtaceae
Name Origin: From the Greek words *eu* (well) and *kalypto* (to cover), referring to how the calyx forms a lid over the flowers in bud.
Species: *cinerea, perriniana, pulverulenta, tetragona;* others
Common Names: eucalyptus; spiral eucalyptus *(E. cinerea);* silver dollar *(E. pulverulenta)*
Availability: Year-round
Description: Most eucalyptus is bluish green to silver-gray. Leaves vary in shape; some are rounded and sessile along tall stems that are useful for establishing line. Other types have oval or slender, elongated leaves. Most have a medicinal or lemon fragrance.
Vase-life: Long lasting, 1 to 4 weeks

EUONYMUS (see Plate 31)

(ew-ON-i-mus)
Family: Celastraceae
Name Origin: From the Latin name for these deciduous and evergreen trees and shrubs.
Species: *japonica, fortunei;* others
Common Name: euonymus
Availability: Year-round
Description: Evergreen foliage is glossy green and variegated. The leaves are oval and densely grouped on woody stems. Useful as a background foliage; shorter stems work well as an accent and filler.
Vase-life: Long lasting, 2 to 3 weeks

EUPHORBIA (see Plate 32)

(ew-FOR-bee-a)
Family: Euphorbiaceae
Name Origin: Named after Euphorbus, the physician to the King of Mauritania, Juba.
Species: *marginata*
Common Name: snow on the mountain, ghostweed
Availability: Year-round
Description: White, yellow, and light green leaves with

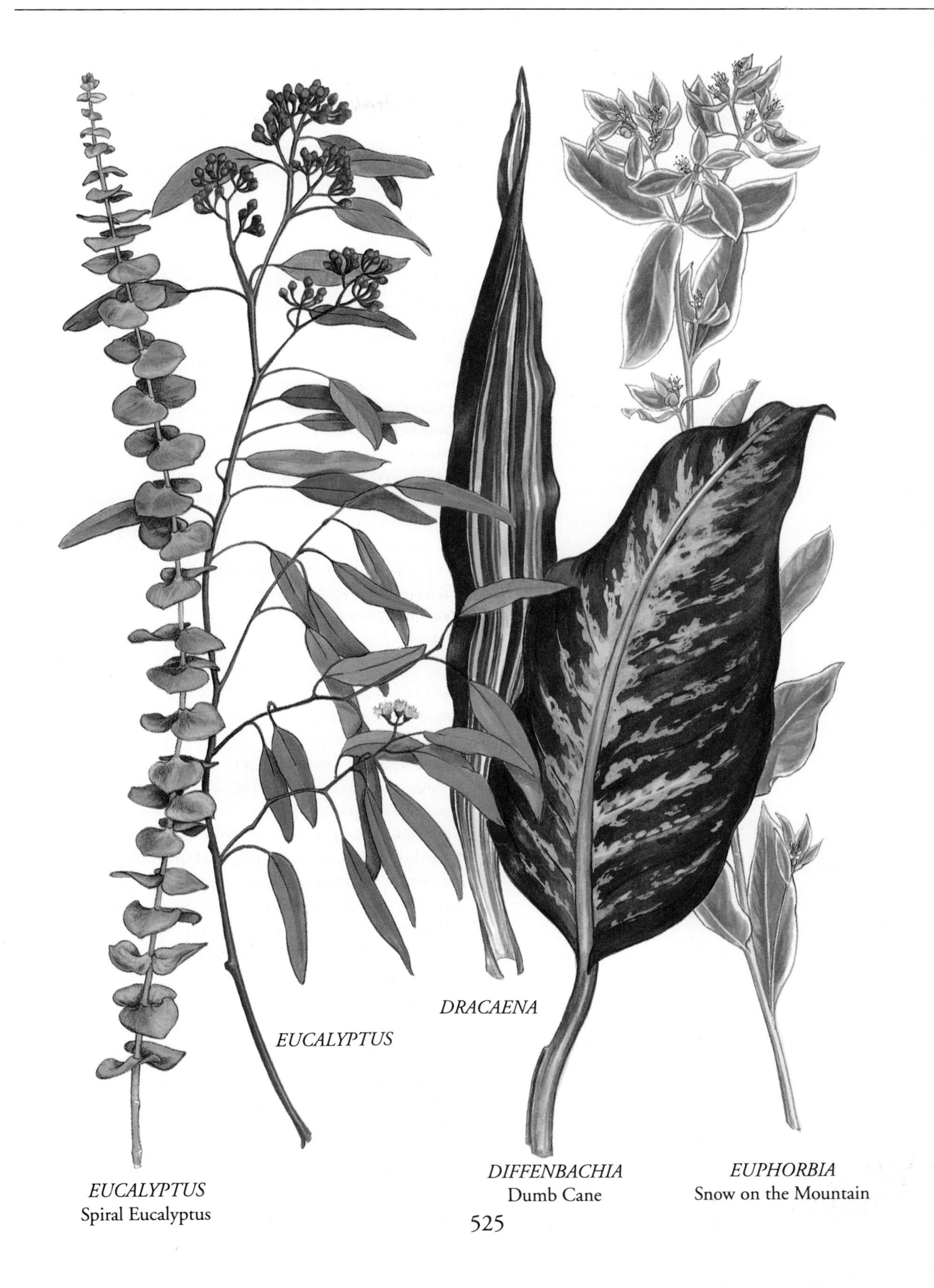
DRACAENA
EUCALYPTUS
EUCALYPTUS
Spiral Eucalyptus
DIFFENBACHIA
Dumb Cane
EUPHORBIA
Snow on the Mountain

contrasting white margin. It exudes a white milky sap when cut that can irritate sensitive skin and cause severe burning or dermatitis. Beautiful accents in foliage arrangements and contemporary designs.
Vase-life: 5 to 7 days; seal stems for increased vase-life; change water often.

F

FATSIA

(FATS-ee-a)
Family: Araliaceae
Species: *japonica* cultivars
Common Name: Japanese aralia
Availability: Mostly summer months
Description: Large, glossy, dark green leaves, deeply lobed like the fingers of a hand. Beautiful additions in foliage arrangements. Useful as background or filler in large arrangements.
Vase-life: Long lasting, 1 to 2 weeks

FATSIA (Japanese Aralia)

FERN see *Nephrolepis*
FIR see *Abies*
FLAT FERN see *Nephrolepsis*
FLAX see *Phormium*
FURZE see *Ulex*

G

GALAX (see Plate 33)

(GAY-lax)
Family: Diapensiaceae
Name Origin: From the Greek word *gala* (milk), referring to the white flowers.
Species: *urceolata*
Common Name: galax
Availability: Year-round; limited supplies in May and June
Description: Single, heart-shaped or rounded leaves on short stems. Generally dark green; however in the autumn and winter months Galax is reddish green. Galax is useful as a background foliage for sympathy pieces and wreaths. Works well as an accent foliage in contemporary and oriental design styles.
Vase-life: 1 to 2 weeks

GAULTHERIA (see Plate 37)

(gawl-THE-ree-a)
Family: Ericaceae
Name Origin: Named after a Canadian botanist and physician, Dr. Gaulthier (1708-58).
Species: *shallon*
Common Names: salal, lemonleaf, shallon
Availability: Year-round, with lesser quantities in July
Description: Large, medium to dark green leaves on long woody branches. Useful as a background foliage in large arrangements; smaller tips are useful as a filler. Individual leaves are useful in corsage and boutonniere designs.
Vase-life: Long lasting, 3 weeks

GIANT PALM see *Chamaedorea*
GOOD LUCK PLANT see *Cordyline*
GORSE see *Ulex*
GREENBRIAR see *Asparagus asparagoides*
GROUND PINE see *Lycopodium*

PANDANUS
PHORMIUM
New Zealand Flax
ULEX
Gorse
ILEX
Holly
GALAX
VACCINIUM
Huckleberry

H

HALA see *Pandanus*
HEART OF JESUS see *Caladium*

HEDERA (see Plate 34)

(HED-er-ah)
Family: Araliaceae
Name Origin: From the Latin name for these evergreen climbers.
Species: *helix, canariensis, colchica; cultivars*
Common Names: ivy; English ivy *(H. helix);* Algerian ivy *(H. canariensis);* Persian ivy *(H. colchica)*
Description: Green or variegated leaves, spaced on flexible stems. Ivy has a graceful trailing quality. Useful as hanging and curving line foliage, and as filler foliage.
Vase-life: About 5 days

HOLLAND RUSCUS see *Ruscus*
HOLLY see *Ilex*
HORSETAIL see *Equisetum*
HUCKLEBERRY see *Vaccinium*

I

ILEX (see Plate 33)

(EYE-leks)
Family: Aquifoliaceae
Species: *aquifolium;* others and cultivars
Common Name: holly
Availability: Commercially, November and December
Description: Dark green or variegated stiff, spiky leaves on woody stems with red berries. Useful as a winter or holiday accent and filler. Spiky leaf margins add interesting form and texture to designs.
Vase-life: Long lasting, 1 to 3 weeks

INDIAN BASKET GRASS see *Xerophyllum*
ITALIAN RUSCUS see *Ruscus*
IVY see *Hedera*

J

JAPANESE ARALIA see *Fatsia*
JUNIPER see *Juniperus*

JUNIPERUS (see *Conifer* illustration)

(joo-NIP-er-us)
Family: Cupressaceae
Name Origin: From the Latin name for these evergreen conifers.
Species: *communis;* others and cultivars
Common Name: juniper
Availability: October through December
Description: Medium to light green needles on spreading branches with powdery blue berries. Generally used as filler foliage in winter and Christmas arrangements and decorations. Texture of foliage and berries provides accent.
Vase-life: Long lasting, 3 to 4 weeks

K

KNIFEBLADE ACACIA see *Acacia*

L

LACE FERN see *Asparagus*
LEATHERLEAF FERN see *Rumohra*
LEMONLEAF see *Gaultheria*

LYCOPODIUM (see Plate 35)

(lie-koh-POH-dee-um)
Family: Lycopodiaceae
Species: *complanatum, obscurum;* others
Common Name: club moss, ground pine
Availability: Year-round
Description: Bright green scalelike needles on short, stiff, forking branches. The stems are 15 to 20 inches long. Useful in establishing line. Also useful as a filler in contemporary designs.
Vase-life: 7 to 10 days

HEDERA
Ivy
RUMOHRA
Leatherleaf Fern
MAGNOLIA
MAHONIA
Oregon Grape

M

MAIDENHAIR FERN see *Adiantum*

MAGNOLIA (see Plate 34)

(mag-NOL-ee-a)
Family: Magnoliaceae
Species: *grandiflora;* others
Common Name: magnolia
Availability: October through March
Description: Large, glossy, dark green leaves with a leathery texture on woody stems. Useful in large arrangements. Useful as a filler in larger designs.
Vase-life: About 5 days

MAHONIA (see Plate 34)

(ma-HON-ee-ah)
Family: Berberidaceae
Name Origin: Named after American horticulturist Bernart McMahon (died 1816).
Species: *aquifolium;* others
Common Name: mahonia, Oregon grape
Availability: Year-round
Description: Evergreen shrubs with glossy, spiny leaves. Useful as a filler in larger arrangements, or as a background foliage.
Vase-life: 5 to 10 days

MANZANITA see *Arctostaphylos*

MARANTA

(ma-RAN-tah)
Family: Marantaceae
Name Origin: Named after Venetian botanist Bartolommeo Maranti (16th century).
Species: *leuconeura* cultivars
Common Name: prayer plant
Availability: Limited; generally leaves are cut from potted plants.
Description: Light and dark green with variegated color and vein patterns. Useful in adding emphasis and accent.
Vase-life: 3 to 5 days

MING FERN see *Asparagus*
MOCK ORANGE see *Pittosporum*

MONSTERA (Swiss cheese plant)

MONSTERA

(mon-STER-ah)
Family: Araceae
Name Origin: Possibly the name derived from the monstrous appearance and size of the leaves.
Species: *deliciosa*
Common Names: monstera, Swiss cheese plant
Availability: Year-round
Description: Large, leathery, glossy, dark green leaves; irregularly cut and perforated with interesting margins and negative spaces, thus the name Swiss cheese. Useful foliage with exotic and tropical flowers.
Vase-life: Long lasting, 2 to 4 weeks

MYRTLE see *Myrtus*

MYRTUS (see Plate 35)

(MUR-tus)
Family: Myrtaceae
Species: *communis*
Common Name: myrtle

LYCOPODIUM
Club Moss
MYRTUS
Myrtle
ARCTOSTAPHYLOS
Manzanita

Availability: October through March
Description: There are two main types of myrtle. One is 48 inches tall with large, glossy, green leaves. The other type has lighter-colored and smaller leaves on shorter, 12-inch stems. Useful as line foliage in setting the framework of a design; smaller pieces are useful as fillers. Myrtle has a lemon fragrance.
Vase-life: Long lasting, 1 to 2 weeks

N

NARROW PALM see *Chamaedorea*

NEPHROLEPIS

(nef-row-LEP-is)
Family: Polypodiaceae
Species: *exaltata, cordifolia; cultivars*
Common Names: Boston fern, *(N. exaltata* 'Bostoniensis'*)*; Oregon fern, Brake fern, flat fern *(N. cordifolia)*
Availability: Year-round
Description: Boston fern has short, light green leaves alongside a short stem usually no longer than 12 inches; an attractive filler and useful in contemporary designs. Oregon fern is much longer and wider than Boston fern. It can easily set the framework of larger designs and is useful in covering mechanics.
Vase-life: 5 days

NEPHROLEPIS (Boston fern and flat fern)

NEW ZEALAND FLAX see *Phormium*
NOBLE FIR see *Abies*

O

OREGON FERN see *Nephrolepis*
OREGON GRAPE see *Mahonia*
OREGONIA see *Buxus*

P

PALM see *Chamaedorea*
PALM CROWN see *Cyperus*
PALM FAN see *Sabal*
PALMETTO PALM see *Sabal*

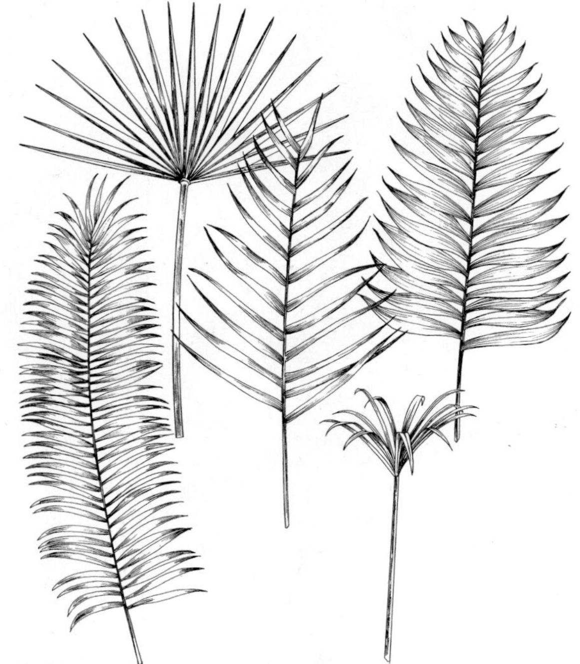

PALMS—from left to right: parlor palm, sabal palm, narrow palm, umbrella palm, and giant palm

PODOCARPUS
Tropical Yew
RUSCUS
Italian Ruscus
RUSCUS
Butcher's Broom
PITTOSPORUM

"It's a type of palm."

Herman © 1989 Jim Unger.
Reprinted with permission of Universal Press Syndicate.

PANDANUS (see Plate 33)

(pan-DAY-nus)
Family: Pandanaceae
Species: *odoratissimus, utilis;* others
Common Names: screw pine, hala
Availability: Year-round
Description: Sword-shaped leaves with yellow striping, 24 to 36 inches long. Pandanus, like New Zealand flax *(Phormium),* makes a dramatic vertical statement. It can also be curled and formed into abstract, geometric patterns.
Vase-life: Long lasting; 2 to 3 weeks

PAPYRUS see *Cyperus*
PARLOR PALM see *Chamaedorea*
PEACOCK LEAF see *Calathea*

PHORMIUM (see Plate 33)

(FOR-mee-um)
Family: Agavaceae
Name Origin: From the Greek word *phormion* (mat), referring to the fiber that is produced from the leaves of *P. tenax.*
Species: *tenax* cultivars
Common Names: flax, New Zealand flax
Availability: Year-round
Description: Long, narrow leaves with variegated green and white and reddish purple patterns. Useful in adding line; may be manipulated into abstract linear patterns. Harmonizes well with tropical flowers.
Vase-life: 5 to 14 days

PINE see *Pinus*

PINUS

(PI-nus)
Family: Pinaceae
Species: *strobus;* many others and cultivars
Common Name: pine; white pine; others
Availability: Commercially, October through February
Description: Long flexible needles on branching stems. Sizes and types vary with species. Useful in setting the skeleton of a design; smaller pieces work well as accents and fillers in winter and Christmas arrangments and decorations.
Vase-life: Long lasting, 3 to 4 weeks

PITTOSPORUM (see Plate 36)

(pi-TOS-poh-rum)
Family: Pittosporaceae
Name Origin: From the Greek words *pitta* (pitch) and *sporum* (seed), referring to the sticky seeds.
Species: *tobira;* others
Common Names: pittosporum, pitt, Australian laurel, mock orange
Availability: Year-round
Description: Medium to light green, thick, leathery leaves including variegated types with cream-colored margins. Pittosporum can easily set the skeleton or background of a design; useful as a filler foliage.
Vase-life: Long lasting, 1 to 2 weeks

PLUMOSA FERN see *Asparagus*

PODOCARPUS (see Plate 36)

(poh-doh-KAR-pus)
Family: Podocarpaceae
Species: *macrophyllus;* others
Common Name: podocarpus, tropical yew
Availability: Year-round
Description: Dark green and coniferlike with long, flexible, flat needles on tall branches. Useful in setting the framework of a design. Shorter branches are useful as

GAULTHERIA
Salal

CYTISUS
Scotch Broom

CORDYLINE
Ti leaves

accents or fillers.
Vase-life: Long lasting, 1 to 3 weeks

POLYSTICHUM

(po-LI-sti-kum)
Family: Polypodiaceae
Name Origin: From the Greek words *poly* (many) and *stichos* (row), referring to the arrangement of the spores on the underside of the fronds.
Species: *munitum;* others
Common Names: western sword fern, shield fern
Availability: Year-round, with lesser quantities in the summer months.
Description: Western sword fern is similar to *Nephrolepis* but has longer and wider leaves on stems that are 20 to 24 inches long.
Vase-life: 10 to 14 days

PRAYER PLANT see *Maranta*

R

RUMOHRA (see Plate 34)

(roo-MOH-rah)
Family: Polypodiaceae
Species: *adiantiformis;* Other genera of the *Polypodiaceae* family *(Dryopteris, Arachniodes)* are often named leather fern, causing some confusion.
Common Names: leather fern, leatherleaf, baker fern
Availability: Year-round
Description: Dark green, triangular-shaped fronds, coarsely toothed. A popular florist foliage; works well in setting the background shape and in covering mechanics. Smaller pieces work well as fillers and in corsages and boutonnieres.
Vase-life: Long lasting, 1 to 3 weeks

RUSCUS (see Plate 36)

(RUS-kus)
Family: Liliaceae
Species: *aculeatus, hypoglossum*
Common Names: butcher's broom, Holland ruscus *(R. aculeatus);* Italian ruscus, smilax ruscus *(R. hypoglossum)*
Availability: Year-round
Description: Butcher's broom has broad green leaves on stems 10 to 20 inches long. Adds an interesting line; also useful as a filler. Italian ruscus has branching stems each with small, green leaves. It resembles string smilax. Provides sweeping curves in designs or can be used as a filler.
Vase-life: 7 to 10 days

S

SABAL

(SAY-bal)
Family: Palmacae (Arecaceae)
Species: *palmetto*
Common Names: sabal, palmetto palm, palm fan, cabbage palm
Availability: Year-round
Description: Palmate leaves, deeply divided and forming a fan about 20 inches wide. Palmetto works well as a background foliage in large designs. May be trimmed for interesting shapes.
Vase-life: Long lasting, 1 to 2 weeks

SAGO PALM see *Cycas*
SALAL see *Gaultheria*
SCOTCH BROOM see *Cytisus*
SCOURING RUSH see *Equisetum*
SCREW PINE see *Pandanus*
SHALLON see *Gaultheria*
SHIELD FERN see *Polystichum*
SILVER DOLLAR see *Eucalyptus*
SMILAX RUSCUS see *Ruscus*
SNAKE GRASS see *Equisetum*
SNOW ON THE MOUNTAIN see *Euphorbia*
SPRENGERI FERN see *Asparagus*

STRELITZIA

(stre-LITZ-ee-a)
Family: Cannaceae, Strelitzaceae
Species: *reginae*
Common Name: bird of paradise
Availability: Year-round
Description: Oval-shaped, light to dark green leaves; up to

STRELITZIA (Bird of Paradise)

20 inches long on sturdy straight stems. Ideal foliage for tall, tropical arrangements.
Vase-life: Long lasting, 2 weeks

STRING SMILAX see *Asparagus*
SPURGE see *Euphorbia*
SWISS CHEESE PLANT see *Monstera*

T

TI LEAF see *Cordyline*
TREE FERN see *Asparagus*
TREE OF KINGS see *Cordyline*
TROPICAL YEW see *Podocarpus*

U

ULEX (see Plate 33)

(EW-leks)
Family: Leguminosae (Fabaceae)
Species: *europaeus*
Common Names: gorse, furze, whin
Availability: June through April
Description: Rigid, dark green, spiny branches. Excellent for adding line; useful as a background element. Provides accent in more contemporary designs.
Vase-life: 5 to 7 days; sensitive to water stress

UMBRELLA PALM see *Cyperus*

V

VACCINIUM (see Plate 33)

(vak-SIN-ee-um)
Family: Ericaceae
Species: *ovatum*
Common Names: huckleberry, huck
Availability: Year-round, with lesser quantities in July
Description: Small ovate leaves densely cover branching stems. Useful as a background foliage in large arrangements. Smaller branches work well as fillers.
Vase-life: Long lasting, 1 to 2 weeks

W

WESTERN SWORD FERN see *Polystichum*
WHIN see *Ulex*
WHITE PINE see *Pinus*

X

XEROPHYLLUM (see Plate 30)

(zer-oh-FIL-um)
Family: Liliaceae
Species: *tenax*
Common Names: bear grass, elk grass, Indian basket grass
Availability: Year-round
Description: Long, thin, curving leaves, about 1/4 inch wide by up to 48 inches long. Bear Grass adds a soft curving line to designs, particularly in otherwise stiff arrangements. Good foliage to add to Waterfall style bouquets.
Vase-life: Long lasting, 1 to 3 weeks

Z

ZEBRA PLANT see *Calathea*

Glossary

abscission. The dropping off of leaves and flowers, which rapidly increases in the presence of ethylene.

abstract design. Freeform design in which plant material and nonfloral items are used in a nonnaturalistic way; emphasis is placed on color, texture, and form.

accent. A subordinate element that enhances the primary structure of a composition. Also, to call attention to a particular location in a floral design; used to create emphasis.

accent cluster. A small floral design added to a larger composition or area, particularly in large funeral pieces.

accessories. Any material or object, other than plant material, such as candles, bows, novelties, and plush toys, added to enhance the theme of a design. Also may refer to plant materials other than flowers and foliage, such as pine cones, sticks, bamboo, and berries.

achromatic. Without color. Any gradation of white, gray, or black.

actual line. Real lines that lead the eye throughout a composition. Curving branches and flower stems are real lines.

advanced design. Refers to floral styles that are complex, requiring specialized skills, mechanics, and techniques.

advertising. Paid promotional activities. Supplying information to the public to induce people to buy a product.

air-drying. A simple method of drying fresh flowers and other plant material. Hanging upside-down, drying upright, or standing in water, allows moisture in the plant to dissipate.

air embolism. The blockage of air emboli or bubbles, that inhibits the uptake of water and nutrients.

aisleabra. Candelabra that are attached to the ends of church pews or chairs on the main aisle for wedding ceremonies.

aisle runner. A white carpet running the length of the center aisle in a church for wedding ceremonies.

alkaline. having a pH greater than 7. The alkalinity of water indicates the water's capacity for pH change with the addition of preservatives and other chemicals.

Allied Florists' Association. A group of growers, wholesalers, and retailers in an area, contributing a percentage of their gross for the advertising and promotion of flowers.

all-sided design. A floral arrangement designed to be viewed from all sides.

alternate complement. A four-color scheme combining a triad and the direct complement of one of the colors.

aluminum sulfate. A white cyrstalline salt used to purify and acidify water for cut flowers that is high in salts.

American Academy of Floriculture (AAF). A division of the Society of American Florists dedicated to service in the floral industry and excellence in leadership.

American Floral Marketing Council (AFMC). A branch of the Society of American Florists that promotes and advertizes the sale of flowers, plants, and floral products throughout the year for nonoccasion days.

American Institute of Floral Designers (AIFD). The premier organization of floral designers established to enact and maintain standards of excellence in the field.

American style. A general description for floral arrangements having a combination of line and mass.

Analysis of Beauty, The. Work published in 1753. William Hogarth theorized that all beauty was based on the serpentine S-line.

analogous color scheme. A color scheme utilizing several adjacent colors on the color wheel, such as yellow, yellow-orange, and orange.

anatomy. In botany, the area that deals with the internal structures of organisms.

anchor pin. Round, plastic holder with four upright prongs. When glued inside the bottom of a container, the prongs hold floral foam in place.

anchor tape. Waterproof tape used primarily to hold floral foam in place.

anther. The pollen-bearing portion of a stamen. In some flowers, such as lilies, the anthers should generally be removed prior to arrangement and delivery because the pollen of the anthers stains clothing..

anti-transpirant. A liquid spray or dip that protects the surface of a flower or foliage and minimizes water loss or transpiration.

arm bouquet. A tied cluster of flowers carried across the forearm for weddings or as a presentation bouquet.

Art Deco. A term derived from the 1925 Paris exhibition *Les Expositions des Arts Decoratifs;* known during the 1920s and 1930s as modernistic. General term describing architecture, furniture, decorative arts, and floral arrangements having strong, streamlined geometric forms, lines and patterns including zigzags, pyramids, and sunburst motifs. Floral designs feature geometrically bold containers, form flowers and foliage. The arrangements are formal linear or high style in appearance.

Art Nouveau. A modern stylistic movement based on the flowing lines of nature that flourished principally in Europe and the United States around 1890-1910. Floral arrangements are often characterized by lavish and cascading asymmetrical waterfall style designs.

asymmetrical balance. Having unequal visual weight on either side of an imaginary center vertical axis.

attar. The fragrant oil in flowers, stored in the epidermal cells of petals or petal substitutes.

auction. A centralized facility, common in Holland, where numerous growers bring their floral products to sell through open-market bidding by distributors, exporters, wholesalers, and retailers.

balance. A design principle. The placement of flowers, foliage, and other objects in strategic locations to create physical and visual stability.

banding. A design technique in which a group of stems is tied or bound together in one or more places using raffia, ribbon, or vines for decorative purposes.

banking pin. Straight silver pins, 1 to 2 inches long. Commonly used to pleat ribbon, or pin flowers on hard foam. The term comes from the early use of such pins for wrapping currency.

baroque period. A powerful and imaginative art and design direction during the seventeenth and eighteenth centuries in Europe characterized by elaborate and massive decorative elements and curved rather than straight lines. A reaction against the severe classic style. Floral arrangements typical of the Baroque period are tightly massed and overflowing, displaying a rhythmic asymmetrical balance.

basing. The application of materials over a design's base or foundation to camouflage floral foam and other mechanics. A variety of techniques may be used for basing arrangements, such as clustering and layering; this technique provides a foundation that adds color and texture to a composition.

basket bouquet. A specialty hand-held bouquet constructed in a basket.

bas-relief. Sculpture in which figures are carved in a flat surface so that they project slightly from the background.

beauty clip. Plastic U-shaped gripping device used to fasten chicken wire to the edge and around the perimeter of a container.

bent neck. A condition found on cut roses, gerbera, and other flowers when water no longer enters the stem, causing the stem beneath the flower to become weak, resulting in nodding flowers with limp stems that bend over.

bible bouquet. A specialty bouquet carried by some brides. A bible or prayer book accented with flowers and ribbon.

Biedermeier. A design style influenced by a period in German and Austrian history. This style utilizes compact concentric circular or spiral patterns in a rounded bouquet.

binding. The process of tying similar materials together in bunches.

binding point. The point or area where all stems come together or intersect, as in a hand-tied bouquet.

biocide. Also called germicide. A general term for a chemical substance that can kill living organisms, especially harmful microorganisms that grow in vase water. An ingredient in floral preservative that inhibits the growth of microorganisms.

blade. The broad, flattened part of a leaf.

blockage. The clogging of water-carrying vessels in flower stems with air, sugars, salts, or bacteria; which inhibits water uptake. Usually occurs at the base of the stem.

blueing. Purplish or bluish coloring which develops on flowers as a result of senescence or cold damage.

bostryx. A cymose inflorescence with successive branches on one side only, normally coiled like a spring.

botanical design. A design style that symbolizes the natural life processes or life cycle of a plant; often features bulb flowers.

botanical name. The two-part Latin name given to plants and flowers, consisting of the genus and species name. These names are used and known worldwide. Often called the Latin name, scientific name, or universal name.

botanicals. Everlasting flowers that are patterned after real flowers and foliage, are botanically correct, and realistic in color, shape, texture, pattern, and size.

botrytis. *Botrytis cinerea* is a gray mold or fungus present in the air that may form on flowers causing irreversible damage. It is encouraged by high humidity and high temperature..

bough pot. A vase for branches or cut flowers. Often refers to the large containers used to hold flowers set in the fireplace during summer months.

boutonniere. A single flower or cluster of flowers worn on a man's lapel.

bract. A modified, usually reduced leaflike structure which subtends a flower or inflorescence in its axil..

broad-leaf. Large or wide leaf, such as the leaves of salal, ivy, and camellia.

broker. An individual who handles functions between growers and wholesalers, including the location of sourcing and coordination of delivery from all over the world.

bud vase. A floral design, usually consisting of one or several flowers, foliage, and fillers, made in a small vertical container vase.

bud cut. Flowers harvested in the bud stage, prior to the petals opening. Many flowers are harvested when the flower is tight in the bud form, such as roses and iris, allowing for a longer vase life.

buffering capacity. The water's ability to resist change with the addition of chemicals, particularly citric acid. Highly buffered water resists change in pH while poorly buffered water is responsive to change.

bulb. A specialized underground plant organ. A general term for flowers that are produced from bulbs, such as tulips, daffodils, and hyacinths.

bunching. Packing flowers in groups of 10, 12, or 25 stems or grouping by the number of blossoms that prepares flowers for shipping.

bundling. The wrapping or tying together of materials.

Byzantine. Referring to *Byzantium,* an ancient city, and the *Byzantine Empire,* in southeast Europe and southwest Asia (a.d. 395-1453). Byzantine floral designs reflect the decorative style of the mosaics. Symmetrical, stylized tree compositions were introduced during this time, as evidenced through mosaics.

cage holder. A stem-anchoring device with numerous openings in the surface, generally surrounding wettable floral foam, that provides a method for supporting cut flower stems.

calyx. Collective term for all the sepals of a flower.

candelabra. A fixture designed to support one or several candles for lighting; often decorated with flowers and ribbon bows.

canopy. An arch or covering used at weddings, often decorated with flowers and used at the altar, the rear of the center aisle, and the reception.

carbohydrate. An organic compound—includes sugars, starch, glycogen, and cellulose. Food that is produced by the plant during growth and serves the plant as a source of energy to keep it alive. Carbohydrates are the primary ingredients in floral preservatives.

care and handling. The set of procedures that involve attention and techniques directed toward increasing the longevity of fresh flowers and foliage. When proper care and handling is carried out at each level of distribution, vase life is increased for the final consumer.

care tags. Paper or plastic tags attached to floral arrangements and plants that provide appropriate information (name of flower or plant, care, etc.) for enhancing quality and increasing longevity.

carpel. One of the flower's female reproductive organs, comprising an ovary and a stigma and containing one or more ovules.

cascade bouquet. A hand-held wedding bouquet style in which the flowers hang down (cascade) below the main portion of the design.

cash and carry. Merchandise paid for with cash and then taken by the customer. No or little added service is given, such as design and delivery.

casket blanket. A blanket of flowers constructed on heavy fabric, such as burlap, and displayed draping over the casket like a blanket.

casket lid inset. A small arrangement of flowers designed to be placed in the opened lid of a casket.

casket scarf. A design of flowers, often attached to fabric, that drapes over a part of the casket (from side-to-side or end-to-end).

casket spray. A floral arrangement that is placed on top of the casket during a funeral service.

catkin. A pendulous, spikelike inflorescence of simple, usually unisexual flowers; found only in woody plants.

chain of life. A marketing and educational program sponsored by the Society of American Florists, specifically focused on proper care and handling at every level (link) in the channel of distribution (chain), ensuring maximum longevity and increased quality for the final consumer.

chaplet. A wreath or garland for the head, customarily made from flowers and foliage. Introduced during ancient periods.

chemical solutions. See *conditioning.*

chenille stem. A straight wire with velvety yarn covering.

chicken wire. A pliable fencing wire used in floral arrangement as an aid for holding flowers or helping heavy, thick-stemmed flowers stay in position while in the floral foam.

chilling injury. Plant disorder caused by low temperatures above freezing. Tropical flowers often experience chilling injury when exposed to temperatures below 47-50°F, flowers fail to open properly, lesions appear on flowers and bracts, and stems and bracts may blacken.

Chinese style. A respected floral art form featuring unstructured, naturalistic designs using seasonal plant material. A rrangements are symbolic and generally include the use of a dominant vertical element and also the use of a more delicate horizontal element.

chlorophyll. The green pigment of plant cells, necessary for photosynthesis.

chroma. A measure of the intensity or purity of a hue. The relative brilliance or dullness of a color.

chuppah. A canopy under which a Jewish couple is married.

cincinnus. A monochasial cymose inflorescence with branches alternating from one side of the vertical axis to the other.

circular form. A spherelike appearance. All materials radiate out from a central area within the container.

citric acid. A natural chemical in citrus fruits. An ingredient of most floral preservatives, a compound used to lower the pH (acidify) of water for cut flowers. Used to create a hydrating solution for flowers.

classical style. A timeless design, generally formal and balanced, restrained and simple.

cleating. Crosswise strengtheners built into fiberboard boxes for shipping cut flowers.

clustering. A technique of gathering like materials together so closely that the quantity and shape cannot be determined; individual materials lose identity and function as one unit, emphasizing color and texture.

clutch bouquet. A simple or casual gathering of flowers tied with ribbon, raffia, or string for a natural, garden-picked appearance.

clutch wiring. A wiring method used to secure clusters of tiny flower stems and fillers; also called *wrap-around* wiring.

collarette. A thin cardboard support placed behind gardenia flowers to protect the fragile petals.

colonial. A term loosely used in referring to the two hundred year period that includes the settlement of the early colonies in America through the Federal period (about 1607-1835). Also called Colonial Williamsburg. Colonial style bouquets are typically rounded and massed, often combining fresh and dried flowers.

colonial bouquet. A hand-held bouquet style that is constructed in a circular shape.

color. The visual interpretation of light waves from the visible spectrum.

color wheel. The standard color wheel or circle has 12 colors, including primary, secondary, and tertiary colors; a simple and useful tool for the designer.

commentator. A person who reports, analyzes, and evaluates floral designs, floral industry products and news at design shows and other industry events.

commercial account. A business account, such as with a hotel, computer company, or bank.

common name. the name by which a flower or foliage is generally known rather than by its botanical, scientific name.

complementary color scheme. Any two hues located opposite each other on the color wheel.

complete flower. In botany, a flower having four whorls of floral parts—sepals, petals, stamens, and carpels.

composition. A grouping or organization of different elements to achieve a unified whole.

compote. A raised, stemmed container.

compound. Consisting of several parts. A leaf with several leaflets, or an inflorescence with more than one group of flowers.

conditioning. The process of preparing flowers for shipping, storage, or arrangement It usually involves treatment of flowers in a conditioning solution such as: preservative water, STS solution, hydrating solution, or pulsing/sugar solution.

cone design. typical to the Byzantine period style. A three-dimensional vertical isosceles triangle design which generally requires a foam base for mass flower insertion.

conifer. A cone-bearing tree, generally characterized by needlelike foliage.

consumption. Use and enjoyment of flowers, floral products, and floral services by the final consumer.

contemporary design/style. Those arrangements that are currently in fashion, popular, and representative of leading trends in creativity. A generic term for whatever is the current trend or on the leading edge in floral design.

continuation. Also referred to as transition, a method of achieving unity in design by planning a gradual change from one element to another, causing continuous eye movement.

cool colors. Blues, greens, and some purples, and colors containing these hues; associated with water and ice; generally restful, peaceful, and soothing. Cool colors are receding and fade into the background.

co-op. The association of several growers, drivers, etc. who offer merchandise or services under one roof or in a joint effort.

corner piece. A small arrangement of flowers, designed to be displayed on the outside corner at the head of the casket.

cornucopia. A basket or other container shaped like a horn or cone overflowing with fruit and vegetables, flowers, foliage, and grain. Introduced during the Greeek period, the cornucopia is known as the symbol for abundance. A lso referred to as a *horn of plenty.* Floral arrangements are often made in cornucopia containers during the autumn, particularly for Thanksgiving.

corolla. All the petals of a flower; normally conspicuously colored.

corsage. A grouping of flowers, ribbon, and other accessories worn by a woman.

corymb. A rounded or flat-topped inflorescence of racemose type in which the lower (outer) flower stalks (pedicels) are longer than the upper (inner) ones so that all the flowers are at about the same level.

customer base. Individuals and commercial accounts who patronize or buy on a regular basis and not just for holidays.

customer bonding. Development of a close friendship between patronizers and the flower shop.

crescent bouquet. A hand-held cascading bouquet in the shape of a distinctive C-shape or quarter moon shape.

crescent form. The organization of floral materials in a C-shape or quarter moon shape.

cultivar. A variety of plant found only under cultivation.

curved line. A swerving line; suggesting a natural, sweeping motion; adds interest and gentleness to floral designs.

custom design/service. Floral arrangements and service designed and provided specifically for the customer.

cyme. An inflorescence in which each terminal growing point produces flowers.

cytokinins. Class of plant growth substances that promote cell division, among other effects; added to some floral preservatives.

delftware. Glazed earthenware, also called delft, usually blue and white in color. It originated in Delft, a city in west Netherlands. A delftware brick is a rectangular box with a perforated top or grill for flowers; originally produced in the late 1500s and then flourished into the mid-1700s. Delft of the early 1600s imitated Chinese porcelain.

delivery co-op. A joint effort by several floral shops to deliver flowers to several areas.

depth The distance from the front to the back of a design and the viewer's awareness of this distance. Depth may be emphasized through various techniques.

desiccant. A substance that absorbs moisture, such as sand, alum, cornmeal, and silica gel, used in drying flowers and foliage.

design. A planned organization of elements to suit a specific purpose.

diagonal line. A slanted line giving the feeling of mobility or instability, creating dynamic movement or tension.

directional facing. Turning or directing flower heads certain ways in order to increase interest and visual movement within a design.

direct mail. Advertising and promotional materials sent by mail to customers.

disc flower. Tubular flowers that compose the central part of a head flower in most *Asteraceae,* such as chrysanthemums, gerberas, etc. contrasted with ray-shaped flowers on the margins of the head flower.

discordant color scheme. Using four colors that are widely separated on the color wheel. Care in color proportion and hue intensity and value, must be taken when using unrelated hues. Examples of discordant schemes include double complement, alternate complement, and tetrad.

dish garden. A container with soil in which several different plants grow together; often decorated with accessories or cut flowers in water tubes, inserted into the soil.

display. To visibly show merchandise to the public.

display cooler. A cooler in the sales area of a floral shop that displays flowers and arrangements for sale.

distribution channel. Route by which floral products move from grower to final consumer. Traditionally the channel of distribution involves the grower, auction, broker, wholesaler, retailer, and final consumer.

dixon pin. Two wooden picks attached on opposite ends of a flexible metal strip; used as a mechanical aid in floral work, particularly large sympathy designs.

domestic market. Products that are produced and consumed within the same country. For example, carnations grown in Colorado, sold and shipped to buyers within the United States and consumed within the U.S. are a domestic product grown for a domestic market.

double complement. A discordant color scheme using two pairs of complementary colors.

double-end spray. A floral spray constructed in an oval or double-triangular shape, used as a funeral design.

dry-pack. Storage or shipping of flowers out of solution (in a dry manner). Proper humidity and temperature conditions are essential for the success of dry-pack storage and shipping.

dry-prone. Flowers and foliage that dry out easily and are subject to wilting, shedding of leaves, needles, and florets. These flowers and foliage must be taken care of first and conditioned quickly upon arrival to the wholesaler or retailer.

Dutch-Flemish style. A style of floral arranging copied from the paintings of the Dutch and Flemish artists of the seventeenth and eighteenth centuries. Typical floral arrangements are massed and overflowing with the use of many varieties and colors of flowers (including tulips) facing in all directions. For authenticity, in recreating this style of design, many accessories are generally placed around a lavish floral bouquet, including fruits, bird nests, and shells.

dynamic line. An active line of continuous movement that counters a design's shape, such as a diagonal line contrasted with horizontal and vertical lines within a rectangular format.

Early-American style. Floral arrangements made in the early American style are simple and charming, using native plant materials, such as wildflowers, weeds, and grains. Representative of the Early American Period (1620-1720), containers are generally simple utility jugs, pitchers, and other kitchenware made from pottery, copper, and pewter.

earthenware. Containers, tableware, and other items made of coarse, brown or red clay. After firing, earthenware is porous and nondurable, unless treated with a glazed finish. Generally, earthenware is coarse-textured and heavy.

easel. A three-legged stand constructed of heavy wire or wood, used to support and display wreaths, other set pieces, and floral sprays, generally at funeral services.

easel spray. A floral arrangement placed on an easel.

eclecticism. In floral arranging, refers to borrowing and mixing styles from various sources and periods and combining them into one style of design with an eye to compatibility.

Egyptian period. Approximately 2800-28 B.C., the floral style was simplistic, repetitious, and highly stylized. Flowers and fruits were placed in carefully alternating patterns. Chaplets, wreaths, garlands, and flower collars were also popular.

emblem tribute. Also called organizational tribute, a funeral design or set piece, ordered by members of a club, business, religion, or school to which the deceased belonged.

emphasis. A design principle synonomous with focal point. The creation of visual importance or accent in a design; the center of interest.

English-Georgian period. Also called *Georgian,* refers to the period in England during the reigns of George I, II, and III (1714-1790). Floral styles during this period included simple hand-held bouquets for which to carry fragrance. Floral arrangements were symmetrical and ranged from small to large with great varieties of fragrant flowers. A variety of containers were used, including *Wedgwood,* metal, and glass.

epergne (eh-PURN). An ornamental stand with several separate dishes or trays used as a table centerpiece for holding fruit and flowers, popular during the Victorian era.

ephemeral. Flowers lasting a very short time, such as cut gardenia blossoms.

equilateral triangle. A design form having the shape of a triangle with all three sides of equal length.

ethylene. Colorless, odorless gas that hastens senescence of flowers. Called the aging hormone, it is often emitted by fruit, foliage, aging flowers, and incomplete combustion of oil and gas in heaters.

ethylene scrubbers. Chemicals used to remove ethylene from the atmosphere.

European style. A loose term generally referring to full, massed bouquets that use a great variety of flowers and colors, in contrast to the *oriental style.*

everlasting flowers. Dried or artificial flowers, contrasted with fresh flowers.

ewer. A large water pitcher with a wide mouth, often used for holding flowers.

exotics. A general term referring to tropical, unsual, and form flowers and foliage.

experimental design. A general term for new, contemporary styles of arrangement, often bizarre in nature.

faience. Glazed earthenware named for Faenze, Italy, where it was originally manufactured.

fan bouquet. A hand-held grouping of flowers, placed on a lace, decorative fan.

fan-shape. A radiating, half-circle design.

feathering. The separation of a flower, especially a carnation, into several small florets for use in a corsage or bridal bouquet.

Federal period. The political, social, and decorative formation era in America following the Revolutionary War (1790-1825). Floral arrangements during this period were styled after ancient classic designs as well as elaborate European massed, symmetrical bouquets.

filament. The anther-bearing stalk of a stamen.

filler. A type of flower, foliage, or accessory, used to fill in spaces in an arrangement, add interest, and complete a design.

filling florist. A florist who receives a flowers-by-wire order and fills it by designing and delivering the flowers to the specified recipient.

final consumer. The last person to receive flowers through the channels of distribution.

flat spray. A showy gathering of flowers and foliage, tied with a ribbon. Used to decorate graves. In regions where flat sprays are popular, many are grouped together to form a larger wall spray.

floral clay. A waterproof, sticky material used in floral arranging for such purposes as fastening stem-anchoring devices, such as needlepoint holders and hard foam to containers.

floral foam. A highly porous material used to absorb water, support stems, and hold flowers, foliage, and accessories in place in a design.

floral holiday. A day that is associated with the giving and receiving of flowers, such as Valentine's Day and Mother's Day, or a day that is associated with using flowers for decorating, such as Christmas and Memorial Day.

floral pomander. A sphere or ball of fragrant flowers that hangs from ribbon.

floral preservative. A chemical mixture of sugar, biocide, and citric acid, added to water to extend the life of cut flowers.

floral tape. A paraffin-coated paper that may be stretched over stems to securely bind them together when creating corsages, boutonnieres, hand-held bouquets, and everlasting designs.

floret. One of the small flowers that make up the total inflorescence.

floriculture. The producing, marketing, sale, and use of floral crops.

Florist Information Committee (FIC). A branch of the Society of American Florists which functions to alleviate any negative press about the floral industry and its products and instead provides positive information to the public.

flower. The reproductive structure of angiosperms; see *complete flower.*

flowers-by-wire. See *wire order.*

foam forms. Hard or wettable foam forms, precut by the manufacturers in popular shapes, such as crosses, hearts, open hearts, and wreaths.

focal area. A part of a design that is emphasized by contrasted materials, the area of focus; generally in the base of a design where all materials meet.

focal point. The center of attraction, usually within the focal area; often highlighted with a distinctive flower or accessory item.

foreign market. Products that are produced in one country and sold to another for consumption. For example, carnations grown in Colombia, sold and shipped to buyers within the United States and consumed within the United States. are a foreign product grown for a foreign market.

form. A design term synonymous with shape or outline; also denotes an unusual or distinctively shaped flower or foliage.

formal. A general term for the classical, symmetrical, or elaborate.

formal-linear design. A high-style design emphasizing shape, form, and line, generally with asymmetrical balance. Minimal materials are used, emphasising individual flowers, stem angles, colors, and textures.

foundation. The base of a design from which elements rise.

foundation colors. The primary colors—red, yellow, and blue—from which all other hues on the color wheel are created

framework. The basic or beginning structure of a floral arrangement, established with stems of flowers and foliage. Framework creates the pattern of design.

framing. A design technique in which the perimeter of the design fully or partially encloses an area and focuses attention to its contents.

free-form design. A design that is not confined to any geometric shape and generally emphasises line and texture.

free-lance floral designer. A knowledgeable and experienced person who designs floral arrangements for individual buyers.

freeze-drying. A method of drying plant materials. Moisture is removed from the cell structure of flowers by mechanical means; flowers retain shape, suppleness, texture, and usually color.

French period. Also known as the *Grand Era,* seventeenth and eighteenth centuries in France during the reign of Louis XIV. This style was influenced by the Dutch Flemish style, with emphasis on classic form, refinement, and elegance, rather than overdone flamboyance. Highly ornamental vases were used to hold tall fan-shaped, rounded, and triangular bouquets. Often termed *French Rococo,* this style was extravagant using symmetry, shells, rocks, and all manner of elaborate decoration.

frond. The leaf of a fern. Also, large, divided leaf.

full-couch. A casket with an undivided full lid (contrasted with a half-couch).

full-service flower shop. A floral shop that provides every floral product and service needed by customers, including delivery and wire service.

fully-open stage. Refers to the state at which some flowers are harvested, at a mature stage before beginning to shed pollen. If cut too early, these flowers do not continue to open to their full potential. Examples include zinnia, sunflower, marigold, and calendula.

funeral basket. An arrangement made in a container of wicker, plastic, papier–mâché, or metal, usually with a handle. It is displayed on the floor or a stand at funeral services.

garland. A wreath, woven chain, or festoon of flowers, leaves, or other materials worn on the head or used as decoration.

genus, *pl.* **genera.** The taxonomic group between family and species that includes one or more species that have certain characteristics in common.

geometric design. A floral arrangement characterized by straight lines, triangles, circles, or similar regular forms.

Georgian period. See *English-Georgian period.*

geotropism. Bending and curving upward against the force of gravity, exhibited by gladiolus and snapdragons.

germicide. Any antiseptic used to destroy germs; see *biocide.*

glamellia. A wired and taped flower constructed from gladiolus petals and formed to resemble a camellia flower; used in corsages and bridal bouquets.

glamour leaves. A ccessories of artificial foliage, made from fabric, net, lace, etc. , used in corsages.

glycerin. A solution used for preserving foliage and some flowers. When properly absorbed in plant stems, it keeps the plant material soft and pliable.

golden anniversary. The 50th wedding anniversary. Floral arrangements for this special occasion are often made in gold containers, incorporate gold and yellow flowers, and often have a gold number 50 insignia accessory.

golden mean. A Greek rule of proportion refering to the division of a line somewhere between one-half and one-third its length that is the most pleasing to the eye.

golden rectangle. A Greek standard for proportion. A rectangle or oblong with its sides in a ratio of 2:3.

golden section. A Greek rule of proportion, based on the golden rectangle. It involves the division of a line or form in such a way that the ratio of the smaller portion to the larger is the same as that of the larger portion to the whole. For example, 2:3 is about the same ratio as 3:5, 5:8, 8:13, and so forth. This rule may be applied to the ratio of the container to the flowers, and the flowers to the entire arrangement.

gradation. A design technique of placing of flowers or foliage in a sequence, from largest to smallest, darkest to lightest, etc.

grades. Standardized measurements of crop quality. Roses are graded according to the length of stem and overall quality.

Grand Era. see *French period.*

Greek period. Floral styles of this time (600-146 b.c.) were garlands and wreaths. Flowers were scattered on the ground during festivals. Fragrance and symbolism was of utmost importance.

greening pin. U-shaped pins similar to hairpins; commonly used to secure moss to foam. Also called philly pins and fern pins.

green tape. See *floral tape.*

grid. Framework of plastic, tape, or wire used at the top of a vase to keep flowers in position without the use of floral foam.

groomsmen. The men who attend the groom at the wedding.

grouping. A design technique characterized by placing like materials in sections or groups with negative space between various groups and some space left between flowers within groups..

grower. The initial level of distribution of cut flowers and foliage and potted plants. Foreign growers sell floral products directly to brokers and wholesalers or take their products to an auction, while domestic growers most often sell to wholesalers. Proper care and handling, or the chain of life, begins at the grower level.

growth regulator treatment. Special hormone-containing solutions frequently used by growers to prevent and inhibit problems, such as leaf yellowing on alstroemerias and lilies.

half-couch. A casket in which the lid is divided in half. The head of the casket is open for viewing, while the foot of the casket is closed.

hamper. A tall container, made of wooden slats or corrugated cardboard, used to ship tall flowers, especially those that are geotropic, such as gladiolus.

hand-tied bouquet. A bouquet that is made in one hand while the other hand places flowers and foliage in a spiral pattern within the bouquet. It is generally tied with ribbon, string, or raffia..

hardgoods. Nonperishable staples or inventory items.

harmony. A pleasing relationship among the parts and elements of a design.

hardening. The conditioning process that allows flowers to become hard or turgid from water uptake.

harvesting. The cutting of flowers and foliage.

head flower. A dense inflorescence of small, crowded, often stalkless flowers; a capitulum.

heika (hay-kah). Vase flowers. General term for naturalistic Japanese arrangements in tall vases.

herbaceous stem. A general term referring to any nonwoody plant. Also referred to as soft stem.

high style. Characterized by bold forms and lines. See *formal linear* and *art deco.*

Hogarth curve. A design shape or prominent line in a design in the shape of the letter S. Named after William Hogarth who theorized that all beauty was based on a serpentine S-line.

holding solution. Preservative solution used in storing cut flowers.

hook-wiring. A method of wiring flat-topped flowers such as chrysanthemums. A U-shaped wire with a hook inserted down into the center of a flower and into the stem.

horizontal form. A low, flat rectangular, oblong arrangement of flowers.

horizontal line. A flat line, parallel to the tabletop or floor. The horizontal lines provide stability and restfulness.

horn of plenty. See *cornucopia.*

horticulture. The science and art of producing and using ornamental plants, fruit, and vegetables.

HQC. 8-hydroxyquinoline citrate. An anti-microbial agent or biocide that kills bacteria and other organisms.

hue. A specific name of a color.

hyacinth stake. A slender, wooden stick used to provide additional support to flowers.

hybrid. The offspring of two parents that differ in one or more heritable characteristics. The offspring of two different varieties or of two different species.

hydrating solution. A citric acid solution that encourages rapid uptake of water and helps flowers recover from dry shipping, dry storage, and water stress.

ikebana (ee-keh-bah-nah). Literal translation: living flowers. General term applied to all Japanese flower arrangements.

Ikenobo (ee-kee-no-bah). Literal translation: hut by the pond. Supposedly referring to the priests' place of retreat in the temple grounds, it is the original term applied to flower arrangements, later used to denote a school. The Ikenobo school, the oldest school of flower arranging in Japan, had its beginnings in the sixth century.

implied line. An imaginary line created by a series of repetitious elements.

inflorescence. Any arrangement of more than one flower on a stem, for example corymb, cyme, panicle, etc.

informal. Casual and relaxed; often asymmetrical and natural.

in lieu of flowers. An addition in death notices asking people not to send flowers to a funeral.

intensity. See *chroma.*

intermediate colors. The secondary colors—orange, violet, and green—formed by mixing two primary colors together.

interpretive design. A floral composition that expresses the designer's feelings and ideas.

inventory. Merchandise and stock on hand. A listing of the stock within the store.

isolation. Separation of a flower or group of flowers from the rest of the arrangement. A method of creating an area of emphasis or a focal point.

isosceles triangle. A triangle form that has two equal sides and one side different in length. Usually a tall, symmetrical triangular floral arrangement.

Japanese design/style. General term for a linear floral design, characterized by three main lines or sections.

jiyu-bana (jee-yoo-bah-nah). Freestyle Japanese arrangement.

juxtapose. To place side-by-side or close together.

kenzan (kehn-zahn). A needlepoint holder used in low bowls.

kubari (koo-bah-ree). A forked twig used as a flower fastener in a taller container used for classical Japanese designs.

lacing. A natural method of crossing stems to form a grid to hold flowers in position within a vase. A technique for making hand-tied bouquets.

landscape design. A floral arrangement that depicts a large area of nature. Flowers, branches, and foliage may represent parts of a natural landscape or groomed garden.

latex. A milky and usually whitish fluid that is produced in the cells of various plants. Bleeding latex of cut flower stems must be cared for properly.

layering. The process of compactly overlapping like materials (usually leaves, such as galax) so that no or little space is left between them. This process produces a scalelike appearance.

leaflet. One of the parts of a compound leaf.

lei. A wreath, garland, or ornamental headdress made of leaves and flowers.

le style 25. See *Art Deco.*

limited-service flower shop. A flower shop characterized by no or little added service and products. For example, a floral department in a grocery store.

liknon. An ancient basket in Greek and Roman times used for holding flowers.

line. A continuous pathway for the eye to follow.

line design. A general term for a floral arrangement characterized by strong vertical, horizontal, diagonal, or curving lines with negative space between for emphasis. This design contrasts with mass design.

line flower. A spike or spikelike inflorescence with an elongated stem, such as gladiolus, liatris, larkspur, and foxtail lily. Foliage and accessories are often termed "line" because of their tall, elongated shape.

management. The act, art, or manner of managing, or handling, controlling, directing employees, designs, money, the business affairs, etc.

marketing. All business activity involved in the moving of goods from the grower or producer to the final consumer, including selling, advertising, packaging, etc.

markup. A system for determining the sale price of an item based on the cost of merchandise.

mass design. A general term for a full, over-abundant floral arrangement. This design contrasts with line design.

mass flower. A single, rounded flower at the top of a peduncle. Used for adding mass to a design. Foliage and accessories may also be termed "mass" when used to cover mechanics, or add mass.

mass merchandiser. One who promotes, advertises, and organizes the sale of flowers in large quantities, such as grocery stores for their floral departments.

mechanical aid. Any material used in a container or in the designing of an arrangement to ease and expedite construction.

mechanical balance. See *balance.*

mechanics. The method of construction of a floral design; the technical aspect.

merchandise. Material offered for sale.

metabolism. The sum of all chemical processes occurring within a living cell or organism.

Middle Ages. The period of European history between ancient and modern times (a.d. 476-1450). Also known as the Medieval period and Dark Ages. Little is known of floral art during this time, however, fragrant flowers were highly favored for strewing on the ground, freshening the air, and for making wreaths and garlands.

mille fleurs design. Literal translation: thousand flowers. General term for a mass floral bouquet with many colors and flowers juxtaposed together.

mirroring. Repetition of the like materials at varying heights and depths. See *shadowing.*

modernistic design. A contemporary floral arrangement expressing current trends and ideas.

monochasium. A cymose inflorescence in which there is a single terminal flower with it a single branch below bearing flowers.

monochromatic color scheme. Uses one hue from the color wheel and may include its tints, tones, and shades.

moribana (moh-ree-bah-nah). Literal translation: piled-up flowers. General term applied to naturalistic Japanese arrangements in shallow or low, flat containers.

muff bouquet. A specialty bouquet in which a small floral arrangement is pinned to a muff for winter weddings.

nageire (nah-gay-reh). Literal translation: Thrown in. A loose, naturalistic style of Japanese arrangement of Ikenobo. General term for arrangements in tall vases or baskets.

naturalistic design. A design style that emphasizes the beauty of flowers and plant materials without manipulation.

needlepoint holder. See *pinholder.*

negative space. Empty areas between flowers or materials.

networking. The exchanging of information with others in an informal way to further business.

neutral colors. Colors that are not located on the color wheel but which influence those that are. Generally black, white, and gray. Other neutral colors include tan and ivory.

new convention design. A design style characterized by only vertical and horizontal placements of materials, all of which are at a ninety degree angle to each other.

new wave design. A style of design in which materials are used in a nonrealistic way. Materials are painted, folded, and manipulated, while lines cross and zigzag throughout the design.

node. The part of a stem, such as on a carnation, where one or more leaves are attached.

nomenclature. The system of naming plants and flowers.

nosegay. A tight grouping of flowers, herbs, and foliage in a hand-held bouquet. Also called a tuzzy-muzzy and posie bouquet.

novelties. Unusual articles that are chiefly decorative and often used with flowers to suggest a floral holiday, season, or theme.

object line/stem. Designation applied to earth line in the Ohara school.

Ohara. Twentieth century school of Japanese flower arrangements. Originator of the moribana style; interprets the moribana and nageire styles.

olfactory. Of the sense of smell.

one-sided design. A floral arrangement designed to be viewed from one side only.

open balance. A type of balance often employed in some contemporary design styles. A relaxed and unstructured balance; a feeling of stability resulting from no formal rules of construction.

open house. Promotion to get the general public to visit the floral shop to see the displays, usually prior to a major floral holiday.

open order. An order for a floral arrangement in which the designer decides both the contents, style, and price.

order-taker. A passive salesperson who simply writes up the orders for the customers without any salemanship involved.

oriental style. A loose term referring to line designs including both the Chinese and Japanese styles, as well as designs that resemble these styles. The oriental style uses few materials and emphasizes simplicity, form, line, and texture.

ovary. An enlarged basal portion of a carpel or of a gynoecium composed of fused carpels. The ovary becomes the fruit.

overhead. Expenses and the general cost of running a business.

packaging. Wrapping or boxing flowers, plants, and floral products for sale and delivery.

packing. The large scale, comercial processing and packaging of flowers and floral products in boxes for shipping.

paddle wire. Continuous wire wrapped around a small length of wood.

pall. A fabric piece placed on a casket. A floral design placed on a casket.

panicle. A branched raceme with each branch bearing a further raceme of flowers. More loosly applies to any complex, branched inflorescence.

pan-melt glue. Small pillows or pellets of glue that are melted down in an electric pan and used to secure mechanical aids, etc. in designs.

parallelism. Placing two or more stems, lines, etc. in the same direction (usually in a vertical format) where the lines never meet together.

parallel systems design. A style of design in which two or more (usually vertical) groupings are placed in a container and negative space surrounds the groups. Each group generally uses like materials.

parasol bouquet. A specialty bouquet in which flower clusters adorn an open or closed parasol (decorative, lacy umbrella).

pavé. A style of design or a technique in which materials are placed closely together in a cobblestone effect. The term is borrowed from jewelry-making, which refers to gems that are placed closely together with little or no metal showing between them.

pedicel. The stalk of a single flower in an inflorescence.

peduncle. The stalk of an inflorescence or of a solitary flower.

perianth. The petals and sepals combined.

period style. A term used to designate a single item or a complete arrangement style prevalent in a specific country at a particular time in history.

perishable. Subject to decay, short-lived.

petal. A flower part, usually conspicuously colored. One of the units of the corolla.

petiole. The stalk of a leaf.

pH. A symbol denoting the relative concentration of hydrogen ions in a solution. pH values run from zero to fourteen and the lower the value, the more acidic a solution. For example, pH 7 is neutral, less than 7 is acidic, more than 7 is alkaline.

phloem. Food conducting tissue.

phoenix design. A style of design in which tall materials burst out from the center of a rounded design. Named for the ancient mythological Egyptian bird.

photosynthesis. The conversion of light energy to chemical energy. The production of carbohydrates from carbon dioxide in the presence of chlorophyll by using light energy.

phototropism. Turning or bending in response to light. Many cut flowers curve towards a light source.

physical balance. See *balance.*

physiology. The study of the activities and processes of living organisms.

pierce-wiring. A method of wiring flowers using a straight wire; the usual method for adding a wire to roses and carnations.

pillowing. A clustering technique used at the base of a design in which pillows or rolling hills and valleys are formed by the groups of flowers.

pigment. In plants, the substance that absorbs light, expressing colors.

pinholder. A holder of steel needles used to hold flowers and other plant materials in place in a low container.

pistil. Central organ of flowers typically consisting of ovary, style, and stigma. A pistil may consist of one or more fused carpels.

plugging. A general term referring to impeded uptake of stems caused by bacteria, salts, air, etc.

pollen. A collective term for pollen grains.

polyfoil. Colored aluminum foil used to wrap potted plants.

positive space. Area occupied by flowers, foliage, and other materials.

postharvest. The period following harvest of a crop.

postharvest physiology. The study of the activities and processes of harvested or cut flowers and foliage.

posy holder. A hand-bouquet holder made from various materials. Manufactured to hold tied nosegays, making them less cumbersome and longer lasting.

pre-cooling. The practice of cooling flowers after boxing by forcing cooled air through holes in the walls of the box.

pre-green. Containers filled with wet floral foam and greenery, that only need to have flowers added. Containers are often pre-greened in mass quantities prior to floral holidays.

presentation bouquet. See *arm bouquet.*

preservative solution. A mixture of floral preservative with water for increasing the life of cut flowers.

preserved flowers. Real flowers that have been dried and preserved by one of a number of drying methods.

pressing. A method of drying flowers and foliage, resulting in flat forms.

pretreatment. Special treatments used prior to using floral preservative in the care and handling process, for example, STS treatment for ethylene-sensitive flowers.

primary colors. Red, yellow, and blue. These three colors are equidistant on the color wheel.

processing. The methods of care and handling, specifically recutting and conditioning flowers and foliage.

processional. The ceremonial proceeding of the wedding party up the aisle prior to the wedding ceremony.

Professional Floral Commentators International (PFCI). A division of the Society of American Florists, established to promote communication and understanding within all segments of the floral industry.

promotion. All advertising, publicity, and personal selling activities leading to public recognition.

proportion. A principle of design. The comparative size relationship between the parts of a floral arrangement to each other and the parts to the whole; flowers, foliage, container, and the surrounding space all relate to proportion.

prospects. Potential customers or flower buyers.

proximity. Near in space.

psychic line. A line that is felt between two elements, created by placing flowers and materials in a way that directs the eye.

pubescent. Covered in soft, short hairs. Coarse in texture adding interest to designs.

publicity. Free promotional activities resulting in public notice.

public relations. The act of promoting good will between a business or person and the public.

pulsing. A postharvest technique used to load flowers with sugar and other chemicals prior to shipment.

queen's bouquet. See *arm bouquet.*

raceme. An inflorescence consisting of a main axis, bearing single flowers alternately or spirally on stalks (pedicels) of approximately equal length.

radial balance. A type of balance or feeling of stability created by all elements radiated or circling out from a common central point like the spokes of a wheel or the rays of the sun.

raffia. Fiber from the palm tree *Raphia ruffia,* used as string.

ray flower. See *disc flower.*

receptacle. That part of the axis of a flower stalk that bears the floral organs.

receiving line. The line in which the bridal party members and parents stand to receive guests' congratulations and greetings after the ceremony or at the reception.

recessional. The ceremonial proceeding of the wedding party down the aisle after the wedding ceremony.

recut. The process of removing the lower one-half to one inch of stem with a sharp blade, shears, or underwater-cutter.

Redbook Master Consultants (RMC). A national organization comprised of florists and industry people interested in advancing and improving the floral industry through education.

relative humidity. The amount of water vapor actually present in the air at a given temperature as compared to the maximum amount the air could hold at that certain termperature. Warm air has the capacity to hold more moisture than cold air.

Renaissance. A period in Europe after the Middle Ages. Beginning in Italy in the fourteenth century, it was marked by a humanistic and classical period in which an unprecedented flourishing of the arts occurred. Floral arrangements characteristic of the Renaissance period were massed in tight symmetrical shapes. Colorful flowers were often combined with fruits and vegetables. Many types of containers were used including urns, jugs, and bowls. A lso well known was a single stem of white lily (Madonna lily) in a simple container. Flowers were symbolic with religious themes.

repetition. A method of obtaining rhythm by repeating similar elements throughout a design.

respiration. An intracellular process in which food is oxidized with the release of energy. The complete breakdown of sugar or other organic compounds to carbon dioxide and water.

retailer. In the channels of distribution, the level between wholesaler and final consumer; the traditional flower shop or one who sells flowers to consumers.

right triangle. A form of arrangement. An asymmetrical triangle with the vertical line being perpendicular to the horizontal line, forming a ninety degree angle.

rikka (ree-kah). Literal translation: standing flowers. One of the oldest forms of Japanese flower arrangement, introduced about 1470. A large, complex, form of arrangement with a kaleidoscopic view of a landscape, often used to adorn temple altars.

Roman period. During this time (28 B.C. - A.D. 325) flowers were used to make garlands and wreaths. The use of plant material was more elaborate than in the previous Greek and Egyptian periods. Fragrance and bright colors were important for flowers.

Romantic era. See *Victorian era.*

round-mound. Also called roundy-moundies, referring to circular, rounded mass bouquets. A common everyday design.

saddle. A container, available in different styles and sizes, made to fit the curvature of a casket lid to hold the foundation of the casket spray design.

salesmanship. The art of selling.

sanitation. Cleanliness. The practice of keeping flower handling facilities, tools, buckets, and equipment clean.

satin acetate. Waterproof utility ribbon, available in a wide assortment of colors and widths.

satin forms. Pre-made satin shapes ready for floral accents to be added. Available in a variety of popular forms, such as pillows, hearts, and crosses.

scale. Another word for size. The ratio or proportion of an arrangement to the surrounding area in which it will be placed.

scalene triangle. An asymmetrical triangle in which all three sides are unequal in length. Scalene triangle designs usually have a prominent vertical line with a wide angle to the downward diagonal line of the design.

scapose. A solitary flower on a leafless peduncle or scape, such as a cut tulip or daffodil, or other inflorescence that does not have ordinary green, leafy foliage on the stem, such as a cut gerbera, agapanthus, or anthurium.

scientific name. See *botanical name.*

scorpioid. A cymose inflorescence; curved to one side like a scorpion's tail.

script. Gold lettering that is attached to a ribbon; often used for funeral designs, and denotes the sender's relationship to the deceased, such as "Loving Grandmother" or "Beloved Father."

sealer. An antitranspirant for fresh flowers (keeps flowers fresher longer) or a finishing spray on dried flowers (keeps flowers intact).

seasonal availability. A general term referring to the peak supply times for flowers and foliage due to the plant's natural growth and flowing cycles.

secondary colors. Orange, violet, and green. Formed by mixing two primary colors together.

secondary stem. Term used to designate *man* line in the Ohara school.

seika (say-kah). Name given by some schools to the classical forms of Japanese arrangement. (Same as shoka in the Ikenobo school.)

sending florist. A florist who takes an order and transmits it to another florist in another city.

senescence. The aging process resulting in wilting and death of a cut flower, foliage, or plant.

sequencing. A design technique in which flowers and materials are placed in a gradual or progressive change, through the gradation of color, the transition of size, etc.

sepal. A floral leaf or individual segment of the calyx of a flower; usually green. The outermost flower structure that usually encloses the other flower parts in the bud.

sessile. Attached directly to the stem; referring to a leaf lacking a petiole or to a flower lacking a pedicel.

set piece. A funeral design that has a certain shape, usually made on a foam form, decorated with flowers, and placed on an easel. Examples include hearts, crosses, and organizational emblems.

shade. Any hue to which black has been added. A dark-colored hue.

shadowing. A design technique in which like materials are placed in pairs, one directly beneath or behind the other to form the appearance of a shadow. Increases the feeling of depth.

shattering. The falling off of florets, leaves, or petals, especially with chrysanthemums suffering from mechanical damage.

sheaf. A cluster of cut flowers, wheat, foliage, etc., tied together at one end with a ribbon bow, raffia, or other tying material.

sheltered design/sheltering. A design style or technique in which the floral arrangement is "protected" and contained beneath the container, or branches and other materials are placed over the arrangement, giving a feeling of protection and recluse.

shin (sheen). Applied by most schools to the main branch, commonly called heaven, in a Japanese flower arrangement. Also used to designate the formal arrangement of several classical methods.

shipper. A firm that specializes in the assembly of flowers, packing in boxes, and shipping them to other buyers.

shoka (sho-kah). The latest of orthodox styles of classical arrangements. The three-branch asymmetrical, classical form of flower arrangement as practiced by the Ikenobo school. It replaced rikka in popularity in the late 18th century.

silica gel. A drying compound (desiccant) for use in drying flowers.

silver anniversary. The 25th wedding anniversary; floral arrangements for this special occasion are often made in a silver-colored container with silver ribbon and a number 25 insignia accessory incorporated with the flowers.

silver conditioning. See *silver thiosulfate.*

silver thiosulfate (STS). A silver compound readily taken up into the flower. Conditioning of ethylene-sensitive crops in this solution helps protect them aginst the effects of ethylene during handling and shipping, and increases vase life.

skeleton flowers. The primary flowers used in setting up the framework of a design, establishing the pattern or outline of the design. Foliage may also be used in setting the framework.

sleepiness. An ethylene induced disorder in flowers, characterized by an inward bending, closing, or wilting of the petals. Flowers appear wilted and limp.

Society of American Florists (SAF). A national trade organization representing the needs and interests of the entire floral industry..

soe (so-eh). Literal translation: harmonizer. Applied to secondary branch, commonly called *man,* in most methods of Japanese flower arrangement. The supporting branch in rikka arrangements.

Sogetsu (so-ghet-soo). Modern school of Japanese flower arrangement founded in 1926.

solitary flower. A single flower on a stem, such as a tulip.

space. An element of design. The three dimensional area in and around a design.

spacing. A design technique in which flowers or other materials are placed closely together to create an area of emphasis, in contrast to materials that have more negative space between them.

spadix. A spike of flowers on a swollen, fleshy axis, usually surrounded by a colorful bract, as with an anthurium.

spathe. A large bract subtended and often surrounding a spadix inflorescence.

specialty bouquet. A novelty bouquet that is unusual or out of the ordinary.

specialty flower shop. A flower shop that targets particular floral needs, such as weddings, high style designs, or everlasting designs.

species. A kind of organism. Species are designated by binominal names written in italics.

spike. An inflorescence in which the main axis is elongated and the flowers are sessile.

split-complementary color scheme. A color scheme using one hue together with the two colors that are adjacent to the direct complement.

stacking. The placing of one material on top of another.

stalk. Generally refers to a stem or peduncle.

stamen. The plant's male reproductive organ. The part of the flower producing the pollen, composed usually of anther and filament. Collectively, the stamens make up the androecium.

state florists' association. A trade association representing the needs and interests of floral industry members within a state.

static line. Lines in a composition that are parallel to the main lines of the design. For example in a design that has predominate vertical and horizontal lines, or a rectangular format, repeating lines of vertical and horizontal display lack of visual energy and are termed static and unmoving.

stigma. The region of a carpel serving as a receptive surface for pollen grains and on which they germinate.

stipule. A leafy appendage, often paired and usually at the base of the leaf stalk.

stitch-wiring. A method of wiring broad leaf foliages.

stoma, *pl.* **stomata.** The pores that occur in large numbers in the epidermis of plants (stems and leaves) and through which gaseous exchange takes place.

storage cooler. An enclosed refrigeration unit in which the bulk of the flowers and foliage are stored.

style. Slender column of tissue which arises from the top of the ovary and through which the pollen tube grows. Also, a recognizeable form of design or school of thought; the end result.

stylized design. A general term referring to a vertical, linear, high-style design.

subject line/stem. Designation of the *heaven* line in the Ohara school.

succulents. Plants with fleshy, water-storing stems and leaves.

symmetrical balance. Equal in visual weight on both sides of a central vertical axis.

table skirt decoration. Floral clusters or small floral accents with ribbon bows that are pinned to the skirt of the tablecloth.

tactile. Referring to the sense of touch.

tai (tye). Literal translation: material substance. Term applied by most teachers to *earth* line in classical forms and some naturalistic arrangements. The tertiary branch of the asymmetrical shoka style.

tailoring. Refers to trimming materials (especially leaves) to give them a sculptured, fitted, or abstract appearance.

technique. The application of design methods. A means to an end.

tension. The quality in a design created by opposing elements that offer resistence in visual flow or eye movement. For example, using discordant colors or crossing stems creates visual tension.

tepal. A perianth segment that is not clearly distinguishable as being either a sepal or petal, such as with the sepals and petals of tulips and lilies.

terracing. The placement of like materials on top of each other, but divided by space in between, giving a stair step appearance.

terra cotta. A hard, brownish-red, usually unglazed earthenware used for containers and sculpture.

tertiary colors. Third colors created by mixing a primary with an adjacent secondary color, named for the parent colors, such as blue-green or red-orange.

tetrad color scheme. A four-color scheme in which the four colors are equidistant from each other on the color wheel.

texture. An element of design that refers to the surface qualities or characteristics, both seen and felt, of flowers, foliage, containers, etc.

theme. An overall feeling, style, or message that a floral design suggests, such as a floral design with a baby girl theme. All parts are harmonized to create the intended theme.

tint. Any color to which white has been added creating light colors and pastel hues.

tipping. A general term for spray painting the edges of flower petals.

tokonoma (toh-koh-noh-mah). Ornamental alcove in a Japanese room used to exhibit art objects and flower arrangements.

tone. Any color to which gray has been added; tones often have a gray, dusty, dull, or frosty appearance.

topiary. An evergreen tree or shrub that has been trimmed or trained to an unnatural shape. A floral design to create this appearance, such as a topiary ball or cone.

toss bouquet. A small bouquet made for the bride to throw or toss to the unmarried females after the wedding or at the reception.

total dissolved solids (TDS). Water salinity or a measure of total soluble elements in the water. A water quality-characteristic, usually given in parts per million (ppm).

trade association. An organization or group that is set up by individuals, merchants, or business firms for the unified promotion and sale of flowers and floral products.

trade publication. Printed materials, such as journals, magazines, newsletters, etc. offering information of interest to all segments of the floral industry on a weekly, monthly, bimonthly, or quarterly basis.

traditional design/style. See *classical style.*

transition. A method of achieving visual rhythm. See *sequencing* and *gradation.*

transpiration. The loss of water vapor by plant parts. Most transpiration occurs through stomata.

transporter. The indivual or method of shipping floral products between each level of distribution.

triadic color scheme. A color scheme using three colors that are equidistant from each other on the color wheel. The combination of red, yellow, and blue is a triadic scheme.

tropism. A response to an external stimulus in which cut flowers curve or bend towards. See *geotropism* and *phototropism.*

tufting. A design technique of clustering or bunching materials closely together at the base of a design to emphasize color and texture. See *clustering* and *pillowing.*

tulle. A type of decorative netting used as an accessory in corsage and wedding designs.

turgid. Full of water, swollen, distended, referring to a cell that is firm due to water uptake.

tussie-mussie. A small hand-held fragrant bouquet sometimes spelled tuzzy-muzzy. The word tuzzy refers to the old English word for a knot of flowers. Originally flower cluster stems were tied together. Later, holders were manufactured for ease of carrying and displaying these tight little bouquets. During the Victorian era, flowers were symbolic and tussie-mussies conveyed sentimental messages to loved ones.

umbel. An umbrella-shaped inflorescence with all the stalks (pedicels) arising from the top of the main stem.

underwater-cutter. Apparatus specially designed to cut flower stems under water, helping them to hydrate rapidly and last a longer time.

unity. A principle of design. The relationship of individual elements or parts to each other which produces a unified whole. The effect created by the cohesive placement and use of materials where the whole is greater than its parts.

urn. A vase of classical shape with a base, often wide-mouthed, with two handles..

value. Refers to the lightness or darkness of a color.

vase life. The useful life of a cut flower after harvest, especially the lasting time and quality for the final consumer.

vegetative design. A naturalistic design style in which flowers and plant materials are placed as they would grow in nature. Materials used are found in nature together with emphasis placed on seasonal compatibility.

venation. The arrangement of veins in a leaf.

vertical form. A design that is taller than it is wide.

vertical line. A line that has a ninety-degree angle from the horizon, emphasizing strength.

vestibule. The area in a church between the outer door and the interior part or chapel of the building.

Victorian era. This period is named for Queen Victoria, who reigned in England from 1837 to 1901. Floral arrangements are characterized as being massive, overdone, and flamboyant. Containers were highly decorative and gaudy.

vignette Refers to displaying or grouping similar types of merchandise for maximum visual appeal.

visual balance. See *balance.*

visual merchandising. Efforts to attract customers to the shop and create interest in the flowers and merchandise so customers will want to buy.

void. A connecting space within a design.

wall pockets. Wall containers of various shapes for flowers.

wall spray. A composite spray made from the grouping of flat sprays and placed against the wall at funeral services.

warm colors. Colors (hues) composed of yellow, orange, and red hues. Associated with warm things like the sun, and heat. Also refererred to as advancing colors..

water quality. The characteristics of water which influence its reactivity with chemicals, particularly floral preservative, and its effect on cut flowers and foliage.

waterfall design. A design style that has many cascading layers forming a pendulous and flowing appearance.

wedding consultant. A knowledgeable and experienced designer who specializes in planning wedding flowers for the bride.

Wedgewood. A fine ceramic ware popular during the English-Georgian period, named after the English potter Josiah Wedgewood. It is depicted ancient Greek and Roman designs and manufactured with special holes and openings for stems, specifically to hold flowers.

western line design. A triangular, L-shaped design with a focal point near the base from which all stems radiate. A prominent vertical line with an opposite downward sweeping line, with open space between the lines.

wetting agent. A chemical added to some floral preservatives to make water "wetter" helping to hydrate flowers quickly.

wholesaler. The traditional middleman in the channel of distribution who buys flowers and floral products from growers and brokers and sells them to retailers. Key functions are to break bulk quantities and sell smaller lots or bunches to retailers.

wilt sensitive. Flowers and foliage that are particularly sensitive or prone to water loss and wilt easily.

wire gauge. The measurement used to determine the thickness of wire.

wire order. Customer's orders which are sent by one florist to another by telephone, computer, or FAX machine.

woody stem. A tougher or harder stem as opposed to a herbaceous stem, such as the stems of heather, leptospermum, and flowering branches.

working capital. The part of a company's capital readily convertible into cash and available for paying bills, wages, etc.. The money needed to run a business.

wrap-around wiring See *clutch wiring.*

wrapping. A design technique in which fabric, ribbon, raffia, etc. Are used to cover, coil, or twine a single stem or group of materials to achieve a decorative effect.

wreath bouquet. A specialty bouquet made in the form of a wreath carried by brides and her attendants.

wristlet. A device added to corsages for wearing on the wrist.

xylem. A complex vascular tissue through which most of the water and minerals of a plant are conducted.

zoning. A technique of restricting the numbers and types of materials used in certain larger areas.

Bibliography

Floral Arrangement History

Bayer, Patricia. *Art Deco Source Book.* Oxford: Phaidon Press Limited, 1988.

Benz, M. *Flowers: Free Form—Interpretive Design.* Houston, TX: San Jacinto Publishing Co., 1960.

Berrall, Julia S. *A History of Flower Arrangement.* New York: The Viking Press, 1968.

Coats, Peter. *Flowers in History.* New York: The Viking Press, Inc., 1970.

Gowing, Lawrence. *Hogarth.* London: Balding & Mansell, 1971.

Maas, Jeremy. *Victorian Painters.* New York: G. P. Putnam's Sons, 1969.

Marcus, Margaret Fairbanks. *Period Flower Arrangement.* New York: M. Barrows & Company, Inc., 1952.

Mitchell, Peter. *European Flower Painters.* Schiedam, The Netherlands: Interbook International B. V., 1973.

Newdick, Jane. *Period Flowers.* New York: Crown Publishers, Inc., 1991.

Paulson, Ronald. *Hogarth: His Life, Art, and Times.* Hartford, CT: Connecticut Printers, Inc., 1971.

Schmutzler, Robert. *Art Nouveau.* New York: Harry N. Abrams, Inc., 1962.

Warren, Geoffrey. *All Colour Book of Art Nouveau.* London: Octopus Books Limited, 1972.

Floral Arrangement and Design

Allen, Judy, C. Edwards, I. Finlayson, A. Johnson, and R. Lamont. *Flower Arranging.* London: Octopus Books Limited, 1979.

Allen, Phyllis Sloan and Miriam F. Stimpson. *Beginnings of Interior Environment.* New York: Macmillan Publishing Company, 1990.

Benz, M. *Flowers: Geometric Form.* College Station, TX: San Jacinto Publishing Co., 1986.

Better Homes & Gardens Flower Arranging. New York: Meredith Press, 1965.

Blacklock, Judith. *Flower Arranging.* Teach Yourself Books. Lincolnwood, IL: NTC Publishing Group, 1975.

Bridges, Derek. *Derek Bridges' Flower Arranger's Bible.* London: Century Hutchinson Ltd., 1990.

Brinton, Diana. *The Complete Guide to Flower Arranging.* London: Merehurst Limited, 1990.

Conder, Susan, Sue Phillips, and Pamela Westland. *The Complete Flower Arranging Book.* London: Octopus Illustrated Publishing, 1992.

Contemporary Floral Designs. Boston, MA: The Rittners Floral School, 1983.

Decorating with Plants and Flowers. Des Moines, Iowa: Creative Home Library, Meredith Coroporation, 1972.

Dreams Come True. Lansing, MI: The John Henry Company, 1991.

Fitch, Charles Marden. *Fresh Flowers Identifying, Selecting, and Arranging.* New York: Abbeville Press, Inc., 1992.

Floral Design Techniques. Lansing, MI: The John Henry Company, 1987.

Forsell, Mary. *The Book of Flower Arranging.* New York: Michael Friedman Publishing Group, Inc., 1988.

Gatrell, Anthony. *Dictionary of Floristry and Flower Arranging.* London: The Bath Press, B. T. Batsford Limited, 1988.

Hillier, Florence Bell. *Basic Guide to Flower Arranging.* New York: McGraw-Hill Book Company, 1974.

Hillier, Malcolm. *The Book of Fresh Flowers: A Complete Guide to Selecting & Arranging.* New York: Simon and Schuster, 1988.

Hillier. *Flower Arranging.* Reader's Digest Home Handbooks. Pleasantville, NY: The Reader's Digest Association, Inc., 1990.

Lauer, David A. *Design Basics.* Fort Worth, TX: Holt, Rinehart and Winston, Inc., 1990.

Love, Daphne and Sid Love. *Flower Arranginng from the Garden.* London: Cassell Educational Limited, 1989.

Mann, Pauline. *The Flower Arranger's Workbook.* London: B. T. Batsford Ltd., 1989.

McDaniel, Gary L. *Floral Design and Arrangement.* Englewood Cliffs, New Jersey: Prentice-Hall, Inc., 1989.

Mitchell, Herbert E. *Design With Flowers.* Woodland Hills, CA: CRB Publishing, Inc., 1991.

Newdick, Jane. *Book of Flowers.* New York: Prentice Hall, 1989.

Ortho Books. *Arranging Cut Flowers.* San Ramon, CA: Chevron Chemical Company, 1985.

Parkin, Beverley. *Say it with Flowers.* Oxford: Lion Publishing, ISIS Large Print Books, 1983.

Piercy, Harold. *The Constance Spry Handbook of Floristry.* London: Christopher Helm Ltd, 1984.

The Professional Floral Design Manual. The AFS Education Center, American Floral Services, Inc. Oklahoma City, OK: Times-Journal Publishing Company, 1990.

Redbook Florist Services. *Advanced Floral Design.* Leachville, Arkansas: Printers and Publishers, Inc., 1992.

Redbook Florist Services. *Basic Floral Design.* Leachville, Arkansas: Printers and Publishers, Inc., 1991.

Rulloda, Phillip M. *Tropical & Contemporary Floral Design.* Phoenix, AZ: Phil & Silverio, Inc., 1990.

Redbook Florist Services. *Selling and Designing Wedding Flowers.* Leachville, Arkansas: Printers and Publishers, Inc., 1991.

Redbook Florist Services. *Selling and Designing Sympathy Flowers.* Leachville, Arkansas: Printers and Publishers, Inc., 1992.

Ryan, Tamaris. *At Home With Flowers.* New York: Gallery Books, 1990.

Sincere Sympathy. Lansing, MI: The John Henry Company, 1988.

Webb, Iris. *The Complete Guide to Flower & Foliage Arrangement.* Garden City, NY: Doubleday & Company, Inc., 1979.

Holiday and Special Occasion

Ainsworth, Catherine Harris. *American Calendar Customs, Volume I.* Buffalo, NY: The Clyde Press, 1979.

Ainsworth. *American Calendar Customs, Volume II.* Buffalo, NY: The Clyde Press, 1980.

Amer, Jean B. *Flower Arrangements for Special Occasions.* Nashville, TN: Allied Publications, Inc., 1962.

Chase's Annual Events. Chicago: Contemporary Books, Inc., 1989.

Douglas, George William. *The American Book of Days.* New York: The H. W. Wilson Company, 1948.

Fendelman, Helaine and Jeri Schwartz. *Official Price Guide Holiday Collectibles.* New York: House of Collectibles, 1991.

Redbook Florist Services. *Floral Design for the Holidays.* Leachville, Arkansas: Printers and Publishers, Inc., 1991.

Floral Fragrance

Bonar, Ann. *Gardening for Fragrance.* London: Ward Lock Limited, 1990.

Duff, Gail. *Natural Fragrances.* Pownal, Vermont: Storey Communications, Inc., 1989.

Lacey, Stephen. *Scent in Your Garden.* Boston: Little, Brown & Company Limited, 1991.

Ohrbach, Barbara Milo. *The Scented Room.* New York: Clarkson N. Potter, Inc., 1986.

Taylor, Jane. *Fragrant Gardens.* London: Ward Lock Ltd., 1991.

Nomenclature and Postharvest Physiology

Bailey, Liberty Hyde and Ethel Zoe Bailey. *Hortus Third.* New York: MacMillan Publishing Co., Inc., 1976.

Benson, Lyman. *Plant Classification.* Boston: D. C. Heath and Company, 1957.

Capon, Brian. *Botany for Gardeners.* Portland, OR: Timber Press, 1990.

Coombes, Allen J. *Dictionary of Plant Names.* Portland, OR: Timber Press, 1987.

Heywood, V. H. *Flowering Plants of the World.* New York: Mayflower Books, Inc., 1978.

Kays, Stanley J. *Postharvest Physiology of Perishable Plant Products.* New York: Van Nostrand Reinhold, 1991.

Lawrence, George H. M. *An Introduction to Plant Taxonomy.* New York: The Macmillan Company, 1956.

Porter, C. L. *Taxonomy of Flowering Plants.* San Fransisco: W H. Freeman and Company, 1967.

Radford, Albert E. *Fundamentals of Plant Systematics.* New York: Harper & Row, Publishers, Inc., 1986.

Raven, Peter H. and Ray F. Evert. *Biology of Plants, 3rd Edition.* New York: Worth Publishers, Inc., 1981.

Salisbury, Frank B. and Cleon W. Ross. *Plant Physiology, 3rd Edition.* Belmont, CA: Wadsworth Publishing Company, 1985.

Seymour, Edward L. D. *The Wise Garden Encyclopedia.* New York: Harper Collins Publishers,1990.

Care and Handling

Graber, Debra Terry. *Fresh Flowers Book 2.* Lansing, MI: The John Henry Company, 1989.

Holstead-Klink, Christy. *Care and Handling of Flowers and Plants.* Alexandria, VA: Society of American Florists, 1985.

Redbook Florist Services. *Purchasing and Handling Fresh Flowers and Foliage.* Leachville, Arkansas: Printers and Publishers, Inc., 1991.

Vaughan, Mary Jane. *The Complete Book of Cut Flower Care.* London: Christopher Helm Ltd., 1988.

Everlasting Flowers

Conder, Susan. *Dried Flowers.* Boston: David R. Godine, Publisher, Inc., 1988.

Condon, Geneal. *The Complete Book of Flower Preservation.* Englewood Cliffs, NJ: Prentice-Hall, Inc., 1972.

Foster, Maureen. *The Flower Arranger's Encyclopedia of Preserving and Drying.* London: Blandford, 1988.

Hillier, Malcolm and Colin Hilton. *The Book of Dried Flowers.* New York: Simon and Schuster, 1986.

Penzner, Diana and Mary Forsell. *Everlasting Design.* Boston: Houghton Mifflin Company, 1988.

Petelin, Carol. *The Creative Guide to Dried Flowers.* London: Webb & Bower Limited, 1990.

Oriental Style of Design

Berrall, Julia S. *A History of Flower Arrangement.* New York: The Viking Press, 1968.

Conway, J. Gregory. *Flowers: East - West.* New York: Alfred A. Knopf, Inc., 1938.

Davidson, Georgie. *Classical Ikebana.* New York: A. S. Barnes & Company, 1970.

Kawase, Toshiro. *Inspired Flower Arrangements.* New York: Kodansha International Ltd., 1990.

Marcus, Margaret Fairbanks. *Period Flower Arrangement.* New York: M. Barrows & Company, Inc., 1952.

Sparnon, Norman J. *Japanese Flower Arrangement.* Tokyo: Charles E. Tuttle Company, 1960.

Webb. Lida. *An Easy Guide to Japanese Flower Arrangement Styles.* New York: Hearthside Press, Inc., 1963.

Wood, Mary Cokely. *Flower Arrangement Art of Japan.* Tokyo: Charles E. Tuttle Company, 1959.

Harvest and Distribution

Graber, Debra Terry. *Fresh Flowers Book 2.* Lansing, MI: The John Henry Company, 1989.

Guide to Floral Industry Transportation. Alexandria, VA: Society of American Florists, 1988.

Redbook Florist Services. *Purchasing and Handling Fresh Flowers and Foliage.* Leachville, Arkansas: Printers and Publishers, Inc., 1991.

The Retail Flower Shop

Cavin, Bram. *How to Run a Successful Florist & Plant Store.* New York: David McKay Company, Inc., 1977.

McDaniel, Gary L. *Ornamental Horticulture.* Reston, VA: Reston Publishing Co., 1979.

Pfahl, Peter B. and P. Blair Pfahl, Jr. *The Retail Florist Business.* Danville, IL: The Interstate Printers & Publishers, Inc., 1983.

Redbook Florist Services. *Retail Flower Shop Operation.* Leachville, Arkansas: Printers and Publishers, Inc., 1991.

Sullivan, Glenn H., Jerry L. Robertson, and George L. Staby. *Management for Retail Florists.* New York: W. H. Freeman and Company, 1980.

Cut Flowers and Foliage

Bailey, Liberty Hyde and Ethel Zoe Bailey. *Hortus Third.* New York: MacMillan Publishing Co., Inc., 1976.

Bianchini, Francesco and Azzurra Carrasa Pantano. *Guide to Plants and Flowers.* New York: Simon & Schuster Inc., 1974.

Bloemen Bureau Holland. Leiden, The Netherlands: Flower Council of Holland, 1992.

Fitch, Charles Marden. *Fresh Flowers Identifying, Selecting, and Arranging.* New York: Abbeville Press, Inc., 1992.

Graber, Debra Terry. *Fresh Flowers Book 2.* Lansing, MI: The John Henry Company, 1989.

Halpin, Anne. *The Naming of Flowers.* New York: Harper & Row, Publishers, Inc., 1990.

Hay, Roy and Patrick M. Synge. *The Color Dictionary of Flowers and Plants for Home and Garden.* New York: Crown Publishers, Inc., 1991.

Heywood, V. H. *Flowering Plants of the World.* New York: Mayflower Books, Inc., 1978.

Hodgson, Margaret, Roland Paine, and Neville Anderson. *A Guide to Orchids of the World.* Prymble, Australia: Collins Angus & Robertson Publishers, 1991.

Holstead-Klink. *Care and Handling of Flowers and Plants.* Alexandria, VA: Society of American Florists, 1985.

Larson, Roy A. *Introduction to Floriculture.* San Diego, CA: Academic Press, Inc., 1992.

Laurie, Alex, D.C. Kiplinger, and Kennard S. Nelson. *Commercial Flower Forcing.* New York: McGraw-Hill Book Company, 1979.

New Cut Flower Crops. Grower Guide, No. 18. London: Grower Books, 1982.

New Pronouncing Dictionary of Plant Names. Chicago: Florists' Publishing Company, 1990.

Raven, Peter H. and Ray R. Evert. *Biology of Plants, Third Edition.* New York: Worth Publishers, Inc., 1981.

Reader's Digest Encyclopaedia of Garden Plants and Flowers. London: The Reader's Digest Association Limited, 1975.

Seiden, Allan. *Flowers of Aloha.* Aiea, Hawaii: Island Heritage Publishing, 1990.

Seymour, Edward L. D. *The Wise Garden Encyclopedia.* New York: HarperCollins Publishers, 1990.

Vaughan, Mary Jane. *The Complete Book of Flower Care.* London: Christopher Helm Ltd., 1988.

Periodicals

Dateline: Washington. Alexandria, VA: Society of American Florists.

Design With Flowers. Costa Mesa, CA: Herb Mitchell Associates, Inc.

Floral Finance. Tulsa, OK: American Floral Services, Inc.

Floral Management. Alexandria, VA: Society of American Florists.

Floral Mass Marketing. Chicago, IL: Cenflo, Inc.

Florist. Southfield, MI: Florists' Transworld Delivery Association.

Florists' Review. Topeka, KS: Florists' Review Enterprises, Inc.

Flowers &. Los Angeles, CA: Teleflora.

Flower News. Chicago, IL: Cenflo, Inc.

Holland Flower. New York: The Flower Council of Holland.

PFD (Professional Floral Designer). Okalahoma City, OK: American Floral Services, Inc.

The Retail Florist. Oklahoma City, OK: American Floral Services, Inc.

Supermarket Floral. Overland Park, KS: Vance Publishing.

Index

Bold entries refer to information in figures.

C

D

E

boot - Lyal Nickals Style

leather leave

baby's breath

Bow

Eucalypts